AAPS Advances in Pharmaceutical Sciences Series

The AAPS Advances in the Pharmaceutical Sciences Series, published in partnership with the American Association of Pharmaceutical Scientists, is designed to deliver well written volumes authored by opinion leaders and authoritarians from around the globe, addressing innovations in drug research and development, and best practice for scientists and industry professionals in the pharma and biotech industries. For more details and to see a list of titles in the Series please visit http://www.springer.com/series/8825

Michael A. Repka • Nigel Langley
James DiNunzio
Editors

Melt Extrusion

Materials, Technology and Drug Product Design

aapspress

Springer

Editors
Michael A. Repka
Department of Pharmaceutics
Pii Center for Pharmaceutical Technology
School of Pharmacy
The University of Mississippi
University, MS
USA

James DiNunzio
Hoffmann-La Roche, Inc. Bridgewater
USA

Nigel Langley
Pharma Ingredients & Services
BASF Corporation
Florham Park, NJ
USA

ISSN 2210-7371
ISBN 978-1-4614-8431-8 ISBN 978-1-4614-8432-5 (eBook)
DOI 10.1007/978-1-4614-8432-5
Springer New York Heidelberg Dordrecht London

Library of Congress Control Number: 2013948475

Printed on acid-free paper

Springer is part of Springer Science+Business Media (www.springer.com)

...To My wife Staci, for her unwavering devotion, and love for our family. My mother, for leading her children into intellectual pursuits. My father, for his unselfish support and guidance. My brother, for his uncompromising principles and loyalty. My children, Michael, Jonathan, Andrew, Jordyn, and Walker, for making everything worthwhile.

Michael A. Repka

...To Lisa for her constant love, encouragement, and support.

James DiNunzio

...To my wife Tomoko for all her love and support and to my children Tom, Sean and Leo.

Nigel Langley

Preface

In the quest to develop new and better therapies to improve the quality of patients' lives, the pharmaceutical industry has relied on a combination of internal innovation and adaptation of external technologies to progress molecules to medicines. Today, melt extrusion stands as one of the several significant adaptations that have enabled therapies and produced novel drug products. The technology currently supports over a dozen commercial products and a range of novel compounds are currently in development using extrusion.

Having a lineage dating back to Archimedes, the concept of extrusion has progressed significantly over the centuries. The first modern designs for the twin screw extruder date back to the 1930s and with the development of the Erdmenger designs to achieve intermeshing and self-wiping in the 1950s, the technology has demonstrated utility and versatility. As an industrial process, the technology has supported a range of products, covering everything from space shuttle components to trash bags and wine corks. Serving as a low-cost production platform, the technology has penetrated a number of fields. Most recently, the technology has gained significant traction in the pharmaceutical space. Surprisingly to many, the technology traces its history back more than 30 years to the approval of Lacrisert, the first melt-extruded pharmaceutical product launched by Merck in 1981. Other major milestone products in the pharmaceutical space manufactured with hot-melt extrusion have included Rezulin, Kaletra, Nuvaring, and Ozerdex. Today, the technology is poised for an explosion as pharmaceutical applications extend into continuous processing, controlled release, and advance drug delivery devices.

It is also not surprising that interest in melt extrusion and the continued interest in solid dispersion technology has been supplemented by a wealth of publications. Within this space, *Melt Extrusion: Materials, Technology and Drug Product Design* has been developed to provide a definitive source on melt extrusion technology in the pharmaceutical arena. This text covers the history of and current technology for hot-melt extrusion. It also provides unique insight from excipient developers whose materials provide the basis for the production of solid dispersion products prepared using hot-melt extrusion. Fundamental overviews of formulation design and characterization are also presented and supplemented with unique industrial perspectives on modern applications of pharmaceutical hot-melt extrusion. The different viewpoints expressed by the authors and their respective organizations highlights the versatility of extrusion technology and points to the future path of the technology within the industry. As editors we wish to acknowledge and thank the authors, for without their contributions and valuable insight this text could not have been possible. It is through their collective efforts that such a comprehensive and valuable text was created and it is hoped that this text will aid in the continued growth of pharmaceutical hot-melt extrusion.

Dr. Michael A. Repka
Dr. Nigel Langley
Dr. James DiNunzio

Contents

Contributors

Fernando Alvarez-Núñez Pharmaceutical R&D, Small Molecule Process and Product Development, Amgen Inc., Thousand Oaks, CA, USA

Jörg Breitenbach Abbott GmbH & Co. KG, Ludwigshafen, Germany

Dominick Daurio Pharmaceutical R&D, Small Molecule Process and Product Development, Amgen Inc., Thousand Oaks, CA, USA

James C. DiNunzio Merck & Co., Inc., Summit, NJ, USA

Hoffmann-La Roche, Inc., Nutley, NJ, USA

Thomas Dürig Ashland Specialty Chemical Co, Wilmington, DE, USA

Jeff T. Gautschi Department of Chemistry, Oregon State University, Corvallis, OR, USA

Agere Pharmaceuticals, Inc., Bend, OR, USA

Costas G. Gogos New Jersey Institute of Technology, The Polymer Processing Institute, University Heights, Newark, NJ, USA

Andreas Gryczke BASF SE, Lampertheim, Germany

Abhay Gupta Division of Product Quality Research, Office of Pharmaceutical Science, US Food and Drug Administration, Silver Spring, MD, USA

Justin M. Keen DisperSol Technologies, LLC, Georgetown, TX 78626, USA

Mansoor. A. Khan Division of Product Quality Research, Office of Pharmaceutical Science, US Food and Drug Administration, Silver Spring, MD, USA

Johannes G. Khinast Institute for Process Engineering, Pharmaceutical and Process Engineering, Graz University of Technology, Graz, Austria

Peter Kleinebudde Institute of Pharmaceutics and Biopharmaceutics, Heinrich-Heine-University, Düsseldorf, Germany

Karl Kolter BASF SE, Ludwigshafen, Germany

Gerold Koscher Institute for Process Engineering, Pharmaceutical and Process Engineering, Graz University of Technology, Graz, Austria

Jay P. Lakshman Novartis Pharmaceuticals Corporation, East Hannover, NJ, USA

Andrew Loxley Particle Sciences Inc., Bethlehem, PA, USA

Daniel Markl Institute for Process Engineering, Pharmaceutical and Process Engineering, Graz University of Technology, Graz, Austria

Charlie Martin Leistritz, Branchburg, NJ, USA

James W. McGinity The University of Texas at Austin, Austin, TX 78712, USA

Dave A. Miller DisperSol Technologies, LLC., Austin, TX, USA

Karthik Nagapudi Pharmaceutical R&D, Small Molecule Process and Product Development, Amgen Inc., Thousand Oaks, CA, USA

Sakaé Obara Specialty Chemicals Research Center, Shin-Etsu Chemical Co., Ltd., Niigata, Japan

Elanor Pinto Ashland Specialty Chemical Co, Wilmington, DE, USA

Ali R. Rajabi-Siahboomi Colorcon Inc., Harleysville, PA, USA

Michael A. Repka Department of Pharmaceutics, Pii Center for Pharmaceutical Technology, School of Pharmacy, The University of Mississippi, University, MS, USA

Eva Roblegg Institute for Process Engineering, Pharmaceutical and Process Engineering, Graz University of Technology, Graz, Austria

Ashish Sarode Biomedical and Pharmaceutical Sciences, University of Rhode Island, Kingston, RI, USA

Sejal P. Shah Department of Pharmaceutics, School of Pharmacy, The University of Mississippi, University, MS, USA

Fumié K. Tanno Specialty Chemicals Research Center, Shin-Etsu Chemical Co., Ltd., Niigata, Japan

Daniel Treffer Institute for Process Engineering, Pharmaceutical and Process Engineering, Graz University of Technology, Graz, Austria

Sampada B. Upadhye Colorcon Inc., Harleysville, PA, USA

Patrick Wahl Institute for Process Engineering, Pharmaceutical and Process Engineering, Graz University of Technology, Graz, Austria

Acknowledgements: Some contributions to this book were made possible by Grant Number P20GM104931 from the National Institute of General Medical Sciences (NIGMS), a component of the National Institutes of Health (NIH) and HRSA/OFAM/DGMO Grant Number D1DHP20294.

Part I
Introduction and Equipment

Chapter 1
Melt Extrusion in Drug Delivery: Three Decades of Progress

Sejal Shah and Michael A. Repka

Abstract This chapter appraises the role of melt extrusion as a solubilization and bioavailability enhancement technique. The introductory chapter highlights major aspects of hot melt extrusion (HME) technology as applied in the pharmaceutical industry, particularly processing techniques, material considerations, recent innovative applications of melt extrusion in drug delivery system design, and a review of current HME-based formulations (marketed or under commercial development). The chapter also focuses on key development aspects of HME processes, such as material sparing screening approaches, process formulation relationships, and stability evaluation of prototype formulations, which emphasize the clinical and biological significance of this technique. In addition, it displays the journey and evolution of this important processing technology into an established pharmaceutical manufacturing platform. The chapter describes several case studies wherein melt extrusion has been utilized to develop commercial drug products.

1.1 Introduction

Although hot melt extrusion (HME) has been a workhorse technology in the plastics industry since the 1930s, research and development within the pharmaceutical manufacturing industry over the past two decades has propelled HME as an alternative "platform technology" for solid dosage form development. Over recent years, several studies have been published describing the use of HME as a technique of choice to address the formulation challenges of new drug molecules. Moreover, several aspects of HME have been extensively reviewed time and again (Breitenbach 2002; Crowley et al. 2007; Repka et al. 2007, 2008, 2012; Shah et al. 2013). Additionally,

M. A. Repka (✉)
Department of Pharmaceutics, Pii Center for Pharmaceutical Technology,
School of Pharmacy, The University of Mississippi,
University, MS, USA
e-mail: marepka@olemiss.edu

S. Shah
Department of Pharmaceutics, School of Pharmacy,
The University of Mississippi,
University, Oxford, MS, USA

M. A. Repka et al. (eds.), *Melt Extrusion*,
AAPS Advances in the Pharmaceutical Sciences Series 9,
DOI 10.1007/978-1-4614-8432-5_1,

the total number of HME-based patents (in comparison to patents granted for other formulation development techniques) has been on a steady rise worldwide.

This introductory chapter highlights major aspects of HME technology as applied in the pharmaceutical industry, particularly HME processing techniques, material considerations, recent innovative applications of HME in drug delivery system design, and a review of current HME-based formulations (marketed or under commercial development). Furthermore, the chapter introduces some of the important topics discussed in the subsequent chapters in this textbook and provides a perspective on the future of this important technique in the pharmaceutical scenario.

1.2 HME as a Drug Delivery Technology

In the HME process, the drug becomes embedded in a carrier system, usually consisting of one or more thermoplastic polymers (Prodduturi et al. 2007; Özgüney et al. 2009; Ghalanbor et al. 2010; Schilling et al. 2010b), low-melting waxes (Liu et al. 2001), sugar alcohols (Ndindayino et al. 2002a), or starch (Bialleck and Rein 2011) . Molten polymers or waxes function as thermal binders during the extrusion process and upon cooling and solidification, act as drug depots and/or drug release retardants. Additionally, functional excipients, such as plasticizers (Repka et al. 1999; Crowley et al. 2002; Wu and McGinity 2003; Crowley et al. 2004; Verreck et al. 2006; Schilling et al. 2007; Thumma et al. 2008a; Schilling et al. 2010a), diluents (De Brabander et al. 2000; Özgüney et al. 2009), pH and release modifiers (Verhoeven et al. 2006; Schilling et al. 2008), stabilizers (Thumma et al. 2008b), surfactants (Ghebremeskel et al. 2006; Thumma et al. 2008b), antioxidants (Crowley et al. 2002; Wu and McGinity 2003), and processing aids (Zhou et al. 1996; Liu et al. 2001) can also be incorporated in the HME process to enhance its efficiency and overcome process limitations on a case-by-case basis.

HME offers some distinct advantages over traditional pharmaceutical formulation techniques. Namely, it is a solvent-free technique, entails a continuous operation (necessitating fewer processing steps), does not require major downstream processing such as compression, and is known to improve bioavailability due to molecular dispersion of the drug in the final dosage form (Forster et al. 2001; Ndindayino et al. 2002; Breitenbach and Magerlein 2003). High processing temperatures, however, tend to limit the applicability of HME in processing thermolabile compounds. However, the combination of HME with other technologies, such as nanotechnology (Miller et al. 2007), powder coating (Sauer et al. 2007) and complexation (e.g., cyclodextrins) (Fukuda et al. 2008; Upadhye et al. 2010) has demonstrated the versatility and inclusiveness of HME.

The end result of HME technology has been a wide array of pharmaceutical dosage forms, such as pellets (Bialleck and Rein 2011) , granules (Liu et al. 2001), immediate and modified release tablets (Crowley et al. 2002; Gryczke et al. 2011), oral fast dissolving systems (Gryczke et al. 2011), transdermal (Repka et al. 1999; Repka and McGinity 2001), transmucosal delivery systems, transungual delivery systems

Table 1.1 Currently marketed and developed drug products produced utilizing hot melt extrusion (HME) technology. (Adapted with permissions from DiNunzio (2012))

Product	Indication	HME purpose	Company
Lacrisert® (Opthalmic Insert)	Dry eye syndrome	Shaped system	Merck
Zoladex™ (Goserelin Acetate Injectable Implant)	Prostate cancer	Shaped system	AstraZeneca
Implanon® (Etonogestrel Implant)	Contraceptive	Shaped system	Organon
Gris-PEG (Griseofulvin)	Antifungal	Crystalline dispersion	Pedinol Pharmacal Inc.
NuvaRing® (Etonogestrel, Ethinyl Estradiol depot system)	Contraceptive	Shaped system	Merck
Norvir® (Ritonavir)	Antiviral (HIV)	Amorphous dispersion	Abbott Laboratories
Kaletra® (Ritonavir/Lopinavir)	Antiviral (HIV)	Amorphous dispersion	Abbott Laboratories
Eucreas® (Vildagliptin/ Metformin HCl)	Diabetes	Melt granulation	Novartis
Zithromax® (Azythromycin enteric-coated multiparticulates)	Antibiotic	Melt congeal	Pfizer
Orzurdex® (Dexamethasone Implantable Device)	Macular edema	Shaped system	Allergan
Fenoglide™ (Fenofibrate)	Dyslipidemia	MeltDose® (Solid dispersion)	Life cycle Pharma
Anacetrapib (Under Development)	Atherosclerosis	Amorphous dispersion	Merck
Posaconazole (Under Development)	Antifungal	Amorphous dispersion	Merck

(Mididoddi et al. 2006; Mididoddi and Repka 2007), and implants (Ghalanbor et al. 2010).

In addition to the versatility (array of dosage forms), this technology offers several advantages in terms of varied application, such as bioavailability enhancement, controlled release, taste-masking, abuse deterrent (Bartholomaeus et al. 2012), and shaped delivery (direct shaping, powder, granules, spheres, films, and patches). Moreover, being a continuous process it has advantages of high throughput, online monitoring, less processing, and minimal process variables.

To date, there are several commercial pharmaceutical products in development using melt extrusion technology (Table 1.1; DiNunzio 2012) demonstrating the production and scale-up feasibility of melt extrusion. In addition, melt extrusion is also developing as an alternative formulation process for drugs in clinical trials.

1.3 Development of Hot Melt Extruded Products

Over the years, melt extrusion has seen a subtle transition from being a novel formulation technique to an essential platform technology in the drug development process. This paradigm shift is due to an overwhelming number of lipophilic drugs entering the development cycle. Melt extrusion finds two distinct roles in the drug development process. The first being solubility and bioavailability enhancement of new molecular entities and risk mitigation strategies for BCS class II drugs and second by life cycle management (LCM) of already commercialized drug products.

LCM is a successfully adopted, innovative, and preemptive strategy to help sustain the market share against strong competition from generic manufacturers or superior products in development. Most companies resort to reformulation or formulation changes as an alternative LCM strategy for blockbuster molecules. LCM through reformulation or by developing enhanced drug delivery systems encompasses a spectrum of innovative delivery technologies, not limited to modified-release for oral delivery, taste-masking, orally disintegrating tablets (ODTs), depot formulations, high-strength parenteral, inhalation, emerging technologies for bioavailability enhancement (melt extrusion, spray drying, and other solubilization techniques) as well as others. An inherent advantage of this approach leads to improvement in the product's therapeutic benefits, and patient's convenience, as well as compliance, thereby extending a product's profitable life. However, for the purpose of this chapter we would discuss melt extrusion as an independent formulation development strategy as applied in early- and late-stage pharmaceutical product development processes.

1.3.1 Early-stage Development

Early-stage development in melt extrusion encompasses various critically interdependent areas involving process considerations, stability assessment of prototype formulations, and performance evaluation (with respect to intended applications) of prototype formulations.

1.3.1.1 Processing Considerations

Processing considerations is a rather broad terminology covering material properties, instrument considerations, and process-formulation interplay. Systematic research over the last couple of decades has revealed that critical product quality attributes are directly dependent on both "formulation" and the "process" employed. It is important to note that the interplay between these determines the finished product attributes.

Material Properties

All of the materials used in melt extrusion (drugs, carriers, processing aids, release modifiers, etc.) should meet certain minimal pharmaceutical criteria, which includes well-characterized safety, and toxicological properties. Thermal stability of the individual components is a prerequisite for the extrusion process, although the short processing times encountered in HME also permit its applicability to thermolabile compounds. The incorporation of plasticizers may lower the processing temperatures encountered in HME, thus reducing the drug and carrier degradation. Incorporating various release-modifying agents can also modulate drug release from extruded systems.

Besides these, the active pharmaceutical ingredient (API) and carrier-related physicochemical properties such as melting point (T_m), glass transition temperature (T_g), melt viscosity, molecular weight, ionic nature, partition coefficient, chemical structure, stability and solubility (pH dependent if any), solubility parameters, number and type of hydrogen bond accepting or donating groups, physical state, hygroscopicity, lipophilicity, and others are key preformulation parameters. The API-related physicochemical properties as determined during preformulation studies guide the preliminary selection of carriers. Carriers are broadly classified as polymeric or nonpolymeric and its selection is based on the intended application. Table 1.2 provides a comprehensive list of carriers used in melt extrusion and its corresponding T_g and T_m values (Repka et al. 2012). The bottlenecks in employing melt extrusion as a processing technology are predominantly, very high-melting temperature of the API, thermal instability of drug and polymer, and high melt viscosity of the drug-polymer mixture. Hence, depending on the nature of the problem encountered, the development strategy is appropriately modified to be amenable to the melt extrusion process.

Screening Criteria and Selection

The pursuit to develop a melt extrusion-based prototype formulation, wherein the intended application is solubilization of the already identified drug-polymer combinations, can be further narrowed down by applying miniaturized (material sparing) screening methods. Such methods determine the drug-polymer miscibility or solubility and stability and employ a screening method coupled with a medium throughput analytical characterization tool. The screening method consists of films, quench cooled melts, and drug-excipient blend whereas the analytical tools generally consist of microscopy, spectroscopy, and calorimetric methods. A collective assessment of these miniaturized-screening experiments would assist in selecting the prototype drug-polymer combinations and drug loads at which the system is stable. Dai et al. (2008) present a comprehensive overview of the screening assays to rapidly identify solubility-enhancing formulations (Dai et al. 2008). Their review addresses three important facets of screening assays high-throughput nature (96 well formats), miniaturization (material sparing; small sample size), and automation (minimal manual intervention).

Barillaro et al. (2008) describe a high-throughput approach for evaluation of phenytoin solid dispersion. Their approach utilized automated solvent casting and subsequent dissolution testing as a screening method (Barillaro et al.

Table 1.2 Carriers used to prepare hot melt extruded dosage forms. (Adapted with permissions from Repka et al. (2012))

Chemical name	Trade name	T_g (°C)	T_m (°C)
Ammonio methacrylate copolymer	Eudragit® RS/RL	64	–
Poly(dimethyl-aminoethylmethacrylate-co-methacrylic esters)	Eudragit® E	50	–
Poly(methacrylic acid-co-methyl methacrylate) 1:2	Eudragit® S/L	160	–
Cellulose acetate phthalate	–	165	192
Poly(vinyl pyrrolidone)	Kollidon®	90–156	–
Poly(vinyl acetate)	Sentry® plus	35–40	–
Hydroxypropyl methylcellulose phthalate	–	137	150
Polyvinylpyrrolidone-co-vinyl acetate	Kollidon® VA64	101	–
Polyvinyl caprolactam-polyvinyl acetate-polyethylene glycol graft copolymer	Soluplus®	70	–
Hydroxypropyl methylcellulose	Methocel®, Benecel®	160–210	–
Hydroxypropyl methylcellulose acetate succinate	Aqoat-AS®	~ 120	–
Ethyl cellulose	Ethocel® Aqualon® EC	130–133	
Hydroxypropyl cellulose	Klucel®	Softens at 130 °C	Chars at 260–275 °C
Polyethylene glycol	Carbowax®	–17°C for MW 6000	37–63°C
Polyethylene oxide	PolyOx® WSR	–57 to –50 °C	62–67 °C
Polymethacrylates	Eudragit ® RSPM Eudragit ® E	52, 40 °C	–
Carnuba wax	–		81–86 °C
Glyceryl palmitostearate	Precirol ATO 5®	–	52–55 °C
Glyceryl trimyristate	Dynasan 114®	–	55–58 °C
Triglyceride tripalmitin	Dynasan 116®	–	61–65 °C

MW molecular weight

2008). Mansky et al. (2007) have extensively described screening methods addressing drug-polymer combinations particularly solubility-enhancing applications (Dai et al. 2007; Shanbhag et al. 2008). Figure 1.1 illustrates the process flow of one such screening method, describing the strategy to determine ideal drug-polymer combinations displaying improved drug solubilization.

One of the major limitations of the above-mentioned screening methods is lack of consistent predictability at a large scale, i.e., findings of such high-throughput screening methods may not hold essentially true for large-scale processing methods

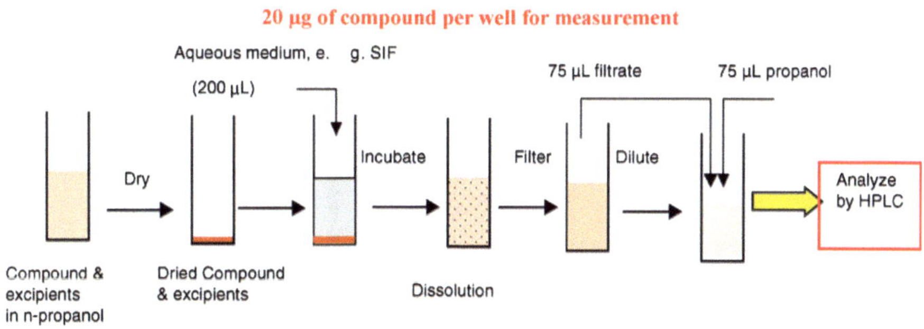

Fig. 1.1 Schematic diagram illustrating formulation screening process workflow. (Adapted with permission from Dai et al. (2007))

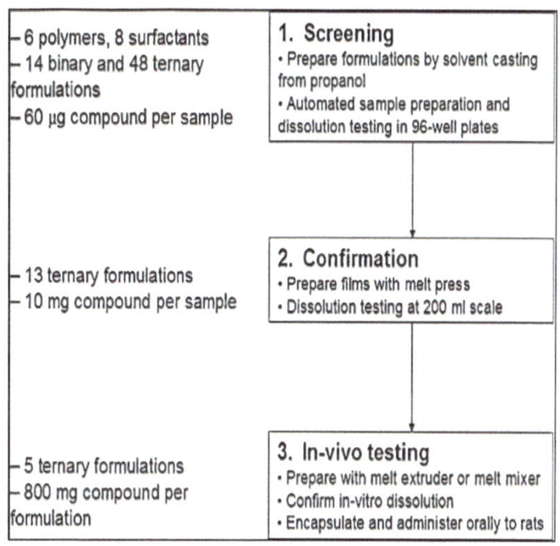

Fig. 1.2 Schematic illustration of the different stages of experimentation. At each subsequent stage, fewer samples are examined; the samples are larger and more compound is used per sample; and the formulation preparation and characterization methods become more relevant to traditional-scale formulation development work. (Adapted with permission from reference Shanbhag et al. (2008))

such as spray drying or melt extrusion. Shanbhag et al. (2008) developed a modified screening method for evaluating drug or polymer solid dispersion formulations. The method employs melt press as an additional "confirmatory step" to identify "hits," which is amenable to melt extrusion as a scale-up processing method (Fig. 1.2).

The authors evaluate the predictive value of the solvent casting-based screening method by selecting 13 hits from the screening stage, further processing them by melt press and testing the dissolution *in vitro*. The screening method successfully identified formulations, which upon melt processing (by melt press), demonstrated better dissolution profiles (Fig. 1.3).

Subsequently, five hits from the melt processing stage were selected for processing at a larger scale by melt extrusion. The oral bioavailability of all five formulations (hits) as evaluated in rats, exceeded that of the unformulated compound by a factor of about 20 (Fig. 1.4).

Polymer \ Surfactant	None	SLS	Poloxamer 407	Crodesta F110	Crodesta F160	Tween 80	Vitamin E TPGS	Volpo 10	Cremophor EL
None	0	0	1	8	31	56	85	87	89
Eudragit RS100	0	5	1	0	0	3	1	3	13
Kollidon VA 64	0	48	21	31	70	32	52	64	18
Plasdone K29/32	0	17	20	14	59	72	56	66	79
HPC SL	4	65	0	22	36	4	54	44	17
Eudragit L100	17	4	59	40	56	50	59	62	70
HPMCP 50	20	5	1	18	39	8	43	64	38

Fig. 1.3 Summary of the results of the screening experiments. The number in each cell is the average value of % dissolved after 1 h of incubation in SIF ($n = 3$ or 6). The color of the cells indicates whether % dissolved was < 25 % (*orange*), between 25 and 50 % (*yellow*), or > 50 % (*green*). The *top* row contains the results for surfactant-only formulations; the *left* column contains the results for polymer-only formulations; the *upper left* corner contains the results for the unformulated compound (no excipients), which was processed by solvent casting in an otherwise identical manner to the formulations; and the remaining cells contain the results for polymer or surfactant formulations. The 13 formulations that were scaled up using the melt press method are identified by the use of a **bold** or *underlined* font (e.g., 79) for % dissolved. (Adapted with permission from reference Shanbhaget al. (2008))

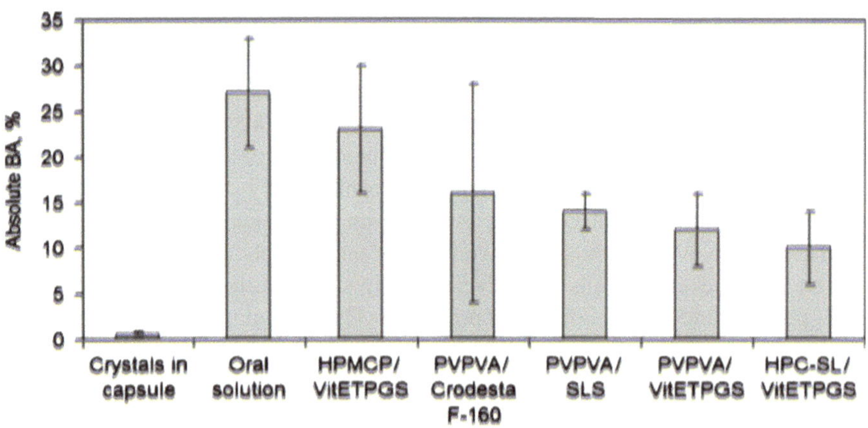

Fig. 1.4 Oral bioavailability of different formulations in rats ($n = 6$). (Adapted with permission from reference Shanbhag et al. (2008))

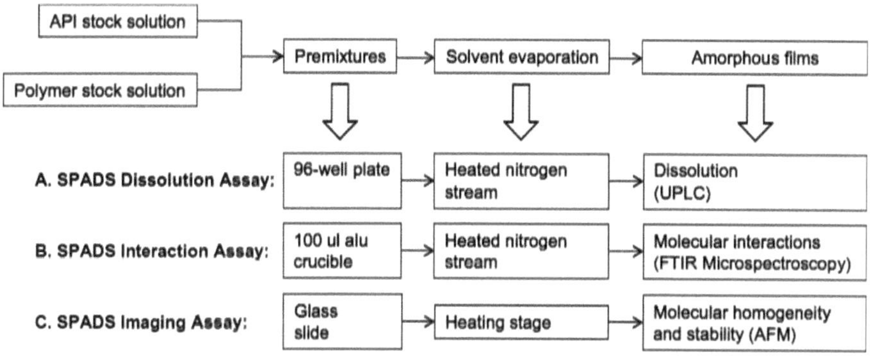

Fig. 1.5 Schematic representation of the experimental procedure of the different SPADS assays. (Adapted with permission from reference Wyttenbach et al. (2013))

While the earlier-described approaches focus on the assessment of the supersaturation potential of the polymer, there are some methods describing the evaluation of amorphous drug stabilization in the solid state (van Eerdenbrugh and Taylor 2010; Lauer et al. 2011; Weuts et al. 2011).

Wyttenbach et al. (2013) present an interesting strategy to identify amorphous solid dispersions (ASD) with maximum supersaturation and solid-state stability. The authors employed three different miniaturized assays (SPADS dissolution assay, FTIR microspectroscopy-based SPADS interaction assay, and atomic force microscopy-based SPADS imaging assay), combined in a two-step experimental flow to determine both the supersaturation potential and the stability of amorphous compositions thus formed with different drug-polymer combinations (Fig. 1.5; Wyttenbach et al. 2013).

The next step in the early-stage development is to extrude different preselected compositions by using scaled-down material-sparing extruders. This step would confirm its extrusion processability and determine the further need of processing aids, etc, which ultimately leads to a prototype formulation.

Stability Evaluation of Prototype Formulation

Stability is assessed during screening, and it is imperative to evaluate physical and chemical stability at all stages of development . The stability of developed melt extruded prototype formulations is assessed by following standard protocols and industry practice.

Stability evaluation is very critical particularly in the case of less stable ASD, at times resulting in the recrystallization of drug from solid dispersions during the manufacturing process, and subsequently during storage (Vasconcelos et al. 2007). The solubility and miscibility of drug in the polymer is directly related to the stabilization of an amorphous drug against crystallization (Qian et al. 2010).

Physical stability of ASD could be improved by the antiplasticization effect of polymers (increasing the viscosity of the binary system and decreasing the diffusion of drug molecules) that would raise the glass transition temperature of the system (Van den Mooter et al. 2001; Kakumanu and Bansal 2002; Sathigari et al. 2012). Hydrogen bonding and hydrophobic interactions between the drug and polymer are the primary driving forces for the formation of solid dispersions during melt extrusion, inhibition of drug crystallization during subsequent storage of melt exudates, and achievement and sustainment of supersaturation in the GI tract.

Storage temperature (Taylor and Zografi 1997; Matsumoto and Zografi 1999; Khougaz and Clas 2000; Miyazaki et al. 2004; Konno and Taylor 2006) and presence of moisture (inherent, during processing or storage) (Rumondor et al. 2009; Marsac et al. 2010) is an important factor resulting in recrystallization and amorphous-amorphous phase separation (Rumondor et al. 2011). The primary packaging component also needs to be properly designed to minimize the water permeation. Thorough characterization of the physicochemical properties of ASD and their corresponding *in vivo* behavior is required for the rational application of these systems in the pharmaceutical industry.

1.3.1.2 Instrument Considerations

Pharmaceutical melt extruders are specifically configured to meet current regulatory norms of manufacturing dosage forms. Extruders are available as single (smooth or grooved barrel), twin (corotating or counter rotating with intermeshing or nonintermeshing types), or multiscrew extruders (static or rotating central shaft). Single screw extruders (SSE) essentially consist of a one-piece screw, which continuously rotates within a barrel developing a good quality melt and generates enough pressures for extrusion. Relatively simple engineering design, combined with low cost and maintenance, make it the machine of choice for the production of virtually all extruded products. On the other hand, it faces limitations of high-pressure compression of dispersed particulates during melting, which leads to agglomerate formation and then insufficient shear deformation further results in poor mixing characteristics.

However, recently, Costeux et al. (2011) proved in 2011 that the SSE could have dominant elongational flow where melting occurred before compression by incorporating a series of spiral flow elongational mixers (SFEM) on to the screw. Due to its elongationally dominant feature, it breaks down blends of high viscosity ratios that cannot be dispersed by shear alone, hence, obviating the need to break the agglomerates. Unlike the twin-screw extruders (TSEs), all of the material can consistently pass through the elongational mixers thereby embedding single heat history. Melting and mixing mostly occurs near the hopper so that a significant part of the total length of the SSE plays a role of mixer. TSEs have overtaken SSEs in pharmaceutical processing and have become the dominant continuous compounding mixer for drug-polymer blends. The TSE works on a fundamentally different and superior principle that is not shear dominated. It melts the blend prior to the final compression of the melted blend, essentially preventing agglomeration of the ingredients.

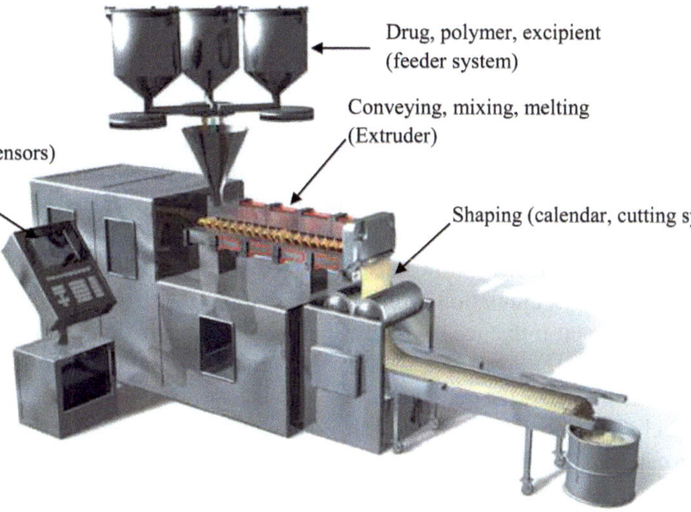

Fig. 1.6 Schematic representation of a typical pharmaceutical twin-screw extruder. (Adapted with permission from Breitenbach (2002))

A majority of the extruders manufactured for pharmaceutical applications are of twin-screw, corotating, and intermeshing types (Repka et al. 2012). As mentioned earlier, TSE overcomes the agglomeration limitation of SSE, additionally, it also offers better conveying, transport mechanisms of the feed, and provides intense mixing of the components. The rotational motion of a twin screw creates an environment of controlled temperature and pressure inside the barrel. High-capacity extruders are designed with temperature sensors and independent heating or cooling units in the barrel that efficiently maintain the individual zones at preset temperatures. The pressure arising from the friction of the moving material against the barrel walls eventually results in the ejection of material through the die cavity (Crowley et al. 2007).

The twin screws can orient in varying configurations depending on the desired level of shear and the speed of mixing or operation (Mollan 2003). Due to their efficient engineering design, adequate kneading, dispersion potential, and shorter and constant residence time (important for heat-sensitive feed material), TSEs with corotating intermeshing screws find widespread applications in pharmaceutical processing.

Although melt extrusion is considered as a unit operation, it consists of series of subprocesses as material feeding, powder conveying and degassing, melting and mixing, melt conveying and venting, and pumping, shaping, and cooling (Fig. 1.6).

Material Feeding: Extruder feeding systems mostly control the homogeneity of the product. Gravimetric (loss in weight) or volumetric feeders are generally used for pharmaceutical extruders. A volumetric feeder that operates by the principle of volume displaced by a pumping mechanism is most suited for preblends with good flow properties, while a gravimetric feeder employs transducers that measures loss in weight and quite consistently generates constant flow rates (Rauwendaal 2001).

Feeding of extruders can be either in the "starve-fed" or the "flood-fed" mode. For pharmaceutical processing, feeding is commonly conducted in a "starve-fed" mode, which results in efficient mixing of the feed material as opposed to flood feeding. Starve feeding uses gravimetric or volumetric feeders to dispense the material directly into screws, that prevents the accumulation of the feed material at the feed zone and thus the mass flow rate is independent of screw speed. At steady state, in a starve-fed mode, the mass flow rate at the feed zone is equal to the mass exiting the barrel and thus accumulation in the barrel is negligible. However, screw speed can have a significant influence on the residence time distribution of the feed material (Rauwendaal 2001). Feed rate, feed type (preblend or multiple), and pulsations in feeding rate influence the degree of fill, which in turn affect the homogeneity, thermal and mechanical energy input into the formulation. Additionally, the side-stuffing option can be employed for predensification of a low-bulk density powder to achieve better throughput. In case of liquid injection, a continuous stream could be achieved by maintaining sufficient backpressure to prevent clogging and variability. Moreover, feed locations on the length of barrel would influence the shear stress, temperature, and mixing experiences of the feed material. For instance, a heat-sensitive material can be added downstream to prevent thermal degradation or excessive shear stress; however, this technique may compromise its mixing capability (Schenck 2010).

Conveying and Venting: As the name indicates, conveying elements move the material from the feed section to further downstream regions in the forward flow direction. Conveying efficiency can be improved by altering certain characteristic geometric features of the conveying elements, such as flight width, pitch, and angle of helix. In addition, the internal to external diameter (Di/Do) ratio, which determines the extruder-free volume often limits the maximum feed rate, throughput, and torque attained. Sufficient venting in the feed section is essential to limit the detrimental effect on throughput due to entrained air and moisture from the feed material (Todd 1998).

Melting and Mixing: Melting of the feed material occurs by conductive thermal energy input via the heated barrel surface and by mechanical energy input supplied by the screws. The barrel heat melting process is likely to be influenced by factors such as uniform product temperature, poor thermal conductivity of the polymers, and volumetric scale-up. About 80–90 % of melting is achieved by viscous dissipation via frictional forces (including interparticle, material/wall, and material/screw friction) (Tadmor and Klein 1970; Todd 1993). The mechanical energy is mainly dissipated in three different ways: frictional energy dissipation (FED) from the frictional movement of polymer solid particles, plastic energy dissipation (PED) from the irreversible deformation of solid particulates, and viscous energy dissipation (VED) from the irreversible deformation, i.e., flow of the polymer melt. PED is essentially the energy dissipated during large and repeated plastic deformations of compacted feed stock particulates while still in the solid state and is much higher than the VED source of polymeric melts. The melting phenomenon is best described by following

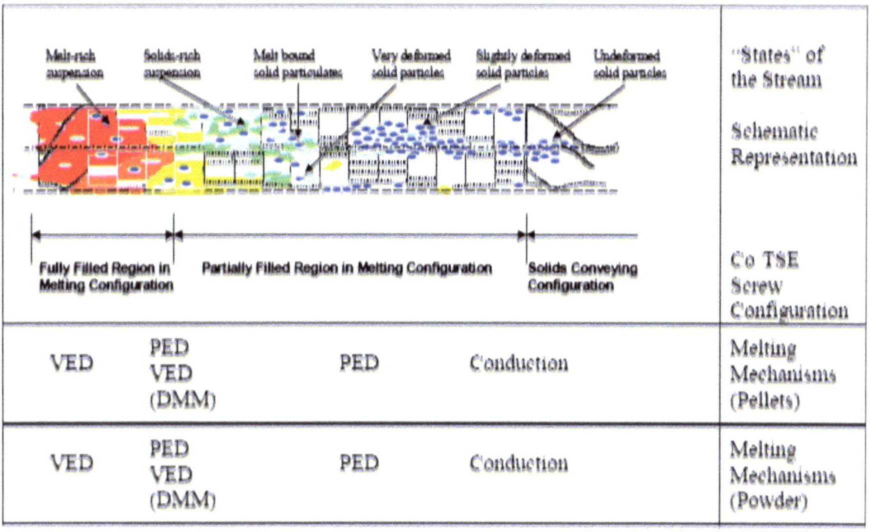

Fig. 1.7 Schematic representation of the evolution of melting of plastic pellets or powder in a co-TSE. (Adapted with permissions from Gogos (1998))

three perspectives: degree of fill; mode of conveying; and structural states of the change as it is being transformed from loose particulates to melt-rich suspensions or fully melted streams as depicted in Fig. 1.7 (Gogos 1998).

Specific energy (ratio of mechanical energy-drive motor reading in kW to feed rate, kg/h) describes the mechanical energy input to the material by the screws per unit mass and plays an important role in scale-up and optimization of the formulation. Various screw designs and configurations directly influence the specific energy, residence time distribution, and maximum shear stress imparted among most process responses. As the material transitions from solid to melt, a distinct change in flow characteristics is observed as the result of the temperature attained, which is greater than the glass transition temperature of the one or two components of the feed material (e.g., mostly polymers). This point marks the beginning of the melt residence time of the material in the barrel. While the fluidity of the polymers accelerates the dissolution of high-melting drugs across the length, it may also affect the degradation of some heat-sensitive compounds. Thus, controlled barrel temperature and effective screw profile and screw speed may result in increased heat transfer from the system and lower localized temperature ultimately leading to desired quality attributes of the extrudates. Generally, high-pressure builds are observed at the melt/mixing sections of the extruder due to the viscous nature of the melt and minimal conveyance afforded by screw geometry that promotes back mixing and delineates other unit operations along the process length (Todd 1998).

Melting and mixing of the feed stock is a result of the combination of material characteristics (viscoelasticity), equipment parameters (screw design- pitch, number of flights, channel depth, flight width, barrel clearance, design and number of

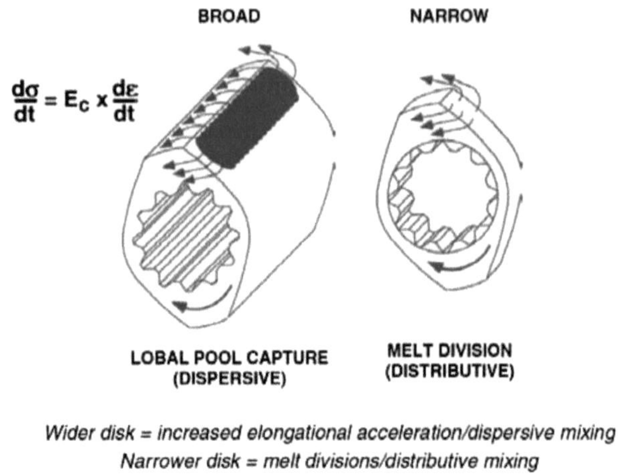

Fig. 1.8 Basic kneading section for dispersive and distributive mixing. (Adapted with permissions from Thiele (2003))

kneading paddles, length and number of mixing sections), and operational parameters, such as screw speed, feed rate, barrel temperature, and temperature imparted through viscous heat generation.

Corotating TSEs have the ability to mix the material longitudinally as well as transversely. The self-wiping nature of the two screws during rotation ensures that intermeshing TSE is self-cleaning. A screw configuration containing only conveying screw elements would mostly move material through the extruder via drag flow with minimal laminar mixing, hence, mixing or kneading elements become an essential component of screw design to attain good content uniformity. Primary mixing for melting and melt dispersion occurs in the kneading blocks or mixing elements, where alternating cycles of constant compression and expansion of the material are very conducive to supplying the forces required for rapid melting and for elongational flow of melts for both dispersive and distributive mixing. Distributive mixing is a type of mixing, wherein the material is divided and recombined in order to achieve better compositional and thermal homogeneity without distorting the individual morphological components. Distributive mixing is achieved using interrupted screw mixing elements (devices promoting division and reorientation of flow elements) and gear mixers or by using paddles with a narrow axial width (Fig. 1.8). The intense shear and shear stress facilitated by wider kneading elements mostly supports dispersive mixing with reduction in the size of morphological components and ultimately leads to molecular dispersion of the miscible components (Thiele 2003).

Melt Conveying and Venting: Residence time and residence time distribution (DiNunzio 2012) are important parameters and have an influence on the quality of the obtained extrudates. Residence time for a given process varies with change in screw speed or feed rate. However, screw design, temperature, and melt viscosity of the blend may also influence the residence time distribution, significantly influencing product attributes such as homogeneity and degradation. Similarly, melt residence time, i.e., the time from which material exists in the molten state across the length of the barrel, will have implications on product attributes during scale-up.

Venting is an essential step further downstream to the mixing section, to remove residual moisture or gas formation occurred during intense melting or the mixing process. Venting or degassing can be achieved by opening the top barrel section over the conveying section often assisted by vacuum to prevent bubbling or foaming of the extrudate.

Pumping, Shaping, and Cooling: The next stage in the extrusion process is to pump the molten extrudate through a die and thereby impart a definite shape for further downstream processing. Die geometry precision control may play a role in the final product with intended applications (e.g., transdermal films), which would require a slit die, annular dies for medical tubing formation. The molten extrudate may also be processed downstream via conventional unit operations (i.e., milling and compression) and in this case precise die geometry is not critical. Mostly circular dies with multiple strands are employed for rapid quench cooling. Pellets can be produced for multiparticulate dosage forms by passing the extrudate strand through a die face-cut pelletizer (Young et al. 2002).

The extruder die is also one of the high-pressure build-up sections of the barrel with nearly 100 % screw fill. Die geometry and the viscoelastic nature of the melt determine the increase in pressure, resistance, and temperature due to viscous heat dissipation resulting in maximum product temperature. Thus, changes to die design are warranted to minimize the pressure build-up for heat-sensitive or pressure-sensitive formulations.

The molten extrudates are often cooled using a conveyor belt with compressed air, or feeding through chilled stainless steel rolls. Cooled extrudates can be further milled into powder, which is either compressed into tablets or filled into capsules. Alternatively, final shaped dosage forms can be obtained from calendaring or injection molding of the melt. These molds yield the classic tablet, capsule shapes, or custom-designed shapes to suit various applications such as denture adhesives, vaginal rings, ear inserts, or pediatric friendly (enhance the esthetic appeal of the product) designs. Some of the dosage forms (made with melt extruded material) that have been previously characterized are films (Repka and McGinity 2001; Trey et al. 2007), pellets, spherical pellets (Young et al. 2002), punched tablets (Fukuda et al. 2006), injection-molded tablets (Quinten et al. 2009), rods, and granules (Robinson and Mcginity 2000) .

1.3.2 *Process-formulation Interplay*

While continuous processing is a salient feature of the melt extrusion technology, it is imperative to assess the influence of engineering aspects on the product quality. Simply said, it is important to understand the complex interplay between formulation and process during melt extrusion to obtain the desired product attributes. For melt extrusion, processing conditions have a direct influence on the product quality and performance for the intended application. Several aspects of the formulation, for instance, melt viscosity of the blend, solubility, heat and pressure sensitivity are directly influenced by processing parameters such as residence time distribution, feed rate, die design, screw configuration, and screw speed (Repka et al. 1999; Schilling et al. 2007; DiNunzio et al. 2010a). A systematic development approach regulating the key parameters that influences the critical quality attributes (CQAs) of the product would further assist in late-stage development with full optimization and scale-up using QbD principles. Hence, early development and later-stage optimization are both governed by the complex formulation and process interplay.

Lowinger (2011) extensively describes the influence of various process or engineering-related factors on the product quality. Modularity in extruder screws enabling change in screw profile may affect factors such as mechanical shear and residence time. Feed rate, type (preblend or multiple feeds), and pulsations in feeding rate, influence the degree of fill, which in turn affects the homogeneity, thermal and mechanical energy imparted into the formulation.

In addition, there are several studies describing the effect of process variables such as screw speed, barrel temperature, residence time on the product quality, and performance (Nakamichi et al. 2002; Verreck et al. 2003; Lyons et al. 2008) of melt extruded products. Liu et al. (2010) studied the dissolution behavior of indomethacin (melting point 162 °C) in Eudragit® EPO ($T_g = 48$ °C) matrices, processed using a batch mixer. The investigator identified the barrel set temperature, counterrotating twin-rotor screw speed, and residence time as important parameters that affected the dissolution behavior of the indomethacin. In addition, the study also revealed that for successful solubilization of the drug, the typical residence time for a particular process should be greater than the time needed for the drug to dissolve in the polymer melt. Both the barrel set temperature and screw speed increased the dissolution rate, which can be explained by the Noyes–Whitney equation (Eq. 1.1).

$$\mathrm{dC/dt} = \mathrm{DA/h}^*(\mathrm{C_s} - \mathrm{C_t}) \tag{1.1}$$

Their findings indicated that dissolution rates (dC/dt) of crystalline indomethacin in the molten EPO matrix could be increased by raising the temperature of the system (i.e., processing above the glass transition temperature of the carrier phase). Therefore increased equilibrium solubility (C_s) and diffusivity (D) of the drug in the molten carrier matrix was achieved due to reduced viscosity of the matrix at elevated temperatures. In addition, increased screw speed enhanced the available particulate surface area (A) of the API and decreased the boundary layer thickness (h), thereby contributing to increased diffusivity (D) and dissolution rate (dC/dt). Thus, each of the terms in the above equation (Eq. 1.1) can be altered to ultimately achieve high dissolution rates (dM/dt). Surface area (A) of the drug can be increased by micronization (Hughey et al. 2010), while increased solubility (C_s) and diffusivity (D) could be achieved by raising the processing temperature or by addition of cosolvents or plasticizers. Applying more shear (suitable screw configuration and screw speed) to the system would result in further reduction of the viscosity (decrease in boundary thickness, h) and thereby enhance the diffusivity and dissolution rate of the drug in the carrier phase. In addition, role of screw configuration in influencing the formation

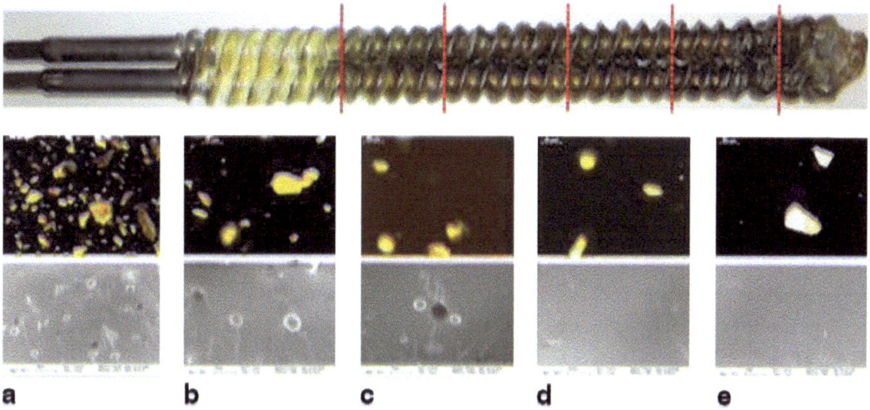

Fig. 1.9 The evolution of morphology along screw M0: polarized light micrographs (*top* row) and scanning electron micrographs (*bottom* row): **a** 8th lobe, **b** 13th lobe, **c** 19th lobe, **d** 24th lobe, and **e** 28th lobe. (Adapted with permissions from Liu et al. (2012))

of cocrystals and for melt granulations (Dhumal et al. 2010; Mu and Thompson 2012) has been reported. Additionally, Liu et al. (2012) evaluated the effect of four different screw configurations on the extent of dissolution of indomethacin in Eudragit®EPO. These researchers monitored the shifts of the indomethacin's benzoyl C=O stretch peak (1,692 cm^{-1} γ-crystal form and 1,684 cm^{-1} amorphous form) for the different screw configurations. Their findings suggested that the first kneading or mixing zones in the screw configurations promoted the transformation of crystalline indomethacin to the amorphous form. Figure 1.9 indicates that the crystalline indomethacin did not completely transform to the amorphous state until the 19th lobe when the screw without the kneading or mixing zone was used.

However, for the other three screws, which have at least one kneading or mixing zone, the transformation was complete at the 13th lobe (Fig. 1.10). FTIR analysis indicated that the first kneading or mixing zone promoted the complete dissolution of indomethacin into the EPO melt. However, their study does not identify the significance of the second kneading or mixing zone, which warrants further investigation.

Subtle changes in the processing conditions can remarkably alter the physicochemical properties of the formulations. A recent invention by a Roche scientist (Chatterji 2012) describes a novel bottom-up microcrystallization manufacturing process utilizing HME. In this case the drug substance exhibits low solubility and is also subject to extensive degradation and metabolism and hence not amenable to formulation by ASD. The patent illustrates the formation of controlled crystalline solid dispersion of API from its super cooled liquid state. The invention describes a process, wherein the crystalline API is converted to noncrystalline form by application of heat and shear up to one-fourth to three-fourth of the barrel length followed by a recrystallization zone in the remaining barrel length, wherein cooling is applied. The cooling of the barrel initiated API nucleation that promoted crystal growth while the shearing action of the screws evenly distributes the nuclei and hence controls the

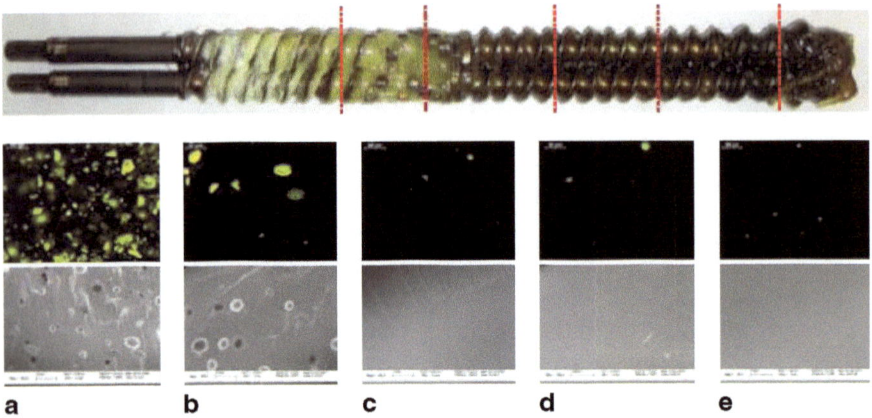

Fig. 1.10 The evolution of morphology along screw M1S: polarized light micrographs (*top* row) and scanning electron micrographs (*bottom* row): **a** 8th lobe, **b** 13th lobe, **c** 19th lobe, **d** 24th lobe, and **e** 28th lobe. (Adapted with permissions from Liu et al. (2012))

mean particle diameter of the newly formed crystalline API. The particle size of newly formed crystalline API is significantly less than the bulk API. Recrystallization of the API is controlled by carrier formulation design and HME process parameters such as barrel temperature and feed rate. The crystalline drug, dalcetrapib, is unstable in its amorphous state; hence, the aforementioned processing technique was found to be a rather suitable method of production. In addition, rapid dissolution was observed as compared to its micronized form. However, this approach faces technical challenges such as maintaining consistent batch-to-batch crystallization (particle size of crystals), and detection of residual amorphous drug content.

1.3.3 Late-stage Development: Scale-up Considerations and QbD-based Approach

In addition to providing a set of potential screening hits, early-stage development in HME also provides an insight into potential problems that may be encountered during scale-up and subsequent commercial-scale processing. Late-stage development of HME formulations particularly focuses on scale-up and adoption or optimization of process with necessary modifications to suit commercial-scale processing. Certain important material characteristics like melt viscosity, thermal sensitivity, and recrystallization potential as evaluated during early-stage development determine the ultimate scale-up strategy adopted during late-stage development.

While several authors have described scale-up approaches for HME-based formulations (Todd 1995; Dreiblatt 2003; Steiner 2003; McKelvey 2008; Schenck 2010; Lowinger 2011; Markarian 2012; Dreiblatt 2012), the present section provides a review of available approaches and describes a few successful case studies employing the listed scale-up approaches (Readers are advised to seek the earlier cited references for more detailed melt-extrusion optimization and scale-up procedures). Additionally, this section discusses the significance of experimental design in scale-up and highlights the importance of novel analytical methods as process analytical technology (PAT) tools in HME processing.

Dreiblatt (2012) reviewed different scale-up strategies for the HME process as listed below:

- Volumetric scale-up (length or diameter ratio, diameter ratio, and screw design)
- Power scale-up
- Heat transfer scale-up
- Die scale-up.

Based on the material properties and processing considerations or constrains any of the earlier-mentioned strategies can be adopted for scale-up. However, power and volumetric scale-up are considered popular scale-up approaches.

Schenck et al. (2008) discussed design space systematic development strategy for the production of solid solutions. The authors describe extrusion as a series of suboperations, material feeding, powder conveying and degassing, melting and mixing, melt conveying and venting, pumping, shaping, and cooling. Further, the authors describe extensive risk assessment on input parameters for each of the earlier-mentioned suboperations that ultimately map a subset of potential critical process parameters (CPPs).

These identified CPPs were expected to have a direct and measurable impact on product quality as described by CQAs. CQAs were defined from an understanding of solid solution product requirements relevant to the patient.

An energy input versus heat history plot for the HME process would result in three distinct zones or areas (Fig. 1.11):

- The inhomogeneous region (operating limit is characterized by low specific energy, short residence time, low product temperature, and/or limited mixing)
- Thermal degradation (upper end of HME processing conditions; can be avoided by use of plasticizers to reduce melt viscosity and the use of antioxidants for sensitive drugs)
- Acceptable extrudate quality (achievable by optimizing processing conditions, identifying CPPs and their effect on CQAs).

Almeida et al. present a case study describing the scale-up and in-line process monitoring of ethylene vinyl acetate hot melt extruded formulations (Almeida et al. 2012). Almeida et al. (2012) evaluated drug release and quality of EVA/drug matrices at different ethylene-vinyl-acetate concentrations (5 and 15 %), manufactured using two different hot melt extruders: a lab-scale mini extruder and a pilot-scale extruder. The process parameters used on both extruders (temperature and screw speed) and drug release from the matrices were compared.

The extrusion temperature for the pilot-scale extruder was modified such that constant pressure was maintained at the extrusion die. In addition, the screw speed for pilot-scale extruder was adjusted such that a balance was achieved between feeding

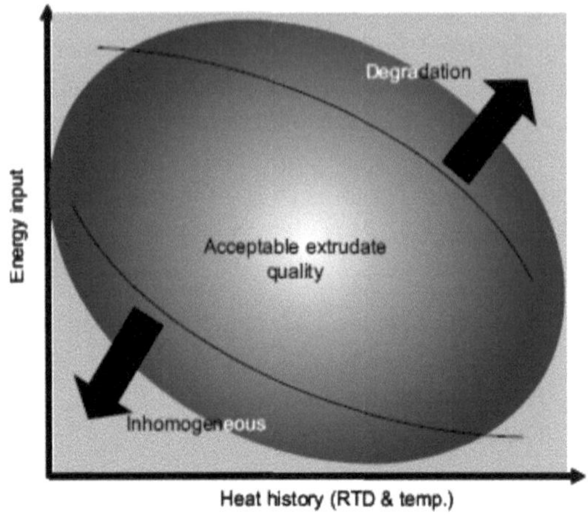

Fig. 1.11 A schematic showing the balance of energy input and heat history, with the extreme combinations of these, resulting in either degradation or inhomogeneous material. (Adapted with permission from Schenck (2010))

rate and screw speed. Despite the differences in extruder diameter, extrusion temperature, and screw speed, the drug release per surface area was similar for material extruded using the lab-scale extruder and pilot-scale extruder, respectively.

Furthermore, the scientists described the application of in-line monitoring tools employing Raman and NIR spectroscopy to evaluate the material behavior at a molecular level in the extrusion barrel as a function of the process settings.

Listro et al. (2012) describe the scale-up of a stable, solid dispersion of Eudragit® E/nifedipine and Eudragit NE 30 D formulation from an 18-mm twin-screw extruder to a 27-mm twin-screw extruder. Scale-up parameters were calculated with software, wherein scale-up parameters on the 27-mm extruder were based on the process parameters from the 18-mm extruder. The extrudates obtained from both extruders were evaluated using PXRD. Crystallinity was not observed in either case, and thus the researchers deduced that processing was comparable for both extruders.

Guns et al. (2012) discussed scale-up aspects of melt extrusion and made a comparative assessment of miconazole and Kollicoat® IR solid dispersion prepared by laboratory scale and pilot-scale extrusion, respectively. The properties of the final dosage forms were influenced by the processing parameters (screw speed, processing temperature, and feed rate) of the melt extrusion process.

Some of the characteristic features essential for successful late-stage development and scale-up include:

- Continuous nature of the process with limited or no manual intervention;
- Ability to characterize the process (online or inline process monitoring with state of the art analytical instrumentation—PAT).
- Scalability—availability of large-scale equipment.

Because of its inherently continuous nature, melt-extrusion-based processes can be successfully scaled up, characterized, and developed into full-scale manufacturing processes employing currently available methodologies (manufacturing equipment scale, process analytical characterization, etc.).

The complexity of analytical testing varies significantly depending on the developmental stage, from simple qualitative testing at an initial developmental stage to comprehensive and quantitative testing at the clinical manufacturing stage. Hence orthogonal testing methods are highly recommended for successful material characterization (Shah et al. 2013).

PAT has gained much attention and application in the pharmaceutical industry (Gupta et al. 2004; Rodrigues et al. 2006; Scott and Wilcock 2006; Gnoth et al. 2007; Qi et al. 2010; Rathore et al. 2010; De Beer et al. 2011; Dierickx et al. 2012a) and has been encouraged by FDA over the last decade (Administration 2004). PAT can be used as a tool to monitor, analyze, and characterize melt extrusion and chemical composition of the products in line to ensure final desired product attributes. Using PAT as an analytical tool, flow properties, polymer structures, polymer-drug interactions as well as concentrations of drugs and additives can be determined immediately downstream of the extrusion process.

Several spectroscopy-based analytical techniques (Raman and Near-infrared (NIR) spectroscopy) have been successfully employed for inline or online analytical monitoring of the melt-extrusion process (Tumuluri et al. 2008; Almeida et al. 2012).

1.4 HME: Applications in Therapeutics

Melt Extrusion (ME) over recent years has found widespread application as a viable drug delivery option in the drug development process. ME applications include trans-drug delivery, medical devices, taste-masking, solid-state stability enhancement, sustained drug release, and solubility enhancement.

1.4.1 Bioavailability Enhancement

With the advancement in molecular modeling and high throughput, drug innovation has led to discovery of numerous new chemical entities as potential therapeutic agents. However, obstacles associated with solubility, bioavailability, and toxicity undermine their development and commercialization. The enhancement of oral bioavailability of poorly water-soluble drugs remains one of the most challenging aspects of the drug development process. ASD containing drug molecules is a proven formulation strategy to improve the bioavailability of BCS class II and IV compounds. Melt extrusion has been successfully applied to enhance solubility of poorly soluble drugs (Lakshman et al. 2008; Tho et al. 2010; Feng et al. 2008b; Liu et al. 2012; Sakurai et al. 2012). Solubilization in the melt-extrusion process occurs through dispersion of a poorly soluble drug(s) in a polymeric (or lipid) carrier matrix. Hydrophilic carriers, including polyvinylpyrrolidone and polyvinylpyrrolidone-co-vinyl acetate, polyethylene glycols, polyethylene oxides, cellulose and derivatives, polymethacrylate derivatives, and a polyvinyl caprolactam-polyvinyl acetate polyethylene glycol graft copolymer have been successfully used to improve the solubility and hence bioavailability of water insoluble compounds using melt-extrusion techniques by the solid dispersion mechanism. The selection of a suitable carrier mainly depends on the following factors: the solubility/miscibility of drug-polymer systems, the polymer physicochemical properties, hydrogen bond interaction between the drug and polymer as well as hydrophobic drug-polymer interaction, stability, as well as the prerequisites of the final dosage forms. Successful solubilization in melt extrusion is determined by several factors, which are material (drug and carrier properties), process (processing temperature, shear, etc.), and equipment (design and operating conditions) related.

During melt extrusion of a drug within a polymer binary mixture, the drug could remain molecularly dispersed within the polymer or exist as an amorphous or crystalline phase. Melt-extrusion processing for solubility enhancement specifically aims towards generation of ASD. ASD not only provide the free energy benefits of an amorphous system, but also offer the maximum specific surface area at the molecular level thereby improving dissolution rates and increasing the apparent drug solubility.

It is essential for many delivery systems to demonstrate a rapid drug release profile and hence potentiate pharmacological activity. Numerous studies have been conducted on the solubility enhancement of water insoluble compounds using polyvinylpyrrolidone (PVP) or its derivatives as a carrier. Fu et al. prepared a nimodipine-PVP VA solid dispersion tablet using HME technology. The dissolution profile of ME tablets was compared with the commercial product Nimotop® (Fu et al. 2010). The melt-extruded tablets exhibited a faster release over Nimotop®, however, the ME tablets demonstrated a reduced dissolution profile after 20–30 min, indicating a recrystallization behavior due to supersaturated nimodipine in the dissolution medium. The bioavailability of ME tablets was further investigated in beagle dogs and the solid dispersion tablets illustrated a similar C_{max} and AUC as Nimotop®. A physically and chemically stable solid dispersion of a new chemical entity manufactured by Novartis Pharmaceuticals Corp. was developed by Lakshaman et al. (2008) using polyvinylpyrrolidone K30 (PVP-K30) as a carrier. The water insoluble API was converted to an amorphous state and the formulation demonstrated a more rapid release compared with the control formulation under nonsink conditions. Furthermore, both of the melt-extruded solid dispersions containing 20 and 30 % drug loading demonstrated a significantly higher bioavailability compared with the control formulation containing 20 % crystalline drug triturated with poloxamer in the dog model. The 20 % drug loading solid dispersion was selected for further studies and was proven as physically and chemically stable at controlled room temperature. A comparative study described the use of three HPMC-based polymers: HPMC 3cps, HPMC phthalate (HPMCP), and HPMC acetyl succinate (HPMCAS) as carriers in ME-based solid dispersions. Solid dispersions of a water insoluble drug, NVS981 were prepared using the HPMC based polymers at drug loadings of 20 and 50 %, respectively. The HPMC 3cps and HPMCAS demonstrated better extrudability compared with HPMCP (partially due to the lower T_g and viscosity). However, the NVS981-HPMCP exhibited a considerably higher release rate, even under nonsink conditions, which was attributed to the possible strong interactions between the polymer and the drug (Ghosh et al. 2011).

Soluplus®, a polyvinyl caprolactam-polyvinyl acetate-polyethylene glycol graft copolymer, was specifically designed and developed for melt extrusion. This polymer has both hydrophilic and hydrophobic elements, which therefore enhances the ability of forming a solid solution as well as increasing solubilization. Therefore, Soluplus® performs very well as a solubilizing carrier for HME technology. In addition, Soluplus® provides a wide range for potential candidates, from low-melting to high-melting drugs, including heat-sensitive APIs, due to its comparatively low T_g (70 °C). Linn and coresearchers studied the capacity of Soluplus® to increase intestinal absorption of three BCS II drugs, danazol, fenofibrate, and itraconazole (Linn et al. 2012). All of the extrudates containing the three drugs demonstrated significant increases in the plasma AUC compared with the crystalline pure drug or the physical mixture. In addition, the results of *in vitro* Caco-2 transport experiments confirmed the strong effect of Soluplus® on the enhancement of absorption behaviors of the model drugs. These data demonstrated the much-needed addition of an effective, extrudable polymer in the pharmaceutical formulators' toolbox. Amorphous itraconazole melt-extruded solid dispersion of HPMC increased drug release up to 90 % in 2 h as compared with 2 % release by powder blend containing crystalline itraconazole. No chemical degradation and no crystallization were observed when the solid dispersion packaged in an aluminum-polyethylene laminated pouch was placed at 25 °C/60 %RH and 40 °C/75 % RH for 6 months.

The drug molecules released from the amorphous dispersion exist as free drug or amorphous drug-polymer nanostructures or aggregates, precipitates, dissolving solid dispersions or bile salt micelles. Each of these forms attains pseudoequilibrium with free drug in solution, with free drug being absorbed as the system approaches thermodynamic equilibrium. Thus, maximizing free drug concentration and sustaining the elevated concentration through solid dispersion contribute to increased bioavailability of the amorphous drug.

1.4.2 *Taste-masking*

Development of palatable taste-masked formulations remains a challenging task in cases of pediatric and geriatric oral formulations. A significant challenge in pediatrics is the lack of appropriate formulations that have been designed to enhance compliance and ensure accurate dosing. As highlighted in the Best Pharmaceuticals for Children Act (BPCA), there is a need in the pediatric sector to develop technologies for the production of solid oral dosage forms that enable accurate dosages of critical medicines to be administered and are orally dissolvable. Several taste-masking approaches have been reported in the literature including dosage forms such as ODTs, quick dissolving films/strips, or pellets. Melt extrusion has been successfully employed in taste-masking applications by efficient entrapment of the bitter tasting

API in nanopolymeric matrices or through molecular interactions of bitter functional groups with the polymeric subgroups.

Gryczke et al. (2011) demonstrated the use of HME as an effective tool to mask the bitter taste of verapamil hydrochloride and Ibuprofen by forming drug-polymer complexes utilizing different grades of Eudragit® polymers. Gryczke et al. (2011) utilized Eudragit® EPO to design fast disintegrating taste-masked tablets with improved palatability and superior patient compliance utilizing HME techniques. Ibuprofen (a bitter drug) and Eudragit® EPO were coprocessed by HME to produce a solid dispersion and subsequently formulated as ODTs, which had physical properties similar to the marketed Nurofen® Meltlet fast disintegrating tablets (FDTs). The formulation demonstrated superior taste-masking ability with zero degree of bitterness and smoother mouth feel compared with the moderate roughness levels post administration of Nurofen®FDTs. Overall, the HME formulations, with similar disintegration properties and crushing strengths as that of the marketed formulation, had faster drug release rates and could mask the bitter taste effectively. In a US patent application by Sherry et al. (2007) ODTs of a nonsteroidal antiinflammatory drug (NSAID), paracetamol, were prepared by dry blending the drugs with sugar alcohols (xylitol, mannitol, and sorbitol) and subsequently melt extruding the mixture by heating to a temperature above the melting point of the sugars.The extrudates obtained after cooling and solidification were milled utilizing a cone mill. The milled extrudates or granules were mixed with other formulation components and compressed into tablets. The ODTs obtained by complete melting of xylitol (low-melting sugar alcohol) in the physical mixture were reported to be more robust compared with those produced by conventional dry blending processes (Sherry 2007). Recently, Maniruzzaman et al. (2011) processed four different bitter cationic APIs (propranolol HCl-PRP, diphenhydramine HCl-DPD, cetirizine HCl-CTZ and verapamil HCL-VRP) with two different grades of anionic polymers (Eudragit® L100 and Eudragit® L100–55) (Acryl EZE®). In addition, a comparative taste-masking study of extruded paracetamol with Eudragit® EPO and cross-linked polyvinylpyrrolidone (Kollidon® VA64) at different drug loadings (ranging from 30 to 60 %) was also conducted by the same group. The taste evaluation of the developed formulations was carried out by using an Astree e-tongue equipped with seven sensor tests and the generated data were analyzed using multidimensional statistics. Analysis of the data showed significant suppression of the bitter taste of all of the APIs with all of the polymers included in the study.

1.4.3 Enteric Release Systems

Employing HME technology, Andrews et al. (2008) designed enteric matrix tablets as an alternative to the traditional enteric coating process. Using 5-amino salicylic acid as a model drug, the enteric matrix tablets were produced using triethyl citrate (TEC) preplasticized Eudragit® L100–55 (enteric polymer) and citric acid (at 17 %w/w) (solid-state plasticizer). Polyvinyl pyrrolidone K30/Carbopol® 971P were employed as optional gelling agents. The extrudates, produced as cylinders were cut into tablets, referred to as "enteric tablets." The drug release from these enteric tablets depended upon the concentration of plasticizer, inclusion of citric acid or the gelling agent in the formulation, and the pH of the dissolution medium. The tablets showed excellent resistance to the acidic environment, releasing less than 10 % of the drug.

Schilling et al. (2010a) developed enteric matrix pellets utilizing a single-step HME-based method. Eudragit® grades L100–55, L100, and S100, and hydroxyl propyl methylcellulose acetate succinate (Aqoat) grades-LF and HF were employed as possible enteric carriers. Theophylline (model drug) pellets containing Eudragit® S100 as the enteric polymer were successfully extruded; these pellets released less than 10 % of the drug in the acidic medium. The researchers evaluated five different plasticizers as extrusion processing aids. PEG 8000 and citric acid, due to their inherent hydrophilic nature, did not promote retention of the enteric properties of polymers and resulted in drug leaching from the porous matrix. Presence of acetyl tributyl citrate (ATBC) and citric acid in formulations promoted partial drug solubilization but required higher extrusion temperatures and ultimately led to drug recrystallization. Triethyl citrate and methyl paraben identified as efficient plasticizers, had a significant impact on the microstructures of the extruded pellets as evaluated by SEM and PXRD studies. Both of these plasticizers produced dense pellets with the drug being homogeneously distributed in its original polymorphic form. The plasticized enteric matrix pellets containing 30 % drug and Eudragit® S100 demonstrated superior gastric protection of the drug and an acceptable extrudability.

Schilling and McGinity (2010c) identified a potential advantage of using HME in processing polymer-coated soft drug granules. Their study highlighted the applicability of HME in development of monolithic matrices containing enteric-coated particles. Spheronized pellets of Theophylline (model drug) obtained by wet extrusion, were enteric coated with Eudragit® L30D-55. The coated drug particles were further extruded with six different hydrophilic matrices. *In vitro* studies revealed that the release properties of the drug-incorporated particles were independent of their tensile strength (mechanical) and rather dependent on the type of polymeric matrix (chemical nature). A good correlation was observed between miscibility of the carrier, the coating polymers, and the drug release in acidic media. Among all the matrices investigated, poloxamer 407 incorporated about 40 % of enteric-coated pellets. The poloxamer 407-based matrices meet USP dissolution requirements for delayed-release dosage forms. Poloxamer 407 was least miscible with Eudragit® L30D-55.

1.4.4 Targeted Release Systems

The primary advantage of a targeted drug delivery system is the reduction of a drug's dosage (minimal amount required to elicit a pharmacological response) and thereby minimize or prevent the side effects associated with a high drug dose. Miller et al. (2008) demonstrated a marked improvement in the oral absorption of Itraconazole by the targeted intestinal delivery of supersaturated drug utilizing HME. ASD of Itraconazole were prepared using HME with Eudragit® L100–55 as a carrier matrix, and Carbopol® 974P, at 20 and 40 % of polymer weight, as a stabilizing agent. Carbopol® prolonged the release of the drug from the drug-polymer matrix *in vitro*. While the *in vivo* evaluation in Sprague-Dawley rats revealed that Carbopol® considerably decreased the intestinal absorption variability otherwise seen with the Eudragit® L100–55 based delivery system. In addition, a five-fold improvement in the intestinal absorption was observed with dispersion containing 20 % Carbopol® in comparison to the supersaturated drug dispersion compositions, which limited absorption mainly from the stomach.

HME-based techniques have identified newer applications of colon-specific targeting. An Eudragit®-based HME delivery system based on a photosensitizer was developed as a novel treatment option for curing serious infections of clinical significance caused by multidrug-resistant organisms such as *Enterococcus faecalis* and *Bacteroides fragilis* (Cassidy et al. 2011).

1.4.5 Innovative Techniques

1.4.5.1 Coextrusion

Coextrusion is the process of extruding two or more materials through a single die with two or more orifices arranged in a manner such that the extrudates merge and coalesce together into a laminar-layered structure before cooling and solidification. Each material is fed to the die from a separate extruder, but the orifices can be arranged so that each extruder supplies two or more layers of the same material. The final product obtained is in the form of a laminar structure with multiple layers, which may offer several advantages due to material and composition characteristics imparted by the individual extruded polymer. The die-design plays an important role in the shaping of the coextrudates. Multiple manifold dies are mostly used for flat-die extrusion, whereby both the layers are separately extruded and then combined inside the extruder to facilitate adhesion between the two layers with uniform distribution and minimal migration of the components of both the layers. While multilayer coextrusion is also possible using single manifold dies with the help of a feed-block (Toensmeier 2000; Giles 2005). In addition, various shaped dies such as flat dies, split dies, annular dies, side fed, and spiral mandrel dies can be used depending on the desired final application (Perdikoulias 2003; Han 2012).

An example of the coextrusion technology is the production of an intravaginal contraceptive thermoplastic ring currently on the market—NuvaRing®. This product contains the drug within a poly (ethylene vinyl acetate), pEVA reservoir, which forms the core while the crystalline pEVA (i.e., lower vinyl acetate fraction) on the exterior serves as a rate-limiting membrane to control drug release (van Laarhoven et al. 2002a, b).

The vaginal ring comprising two steroidal contraceptives (etonogestrel and ethinyl estradiol) is present in the molecularly dissolved state in a coaxial fiber consisting of two types of polyethylene vinyl acetate (EVA) copolymers. The coaxial fiber comprised a core polymer (EVA 28; 28 % vinyl acetate; high solubility and permeability of steroids) with the two steroids incorporated and enveloped within a thin polymer membrane (EVA 9; 9 % vinyl acetate; lower solubility and permeability of the steroids). This study demonstrated the application of HME in manufacturing structurally and functionally intricate dosage forms such as this vaginal ring, incorporated with more than one drug.

Intravaginal rings (IVRs) also find application in prevention of sexual transmission of the human immunodeficiency virus. Sustained release, dual segmented polyurethane intravaginal rings have been fabricated for two antiretroviral drugs with disparate hydrophilicity, dapivirine, and tenofovir (Johnsonet al. 2010). Drugs were individually formulated using appropriate polymers with due consideration of both drug and polymer hydrophilicities using a solvent-casting methodology followed by HME. The resulting drug-loaded rods were butt-welded to form dual-segment IVRs that were mechanically comparable to NuvaRing® IVR that were already in widespread therapeutic use. In addition, the formulated ring exhibited sustained release of medicament for a period of 30 days while it was found to be chemically and physically stable under accelerated stability conditions for a period of 90 days.

Coextrusion also finds application in design of combination dosage forms wherein two or more drugs are administered as a single dosage form to treat various illnesses such as infectious diseases, HIV, and cancer. The combination therapy may result in lower treatment failure rates and slower development of resistance. However, this methodology can have several limitations, such as physical/chemical incompatibility among excipients and/or drugs, inconsistent release rates, as well as inadequate or nonreproducible pharmacokinetic profiles.

Coextrusion has been used to develop a multilayer (core or coat) dosage form (Dierickx et al. 2012a). In this case, the core provides sustained drug release and the coat imparts an immediate API release. Dierickx et al. (2012a) carefully selected such polymers, which could be combined in a coextruded dosage form. A combination of polycaprolactone (core) and polyethylene oxide (coat) was selected for coextrusion trials, after consideration of their drug release profiles and extrusion temperature. Metoprolol tartrate (MPT) and hydrochlorothiazide (HCT) were used as sustained and immediate release model drugs, respectively. Fixed-dose combination minitablets with good *in vitro* and *in vivo* performance were successfully developed by means of coextrusion, using a combination of polycaprolactone and polyethylene oxide. The concept of coextrusion is promising for many novel dosage forms, including oral, transdermal, and implant applications.

1.4.5.2 Twin-screw Melt Granulation

Melt granulation is a thermal agglomeration process of powder blends containing binders, which softens/melts at an elevated temperature producing granules and solidifies when cooled to room temperature. Traditionally, melt granulation is carried out in a heat jacketed high-shear granulator or fluidized bed granulator. Recently, twin-screw extrusion has emerged as a platform for granulation, due to its favorable attributes of continuous manufacturing, modular configuration, energy efficient process, and yielding consistent product quality. The flow mechanics and heat transfer within an extruder are significantly different from a high shear mixer or fluidized bed in that it provides much more localized and controlled, scalable, efficient and uniform heat transfer. Moreover, the mechanism of granule formation is different for individual processes, including either or both as immersion or distribution and coalescence mechanisms, based on the binder droplet size and its method of addition (Abberger et al. 2002; Vilhelmsen and Schaefer 2005). However, if the binder droplet size is larger than the powder size, the immersion mechanism will be dominant and would result in denser granules and thereby influencing the drug release. Melt granulation using TSE utilizes hydrophilic or hydrophobic binders (5–30 %) with relatively low-melting temperatures in the range of 40–70 °C. Selection of the binders is based on drug-excipient compatibility and desired drug-release profiles.

Melkebebeet al. (2006) demonstrated the utility of twin-screw granulation for increasing the wetting and dissolution of a BCS class II compound, employing polyethylene glycols (PEG 400 and 4000) as binders. The veterinary drinking water formulation thus developed, had immediate drug release profile. Formulation and processing variables such as drug content, PEG 400/4000-ratio, surfactant (type and concentration) and granulation temperature had a profound influence on granule properties and dissolution characteristics. TSE granulation ensured the fine dispersion of drug into the PEGs, creating a microenvironment around the drug particles enhancing the dissolution rate. Complete drug release was obtained within 10 min for a formulation containing 10 % drug and 2 % (w/w) of surfactant (polysorbate 80 or Cremophor® RH40).

Lakshman et al. (2008) demonstrated the utility of the twin-screw extruder in a melt granulation process to enhance the compactibility of Metformin HCl, a poorly compactible high-dose drug. Tablets produced through melt granulation were compared with modified wet granulation and solvent granulation processes. Extruder processing parameters, such as barrel temperature, was above the glass transition temperature of the binder, HPC (130 °C), and well below the melting temperature of Metformin HCl (224 °C). In addition, high feed rate, and low screw speed were optimized to yield desired granules attributes. Thus, process optimization assisted the uniform and intimate mixing of the binder with the API leading to better compressible granules. The tablets prepared using melt granulation exhibited superior hardness and friability properties; moreover, the tablets were not sensitive to changes in atmospheric moisture levels. The process of melt granulation can successfully reduce the tablet sizes of high-dose medications (Metformin HCl) in this case and combination products by eliminating the need to add relatively large amount of excipients.

Similarly, Vasanthavada et al. (2011) applied melt extrusion in developing a modified-release tablet of a high-dose drug, imatinib mesylate. Tablets demonstrated excellent physical attributes and remained stable after six-month storage under stressed conditions. Recently, Mu and Thompson (2002) studied the mechanism of melt granulation inside the twin-screw extruder. The influence of process and formulation parameters, screw configuration, binder viscosity, particle size, and process temperature on granule development was studied. The study concluded that the immersion nucleation mechanism was dominant as construed from the characterization of granule size, granule strength, and binder concentration in the final product.

1.4.5.3 Cocrystal Formation using Melt Extrusion

The formation of cocrystals is considered as one of the potential techniques that can achieve the desired API properties of interest, such as improved dissolution rate, stability, and bioavailability (Smith et al. 2011). Cocrystals are defined as nonionic, multimolecular crystalline complexes. The crystal lattice of cocrystals is mainly composed of an active component and cocrystal former in a stoichiometric ratio, which is bound together through noncovalent interactions, primarily hydrogen bonding. The cocrystals formed might exhibit entirely different physical properties than the pure components. Many studies have proved that cocrystals are beneficial in improving the solubility, stability, bioavailability, and mechanical properties of the active drug. Cocrystals are mainly formed by solvent-based crystallization such as slurry, conversion, evaporation, antisolvent addition, and solid-based techniques such as cogrinding and melt extrusion. Of all the above-mentioned techniques, slow evaporation and cogrinding are the most widely used; however, both of these techniques possess scale-up limitations. Therefore, in recent years, significant progress has been made to use HME technology to produce pharmaceutical scalable, solvent-free and cost-effective cocrystals.

Dhumal et al. (2010) have successfully prepared cocrystals of Ibuprofen and Nicotinamide (1:1) utilizing melt-extrusion technology. In addition, these researchers have studied the effect of different process parameters such as screw configuration (low shear, medium shear, and high shear), screw speed (20, 30, and 40 rpm), and extrusion temperature (70, 80, and 90 °C) on formation of cocrystals and their purity. Preliminary DSC studies have reported that the eutectic temperature for Ibuprofen and Nicotinamide was 74 °C, therefore extrusion temperatures above and below the eutectic temperature were selected to study their effect on cocrystallization. PXRD and DSC studies were performed to characterize the crystallinity of the final products produced from ME. The formulations extruded at temperatures above the eutectic temperature (80 and 90 °C) were found to have high cocrystal purity. In addition, at the studied temperatures above the eutectic temperature, screw configuration exhibited a major role in the cocrystal formation and its purity. Residence time decreases as the screw speed increases, therefore the cocrystals produced at slower speed demonstrated higher cocrystal purity than those produced at higher screw speeds. Even though both the batches of cocrystals produced at 80 and 90 °C were above the eutectic temperature, the cocrystal produced at 90 °C showed higher purity than that at 80 °C. Of all the formulations, the cocrystals produced at high shear, higher temperature (90 °C), and lower screw speed demonstrated pure cocrystals with no traces of unreacted components. Thus, the authors further studied dissolution and stability on the cocrystals with high purity. The dissolution studies indicated that the cocrystals showed improved release profiles compared with pure Ibuprofen, micronized

Ibuprofen, and the physical mixture of Ibuprofen with Nicotinamide. In addition, these cocrystals were found to be physically stable for up to 6 months when stored at ambient conditions.

In another study, Liu et al. (2012) investigated the *in situ* formation of cocrystals in a single-step, utilizing ME technology, in order to achieve minimal thermal degradation of the active by extruding the heat-sensitive drugs at lower processing temperatures. In this study, Carbamazepine (CBZ) and Nicotinamide were used as a model drug and cocrystal former, respectively, while polyvinylpyrolidone/vinylacetate (PVP/VA 64), Soluplus® and HPMC were used as different carrier matrices. The active, coformer and the polymeric matrix were blended and fed through the extruder to obtain a solid dispersion with *in situ* cocrystallization of API during the process (Liu et al. 2012).

The authors noticed the possibility of the drug cocrystallization with Nicotinamide in PVP/VA 64 matrices during DSC characterization of the drug-polymer physical mixtures. Hot-stage microscopy studies revealed that the cocrystals were formed completely *in situ* during the extrusion process, in both Soluplus® and PVP/VA 64 matrices (a single melting point lower than the API); however, only partial cocrystallization occurred in HPMC matrices which was evident with melting of crystals at both 165 °C (melting temperature of the cocrystal) and 190 °C (melting temperature of Carbamazepine). These results were also confirmatory with FTIR and XRD data, which further demonstrated that cocrystal formation is instantaneous during melt-extrusion processing in PVP/VA 64 matrices. In addition, dissolution data in PVP/VA 64 matrices revealed that the release from cocrystals was relatively faster than compared with the pure API. The drug release from the physical mixtures and the drug-polymer melt extrudates was less than 80 % in 2 h, and about 100 % in 1 h, respectively, while the solid dispersion containing API-Nicotinamide-PVP/VA64 released 100 % API in just 20 min. These data demonstrated the influence of Nicotinamide as a hydrotropic agent and its formation of cocrystals simultaneously during the extrusion process. Similar dissolution results were observed for the extrudates formed from Soluplus® and HPMC matrices. Overall, the melting point of the API was drastically reduced due to the formation of the cocrystal, which helped in extruding the drug incorporated polymeric solid dispersions at lower temperatures. The processing temperatures could also be adjusted based on choice of the coformer employed during cocrystallization; however, the effect of coformer on drug recrystallization, physical stability, moisture sorption properties, as well as other parameters, should be considered for the production of a stable formulation.

Another example of cocrystal complexes that were successfully formed utilizing ME technology are Ibuprofen-Nicotinamide (Dhumal et al. 2010). In addition, Daurio et al. (2011) describe twin-screw extrusion as an effective, environment-friendly, and scalable method for producing cocrystals of Caffeine-Oxalic acid, Nicotinamide trans Cinnamic acid, Carbamazepine-Saccharin, and Theophylline-Citric acid. The authors identified this neat and liquid-assisted melt-extrusion technique as a viable alternative to the solution crystallization for formation of *in situ* cocrystals, where the temperature and the extent of mixing were indicated as the primary parameters that influenced cocrystal formation (Daurio et al. 2011).

1.5 Clinical and Therapeutic Potential

Development of solid dispersion-based formulations is a popular means to improve bioavailability of investigational drugs. HME of candidate molecules commonly results in ASD thereby improving their bioavailability.

This section describes the pharmacokinetics and bioavailability aspects of some of the marketed (and developmental) products developed using HME technology, highlighting the clinical and therapeutic significance of this platform technology. In addition, the section also highlights existing limitations in the comparative evaluation (*in vivo* evaluation models, *in vitro–in vivo* correlation) of melt-extruded/solid dispersion-based products and corresponding traditional dosage forms. Finally, the section discusses the need for standardizing the evaluation parameters (*in vitro* and *in vivo*) for solid dispersion-based therapeutics.

Kaletra® represents an excellent example of the successful application of melt-extrusion technology in the design of patient centric formulations and improvement of the overall therapeutic efficacy of the drug (Breitenbach 2006). Significant improvements of product attributes include improved storage stability—eliminating the need for refrigeration, reduced dosage, and reduced dose frequency). Kaletra® is an antiviral drug product containing HIV protease inhibitors lopinavir and ritonavir. Ritonavir by itself has minimal antiviral effects at the dosage levels used in the formulation, however, the coformulation of lopinavir with subtherapeutic concentrations of ritonavir (a potent inhibitor of cytochrome P450 3A4), ensures ideal pharmacokinetic profiles and enhances the bioavailability of lopinavir.

The very first form of Kaletra® was available as a soft-gelatin capsule (SGC). SGCs required refrigeration to maintain stability; the recommended adult daily dose comprised six capsules and had to be administered with food for optimal bioavailability of lopinavir. All of these factors together (storage, food, and multiple daily doses) complicated the anti-HIV therapy from the patient's perspective. Managing such a complicated therapy became particularly challenging in the worst affected African continent wherein refrigerated storage and an effective supply chain system (limited infrastructure) are not readily available.

The application of the Meltrex® melt-extrusion technology to Kaletra® has revolutionized anti-HIV therapy involving the lopinavir or ritonavir drug combination. The new form of Kaletra® is available as a tablet (as opposed to SGCs marketed initially) and does not require refrigerated storage. The total number of daily doses has been reduced to four tablets as opposed to six SGCs. The plasma concentrations of both drugs were more consistent independent of fed conditions, eliminating the requirement to administer the medication with food (Breitenbach 2006).

Klein et al. (2007) describe studies wherein the bioavailability of melt-extruded tablets of lopinavir/ritonavir (800/200 mg or 400/100 mg dose) was compared with equal doses of the SGC under moderate-fat meal conditions. The tablet formulation of Kaletra® was reported to be bioequivalent to the SGC post moderate fat meal based on comparable AUC-time curves for lopinavir and ritonavir. Thus, the tablet formulation resulted in consistent lopinavir and ritonavir exposure within and across studies and across meal conditions.

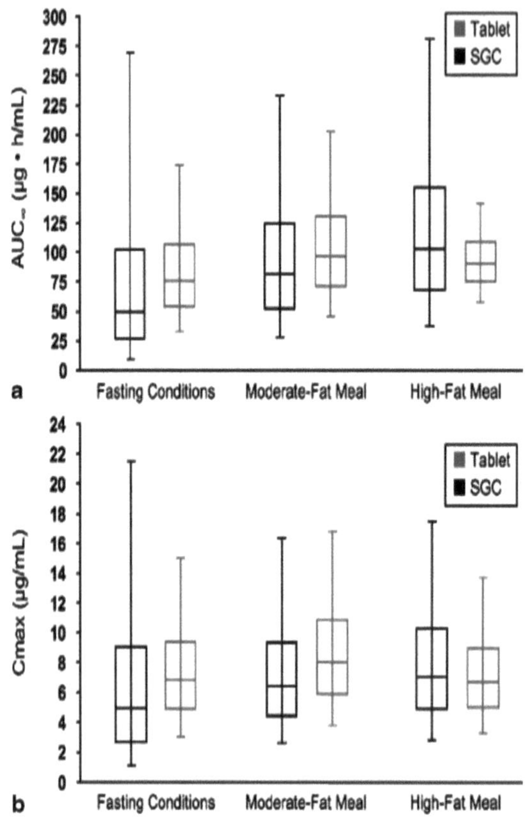

Fig. 1.12 The lower and upper quartiles (*box*), median (*horizontal line* in box), and 5th and 95th percentiles (whiskers) of the AUC (**a**) and Cmax (**b**) of lopinavir from a 400/100-mg dose as the SGC (*black*) or tablet (*gray*) formulation under various meal conditions. (Adapted with permission from (Kleinet al. (2007))

The experimental findings suggest that the melt-extruded tablet administered under different meal conditions (fasting, high fat, or moderate fat meals) provides lopinavir and ritonavir exposures similar to those with the SGC administered with a meal (recommended administration method on the product label).

The melt-extruded tablet showed decreased pharmacokinetic variability (lower %CV), particularly under fasting conditions. This is further confirmed in the probability distributions, wherein the interquartile and 5th and 95th percentile ranges for lopinavir concentrations are narrower for the tablet than for the SGC, regardless of meal status. The central values, interquartile ranges, and 5th and 95th percentile ranges for lopinavir concentrations achieved with the tablet under moderate- or high-fat meal conditions were within the ranges noted for the SGC under moderate- or high-fat meal conditions (Fig. 1.12). The reduced food effect and limited variability of drug plasma concentrations for the tablet result in consistent exposure to lopinavir and ritonavir, thereby minimizing the likelihood of extreme values for high concentrations associated with reduced tolerability (adverse effects) and for low concentrations associated with suboptimal efficacy (ineffective therapy).

Similar to Kaletra®, Meltrex® technology has also been used for designing drug products for Ritonavir (Norvir®), Ibuprofen (Klueglich et al. 2005) and Verapamil tartrate (Isoptin® SR) (Roth et al. 2009), with aims of stabilizing the drug or providing

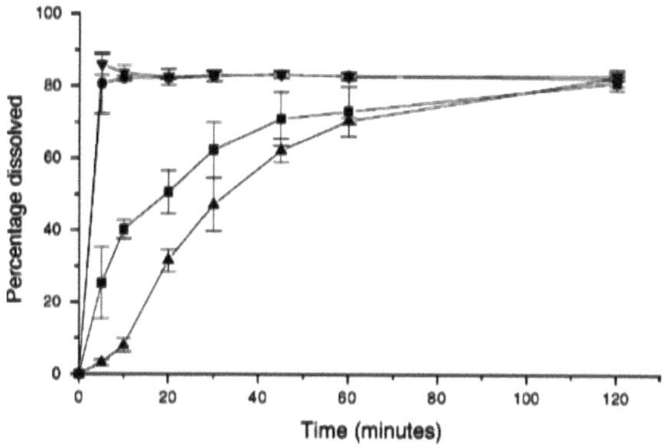

Fig. 1.13 Dissolution profiles of Sporanox® (▲), HPMC extrudate (■), Eudragit® E100 extrudate (▼), and Eudragit® E100-PVPVA64 extrudate (•) in capsules containing 100 mg itraconazole in SGF_{sp}. The dissolution (paddle method) was carriesd out in 500 ml SGF_{sp}, at 37 °C, 100 rpm. Error *bars* indicate the S.D. (Adapted with permission from Six et al. (2005))

better pharmacokinetic profiles for the drugs (the readers are directed to Chap. 18 for further information on Meltrex® technology and the associated drug delivery systems).

Azole antifungals, such as ketoconazole, itraconazole, and posaconazole are weak bases, exhibit a pH-dependent solubility, and have limited bioavailability. While traditional drug delivery approaches have resulted in effective formulations of ketoconazole, itraconazole was one of the first commercially marketed ASD (Sporanox®) (Gilis 1997).

While the ASD (Sporanox®) significantly improved the oral bioavailability of itraconazole, it presented issues such as variable absorption and significant food effects, particularly with acidic beverages (Jaruratanasirikul and Kleepkaew 1997).

Several formulations, including melt-extruded dispersions have been made to improve the oral bioavailability of itraconazole. Six et al. (2003a) describe the preparation of solid dispersion formulations following melt extrusion (Verreck et al. 2003; Six et al. 2004) and their subsequent bioavailability assessment in comparison with Sporanox® following a double-blind single-dose cross-over study in human volunteers (Six et al. 2005).

Solid dispersions of itraconazole (40 % w/w) and HPMC 2910, Eudragit® E100 or a mixture of Eudragit® E100-PVPVA64 were manufactured by HME and filled into gelatin capsules (Six et al. 2005).

The *in vitro* release behavior of the solid dispersions with Eudragit® E100 or Eudragit® E100 and PVPVA64 mixture, showed rapid dissolution of the drug with about 80 % drug being released in the first 10 min. The dissolution of itraconazole from HPMC dispersions or Sporanox® was much slower with about 80 % drug being dissolved after 2 h (Fig. 1.13).

The *in vivo* profiles, however, were significantly different from *in vitro* release behavior indicating a poor *in vitro/in vivo* correlation. The mean AUC values (0–72 h) after administration of Sporanox® and the HPMC formulation were comparable and were higher than that for the Eudragit® E100 and Eudragit® E100-PVPVA64 formulations. In fact, the *in vivo* profiles for formulations based on Eudragit® E100 and Eudragit® E100-PVPVA64 showed the lowest mean AUC and C_{max} respectively. The authors attributed the observed dissolution behavior to slow dissolution kinetics of HPMC as compared with that of Eudragit® E100 or Eudragit® E100 and PVPVA64 mixtures. Furthermore, based on the poor IVIVC, it was hypothesized that the rapid diffusion or dissolution of the polymer into the bulk phase, renders the drug unprotected and prone to precipitation.

Miller et al. (2008) present an interesting supersaturation-based strategy for delivery of itraconazole. The strategy is to control drug release so as to retard its precipitation as pH increases (acidic-to-neutral pH transition as the drug moves into the small intestine), thereby extending the absorption window of itraconazole.

Their findings confirmed the hypothesis that supersaturation of itraconazole following an *in vitro* pH transition correlates directly to *in vivo* absorption (Miller et al. 2008). It was observed that the polymeric carriers Methocel™ E50 and Eudragit® L 100–55 for amorphous itraconazole solid dispersions produced substantially greater *in vivo* absorption due to improved supersaturation of itraconazole in the small intestine (Miller et al. 2008). Of the two polymeric carriers, Methocel™ E50 was preferred due to reduced variability. The authors also proposed incorporation of Carbopol® 974P into the Eudragit® L 100–55-based solid dispersion system for stabilization of itraconazole supersaturation and the enhancement of oral absorption.

Similar to itraconazole, posaconazole dissolves in acidic conditions but rapidly precipitates in the environment of the upper small intestine. Posaconazole is commercially marketed as a conventional suspension formulation (Noxafil®) and exhibits highly variable bioavailability with a significant food effect. Fang et al. (2009) describe a novel posaconazole formulation developed using HME. The formulation is essentially a solid dispersion of posaconazole with hypromellose acetate succinate (HPMCAS). Since posaconazole degrades rapidly at temperatures above 160 °C, the process is designed such that posaconazole is solubilized at temperatures below its melting point. Such a uniquely processed formulation is stable and has improved bioavailability. Extrudates are finally prepared as a tablet(s) or capsule containing a 100-mg dose of posaconazole (Krishna et al. 2012).

Bioavailability studies indicate that under fasted conditions the mean plasma concentrations of posaconazole were significantly higher for the solid dispersion formulations (tablet or capsule) compared with those for the posaconazole oral suspension. Similar to the fasted conditions, under fed conditions, posaconazole solid dispersion formulations also demonstrated higher mean plasma concentrations for posaconazole compared with the oral suspension.

The plasma concentrations for posaconazole were considerably higher for the oral suspension (2.5- to 3-fold) under fed conditions; however, the drug plasma concentrations for tablet and capsule formulations remained significantly unaffected by food. Additionally, the tablet and capsule formulations demonstrated less intersubject variability in peak plasma concentrations and total exposures (AUC) compared with the oral suspension under fasted conditions.

Cholesteryl Ester Transfer Protein (CETP) inhibitors are a class of poorly soluble compounds under clinical development for reducing the risk of atherosclerosis. Pfizer (Torcetrapib), Merck (Anacetrapib), and Hoffmann-La Roche (Dalcetrapib) have reported data for formulations of CETP inhibitors and their use in preclinical and clinical trials. Based on clinical findings up to January 2013, the development of

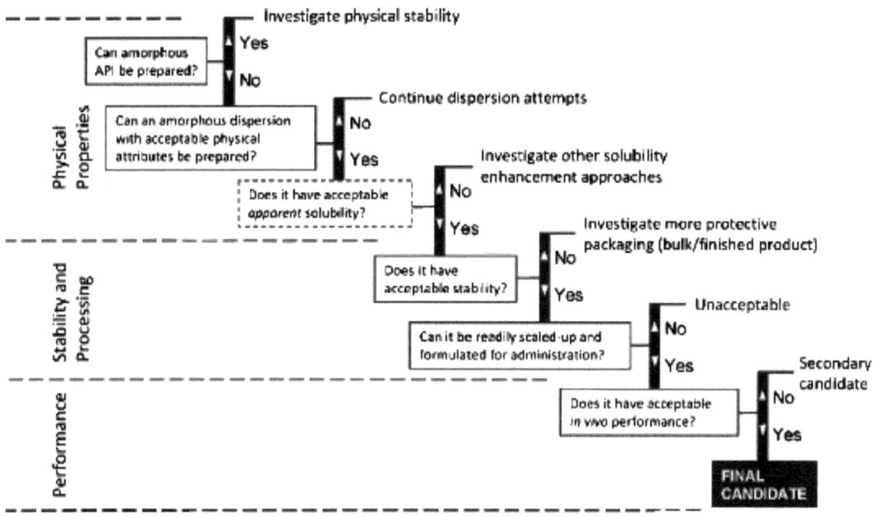

Fig. 1.14 Decision tree for dispersion selection. (Adapted with permission from Engers et al. (2010))

most of these drugs has been terminated, potentially due to serious adverse events (Torcetrapib) or lack of meaningful clinical improvement despite HDL increases (Dalcetrapib).

Despite limited clinical success with CETP inhibitors, design of solid dispersion-based formulations (either by spray drying or HME) has become a popular strategy for formulating these poorly soluble molecules (Dwayne 2012; Geers 2010; Miller 2011; Chatterji 2012). In most cases the solid dispersion-based formulations serve as "biosuperior/biobetter" drug products at later stages (after a successful phase I evaluation of liquid filled capsule (LFC) or a similar drug product) in the clinical development cycle. This strategy is very similar to the LCM strategy adopted for previously approved drugs. In fact, an anacetrapib-incorporated HME tablet was introduced in phase III following limited phase I evaluation in comparison to an LFC formulation (Krishna et al. 2011).

Designing ASD (by spray drying or melt extrusion) undoubtedly remains a popular strategy for development of drugs wherein traditional formulation methods have not proved successful. Engers et al. describe a decision tree approach for design of ASD (Fig. 1.14) (Engers et al. 2010).

However, there are some areas, which need to be carefully evaluated before considering the design of solid dispersions. Poor *in vitro/in vivo* correlation (IVIVC) and lack of standardized *in vitro* evaluation methods are a major concern in the evaluation of ASD. Newman et al. (2012) have carefully reviewed important issues related to the assessment of ASD. Findings from about 40 studies reporting dissolution and bioavailability studies of drugs were reviewed. These investigators found that amorphous dispersions improved bioavailability in about 82 % of the cases, amorphous dispersions possessed lower bioavailability than the reference material in about 8 % of the cases and in about 10 % of cases the amorphous dispersions had similar bioavailability as the reference material. A careful analysis of these studies revealed several *in vitro* (dissolution testing and equipment, content and amount of dissolution media, sink or nonsink conditions, agitation rates, media pH, dissolution characteristics of the polymer, and dispersion particle size) and *in vivo* (animal species utilized, fasting versus fed conditions, and regional differences in gastrointestinal (GI) content and volume) variables that could have influenced the results. Their findings elicited several recommendations, emphasizing the need for standardization of *in vitro* and *in vivo* evaluation parameters.

In addition to the example of itraconazole (discussed earlier), poor IVIVC has been a pressing concern in evaluation of HME-based drug products and has been highlighted in several independent studies.

Zheng et al. (2007) investigated the *in vitro* dissolution properties and oral bioavailability of three solid dispersions of nimodipine in beagle dogs. The solid dispersions were compared with pure nimodipine, their physical mixtures, and Nimotop®. Nimodipine solid dispersions were prepared by a HME with hydroxypropyl methylcellulose (HPMC, Methocel E5), polyvinylpyrrolidone/vinyl acetate copolymer (PVP/VA, Plasdone S630®), and ethyl acrylate, methyl methacrylate polymer (Eudragit® EPO).

Bioavailability studies suggested that Nimotop® had the highest *C*max followed by the Eudragit® EPO solid dispersion. The mean AUC (0–12hr) for Nimotop® and the Eudragit® EPO solid dispersions were comparable but lower for other formulations (HPMC and PVPVA). This *in vivo* finding is in contradiction to the observed *in vitro* drug dissolution rates wherein the formulations were ranked as PVPVA > Eudragit® EPO > HPMC solid dispersions. *In vitro* dissolution studies in a biorelevant media containing 0.1 N hydrochloric acid and 0.05 % (w/v) sodium dodecyl sulfate (SDS) showed that the solid dispersion with Eudragit® EPO dissolved significantly faster (due to the acidic pH) than all formulations except Nimotop®, supporting the observed *in vivo* data for Eudragit® EPO solid dispersions.

Interestingly, there are several instances (particularly implants) where the *in vivo* drug release from melt-extruded systems has been fairly well characterized and suitable *in vitro* predictive models have been developed. Rothen-Weinhold et al. describe an implant for the controlled release of Vapreotide pamoate (a somatostatin analogue). Vapreotide implants were prepared using poly(lactide)-co-glycolide (PLGA), a biodegradable carrier, by an extrusion method. The *in vivo* release properties of Vaperotide as assessed in a rat model were shown to be affected by processing/formulation-related factors such as drug loading, polymer molecular weight, copolymer composition, and end group modifications.

Amann et al. (2010) describe the *in vivo* evaluation of a biodegradable implant containing the antipsychotic drug risperidone consisted of 60 % of PLGA prepared by HME. The authors were able to demonstrate a consistent *in vitro/in vivo* correlation for the developed implants in a rat model.

Melt extrusion as a technology is not just limited to design of delivery systems for synthetic drugs discovered by traditional chemical means but finds interesting applications with Chinese polyherbal/herbal formulations. Zhang et al. (2009) have reviewed technologies employed in the recent past for improving bioavailability of Chinese herbal medicines.

Luo et al. (2013) describe an application wherein the dissolution and bioavailability of Ginsenosides is improved by melt extrusion and cogrinding. These findings are indicative that in the future, melt extrusion will find broader applications to improving herbal drug attributes and not just be limited to areas such as moisture proofing of herbal extracts (Chen et al. 2010).

1.6 Conclusions and Future Perspectives

Over the past three decades, HME has transformed itself from an adapted technology (modified from the plastics industry) into a manufacturing platform within the pharmaceutical industry. HME has not only resulted in several successful drug delivery systems but has also opened new research avenues such as shaped delivery, continuous manufacturing, and online process characterization. While melt extrusion has been successfully applied to mainstream industrial pharmaceutics (e.g., bioavailability enhancement, taste-masking, trans delivery systems, controlled release drug delivery, and stable dosage forms), the future of this technology is also applicable in areas, such as injection molding (for shaped delivery), cocrystal formation or development, coextrusion, melt granulation, and supercritical carbon dioxide-based extrusion (foamed drug delivery systems). In addition to these applications, fundamental research and manufacturing areas such as scale-up technology and analytical characterization development will also be embraced as the "norm" for improvements in product development. In summary, for the future, melt extrusion as a technology will encompass significant applications throughout numerous pharmaceutical areas, which remain yet unresolved by traditional pharmaceutical techniques.

References

Abberger T, Seo A et al (2002) "The effect of droplet size and powder particle size on the mechanisms of nucleation and growth in fluid bed melt agglomeration". Int J Pharm 249(1–2):185–197

Administration F. a. D. (2004).) Process Analytical Technology Initiative, Guidance for Industry, PAT—a framework for innovative pharmaceutical development, manufacturing and quality assurance. U. S. D. o. H. a. H. S. F. a. D. administration

Almeida A, Saerens L et al (2012) "Upscaling and in-line process monitoring via spectroscopic techniques of ethylene vinyl acetate hot-melt extruded formulations". Int J Pharm 439(1–2): 223–229

Amann LC, Gandal MJ et al (2010) "*In vitro-in vivo* correlations of scalable PLGA-risperidone implants for the treatment of schizophrenia". Pharm Res 27(8):1730–1737

Andrews GP, Jones DS et al (2008) "The manufacture and characterisation of hot-melt extruded enteric tablets". Eur J Pharm Biopharm 69(1):264–273

Barillaro V, Pescarmona PP et al (2008) "High-throughput study of phenytoin solid dispersions: formulation using an automated solvent casting method, dissolution testing, and scaling-up". J Comb Chem 10(5):637–643

Bartholomaeus JH, Arkenau ME, Galia E (2012) "Opioid extended-release tablets with improved tamper-resistant properties". Expert Opin Drug Deliv 9(8):879–891
Bialleck S, Rein H (2011).) "Preparation of starch-based pellets by hot-melt extrusion". Eur J Pharm Biopharm 79(2):440–448
Breitenbach J (2002) "Melt extrusion: from process to drug delivery technology". Eur J Pharm Biopharm 54(2):107–117
Breitenbach J, Magerlein M (2003). Melt extruded molecular dispersions. In Ghebre-Sellassie I, Martin C. (eds) Pharmaceutical extrusion technology, vol 133. Marcel Dekker Inc., New York
Breitenbach J (2006) "Melt extrusion can bring new benefits to HIV therapy: the example of KaletraÂ® tablets". Am J Drug Deliv 4(2):61–64
Cassidy CM, Tunney MM et al (2011) "Development of Novel Oral Formulations Prepared via Hot Melt Extrusion for Targeted Delivery of Photosensitizer to the Colon". Photochem Photobiol 87(4):867–876
Chatterji A (2012).) Process for controlled crystallization of an active pharmaceutical ingredient from supercooled liquid state by hot melt extrusion. U. S. P. T. O. (ed), Roche.
Chatterji A, Miller DA, Sandhu H, Shah N (2012).) Process for controlled crystallization of an active pharmaceutical ingredient from supercooled liquid state by hot melt extrusion. www.uspto.gov. U. S. P. a. T. Office, United states of America
Chen H, Shi S et al (2010) "Combined application of extrusion-spheronization and hot-melt coating technologies for improving moisture-proofing of herbal extracts". J Pharm Sci 99(5):2444–2454
Costeux S, Barger M, Luker K, Badami A, Walton K (2011).) "Facile TPO dispersion using extensional mixing". 69th Annual Technical Conference of the Society of Plastics Engineers 2011. ANTEC, vol 1, pp 22–27
Crowley MM, Fredersdorf A et al (2004) "The influence of guaifenesin and ketoprofen on the properties of hot-melt extruded polyethylene oxide films". Eur J Pharm Sci 22(5):409–418
Crowley MM, Zhang F et al (2002) "Stability of polyethylene oxide in matrix tablets prepared by hot-melt extrusion". BioMaterials 23(21):4241–4248
Crowley MM, Zhang F et al (2007) "Pharmaceutical applications of hot-melt extrusion: Part I". Drug Dev Ind Pharm 33(9):909–926
Dai WG, Dong LC et al (2007) "Parallel screening approach to identify solubility-enhancing formulations for improved bioavailability of a poorly water-soluble compound using milligram quantities of material". Int J Pharm 336(1):1–11
Dai WG, Pollock-Dove C et al (2008) "Advanced screening assays to rapidly identify solubility-enhancing formulations: high-throughput, miniaturization and automation". Adv Drug Deliv Rev 60(6):657–72
Daurio D, Medina C et al (2011).) "Application of Twin Screw extrusion in the manufacture of cocrystals, Part I: Four case studies". Pharmaceutics 3(3):582–600.
De Beer T, Burggraeve A et al (2011) "Near infrared and Raman spectroscopy for the in-process monitoring of pharmaceutical production processes". Int J Pharm 417(1–2):32–47
De Brabander C, Vervaet C et al (2000) "Matrix mini-tablets based on starch/microcrystalline wax mixtures". Int J Pharm 199(2):195–203
Dhumal RS, Kelly AL et al (2010) "Cocrystalization and simultaneous agglomeration using hot melt extrusion". Pharm Res 27(12):2725–2733
Dierickx L, Saerens L et al (2012a) "Co-extrusion as manufacturing technique for fixed-dose combination mini-matrices". Eur J Pharm Biopharm 81(3):683–689
DiNunzio JC, Brough C et al (2010a) "Fusion production of solid dispersions containing a heat-sensitive active ingredient by hot melt extrusion and Kinetisol dispersing". Eur J Pharm Biopharm 74(2):340–351
DiNunzio JC, Brough C et al (2010b) "Fusion processing of itraconazole solid dispersions by kinetisol dispersing: a comparative study to hot melt extrusion". J Pharm Sci 99(3):1239–50123
DiNunzio JC, Zhang F, Martin C, McGinity JW(2012) Melt extrusion. In: Williams RO, III, Watts AB, Miller DA (eds) Formulating poorly water soluble drugs, vol 3. Springer, New York, pp 311–362
Dreiblatt A (ed) (2003) Process design. Pharmaceutical extrusion technology. Marcel Dekker Inc., New York

Dreiblatt A (ed) (2012) Technological considerations related to scale-up of hot-melt extrusion processes. Hot-melt extrusion: pharmaceutical applications. Wiley, Chichester

Dwayne T, Friesen D, Douglas KL, Lorenz A, Bruno C. Hancock T, McDermott J, Shanker RM (2004).) Dosage form comprising a CETP inhibitor and an HMG-COA reductase inhibitor. U. S. P. T. O. (ed), United States of America.

Engers D, Teng J et al (2010) "A solid-state approach to enable early development compounds: selection and animal bioavailability studies of an itraconazole amorphous solid dispersion". J Pharm Sci 99(9):3901–3922

Feng J, Xu L et al (2011) "Evaluation of polymer carriers with regard to the bioavailability enhancement of bifendate solid dispersions prepared by hot-melt extrusion". Drug Dev Ind Pharm 38(6):735–743

Forster A, Hempenstall J et al (2001) "Characterization of glass solutions of poorly water-soluble drugs produced by melt extrusion with hydrophilic amorphous polymers". J Pharm Pharmacol 53(3):303–315

Fu J, Zhang L et al (2010) "Stable nimodipine tablets with high bioavailability containing NM-SD prepared by hot-melt extrusion". Powder Technol 204(2–3):214–221

Fukuda M, Peppas NA et al (2006) "Floating hot-melt extruded tablets for gastroretentive controlled drug release system". J Control Release 115(2):121–129

Fukuda M, Miller DA et al (2008) "Influence of sulfobutyl ether beta-cyclodextrin (Captisol) on the dissolution properties of a poorly soluble drug from extrudates prepared by hot-melt extrusion". Int J Pharm 350(1–2):188–196

Geers S (2010) Polymer formulations of CETP inhibitors. U. S. P. T. O. (ed).Merck, United States of America

Ghalanbor Z, Korber M et al (2010) "Improved lysozyme stability and release properties of poly(lactide-co-glycolide) implants prepared by hot-melt extrusion". Pharm Res 27(2):371–379

Ghebremeskel AN, Vemavarapu C et al (2006) "Use of surfactants as plasticizers in preparing solid dispersions of poorly soluble API: stability testing of selected solid dispersions". Pharm Res 23(8):1928–1936

Ghosh I, Snyder J et al (2011) "Comparison of HPMC based polymers performance as carriers for manufacture of solid dispersions using the melt extruder". Int J Pharm 419(1–2):12–9

Giles HFJ, Wagner JRJ, III MEM (2005).) Extrusion: the definitive processing guide and handbook. William Andrew Inc., New York

Gilis PM, De Condé VFV, Vandecruys RPG. (1997).) Beads having a core coated with an antifungal and a polymer. U. S. P. T. Office. United States of America.

Gnoth S, Jenzsch M et al (2007) "Process Analytical Technology (PAT): batch-to-batch reproducibility of fermentation processes by robust process operational design and control". J Biotechnol 132(2):180–186

Gogos CC, Tadmor Z, Kim MH (1998) "Melting phenomena and mechanisms in polymer processing equipment". Adv Poly Technol 17(4):285–305.

Gryczke A, Schminke S et al (2011) "Development and evaluation of orally disintegrating tablets (ODTs) containing Ibuprofen granules prepared by hot melt extrusion". Colloids Surf B Biointerfaces 86(2):275–284.

Guns S, Mathot V et al (2012) "Upscaling of the hot-melt extrusion process: comparison between laboratory scale and pilot scale production of solid dispersions with miconazole and Kollicoat IR". Eur J Pharm Biopharm 81(3):674–682

Gupta P, Kakumanu VK et al (2004) "Stability and solubility of celecoxib-PVP amorphous dispersions: a molecular perspective". Pharm Res 21(10):1762–1769

Han W, Wang X (2012) "Optimal geometry design of the coat-hanger die with uniform outlet velocity and minimal residence time". J Appl Polym Sci 123(4):2511–2516

Hughey JR, DiNunzio JC et al (2010) "Dissolution enhancement of a drug exhibiting thermal and acidic decomposition characteristics by fusion processing: a comparative study of hot melt extrusion and KinetiSol dispersing". AAPS PharmSciTech 11(2):760–774

Jaruratanasirikul S, Kleepkaew A (1997) "Influence of an acidic beverage (Coca-Cola) on the absorption of itraconazole". Eur J Clin Pharmacol 52(3):235–237

Johnson TJ, Gupta KM et al (2010) "Segmented polyurethane intravaginal rings for the sustained combined delivery of antiretroviral agents dapivirine and tenofovir". Eur J Pharm Sci 39(4): 203–212
Kakumanu VK, Bansal AK (2002) "Enthalpy relaxation studies of celecoxib amorphous mixtures". Pharm Res 19(12):1873–1878
Khougaz K, Clas SD (2000) "Crystallization inhibition in solid dispersions of MK-0591 and poly(vinylpyrrolidone) polymers". J Pharm Sci 89(10):1325–1334
Klein CE, Chiu YL et al (2007) "The tablet formulation of lopinavir/ritonavir provides similar bioavailability to the soft-gelatin capsule formulation with less pharmacokinetic variability and diminished food effect". J Acquir Immune Defic Syndr 44(4):401–410
Klueglich M, Ring A et al (2005) "Ibuprofen extrudate, a novel, rapidly dissolving ibuprofen formulation: relative bioavailability compared to ibuprofen lysinate and regular ibuprofen, and food effect on all formulations". J Clin Pharmacol 45(9):1055–1061
Konno H, Taylor LS (2006) "Influence of different polymers on the crystallization tendency of molecularly dispersed amorphous felodipine". J Pharm Sci 95(12):2692–2705
Krishna R, Bergman AJ et al (2011) "Model-based development of anacetrapib, a novel cholesteryl ester transfer protein inhibitor". AAPS J 13(2):179–190
Krishna G, Ma L et al (2012) "A new solid oral tablet formulation of posaconazole: a randomized clinical trial to investigate rising single- and multiple-dose pharmacokinetics and safety in healthy volunteers". J Antimicrob Chemother 67(11):2725–2730
Lakshman JP, Cao Y et al (2008) "Application of melt extrusion in the development of a physically and chemically stable high-energy amorphous solid dispersion of a poorly water-soluble drug". Mol Pharm 5(6):994–1002
Fang LY, Gopal Krishna DH, Allen E, Moton JR, Prestipino RC, Steinman M, Wan J, Waskin HA (2009).) High density compositions containing posaconazole and formulations comprising the same. W. I. P. O. (ed), Schering Corporation, United States of America.
Lauer ME, Grassmann O et al (2011) "Atomic force microscopy-based screening of drug-excipient miscibility and stability of solid dispersions". Pharm Res 28(3):572–584
Linn M, Collnot EM et al (2012) "Soluplus(R) as an effective absorption enhancer of poorly soluble drugs *in vitro* and *in vivo*". Eur J Pharm Sci 45(3):336–343
Listro T, Borek M, Crowley MM, Nollenberger K (2012).) Hot melt extrusion: analytical tools & techniques in hot melt extrusion & case studies on formulation development & process scale-up. Drug Dev Deliv 12(7):36–40
Liu H, Wang P et al (2010) "Effects of extrusion process parameters on the dissolution behavior of indomethacin in Eudragit EPO solid dispersions". Int J Pharm 383(1–2):161–169
Liu H, Zhu L et al (2012) "Effects of screw configuration on indomethacin dissolution behavior in Eudragit EPO". Adv. Polym. Technol 31(4):331–342
Liu J, Zhang F et al (2001) "Properties of lipophilic matrix tablets containing phenylpropanolamine hydrochloride prepared by hot-melt extrusion". Eur J Pharm Biopharm 52(2):181–190
Liu X, Lu M et al (2012) "Improving the chemical stability of amorphous solid dispersion with cocrystal technique by hot melt extrusion". Pharm Res 29(3):806–817
Lowinger M (2011).) "Process development: scaling a melt extrusion process from conception to commercialization" American Pharmaceutical Review, pp 80–84..
Luo Y, Xu L et al (2013) "Improvement of dissolution and bioavailability of Ginsenosides by hot melt extrusion and cogrinding". Drug Dev Ind Pharm 39(1):109–116
Lyons JG, Blackie P et al (2008) "The significance of variation in extrusion speeds and temperatures on a PEO/PCL blend based matrix for oral drug delivery". Int J Pharm 453(1):233–252
Maniruzzaman M BJ, Douroumis D (2011) "Taste masking of bitter APIs by using hot melt extrusion (HME)". AAPS J 13 (S2):T3273–T3273
Mansky P, Dai W-G et al (2007) "Screening method to identify preclinical liquid and semi-solid formulations for low solubility compounds: Miniaturization and automation of solvent casting and dissolution testing". J Pharm Sci 96(6):1548–1563
Markarian J (2012) Scale-up challenges in hot-melt extrusion critical issues that should be considered when scaling up a hot-melt extrusion process. Pharm Technol (Advanstar Communications Inc.) 36:88–92

Marsac PJ, Rumondor AC et al (2010) "Effect of temperature and moisture on the miscibility of amorphous dispersions of felodipine and poly(vinyl pyrrolidone)". J Pharm Sci 99(1):169–185

Matsumoto T, Zografi G (1999) "Physical properties of solid molecular dispersions of indomethacin with poly(vinylpyrrolidone) and poly(vinylpyrrolidone-co-vinyl-acetate) in relation to indomethacin crystallization". Pharm Res 16(11):1722–1728

McKelvey CA, Schenck L, Lowinger M, Gregory M (2008) Applying Extrusion to Pharmaceutical Formulation Design. Society of Plastics Engineers Annual Technical Conference

Mididoddi PK, Prodduturi S et al (2006) "Influence of tartaric acid on the bioadhesion and mechanical properties of hot-melt extruded hydroxypropyl cellulose films for the human nail". Drug Dev Ind Pharm 32(9):1059–1066

Mididoddi PK, Repka MA (2007) "Characterization of hot-melt extruded drug delivery systems for onychomycosis". Eur J Pharm Biopharm 66(1):95–105

Miller DA. (2011) Hot-melt extrusion for improved dissolution properties of poorly watersoluble molecules. U. S. P. a. T. O. (ed). United States of America, F. Hoffmann-La Roche, Inc., United States of America

Miller DA, McConville JT et al (2007) "Hot-melt extrusion for enhanced delivery of drug particles". J Pharm Sci 96(2):361–376

Miller DA, DiNunzio JC et al (2008) "Targeted intestinal delivery of supersaturated itraconazole for improved oral absorption". Pharm Res 25(6):1450–1459

Miyazaki T, Yoshioka S et al (2004) "Ability of polyvinylpyrrolidone and polyacrylic acid to inhibit the crystallization of amorphous acetaminophen". J Pharm Sci 93(11):2710–2717

Mollan M (2003) Historical Overview. Marcel Dekker Inc., New York

Mu B, Thompson MR (2012) "Examining the mechanics of granulation with a hot melt binder in a twin-screw extruder". Chem Eng Sci 81:46–56

Nakamichi K, Nakano T et al (2002) "The role of the kneading paddle and the effects of screw revolution speed and water content on the preparation of solid dispersions using a twin-screw extruder". Int J Pharm 241(2):203–211

Ndindayino F, Vervaet C et al (2002a) "Bioavailability of hydrochlorothiazide from isomalt-based moulded tablets". Int J Pharm 246(1–2):199–202

Ndindayino F, Vervaet C et al (2002b) "Direct compression and moulding properties of co-extruded isomalt/drug mixtures". Int J Pharm 235(1–2):159–168

Newman A, Knipp G et al (2012) "Assessing the performance of amorphous solid dispersions". J Pharm Sci 101(4):1355–1377

Özgüney I, Shuwisitkul D et al (2009) "Development and characterization of extended release Kollidon® SR mini-matrices prepared by hot-melt extrusion". Eur J Pharm Biopharm 73(1):140–145

Perdikoulias J, Dobbie T (2003) Die design. Marcel Dekker Inc., New York

Prodduturi S, Urman KL et al (2007) "Stabilization of hot-melt extrusion formulations containing solid solutions using polymer blends" AAPS Pharm Sci Tech 8(2):5–0

Qi S, Belton P et al (2010) "Characterisation and prediction of phase separation in hot-melt extruded solid dispersions: a thermal, microscopic and NMR relaxometry study". Pharm Res 27(9):1869–1883

Qian F, Huang J et al (2010) "Drug-polymer solubility and miscibility: Stability consideration and practical challenges in amorphous solid dispersion development". J Pharm Sci 99(7):2941–2947

Quinten T, Gonnissen Y et al (2009) "Development of injection moulded matrix tablets based on mixtures of ethylcellulose and low-substituted hydroxypropylcellulose". Eur J Pharm Sci 37(3–4):207–216

Rathore AS, Bhambure R et al (2010) "Process analytical technology (PAT) for biopharmaceutical products". Anal Bioanal Chem 398(1):137–154

Rauwendaal C (ed) (2001) Polymer extrusion. Hanser, Munich

Repka MA, McGinity JW (2001) "Influence of chlorpheniramine maleate on topical hydroxypropylcellulose films produced by hot-melt extrusion". Pharm Dev Technol 6(3):297–304

Repka MA, Gerding TG et al (1999) "Influence of plasticizers and drugs on the physical-mechanical properties of hydroxypropylcellulose films prepared by hot melt extrusion". Drug Dev Ind Pharm 25(5):625–633

Repka MA, Battu SK et al (2007) "Pharmaceutical applications of hot-melt extrusion: Part II". Drug Dev Ind Pharm 33(10):1043–1057
Repka MA, Majumdar S et al (2008) "Applications of hot-melt extrusion for drug delivery". Expert Opin Drug Deliv 5(12):1357–1576
Repka MA, Shah S et al (2012) "Melt extrusion: process to product". Expert Opin Drug Deliv 9(1):105–125
Robinson JR, Mcginity JW (2000).) Ethypharm SA, effervescent granules and methods for their preparation. United States of America
Rodrigues LO, Alves TP et al (2006) "Improving drug manufacturing with process analytical technology". IDrugs 9(1):44–48
Roth W, Setnik B et al (2009) "Ethanol effects on drug release from Verapamil Meltrex, an innovative melt extruded formulation". Int J Pharm 368(1–2):72–75
Rumondor AC, Marsac PJ et al (2009).) "Phase behavior of poly(vinylpyrrolidone) containing amorphous solid dispersions in the presence of moisture". Mol Pharm 6(5):1492–1505.
Rumondor AC, Wikstrom H et al (2011) "Understanding the tendency of amorphous solid dispersions to undergo amorphous-amorphous phase separation in the presence of absorbed moisture". AAPS PharmSciTech 12(4):1209–1219
Sakurai A, Sakai T et al (2012) "Polymer combination increased both physical stability and oral absorption of solid dispersions containing a low glass transition temperature drug: physicochemical characterization and *in vivo* study". Chem Pharm Bull (Tokyo) 60(4):459–464
Sathigari SK, Radhakrishnan VK et al (2012).) "Amorphous-state characterization of efavirenz-polymer hot-melt extrusion systems for dissolution enhancement" J Pharm Sci 101(9): 3456–3464
Sauer D, Zheng W et al (2007) "Influence of processing parameters and formulation factors on the drug release from tablets powder-coated with Eudragit L 100–55". Eur J Pharm Biopharm 67(2):464–475
Schenck L, Troup GM, Lowinger M, Li L, McKelvey C (eds) (2010).) Achieving a hot melt extrusion design space for the production of solid solutions. Chemical engineering in the pharmaceutical industry: R & D to manufacturing. John Wiley, New York
Schilling SU, Shah NH et al (2007) "Citric acid as a solid-state plasticizer for Eudragit RS PO". J Pharm Pharmacol 59(11):1493–1500
Schilling SU, Bruce CD et al (2008) "Citric acid monohydrate as a release-modifying agent in melt extruded matrix tablets". Int J Pharm 361(1–2):158–168
Schilling SU, Lirola HL et al (2010a) "Influence of plasticizer type and level on the properties of Eudragit S100 matrix pellets prepared by hot-melt extrusion". J Microencapsul 27(6):521–532
Schilling SU, McGinity JW (2010b) "Novel application of hot-melt extrusion for the preparation of monolithic matrices containing enteric-coated particles". Int J Pharm 400(1–2):24–31
Schilling SU, Shah NH et al (2010c) "Properties of melt extruded enteric matrix pellets". Eur J Pharm Biopharm 74(2):352–361
Scott B, Wilcock A (2006) "Process analytical technology in the pharmaceutical industry: a toolkit for continuous improvement". PDA J Pharm Sci Technol 60(1):17–53
Shah S, Maddineni S et al (2013) "Melt extrusion with poorly soluble drugs". Int J Pharm 453(1):233–252
Shanbhag A, Rabel S et al (2008) "Method for screening of solid dispersion formulations of low-solubility compounds—miniaturization and automation of solvent casting and dissolution testing". Int J Pharm 351(1–2):209–18
Sherry R (2007).) Reckitt Benckiser Healthcare (UK) Limited assignee. Granules comprising paracetamol, a NSAID and a sugar alcohol made by melt extrusion patent
Six K, Berghmans H et al (2003a) "Characterization of solid dispersions of itraconazole and hydroxypropylmethylcellulose prepared by melt extrusion: Part II". Pharm Res 20(7):1047–1054
Six K, Murphy J et al (2003b) "Identification of phase separation in solid dispersions of itraconazole and Eudragit E100 using microthermal analysis". Pharm Res 20(1):135–138
Six K, Verreck G et al (2004) "Increased physical stability and improved dissolution properties of itraconazole, a class II drug, by solid dispersions that combine fast- and slow-dissolving polymers". J Pharm Sci 93(1):124–131

Six K, Daems T et al (2005) "Clinical study of solid dispersions of itraconazole prepared by hot-stage extrusion". Eur J Pharm Sci 24(2–3):179–186
Smith AJ, Kavuru P et al (2011) "Cocrystals of quercetin with improved solubility and oral bioavailability". Mol Pharm 8(5):1867–1876
Steiner R (2003) Extruder design. Marcel Dekker Inc., New York
Tadmor Z, Klein I (1970) Engineering principles of plasticating extrusion. Van Nostrand Reinhold Co., New York
Taylor LS, Zografi G (1997) "Spectroscopic characterization of interactions between PVP and indomethacin in amorphous molecular dispersions". Pharm Res 14(12):1691–1698
Thiele W (ed) (2003) Twin-screw extrusion and screw design. Pharmaceutical extrusion technology. Marcel Dekker Inc., New York
Tho I, Liepold B et al (2010) "Formation of nano/micro-dispersions with improved dissolution properties upon dispersion of ritonavir melt extrudate in aqueous media". Eur J Pharm Sci 40(1):25–32
Thumma S, ElSohly MA et al (2008a) "Influence of plasticizers on the stability and release of a prodrug of Delta(9)-tetrahydrocannabinol incorporated in poly (ethylene oxide) matrices". Eur J Pharm Biopharm 70(2):605–614
Thumma S, Majumdar S et al (2008b) "Chemical stability and bioadhesive properties of an ester prodrug of Delta 9-tetrahydrocannabinol in poly(ethylene oxide) matrices: effect of formulation additives". Int J Pharm 362(1–2):126–132
Todd D (1993).) "Melting of plastics in kneading blocks". Int Polym Proc 8:113–118
Todd DB (ed) (1995) Introduction to compounding. Plastics compounding: equipment and processing. Hanser, New York
Todd D (1998) "Practical aspects of processing in intermeshing twin screw extruders". J Reinforc Plast Compos 17:1607–1616
Toensmeier PA (2000).) "High-value niche grows in multi-layer die design". Modern Plastics 43:52–54
Trey SM, Wicks DA et al (2007) "Delivery of itraconazole from extruded HPC films". Drug Dev Ind Pharm 33(7):727–735
Tumuluri VS, Kemper MS et al (2008).) "Off-line and on-line measurements of drug-loaded hot-melt extruded films using Raman spectroscopy". Int J Pharm 357(1–2):77–84
Upadhye SB, Kulkarni SJ et al (2010) "Preparation and characterization of inclusion complexes of a hemisuccinate ester prodrug of delta9-tetrahydrocannabinol with modified beta-cyclodextrins". AAPS PharmSciTech 11(2):509–517
van den Mooter G, Wuyts M et al (2001) "Physical stabilisation of amorphous ketoconazole in solid dispersions with polyvinylpyrrolidone K25". Eur J Pharm Sci 12(3):261–269
van Eerdenbrugh B, Taylor LS (2010) "Small scale screening to determine the ability of different polymers to inhibit drug crystallization upon rapid solvent evaporation". Mol Pharm 7(4): 1328–1337
van Laarhoven JA, Kruft MA et al (2002a) "Effect of supersaturation and crystallization phenomena on the release properties of a controlled release device based on EVA copolymer". J Control Release 82(2–3):309–317
van Laarhoven JA, Kruft MA et al (2002b) "*In vitro* release properties of etonogestrel and ethinyl estradiol from a contraceptive vaginal ring". Int J Pharm 232(1–2):163–173
van Melkebeke B, Vermeulen B et al (2006) "Melt granulation using a twin-screw extruder: a case study". Int J Pharm 326(1–2):89–93
Vasanthavada M, Wang Y et al (2011) "Application of melt granulation technology using twin-screw extruder in development of high-dose modified-release tablet formulation". J Pharm Sci 100(5):1923–1934
Vasconcelos T, Sarmento B et al (2007) "Solid dispersions as strategy to improve oral bioavailability of poor water soluble drugs". Drug Discov Today 12(23–24):1068–1075
Verhoeven E, Vervaet C et al (2006) "Xanthan gum to tailor drug release of sustained-release ethylcellulose mini-matrices prepared via hot-melt extrusion: *in vitro* and *in vivo* evaluation". Eur J Pharm Biopharm 63(3):320–330

Verreck G, Decorte A et al (2006).) "The effect of pressurized carbon dioxide as a plasticizer and foaming agent on the hot melt extrusion process and extrudate properties of pharmaceutical polymers" J Supercrit Fluid 38(3):383–391
Verreck G, Six K et al (2003) "Characterization of solid dispersions of itraconazole and hydroxypropylmethylcellulose prepared by melt extrusion–Part I". Int J Pharm 251(1–2):165–174
Vilhelmsen T, Schaefer T (2005) "Agglomerate formation and growth mechanisms during melt agglomeration in a rotary processor". Int J Pharm 304(1–2):152–164
Weuts IF van Dycke F et al. (2011) "Physicochemical properties of the amorphous drug, cast films, and spray dried powders to predict formulation probability of success for solid dispersions: etravirine". J Pharm Sci 100(1):260–274
Wu C, McGinity JW (2003) "Influence of methylparaben as a solid-state plasticizer on the physicochemical properties of Eudragit RS PO hot-melt extrudates". Eur J Pharm Biopharm 56(1):95–100
Wyttenbach N, Janas C et al (2013) "Miniaturized screening of polymers for amorphous drug stabilization (SPADS): Rapid assessment of solid dispersion systems". Eur J Pharm Biopharm 84(3):583–598
Young CR, Koleng JJ et al (2002) "Production of spherical pellets by a hot-melt extrusion and spheronization process". Int J Pharm 242(1–2):87–92
Zhang W, Yang S et al (2009) "Technology for improving the bioavailability of small molecules extracted from traditional Chinese medicines". Expert Opin Drug Deliv 6(11):1247–1259
Zheng X, Yang R et al (2007) "Part I: characterization of solid dispersions of nimodipine prepared by hot-melt extrusion". Drug Dev Ind Pharm 33(7):791–802
Zhou F, Vervaet C et al (1996) "Matrix pellets based on the combination of waxes, starches and maltodextrins". Int J Pharm 133(1–2):155–160

Chapter 2
Twin Screw Extrusion for Pharmaceutical Processes

Charlie Martin

Abstract The use of twin screw extruders (TSEs) to create sophisticated dosage forms is expanding because the continuous melt extrusion (ME) process has been proven to be a consistent and repeatable way to make high-quality products. TSEs are continuous small mass continuous mixers that are the plastics industry's preferred manufacturing methodology for compounding and devolatilization. It is without question that pharmaceutical companies have reaped the benefit of over half a century of technological developments and process refinement in plastics. It is only in the last decade or so that extrusion has emerged as a viable platform for pharmaceutical development.

This chapter will describe the design and functionality of co-rotating and counter-rotating TSEs for pharmaceutical applications. Screw and process design to facilitate effective mixing are addressed. Control parameters and the associated interactions (i.e., screw rpm versus feed rate) to achieve a quality product are proffered. Common TSE terms are defined and examples of useful TSE formulas are presented and explained. Staging of unit operations in a TSE and downstream system functionalities are described. Case studies are also presented to provide insight into how melt extrusion is being applied today. Finally, potential future growth areas are identified for melt extrusion.

2.1 Introduction

The use of twin screw extruders (TSEs) to create sophisticated dosage forms is expanding because the continuous melt extrusion (ME) process has been proven to be a consistent and repeatable way to make high-quality products. TSEs are continuous small mass mixers that are the plastics industry's preferred manufacturing methodology for compounding and devolatilization. It is without question that pharmaceutical companies have reaped the benefit of over half a century of technological developments and process refinement in plastics. It is only in the last decade or so that extrusion has emerged as a viable platform for pharmaceutical development.

C. Martin (✉)
Leistritz, Branchburg, NJ, USA
e-mail: cmartin@alec-usa.com

M. A. Repka et al. (eds.), *Melt Extrusion*,
AAPS Advances in the Pharmaceutical Sciences Series 9,
DOI 10.1007/978-1-4614-8432-5_2,

Fig. 2.1 Twin screw extruder with feeders for processing pharmaceuticals

Driven by FDA's Process Analytical Technology (PAT) initiative, the use of TSEs has been embraced by virtually every major pharmaceutical company. The PAT initiative encourages that new dosage forms be manufactured via continuous processes with in-line monitoring of key parameters—it literally could have been written by an extruder supplier.

TSEs for pharmaceutical class installations, as compared to most plastics installations, must also integrate validation documentation into the scope of delivery. Detailed/project specific document packages for the Factory Acceptance Test (FAT), Installation Qualification (IQ) and Operational Qualification (OQ) add months of time and effort to the installation and commissioning of the equipment. There are also cGMP guidelines for cleaning pharmaceutical class TSE systems. For instance, the equipment must be cleaned at appropriate intervals and written procedures are required that must be specific and detailed. Cleaned equipment must be protected from contamination while in storage and inspected just before using. Records and equipment logs of all cleaning and inspection must be kept, and the time between end of processing and cleaning steps must be recorded (Martin 2012a).

The goal of this chapter is to provide insight into the design of TSEs, and how these devices are and can be applied to manufacture novel dosage forms in a precise and economical manner (Fig. 2.1). Some simple formulas are interspersed within the text to provide insight into the TSE design and process.

2.2 TSE Design and Applications

Let us start with some simple definitional terms. In a TSE there are, obviously, two screws that fit inside a "figure eight" barrel. The outside diameter of each screw is referred to as the OD, and there is an overflight gap between the tip of the screw and

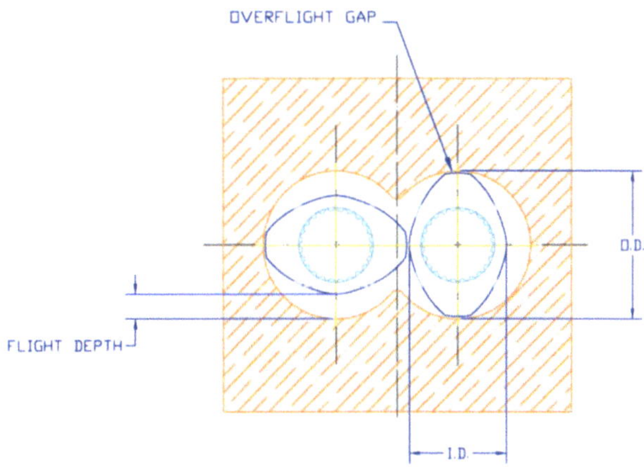

Fig. 2.2 TSE terms: OD, ID, and overflight gap

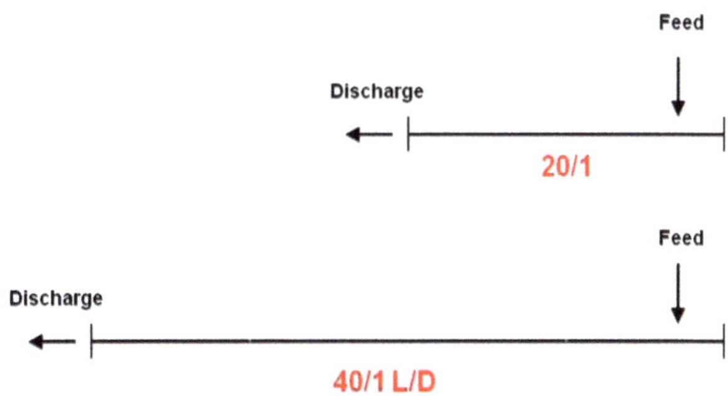

Fig. 2.3 Comparison between 20/1 and 40/1 L/D TSE process section lengths

the barrel wall. There is also an inside diameter (ID) of each screw. In a TSE, the OD and ID are constant throughout the length of screws. OD divided by the ID gives you the OD/ID ratio, which determines the flight depth and the free volume in the process section. See Fig. 2.2 for a schematic that defines some typical TSE terms.

The length of the barrel process section is frequently described in terms of the length to diameter (L/D) ratio, with greater L/D values indicating a longer process section. The L/D ratio is defined by dividing the overall length of the process section by the diameter of the screws. For most pharmaceutical compounding operations L/D values ≤ 40 are used, while reactive and devolatilization extrusion operations may utilize longer, such as 60/1 L/D process sections. Heat- and shear-sensitive products often use shorter process sections, such as a 20/1 L/D process length. It becomes obvious that the longer the L/D the more unit operations can be performed along the length of the process section.

For instance, if the OD of a screw is 20 mm and the length of the process section is 800 mm then the L/D ratio is 800/20 or 40/1. If the length was 400 mm the L/D would be 400/20 or 20/1 (Fig. 2.3).

TSEs perform dispersive and distributive mixing. The intense mixing associated with the short inter-screw mass transfer distances inherent with a TSE results in highly efficient mixing and a more uniform product as compared to large-mass batch

Fig. 2.4 TSE co-rotating intermeshing screw set

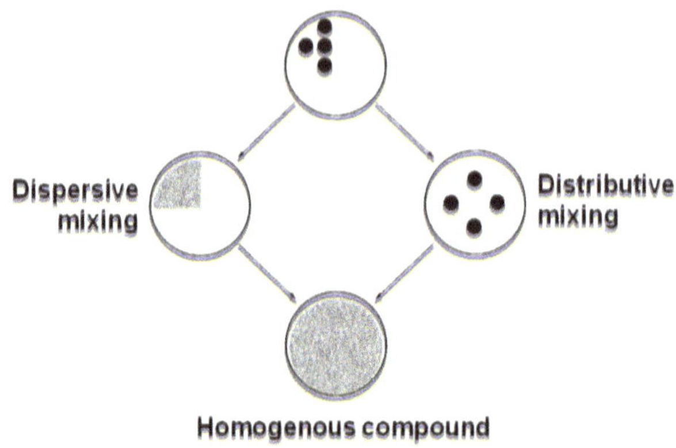

Fig. 2.5 Dispersive and distributive mixing comparison

mixers with much longer mass transfer distances. The short residence time associated with a TSE (typically less than 2 min and as short as 5 s) is particularly beneficial for heat/shear-sensitive formulations, as the TSE can be designed to limit exposure to elevated temperatures to just a few seconds (Fig. 2.4).

In dispersive mixing, agglomerates or liquid droplets held together by interfacial tension are subjected to mechanical stress by the rotating screws to achieve size reduction. The most important flow characteristics to achieve dispersive mixing are the extensional and planar flow fields as generated by TSE mixing elements. In distributive mixing, repeated rearrangement of the minor components without size reduction enhances product homogeneity (Fig. 2.5).

The modular nature of TSEs offers process flexibility with regard to screw design, barrel configuration, and shear intensity. The co-rotating intermeshing TSE mode dominates the market, having captured approximately 90 % of current installations. Co-rotating intermeshing TSEs are very efficient at feeding powders and for staging of downstream unit operations. Counterrotating TSEs are also utilized in the pharmaceutical industry and have shown great promise in many applications. For instance, counterrotating intermeshing TSEs can be designed so that the materials being processed experience a tighter residence time distribution, and can provide more efficient pumping into a die (Martin 2001; Fig. 2.6).

In the plastics and food industries, TSE screw diameters (ODs) range from 5–400+ mm with outputs from a 20 gm to 50,000+ kg/h. Pharmaceutical TSEs are generally 60 mm and below, due to more efficient heat transfer characteristics of smaller extruders and, of course, the smaller batch sizes associated with pharmaceutical as compared to plastic products. Research and development efforts for pharmaceuticals are generally performed on TSEs with screw diameters in the 10–30 mm range.

The TSE utilizes a modular design for barrels and screws (Fig. 2.7). Segmented screws convey and shear the materials in channels bounded by screw flights and barrel

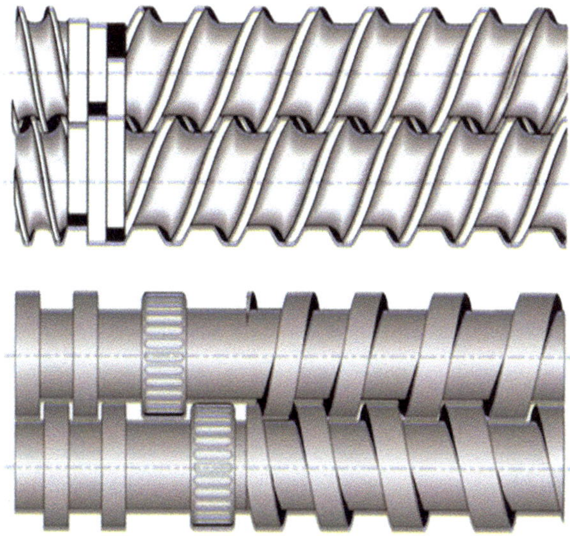

Fig. 2.6 Co-rotating and counterrotating intermeshing TSE geometries

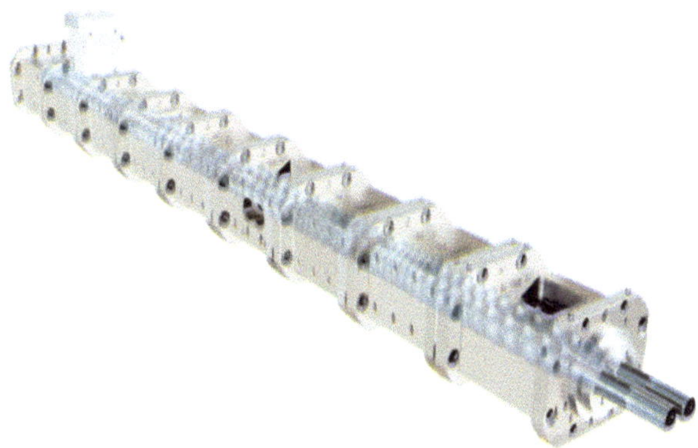

Fig. 2.7 Modular TSE process section

walls, with short mass transfer distances. Solids conveying and melting occurs in the early stages of the process section followed by mixing and devolatilization. Discharge elements then build and stabilize pressure prior to exiting the extruder.

The controlled pumping and wiping characteristics of the TSE screws in combination with a segmented design allows specific screw element geometries to be matched to the unit operations being performed in the TSE process section (Fig. 2.8).

TSE control parameters include screw speed (rpm), feed rate, process temperatures, and vacuum level. Melt pressure, melt temperature, motor amperage, and various in-line sensors also monitor the process to ensure a consistent/quality product. State-of-the-art extruders utilize programmable logic controllers (PLCs) for logic functionality that implements graphical touch-screens, data acquisition, and recipe management as standard features. It is also now routine to integrate subsystem PLCs with a supervisory PLC for integration of complex equipment from multiple suppliers, referred to as "distributed processing" (Fig. 2.9).

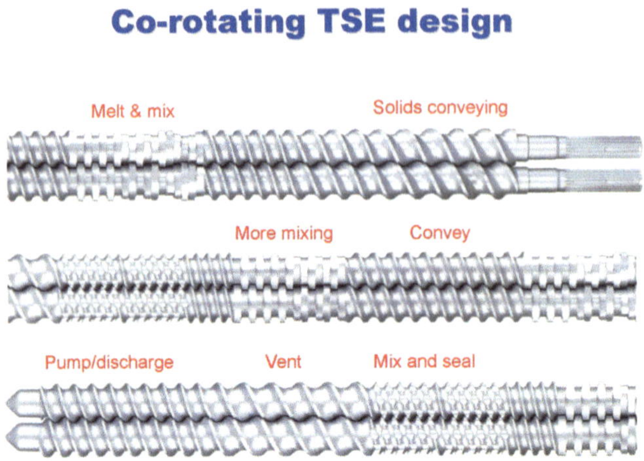

Fig. 2.8 TSE segment screw set with unit operations denoted

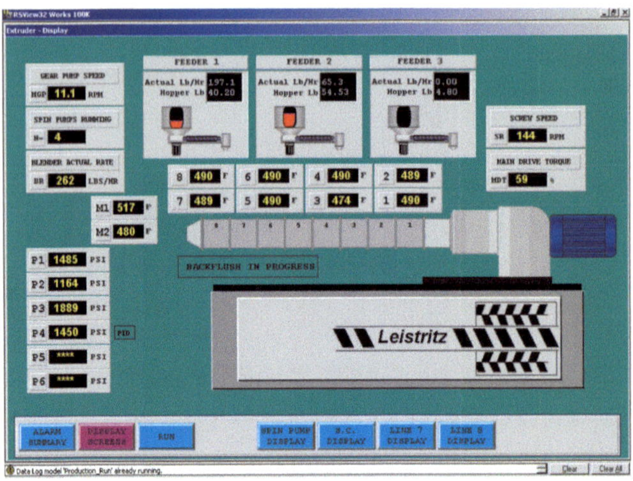

Fig. 2.9 Human Machine Interface (HMI) touch screen

Pharmaceutical class TSEs generally adhere to FDA part 11 of Title 21 of the Code of Federal Regulations, which defines the criteria under which electronic records are deemed trustworthy. Practically speaking, this regulation requires drug makers to implement controls, audits, and system validation for software and systems involved in processing electronic data. Protocols must be followed with regard to limiting system access to authorized individuals, operational checks, device checks, controls over systems documentation, and a plethora of other guidelines. Strict adherence with regard to copies of records and record retention is part of the guideline (Kapp and Palmer 2003).

Starve feeding refers to when the extruder is fed at a rate less than the forwarding efficiency of the screws. TSEs are almost always starve-fed, with the output rate determined by the feeder(s). The TSE screw rpm is independent and used in concert

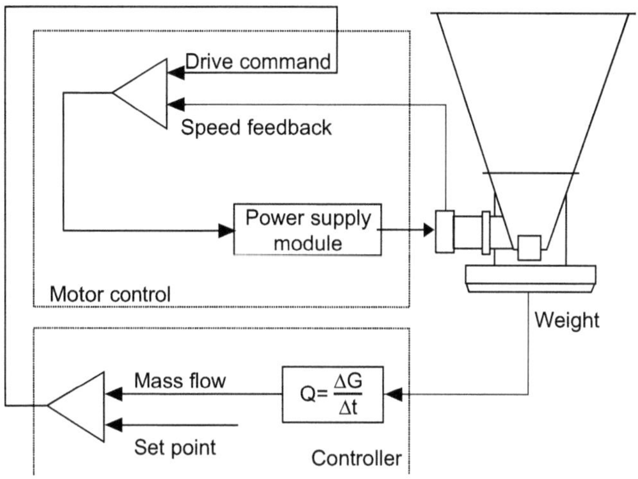

Fig. 2.10 Loss-in-weight feeder control functionality

with the feed rate to optimize compounding efficiencies. Feed rate versus screw rpm, and screw design, impact the average residence time (RT) and residence time distribution (RTD) for the materials being processed. Typical RTs are in the 15 s–3 min range (Martin 2001).

Feeders also maintain formulation consistency and are situated at different locations along the TSE process section to introduce ingredients in the proper order. The accuracy of the feeding system is critical to maintain uniform product quality. For this reason, loss-in-weight (LIW) feeders are normally specified to ensure that the ingredients are delivered at a constant mass flow rate to the extruder (Fig. 2.10). LIW feeders can meter either a premix or individual components to the TSE, as dictated by the formulation and desired level of continuous processing/monitoring. Volumetrically controlled feeders for solids are generally only used with a premix or at very low rates (i.e., 20 g/h), where LIW controls become impractical.

A LIW feeding device consists of the feeding module, the feeder hopper, a refill device, a load cell, and a control system. In an LIW feeder, the mass flow rate is calculated by dividing the weight reduction in the hopper by the time interval. At short/determined time intervals, the weight is measured and transmitted to the controller. The real-time mass flow rate is calculated from the weight reduction per unit time. To compensate for the difference between the set point and the measured value of mass flow, the motor speed is continuously modified. In operation, the feeder, hopper, and material are continuously weighed, and the feeder's discharge rate (which is the rate at which the feeding system is losing weight) is controlled to match the desired feed rate. With this technology, a constant mass flow to the TSE is ensured and verified.

The LIW principle is also utilized for the delivery of the liquids to the TSE. There are a variety of different types of pumps that may be specified based on the viscosity range and changeover requirements for the liquid being introduced (piston pump, gear pump, peristaltic pump, among others). Liquid injection systems can be

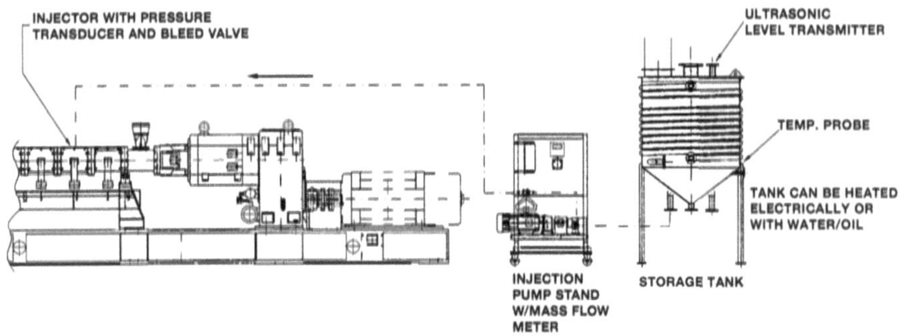

Fig. 2.11 Example of heated liquid feed stream to a TSE

Fig. 2.12 Refill system connected to LIW feeder

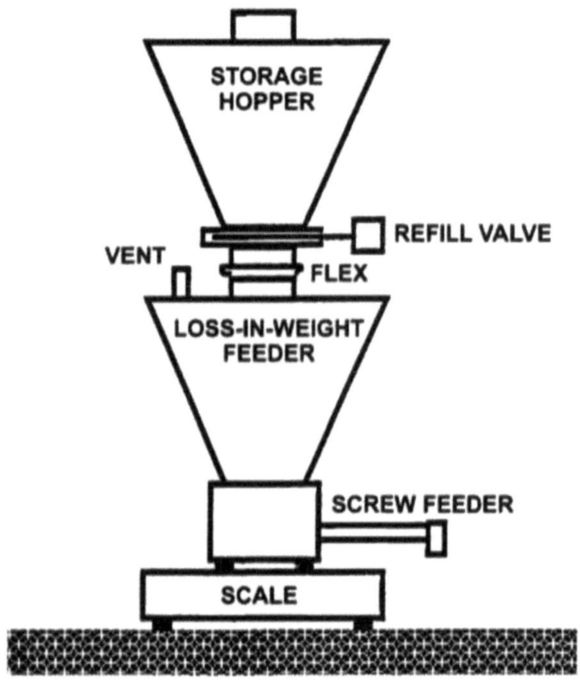

supplied as ambient or heated. The mass flow rate is measured and controlled by a liquid tank placed on load cells with the same LIW control to control the pump motor speed. Alternatively, a mass flow meter can be used to control and monitor the fluid rate to the extruder (Fig. 2.11).

Refilling a LIW feeder is an important part of optimizing feeder technology. In the refill mode, the LIW feeder must maintain a constant flow of material to the TSE. Refill times should be relatively short to allow the feeder to return to a true LIW operation. Additionally, the flow cutoff action to the LIW feeder must be quick and sure. Options include slide gates, flap valves, modulating butterfly valves, and rotary valves. Butterfly valves are often specified due to their easy-to-clean design and utility in a high containment environment (Martin and Nowak 2011; Fig. 2.12).

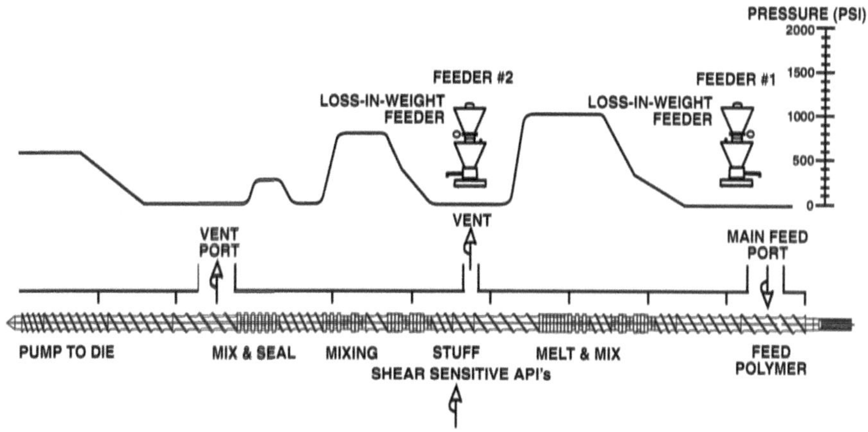

Fig. 2.13 Controlled pressure profile in starve-fed TSE

The pressure gradient in a TSE process section is determined by the selection of screws. Flighted elements are strategically placed so that the screw channels are not entirely filled, which results in zero pressure underneath downstream vent/feed sections. The controlled pressure gradient allows entrapped air, moisture, and other volatiles to be efficiently removed by vents without vent flooding. The zero pressure locations also facilitate the downstream introduction of heat and shear-sensitive active pharmaceutical ingredients (APIs) late in the process section to minimize exposure to shear forces and dwell time in the melt stream (Fig. 2.13).

A "side stuffer" is a device that is often used in a TSE system to introduce shear-sensitive APIs into the process melt stream. A side stuffer utilizes co-rotating, intermeshing, self-wiping screws that "push" materials into an unfilled section of the TSE screws. The side stuffer barrel is jacketed for liquid cooling/heating. Side stuffing is desirable for processing a high percentage of API into the process to facilitate more efficient melting/wetting/mixing by avoiding the plastication zone. A side stuffer is also starve-fed and requires its own LIW metering feeder upstream to set the rate (Martin 2001; Fig. 2.14).

The TSE and related devices are often placed within a process containment isolator when handling high-potency APIs. Sometimes just the API feed system is contained in an isolator unit. Door panels utilize gaskets with compression latches and connections are hermetically sealed. View panels are fitted with glove ports that are sealed, and primary chambers are configured with HEPA-filtered exhaust. As TSE systems are custom designed, isolator systems are built-to-order based on the occupational exposure level (OEL) of the installation, as well as TSE system features, the range of products, and batch size (Fig. 2.15).

The gearbox of a TSE transmits energy from the motor to the screws, and reduces the motor speed to the desired screw rpm while multiplying torque. TSEs typically utilize various failsafes to prevent gearbox damage in an over-torque situation, including both an electronic current limit and mechanical overtorque coupling that automatically disengages the motor in an over-torque situation (Fig. 2.16).

Fig. 2.14 Side stuffer attached to TSE

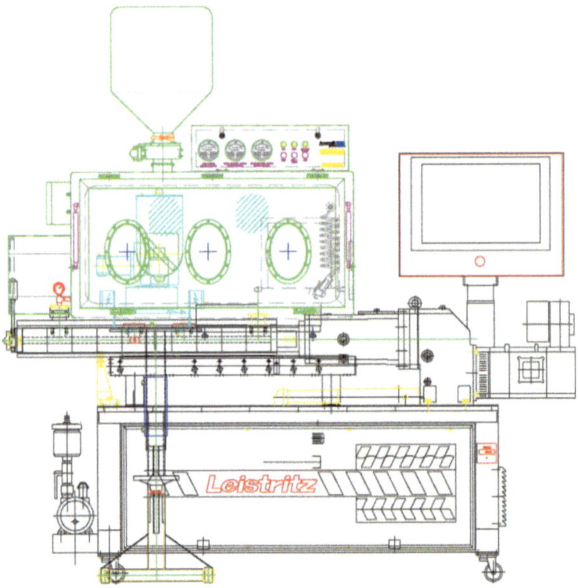

Fig. 2.15 Twin screw extruder with isolator containment system for API feed stream

The TSE motor inputs energy into the process via rotating screws that impart shear and energy into the materials being processed. Alternating current (AC) motors/drives utilize digital communications, and can upload/download drive settings. The percentage motor torque is a particularly critical parameter to be monitored, and is measured as follows:

% Torque This formula indicates the percentage of available rotational force (screws) being used in a process, and is a typical readout denoted as a percentage that is calculated as follows:

$$\% \text{ Torque} = \text{Applied KW/Maximum KW}$$

Fig. 2.16 Water-cooled motor attached to over-torque coupling and TSE gearbox

Unit

Kilowatt (KW) is the electrical current or load from the main motor.

For example, if a TSE has a maximum rating of 100 KW and a process draws 40 KW then the torque is 40 % (40 KW/100 KW).

$$\%\ \text{Torque} = 40/100 = 40\ \%$$

For quality control and process troubleshooting, specific energy (SE) is a particularly important parameter to monitor, and is calculated as follows:

Specific Energy Specific Energy is the amount of power that is being input by the motor into each kilogram of material being processed. It is important that SE records be maintained for quality assurance and troubleshooting purposes. SE can also be used as a benchmark for scale-up and/or comparing different manufacturing operations. SE is calculated in two steps:

Applied power can be calculated as:

$$\text{KW (applied)} = \text{KW (motor rating)} \times \%\ \text{torque} \times \text{rpm running/max. rpm} \times 0.97\ \text{(gearbox efficiency)}$$

Now, the SE can be calculated as:

$$\text{Specific Energy} = \frac{\text{KW (applied)}}{\text{kg/h}}$$

Units

KW	Kilowatts (the motor rating)
% Torque	% used of the maximum allowable torque
Rpm	Screw rotations per minute

SE is denoted in KW per kg/h.

For example:

A 40 mm TSE is processing an HPMCAS formulation at 30 kg/h, running at 100 rpm with 68 % torque. The machine has a 50 KW motor and a maximum screw rpm of 600.

$$\text{KW(applied)} = 50\ \text{KW} \times 0.68 \times 100/600 \times 0.97 = 5.8,\ \text{then}$$
$$\text{SE} = 30/5.8\ \text{or}\ 0.195\ \text{KW/kg/h}.$$

If the 0.195 SE suddenly changed to 0.25 or 0.15 and the process conditions were the same, it would indicate that either the material or hardware had significantly changed, and that the production of that batch should be discontinued until the problem was identified and corrected.

TSEs are available with top screw rpms of 1,200 or higher. The TSE should not be geared substantially higher than the required screw rpm, particularly for TSEs with ODs of 40 mm and larger. Due to the heat/shear-sensitive nature of many pharmaceutical formulations, the motors and gearbox are often specified for top screw speeds of 500 rpm or lower.

The following formula can be used as a guideline on how to configure the motor/gearbox:

Shaft Torque as It Relates to Screw rpm The cross-sectional area of the screw shafts, the shaft design/metallurgy, and the manufacturing method determines the torque that can be imparted into a process. This formula helps determine the proper motor and gearbox configuration for a TSE:

$$\text{Torque} = 9{,}550 \times \text{KW/top rpm}$$

Units

9,550	Constant
Torque	Total torque for both screw shafts, typically denoted in Nm (Newton meters)
KW	Motor rating on the TSE

For example:

If a TSE with 30 mm OD screws has a torque rating of 318 Nm, and therefore uses a 20 KW motor at full torque if geared for 600 rpm. A 40 KW motor would be specified at 1,200 rpm at full torque as indicated in the following comparison:

$$318 = \frac{9{,}550 \times 20}{600} \quad 318 = \frac{9{,}500 \times 40}{1{,}200}$$

If a process has been defined and it has been determined that TSE will never operate above 400 rpm, then it should not be geared for 1,200. Since the torque is a constant, a 30 mm TSE can either be geared at 400 rpm with a 15 kW motor (13 kW applied), or at 1,200 rpm with a 40 kW motor (or with another motor/gearbox combination). In most instances, high rpms will not be specified for pharmaceutical applications, unless the TSE is intended for research and future usage is unknown (Martin 2006).

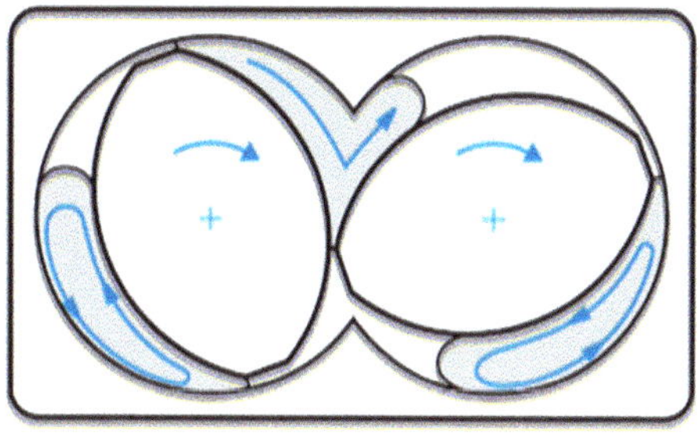

Fig. 2.17 TSE Bilobal co-rotating, intermeshing screws—"self-wiping"

2.3 TSE Process Section: the "Key" to Success

The heart of the TSE is its screws and barrels, referred to as the process section. Screws and barrels are often manufactured of hardenable stainless steels. Nickel-based alloys are specified for corrosive process environments. The metallurgies selected must not be additive, reactive, or absorptive with the materials being processed to be used in a pharmaceutical class environment. Cleaning and storage protocols must be adhered to when handling TSE parts.

In a co-rotating, intermeshing TSE, the screws are termed "self-wiping". The surface velocities of the screws in the intermesh region are in opposing directions, which results in the materials being "wiped" and forced to follow a figure 8 pattern down the length of the screws. Most co-rotating TSEs are bilobal, referring to the number of "lobes" that are possible at a given OD/ID ratio (Fig. 2.17).

The heart of the high-speed TSE is its screws. There are seemingly an infinite number of screw variations possible. There are, however, only three basic types of screw elements: flighted elements, mixing elements, and zoning elements. Flighted elements forward material past barrel ports, through mixers and out of the extruder to pressurize the die. Zoning elements isolate two unit operations. Screw designs can be made shear-intensive or passive, based upon the elements used in the design.

Mixing elements can be dispersive and/or distributive, or a combination thereof. The kneader is the most prevalent mixing element used in a TSE. The wider a kneader is the more dispersive it becomes as extensional and planar shear effects occur as materials are forced up and over the land. Narrower kneaders are more distributive in nature that force high melt division rates with significantly less extensional and planar shear effects. Distributive mixing elements can be particularly useful for mixing heat- and shear-sensitive materials. Kneading elements can be arranged with a forward pitch (less aggressive), neutral, or reverse pitch (most aggressive). High liquid phase mixing generally benefits from specialty distributive elements that prevent "pooling" of the liquids in the TSE process section (Martin 2001; Fig. 2.18).

Counterrotating intermeshing TSEs are also viable for pharmaceutical applications. Looking into the feed throat, the screws rotate outward that results in feeding on both screws. In the screw intermesh region, the flight of one of the screws penetrates the flight depth of the second screw, and the velocity of the screws in the intermesh is in the same direction. This region is referred to as the calender gap. Screw rotation

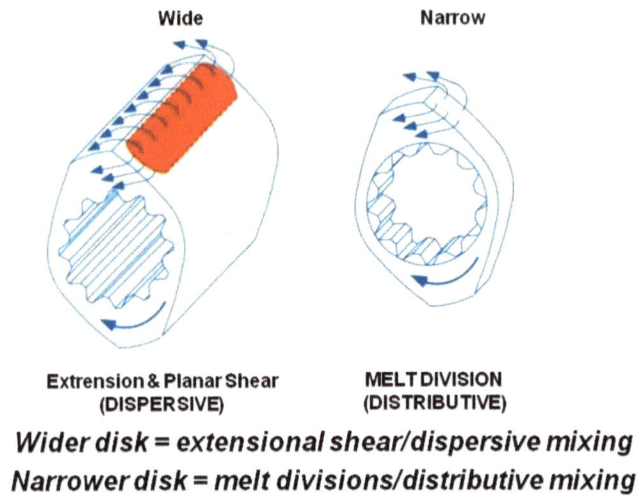

Fig. 2.18 Example of dispersive and distributive mixing elements.

forces materials up and through the calender gap to facilitate melting and mixing as the processed materials experience an extensional mixing effect. Essentially, the entire length of the screw can function as a mixing device as materials continually experience the extensional mixing and shear associated with the calender gap. In addition to calender gap mixing, gear mixers can be utilized for distributive mixing, as well as blister rings for planar shear mixing and/or to provide a seal for vacuum venting. At the discharge end of the screws, positive displacement pumping elements can be specified that pump in a C-locked chamber. Because of screw deflection inherent with the materials being forced through the calender gap, the top screw rpms are typically lower as compared to co-rotating TSEs. In counterrotation, hexalobal mixers are possible at the same flight depth as in bilobal co-rotating TSE designs, which translates into more possible mixing events for each screw rotation (Martin 2005; Fig. 2.19).

TSEs are also efficient as a means to purify the melt stream, referred to as devolatilization which occurs via vents in the TSE process section (Fig. 2.20). Factors that impact devolatilizing efficiency are: Residence time (RT) under the vent or vents (longer the better), surface area of the melt pool (higher the better), and surface renewal (higher the better). Single or sequential atmospheric- and/or vacuum pump-assisted vents can be integrated into the TSE design (Fig. 2.21). The type of pump and vacuum system design depends upon the level and type of volatiles being removed. It is possible to remove more than 20 % volatiles (or more) in a TSE process (an example is the removal of 20 % water from a thermoplastic starch slurry via four vents). Increasing the screw rpm and/or decreasing the rate generally improves devolatilization efficiencies, and the use of stripping agents (i.e., supercritical CO_2) can be utilized to achieve near zero residual levels (Martin 2001).

State-of-the-art TSE barrels are sequential, modular blocks and typically use electric cartridge heaters and internal bores for liquid cooling. Barrels can also use liquid

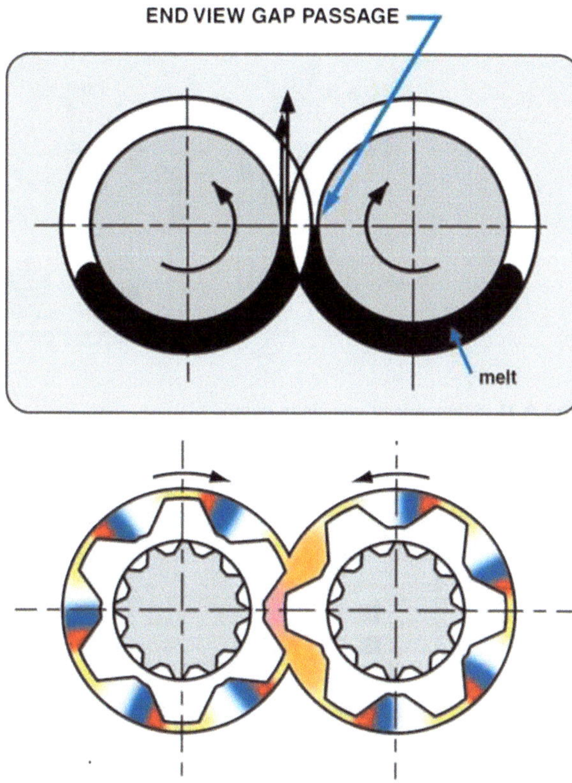

Fig. 2.19 Example of counterrotating calender gap and hexalobal mixing elements

temperature controls. Various types of barrels are available and matched to the unit operations being performed in that region of the process section. Increasing the coolant flow and heat transfer capabilities of the barrels is beneficial for heat-sensitive APIs. Just like screws, hardened stainless steels are adequate for most processes, while nickel-based alloys are specified for corrosive resistance (Fig. 2.22).

Barrel sections are available in two basic configurations: sequential blocks and clam shell, providing unique advantages and disadvantages. Clam shell designs allow for the barrel to be opened, analogous to opening a clam and hence the name of

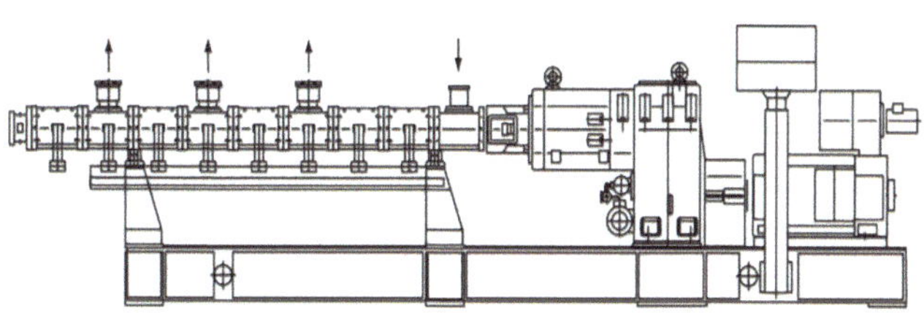

Fig. 2.20 Multistage venting and devolatilization factors

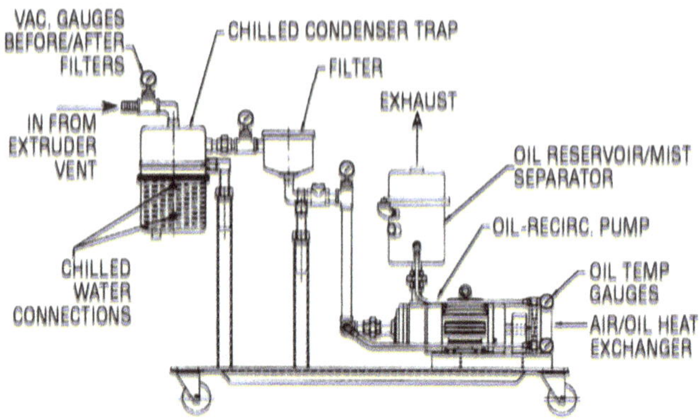

Fig. 2.21 Schematic of a vacuum pump

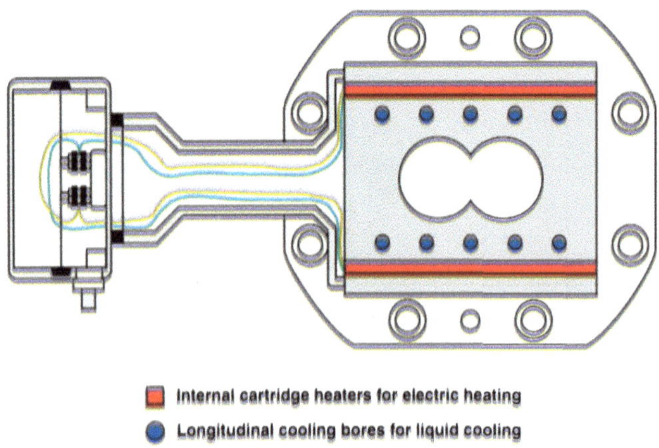

Fig. 2.22 End view of TSE barrel heating/cooling design

the design. This allows users to easily access all points of the barrel for sampling and cleaning activities, which can be advantageous by allowing for regional determination of process performance without removal of the screws from the barrel. Modular interchangeability with these systems is often more limited when compared to the sequential block design. Also, leakage along the barrel seam is possible for high-pressure operations, which is less likely with other barrel geometries. Temperature control is also generally not as precise with the clam shell design. In the case of the sequential block design, modularity of the system and temperature are improved. Locational sampling along the length of the extruder requires specific shut-down, cooling, and screw removal protocols for equivalency to the clam shell design.

As previously stated, the OD/ID ratio of a TSE is defined by dividing the outside diameter (OD) by the inside diameter (ID) of each screw. For instance, a TSE with an OD of 50 mm and an ID of 30 mm would have an OD/ID ratio of 1.66 (50/30). The torque limiting factor for a TSE is the screw shaft diameter and design. Deeper screw

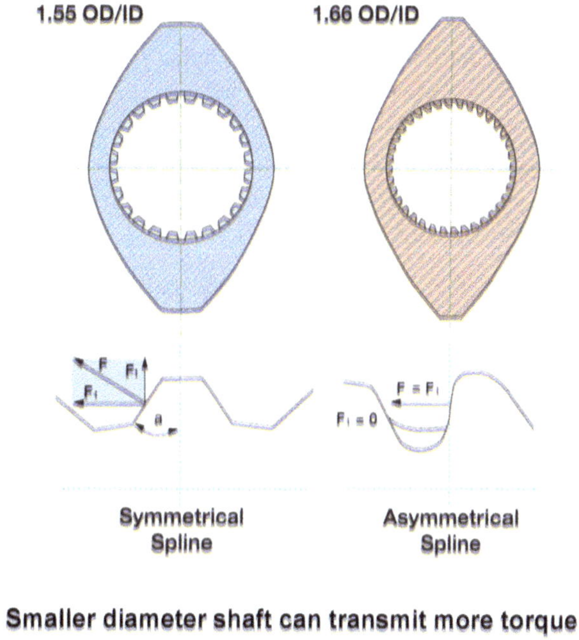

Fig. 2.23 End view of 1.55 and 1.66 OD/ID screw element geometries with comparative symmetrical and asymmetrical shaft design (1 tooth)

flights result in more free volume, but with less torque, since a smaller diameter screw shaft is mandated. Based on the use of a symmetrical splined shaft, a 1.55 OD/ID ratio has been deemed to offer the best balance of torque and volume. The use of asymmetrical splined shafts improves the power transmission efficiency of the shaft so that a smaller diameter shaft can transmit higher torque since the tangential force vector from the shaft tooth is isolated and therefore more efficient in transmitting power from the motor into the shafts/screws/materials being processed. The use of an asymmetrical splined shaft facilitates an increased OD/ID ratio of 1.66/1 with increased torque (Fig. 2.23).

A higher OD/ID ratio (i.e., 1.66/1 vs. 1.55/1 OD/ID) results in both a deeper channel and narrower kneader crest. The materials that pass over the kneader tip experience less RT in planar shear. Both factors contribute to a lower average shear rate inherent with a deeper-flighted TSE, which is often beneficial for processing shear-sensitive formulations (Martin 2012b).

As previously stated, L/D ratios for pharmaceutical processes range from 20/1 for a premix feed to 50/1+ L/D for some injection and solvent extraction processes. Whatever the OD/ID and L/D ratio, it is important to know the specific volume of the TSE:

Specific Volume Specific volume (SV) represents the approximate volume for 1 L/D of the process section, which is useful in a number of other formulas.

$$\text{Specific Volume} = 0.94 \times (\text{OD}^2 - \text{ID}^2) \times \text{OD}/1000$$

Units:

SV is denoted in cc/diameter of length
OD Screw outside diameter (each)
ID Screw inside diameter (each)

For example:

For a HSEI-TSE with a screw OD of 50 and an ID of 30 mm (a 1.66 OD/ID ratio), the approximate SV is as follows:

$$SV = 0.94 \times \left(50^2 - 30^2\right) \times 50/1000 = 75.2\ \text{cc/dia}$$

Residence Time This formula provides the approximate residence time (RT) in the process section. It is important to note that the residence time distribution (RTD) is dependent upon the degree of screw fill. The following formula can be used for RT:
Step 1: Calculate degree of fill in TSE process section:

$$\%\ \text{Fill} = \frac{\text{Rate} \times 0.2777}{\text{FV} \times (\text{Run rpm}/60) \times \text{SG} \times 0.35} \times 100$$

Units:

Rate kg/h
FV Free volume of the extruder, in cc/dia
SG Specific gravity
Run rpm Operating screw rpm

For example:

A 50 mm extruder with a FV of 70 cc/dia., processing 100 kg/h at 200 screw rpm with a 1.0 SG:

$$\%\ \text{Fill} = \frac{100 \times 0.2777}{70 \times (200/60) \times 1.0 \times 0.35} \times 100$$

$$\%\ \text{Fill} = 34\ \%$$

Step 2:

$$\text{Residence Time (RT)} = \frac{\text{L/D} \times 0.28}{\%\ \text{Fill}} + (\text{MP} \times 14.5 \times 0.01)$$

Units:

RT Residence time in seconds
SG Specific gravity
L/D Length/diameter ratio of extruder

0.28 is a composite forwarding efficiency for the screws

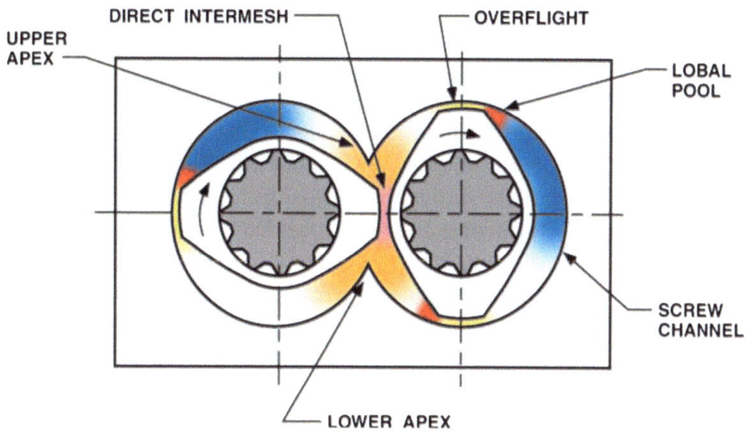

Fig. 2.24 End view of the five TSE mass transfer regions

% Fill Degree of fill, expressed as a decimal (i.e., 40 % = 0.4; dependent on screw design)
MP Melt pressure in Bar

For example:

A 50 mm extruder with a 40 L/D processing 100 kg/h with a 1.0 SG with a 34 % screw fill (0.34), with a 20 Bar outlet pressure, the following applies:

$$\text{RT (average)} = \frac{(1.0 \times 40 \times 0.28)}{(0.34)} + (20 \times 14.5 \times 0.01)$$

$$\text{RT (average)} = 35.8\ \text{s}$$

Note: This a rough approximation and can be made more accurate by adjusting the forwarding efficiency constant based upon experimental results. Also, the residence time distribution (RTD) will be highly dependent upon the degree of screw fill.

In any TSE, there are five shear regions inherent with the screws (Fig. 2.24):

- *Channel region* is a low shear region and is dependent on the degree of screw fill in a starve-fed TSE. The mixing rate in the channel in a TSE is significantly lower as compared to the other shear regions.
- *Overflight gap region* is a high shear region and is independent of the degree of screw fill. This is the area between the screw tip and the barrel wall where the material experiences planar shear.
- *Extensional mixing region* is a high shear region and is independent of the degree of screw fill. This is the area where the material accelerates/experiences a mixing effect from the channel entering into the overflight gap.
- *Apex (upper/lower) region* is a high shear region and is independent of the degree of screw fill. This is where the interaction from the second screw results in mixing due to the associated pressure fields, compression/expansion of the melt, and directional flow changes.

- *Intermesh region* is a high shear region and is independent of the degree of screw fill. This is the area between the screws; in co-rotation the screw surface velocities are in the opposite direction, and in counterrotation the same direction.

It is worth noting that the four high mass transfer regions can, to a certain extent, be considered independent of the degree of screw fill. This partially explains why, in a starve-fed TSE, when the throughput rate is decreased at a constant screw rpm, more mixing occurs, as the materials being processed have a longer RT in the mixing zones. Conversely, as the throughput rate is increased, the low shear channel region dominates more, and the materials being processed will spend less time in the mixing zones (Martin 2001).

The following are some additional formulas that provide insight into any TSE process:

Peak Shear and Shear Stress The peak shear can be critical to achieve dispersive mixing, and can also result in degradation. The following provides a benchmark for the peak shear in a TSE:

$$\text{Peak shear rate} = (\pi * D * n) / (h * 60)$$

Units:

D Screw diameter
n Screw rpm
h Overflight gap

Hence, for a TSE with a 27 mm OD screw and 0.1 mm overflight gap operated at 600 rpm, the peak shear would be calculated as follows:

$$\text{ZSE-27} = \frac{(\pi * 27 * 600)}{(0.1 * 60)} = 8478/\text{s}$$

The formula can be used as a rough estimate to match rpms for TSEs with different diameters, or to estimate the peak shear at a particular rpm. Peak shear rate can be used to calculate shear stress, which is the critical factor to accomplish dispersive mixing:

$$\text{Shear stress} = \text{Peak shear rate} \times \text{Viscosity}$$

In the early stages of the TSE process section, viscosities are higher and high stress rates are produced that cause dispersive mixing (or degradation). In the later stages of the TSE process section, the viscosities typically decrease and result in comparatively lower stress rates. Increased cooling can be used as a tool to increase the viscosity of the melt to achieve dispersive mixing. It must be noted that shear stress calculation does not reflect and is not a measurement of extensional shear, which is much more complicated to calculate and requires modeling, but does provide a benchmark and insight into the dispersive mixing effect.

Fill % Approximation This formula provides the percentage of the available volume in the process section that is being utilized in the starve-fed TSE

$$\%\ \text{fill} = \frac{Q \times 0.2777}{SV \times \text{rpm}/60 \times SG \times Ef}$$

where

- Q Flow rate in kg/h
- Ef Average forwarding efficiency (approx. 35 % for a screw with 1/3 kneaders and 2/3 flighted elements, or 0.35)
- SV Specific volume of extruder in cc/dia
- rpm Screw rpm
- SG Specific gravity of the material being processed

For example, a 40 mm extruder with an SV of 35 cc/dia is running 30 kg/h at 100 rpm,

$$\%\ \text{fill} = \frac{30 \times 0.2777}{35 \times (100/60) \times 1.0 \times 0.35} = 0.41 \text{ or } 41\ \%\ \text{filled}$$

This calculation might show that a process that is devolatilization intensive (i.e., for solvent extraction) is run with a 30 % degree of screw fill because a starved process section increases the surface area of the melt pools under the vent or vents. Another example might demonstrate that a shear-sensitive process might run with a 60 % degree of screw fill.

This formula is a rough estimation and is meant to provide insight into the dynamics of a starve-fed TSE process section, as compared to an absolute value. More advanced/accurate models are available that take into account the specific screw geometries, the viscosity of the melt, and the degree of screw fill.

What is known is that higher fill levels result in a tighter RTD and less shear intensity, with lower fill levels associated with a wider RTD and more shear intensity (Fig. 2.25). Fill level approximations help explain why processes benefit from different feed rates versus screw rpms.

Scale-Up Formula Scale-up is useful for estimating rates for future production TSEs based on lab-scale experiments. The geometries for the two extruders (OD/ID and L/D ratios) should be similar for this equation to be valid. For processes that scale-up volumetrically, the formula is as follows:

$$\text{Scale up: } Q_{\text{target}} = Q_{\text{reference}} \times \left[\frac{OD_{\text{target}}}{OD_{\text{reference}}}\right]^3$$

Units:

- Q Throughput rate (in any units)
- OD Screw outside diameter (each)

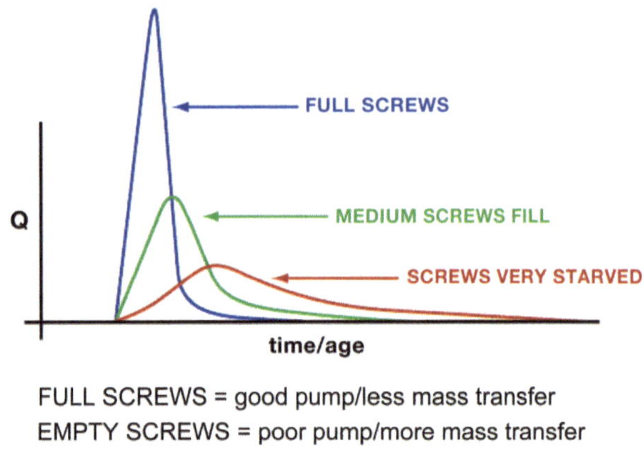

Fig. 2.25 Residence time distribution comparison for various levels of screw fill

For example:

A 18 mm TSE is producing 2 kg/h and the process is not limited by heat transfer or mass transfer boundary conditions. To estimate how much a 40 mm TSE will produce, the following calculations will apply:

$$Q\,(18\ \text{mm}) = 2 \times (40/18)^3$$

$$Q\,(40\ \text{mm}) = 2 \times 2.22^3 = 22\ \text{kg/h}$$

For a heat transfer limited process, the exponent in the scale-up formula is closer to 2. For devolatilization and many pharmaceutical processes, the scale-up exponent is between 2.3 and 2.5. The greater the difference in extruder ODs, the less reliable this calculation becomes. Advanced formulas and computer modeling approaches are also available for more intensive scale-up work (Martin 2006).

How the TSE is configured and operated makes the difference between success and failure. In addition to the process section design (screws/barrels), feed rate, screw rpm, temperatures, and sequence/location of feed streams all play a role in the shear intensity to which the materials are exposed. Table 2.1 provides a brief overview of how a pharmaceutical, or any, TSE process can be managed.

The balance between the rate set by feeders and the screw rpm is obviously important for managing the process intensity and shear history. Process temperatures and the screw design also play critical roles. Everything must work together to optimize the final product (Thiele 2003).

The quality of the product is often affected by the melt temperature of the extrudate, which is often highly impacted by the front-end pressure of the TSE. Front-end management of the melt stream can make the difference between success and failure of a process.

Temperature Rise Caused by Pressure Pressure generation in the extruder front-end results in a temperature rise. The more restrictive the front-end, the higher is the pressure and melt temperature rise, which often adversely effects product quality. The temperature rise formula is as follows:

$$\Delta T\,(^\circ\text{C}) = \frac{\Delta P\ (\text{bar})}{2}$$

Table 2.1 Conditions regulating process intensity

Type of processing	Gentle	Strong
Screw speed	Low	High
Mixing rate in each of the mass transfer along the screws is proportional to rpm		
Screw fill	High	Low
Starving the feed increases remastication and time in mixers to increase process intensity		
Temperature	High	Low
Lowering the temperature increases the controlling modulus and dispersive stress rate		
Extensional/planar mixing	Low	High
Increasing the number of lobal events (i.e., rpms) increases process severity		
Sequential feed	Depends	Depends
Sequential feed may be employed to either utilize or avoid high stress rate exposure		

Units:

ΔT Change in temperature in °C
ΔP Change in pressure (1 bar = 14.503 psi)

For example:

If the die pressure is 40 bar then the associated melt temperature rise can be 20 °C. ($\Delta T = 40/2$)

This formula is meant to be insightful, if not necessarily accurate, as TSE rpm, geometry of the discharge screw elements, and formulation viscosity all play significant factors in the actual melt temperature.

2.4 Downstream Systems

A wide variety of downstream systems are available to extrude an infinite array of pellets and/or shapes. Pelletization is a process where the melt stream is pumped through a die, cooled and formed into a pellet, typically between 0.5 and 5 mm. In strand pelletization, "spaghetti" strands are extruded and cooled on a stainless steel or plastic belt conveyer. The pelletizer feed rolls pull the strands and push them into the cutting assembly. Smaller pellets can be used for direct capsule filling, whereas larger pellets are typically milled.

Die face pelletization is also common, where the pellets are cut at the die face and conveyed/cooled to air quenched chimneys and vibratory towers/trays. The advantage of this system, if workable, is that strand breakage during the cooling phase is

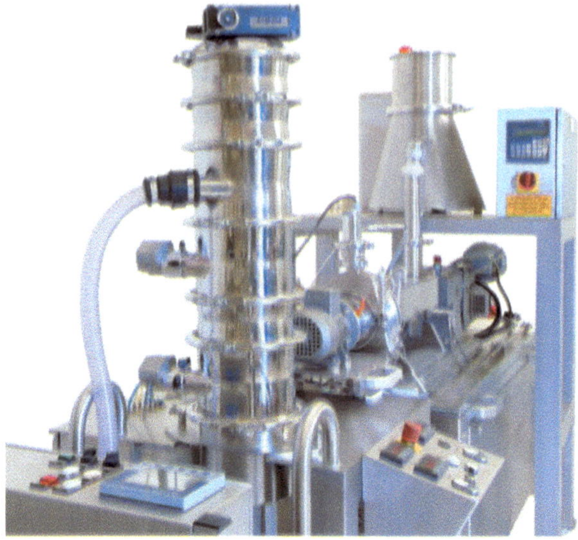

Fig. 2.26 Micro-pelletizer attached to TSE

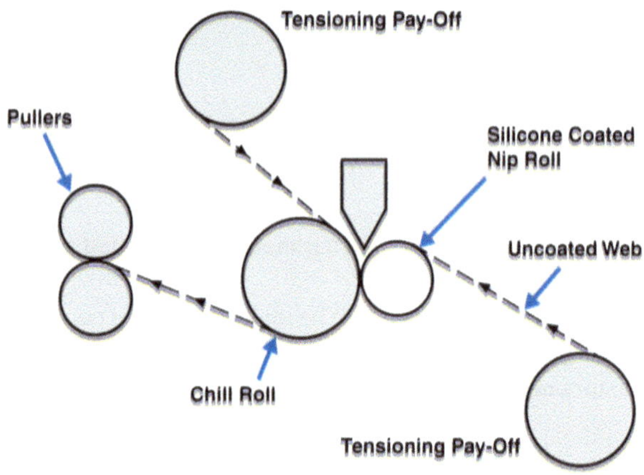

Fig. 2.27 Schematic of lamination system

eliminated. Die face pelletizing is not as wide spectrum with regard to the range of materials that can be processed, and start-up procedures need to be carefully defined and repeated. However, if the process is amenable, die face pelletizing is often preferred to strand systems (Case and Martin 2005; Fig. 2.26).

Film and lamination systems are often used to produce transdermal and dissolvable films (Fig. 2.27). To maintain dimensional tolerances, a gear pump (or screw pump) that builds/stabilizes pressure to the die is often specified. A gear pump is a positive displacement melt delivery device that builds and stabilizes the melt stream from the TSE and into the die. The melt is then distributed in a flat die and cooled on rolls. The roll surface is maintained at the desired temperature by liquid circulating through internal cooling channels. For some flat products the nip force across the roll face is used to "squeeze" the extrudate between the rolls. Unwind stations can be

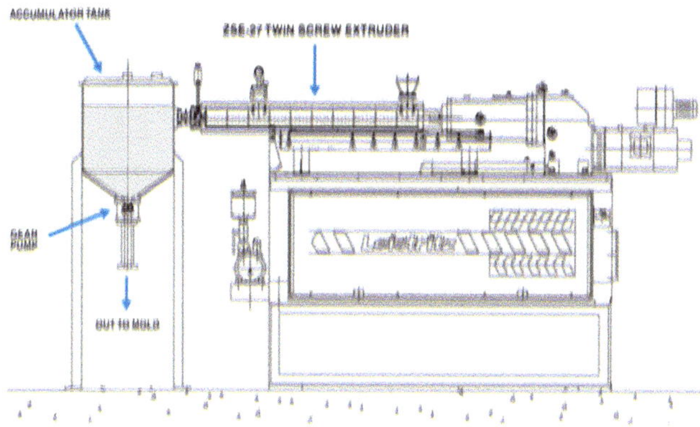

Fig. 2.28 Continuous compounding combined with in-line molding system

utilized to laminate the film onto a substrate. The final product is then either wound or cut-to-length.

Shape extrusion is when the process melt is directly extruded into a part with specific dimensions. Unique shapes are possible. The extrudate can be a simple rod, or complex shape, referred to as a "profile". The extruded profile is formed in the die, sized by calibration tooling, and conveyed and supported through a variety of different types of air cooling devices. A belt puller pulls the extrudate and then feeds it to an on-demand or flywheel type cutter. In this manner, for example, a pre-form tablet might be produced (Elliott 2003).

In-line molding is also possible to produce unique three-dimensional dosage forms (Fig. 2.28). In this operation, the formulation is mixed and devolatilized in a TSE and discharged into an accumulator. A gear pump or screw pump then operates cyclically to fill a mold that determines the final product shape and dimensions. This process is highly formulation-dependent.

2.5 Examples of Pharmaceutical Processes Performed on Twin Screw Extrusion Systems

The following provides a few examples of related processes/applications:

2.5.1 Nano-16 Twin Screw Extrusion System Case Study

A challenge in the early stages of development is that APIs are expensive and only available in limited quantities. Hence, a need arose to process very small samples (between 20 and 100 g) via ME. Therefore, a low volume TSE with 16 mm OD

Fig. 2.29 Micro-plunger feeder attached to low-volume TSE

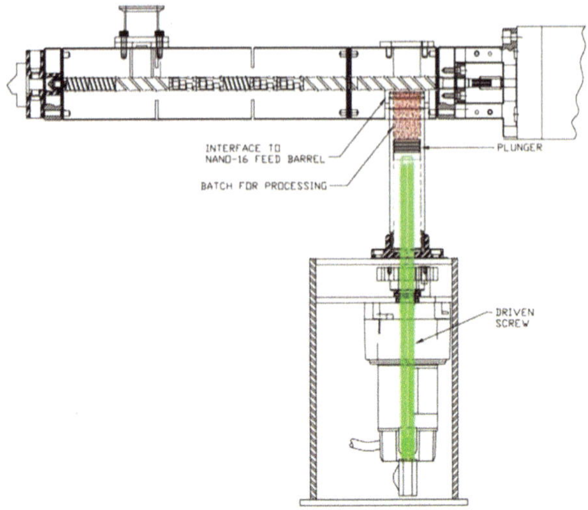

screws and a 1 mm flight depth with a 1.2/1 OD/ID ratio and 1 cc/dia free volume was used to process a series of 50 g batch samples of a HPMCAS polymer with 40 % of a poorly soluble API. The objective was to demonstrate the viability of the extrusion process utilizing a small sample with minimal waste. The materials were premixed and metered by a patented micro-plunger feeder (Fig. 2.29). A 25–1 L/D process section was utilized. The screw design included flighted elements and kneading/shear-inducing elements. An atmospheric vent and a low-volume strand die front-end attachment were used.

Approximately 44 g of the sample was collected and usable for evaluation purposes. Approximately 6 g of material was "lost" in the TSE process section as follows: 1 g at the extruder/plunger interface, 2–3 g on the screws, and 2 g in the die/front end. The temperature profile and screw rpms were selected based upon experience with similar formulations. A PC-based control/data acquisition package allowed for detailed analysis of the run conditions (Fig. 2.30).

The strand was cooled in an air-quenched annular chamber, cut into 1 mm pellets, and collected. The pellets were then milled into a powder and compressed into tablets for dissolution testing. The results indicated that the API was converted from a crystalline to amorphous state during processing. Without extrusion, this API might not have been a candidate for additional development (Fig. 2.31).

Subsequent tests were performed at 2 kg/h (TSE with 18 mm screw diameter) and 6 kg/h (TSE with 27 mm screw diameter) with similar results, confirming the viability of the initial screening device (Martin 2010).

2.5.2 *Melt Granulation*

Melt granulation has traditionally been performed by a high-shear granulator with double-jacketed granulation bowl, where either water or oil is circulated as the

Run Data

Runtime	Z1	Z2	Z3	Z4	Z5	A1	A2	A3	A4	A5
0:00:00	120	125	130	131	131	0	0	86	53	0
0:01:00	120	125	130	131	130	100	2	207	190	0.27
0:02:00	120	125	130	130	130	100	2	138	179	0.45
0:03:00	120	125	130	130	129	100	2	207	203	0.63
0:04:00	120	125	130	130	128	100	2	466	266	0.87
0:05:00	120	125	130	130	127	100	2	742	303	1.17
0:06:00	120	125	130	130	126	100	2	760	314	1.48
0:07:00	120	125	130	130	125	100	2	673	314	1.79
0:08:00	120	125	130	130	124	100	2	742	324	2.1
0:09:00	120	125	130	130	124	100	2	725	324	2.43
0:10:00	120	125	130	130	123	100	2	1019	385	2.79
0:11:00	120	125	130	130	123	100	2	1019	409	3.19
0:12:00	120	125	130	130	122	100	2	1036	401	3.6
0:13:00	120	125	130	130	122	100	2	932	374	3.98
0:14:00	120	125	130	130	122	100	2	1036	390	4.36
0:15:00	120	125	130	130	122	100	2	1088	398	4.77
0:16:00	120	125	130	130	121	100	2	1071	409	5.19
0:17:00	120	125	130	130	121	100	2	1209	443	5.62
0:18:00	120	125	130	130	121	100	2	1226	467	6.07
0:19:00	120	125	130	130	121	100	2	1140	427	6.51
0:20:00	120	125	130	130	121	100	2	1105	417	6.94
0:21:00	120	125	130	130	121	100	2	1157	459	7.37
0:22:00	120	125	130	130	121	100	2	1278	475	7.82
0:23:00	120	125	130	130	121	100	2	1364	496	8.31
0:24:00	120	125	130	130	121	100	2	1519	511	8.82
0:25:00	120	125	130	130	121	100	2	1347	525	9.32
0:26:00	120	125	130	130	121	100	2	1312	469	9.82
0:27:00	120	125	130	130	121	100	2	1450	562	10.31
0:28:00	120	125	130	130	121	100	2	1485	525	10.83
0:29:00	120	125	130	130	121	100	2	1416	506	11.35
0:30:00	120	125	130	130	121	100	2	1416	509	11.86
0:31:00	120	125	130	130	121	100	2	1381	546	12.35
0:32:00	120	125	130	130	121	100	2	1381	509	12.86
0:33:00	120	125	130	130	121	100	2	1329	485	13.36
0:34:00	120	125	130	130	121	100	0	1220	448	13.72

Z1 = Zone 1 Temperature; A1 = Screw Speed
Z2 = Zone 2 Temperature; A2 = Plunger Feeder cc/min
Z3 = Zone 3 Temperature; A3 = Pressure
Z4 = Zone 4 Temperature; A4 = Torque
Z5 = Melt Temp; A5 = TTQ

Fig. 2.30 Nano-16 TSE process conditions/run record

heating medium. The disadvantages of this process include: inefficient processing due to the slow heat transfer from the bowl to the powder blend, difficulty in process scale-up since the area of the heated surface to the volume of powder blend

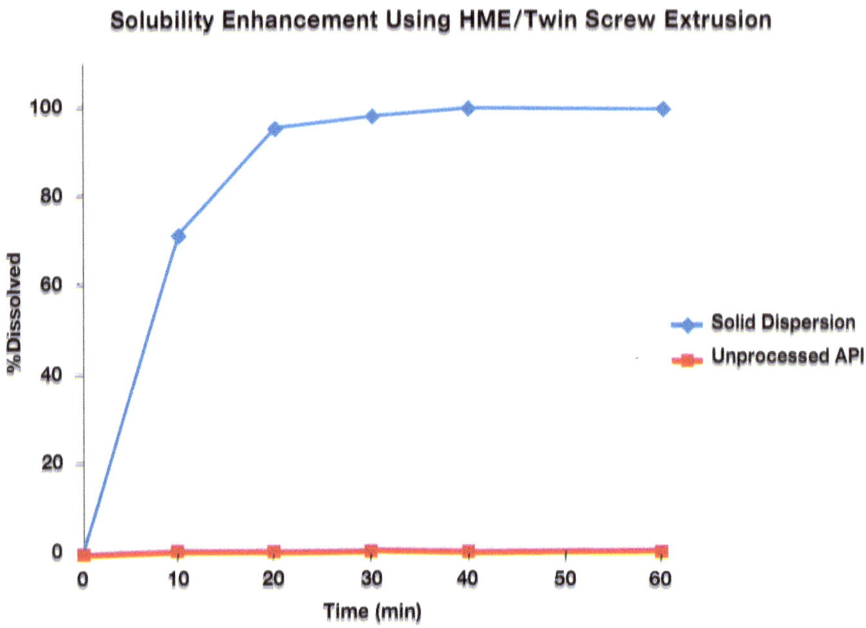

Fig. 2.31 Nano-16 sample results from dissolution testing

decreases with an increase in batch size, and batch-to-batch variability associated with inconsistent heat transfer.

A powdered premix was metered into a TSE with a 16 mm OD. The mechanical energy of co-rotating TSE screws, rather than the heat conducted from the barrels, melted/softened the premix with minimal heat transfer between the extruder barrel and materials. A binder that melted at a relatively low temperature (60–80 °C) was used to achieve the agglomeration of the formulation composition. During extrusion, the formulation was subjected to high pressure which resulted in granules with a higher density and better flowability in comparison with the granules from a high-shear granulator. An air quench front-end with mild compression helped optimize granule formation. The granules were formed after the material exited the die and the molten binders solidified.

Through the use of TSE melt granulation highly compressible granules were produced that yielded tablets with minimal weight variation. It can also be theorized that the TSE melt granulation with controlled/repeatable mechanically driven melting and short mass transfer distances is easier to scale-up as compared to a high-shear granulator (Keen et al. 2012; Fig. 2.32).

2.5.3 Micro-Pelletization for Capsule Filling

A Eudragit™ polymer/API powdered premix was metered by an LIW feeder into a an 18 mm TSE that mixed, devolatilized, and conditioned the materials and pressurized

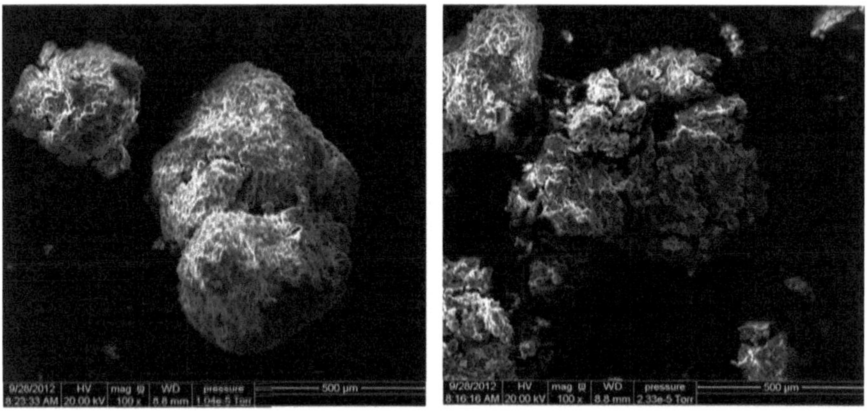

15% tramadol HCl granules obtained by continuous melt granulation with NANO16

15% tramadol HCl granules obtained by melt mixing and grinding through No. 20 mesh

Fig. 2.32 Melt granulation morphology produced from TSE

Fig. 2.33 Eudragit™ 0.5 mm micro-pellets produced from TSE

the inlet of a gear pump front-end attachment that built/stabilized pressure to an annular die with a series of 0.5 mm micro-die holes (Fig. 2.33). A differential pressure before/after the gear pump of 100 bar was required for this particular formulation to keep the micro-die holes from getting blocked. Rotating blades cut the strands at the die face that were then vacuum-conveyed to a cooling chimney which finished the cooling process and discharged the micro-pellets.

It should be noted that die face pelletizing can be problematic because some formulations will smear at the die face during cutting, and pellets may also have a tendency to stick together in the cutting chamber. Micro-pelletizing is particularly challenging as holes tend to get blocked and high pressures are often a requirement. For success, extensive formulation and process development can be expected (Elliott 2003).

Fig. 2.34 End view of coextruded structure rod/profile for sustained release of product

2.5.4 *Thermo-Plastic Urethane (TPU) Coextruded Rod System*

A thermo-plastic urethane (TPU) was premixed with a low percentage of API and metered into a TSE (27 mm OD screws) by an LIW feeder. The TSE process section design facilitated intimate mixing without over-shearing the TPU/API formulation. A gear pump was attached to the TSE to pressurize a coextrusion die. A single screw extruder (SSE) with 25 mm OD melted/conditioned a "virgin" TPU and was mated to and pressurized the same coextrusion die. The SSE is a flood-fed device that is used to melt and pump a precompounded material where the screw rpm determines the rate of the melt stream being fed into the die. The coextrusion die distributed the TPU melt from the SSE so that it completely encapsulated the inner TPU core that contained the active drug.

Since the inner core of the TPU coex structure was encapsulated, it was possible to pass the part through a water trough with guide rolls for more efficient cooling, as compared to air. If it was deemed that the structure could not contact water, then a series of driven/contoured rollers might be specified to cool the product. A precision belt puller pulled the extrudate through the water trough and fed an on-demand cutter that cut the extrudate into approx. 75 mm lengths. Longer or shorter lengths are possible. The parts were candidates to be post welded, as might be used as a vaginal ring. Flywheel type cutters can also be used to produce shorter length parts, i.e., in the 0.5 to 3 mm range (Fig. 2.34).

2.5.5 *Transdermal Sheet/Laminate Extrusion*

Multiple rubber feed streams were conditioned and metered into a TSE at a controlled rate. Rubbers are typically ground and fed into the TSE, or melted and pumped into the TSE. The API was metered by an LIW feeder into a side stuffer that pushed the API into the TSE process section. The materials were then mixed and discharged from the TSE into a gear pump that metered the melt stream to a flexible lip sheet die that distributed material across the die width. The downstream take-off was designed so that the nip force across the roll face "squeezed" the extrudate between the rolls, which were cored for liquid temperature control. Unwind stations, with electronic

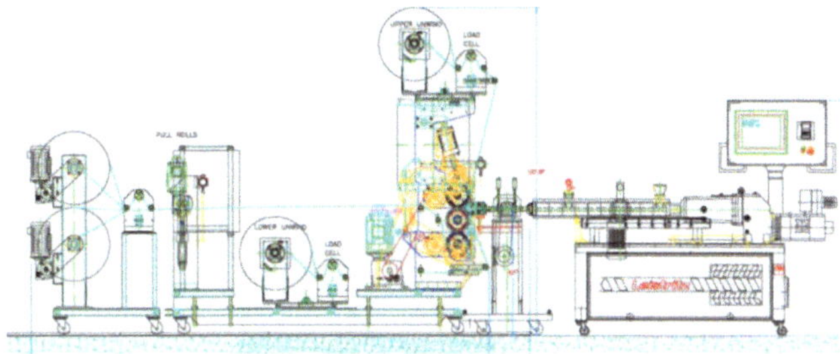

Fig. 2.35 Schematic of TSE transdermal-laminate system

tension control, introduced backing and peelable substrates and the extrudate together at the correct position to facilitate adhesion. An automatic cut-and-transfer winder was used to wind product prior to post-processing operations (DiNunzio et al. 2010; Fig. 2.35).

2.5.6 Extrusion of Foamed Extrudate

A Eudragit™ polymer/API/talc were premixed and metered into a (27 mm OD screws) TSE by an LIW feeder. The materials were melted/mixed in the early stages of the TSE process section and dynamic seals were integrated into the screw design to accommodate high-pressure injection pressures inherent with supercritical CO_2. The supercritical CO_2 injection system was equipped with a precision piston type injection pump and chilled/temperature controlled plumbing connections. The supercritical fluid was injected and intimately mixed with the molten extrudate via high division distributive mixers to minimize viscous heating. The latter part of the TSE process screw design utilized low energy input pumping elements to allow the TSE barrels to serve as a heat exchange device to cool the process melt stream. The extrudate was pumped through a pelletization die, cut at the die face by rotating knives and vacuum conveyed to a cooling tower. The resulting foamed pellets were then milled and pressed into tablets that resulted in faster dissolution rates, indicating that by increasing the porosity of the matrices a faster acting dosage form is possible (Listro et al. 2012; Figs. 2.36 and 2.37).

2.5.7 Production ME System

A PVP/API premix was metered at 100 kg/h by an LIW feeder to a TSE (50 mm OD screws) that melted, intimately mixed, devolatilized, and pumped the formulation to a strand die. Vacuum venting was utilized to remove residual volatiles from the formulation. A 10-hole strand die was specified with an angle discharge onto a

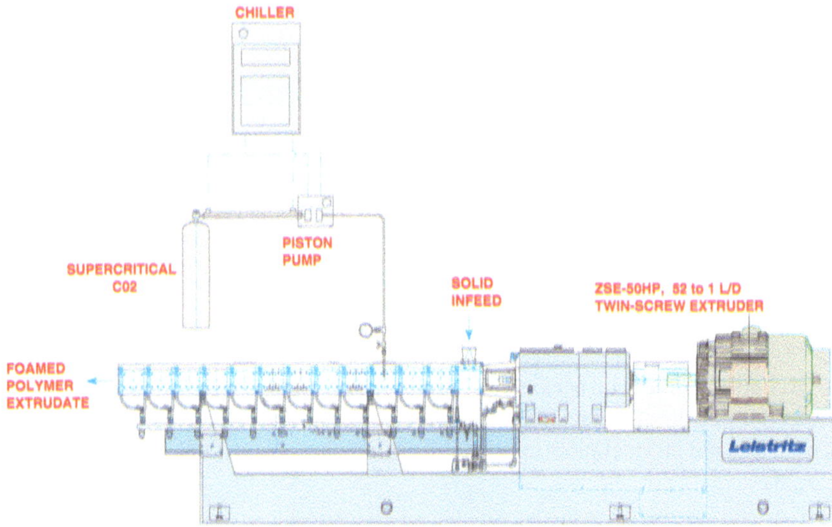

Fig. 2.36 Schematic of attachments of a TSE supercritical injection system

Fig. 2.37 Close-up of attachment "combing" mixing elements

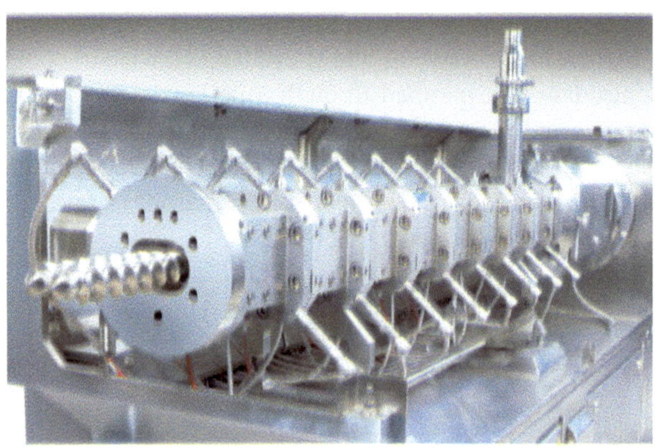

Fig. 2.38 Production class 50 mm TSE

stainless steel conveyor which cooled the strands which were then directly fed into a pelletizer and in-line mill. The temperature and speed of the cooling conveyor was critical so that the strands could be pelletized and cooled in-line (Fig. 2.38).

2.6 Summary/Conclusion

ME continues to evolve and create sophisticated dosage forms utilizing extrusion technologies that have already been developed in the plastics (and food) industries—extrusion is a battle-hardened, well-proven, manufacturing process that has been validated in 24 h/day industrial settings making the plastic products we see and use every day. Now that the pharmaceutical industry has embraced ME, machinery suppliers have downsized and redesigned equipment to meet the requirement of cGMP environments. Additional efforts have also been made to design TSE systems to test early-stage materials available only in minimal quantities. Coextruded structures, utilizing multiple extruders, foamed products and integration with in-line molding are the next generation of development efforts that are foreseen.

The best indication of the future is the past. The plastics industry evolved from batch to continuous processing to make better, more consistent products at lower cost. This is now happening in the pharmaceutical industry. The future is bright for ME, and the path is well-proven.

References

Case C, Martin C (2005) Pelletizing/Compounding, Szycher's Handbook of Polyurethanes, 2nd edn. CRC Press

DiNunzio J, Martin C, Zhang F (2010) Melt extrusion: shaping drug delivery in the 21st century. Pharm Tech, S 30

Elliott B (2003) Film, sheet, & laminates. In: Ghebre-Sellassie I, Martin C (eds) Pharmaceutical extrusion technology. Drugs & the pharmaceutical sciences volume 133. Marcel Dekker Inc, New York, p 225

Kapp S, Palmer P (2003) Controls and instrumentation. In: Ghebre-Sellassie I, Martin C (eds) Pharmaceutical extrusion technology. Drugs & the pharmaceutical sciences volume 133. Marcel Dekker Inc, New York, p 361

Keen JM, Hughey JR, Bennett RC, Martin C, Cuppok Y, Jannin V, Musakhanian J, Marchaud D, McGinity JW (2012) Melt granulation using a twin screw extruder to obtain lipid granules containing tramadol HCl. AAPS, Chicago IL

Listro T, Kane T, Adegoke A, Sarode A, Wang P (2012) Foam hot melt extrusion of an Eudragit—Nifedipine matrix formulation using supercritical CO_2. Controlled Release Society, Portland OR

Martin C (2001) Twin screw extruders. The SPE guide on extrusion technology and troubleshooting, Part I, Chapter 2

Martin C (2005) Counter-rotating twin screw extruders, SPE plastics technician's toolbox: Extrusion

Martin C (2006) Understanding twin screw extruders: the High-Speed Energy Input (HSEI) Concept, Plastics Today Online

Martin C (2010) Hot melt extrusion: new R&D options and the path to production. Controlled Release Society, Portland OR

Martin C (2012a) Twin screw extruder developments to produce consistent/repeatable concentrates, SPE Color RETEC, Louisville, KY

Martin C (2012b) Comparing twin screw extrusion systems to process bioplastics. Polymer Processing Society, Niagara Falls, Ontario

Martin C, Nowak S (2011) Loss-in-weight feeding and continuous extrusion: new Options in pharmaceutical processing, Pharmaceutical Online

Thiele W (2003) Twin-screw extrusion & screw design. In: Ghebre-Sellassie I, Martin C (eds) Pharmaceutical extrusion technology. Drugs & the pharmaceutical sciences volume 133. Marcel Dekker Inc, New York, p 69

Part II
Excipients

Chapter 3
Properties and Applications of Polyvinyllactam Polymers

Karl Kolter

Abstract A number of polymers used in melt extrusion have been available to the pharmaceutical scientists for many decades. However, in one case, a vinylcaprolactam structure (Soluplus®) was developed with the intention of using it in particular for melt extrusion. It was designed to address some of the shortcomings of the available polymers. In addition, the properties and applications of many polyvinyllactam polymers, such as the Kollidons®, are outlined, discussing immediate release to sustained release. These "older and newer" polymers, including their structure and applications, are thoroughly discussed within this chapter, including a historical perspective.

3.1 Introduction

In the manufacture of pharmaceutical polyvinyllactams, the monomer vinylpyrrolidone is mainly used, whereby this is polymerized to the homopolymer polyvinylpyrrolidone (povidone). Alternatively, together with vinyl acetate, it can be polymerized to the copolymer copovidone. However, since 2009, a new polymer has been available in the market; this is based on vinylcaprolactam (Soluplus®) and differentiates itself from vinylpyrrolidone in that it has a larger ring (Fig. 3.1).

Polyvinylpyrrolidones are polymers that have been in use in the pharmaceutical industry for quite some time (Bühler 2008a). First synthesized in the 1930s by W. Reppe, they were initially used as plasma expanders in the Second World War. Since then, however, they have established themselves as very effective binders (Agrawal and Prakasam 1988). In addition, numerous other applications have been found, such as crystallization inhibitors (Simonelli et al. 1970; Ishida et al. 2007), viscosity-increasing agents (Bühler 2008b), film formers (Amnuaikit et al. 2005) solubilizers (Damian et al. 2002; Nair et al. 2002), and complex formers (Garekani et al. 2003; Plaizier-Vercammen 1983; Plaizier-Vercammen and Neve 1983). They are thus multifunctional polymers and as such, due to their many and diverse properties, can be used in the most varied applications (oral, Brouwers et al. 2009;

K. Kolter (✉)
BASF SE, Ludwigshafen, Germany
e-mail: karl.kolter@basf.com

M. A. Repka et al. (eds.), *Melt Extrusion*,
AAPS Advances in the Pharmaceutical Sciences Series 9,
DOI 10.1007/978-1-4614-8432-5_3,

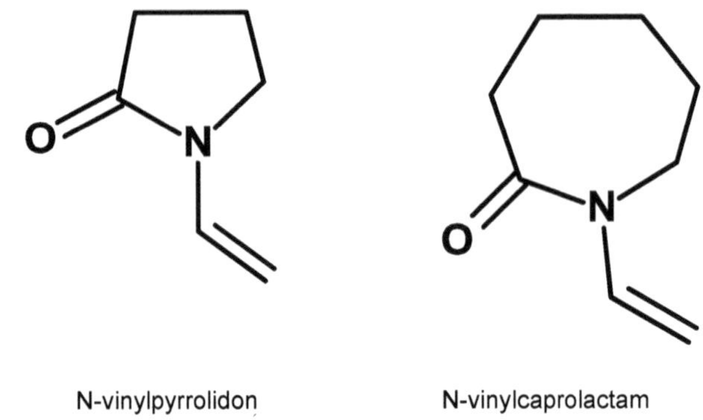

Fig. 3.1 Structure of vinyllactams

He et al. 2008; parenteral, Akers 2002; dermal, Lee and Chien 1995; Ma et al. 1996) and also in different dosage forms (tablets, capsules, oral solutions, injection solutions, etc.) (Tantry et al. 2007; Kuentz and Röthlisberger 2002; Strickley 2004), which these different functions can be optimized for various uses.

Polyvinyllactams are prepared by radical polymerization of the corresponding monomers. An initiator is required to start the reaction; this generates the radicals that then react with the monomers. The reaction continues with further monomer molecules until at some point it ceases. The polymerization reaction can take place in water or a suitable solvent. Depending on the way the reaction is set up and controlled, products with different properties can be obtained. For example, the molecular weight (MW) can be controlled to yield products of MW of approximately 2,500–1,250,000 Da, whereby those properties, such as viscosity, that are dependent on this can vary. If other polymer properties are to be modified significantly, comonomers should be used. Such comonomers have an effect especially on the lipophilic nature of the product, its behavior in various solvents, its film properties and its glass transition temperature. Thus, vinyl acetate can be used as a comonomer to form a polymer (copovidone) that is considerably more plastic, more lipophilic, and that has better film-forming properties. However, this polymer is still completely water soluble. In the alternative, if the proportion of vinyl acetate were significantly increased, a polymer would be produced that is insoluble in water.

A polymer can be even more strongly modified if the monomers are grafted onto another polymer, which is referred to a graft copolymer. In these graft polymers, the individual areas—the graft components and the grafted side-chains—differ significantly in their properties. If these polymer areas differ in terms of lipophilicity/hydrophilicity, they are called amphiphilic polymers.

3.2 Properties of Polyvinyllactam Polymers

3.2.1 Povidones

The various types of povidone differ primarily in their molecular weights, which can be varied over a considerable range (2,500–1,250,000 Da). As molecular weight

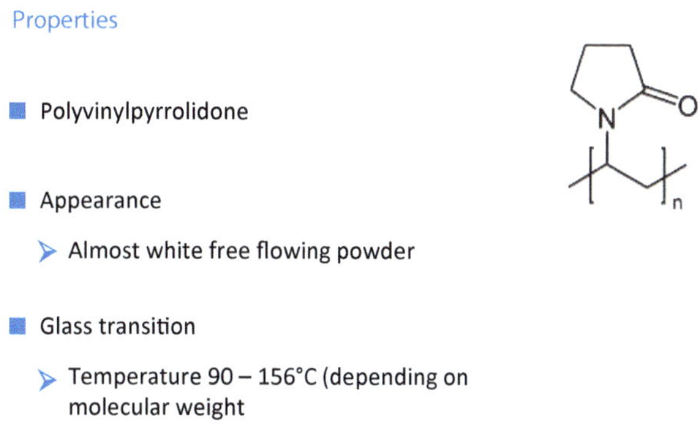

Fig. 3.2 Structure and properties of povidones

Table 3.1 K-values and molecular weights of povidones

Kollidon® grade	K-value (1 or 5 % solutions in water)	M_w (g/mol)
Kollidon® 12PF	11–14	~2500
Kollidon® 17PF	16–18	~9000
Kollidon® 25	24–27	~31000
Kollidon® 30	28–32	~50000
Kollidon® 90F	85–95	~1250000

cannot simply be routinely and reproducibly determined, the so-called K-value is used for characterization. This is calculated from the relative viscosity of a dilute solution of polymer in water (Bühler 2008a) and is suffixed to the name. Of course, there is a correlation between molecular weight and the K-value. The povidone monograph in the pharmacopoeias comprises all polyvinylpyrrolidones of different molecular weight whereby precise limits for the K-values are specified (Fig. 3.2; Table 3.1).

Povidones combine aqueous solubility with solubility in organic solvents of medium lipophilicity (Bühler 2008a). This is significant as it means that povidones can be used in both aqueous and organic media. These polymers are also capable of interacting with both hydrophilic and lipophilic active ingredients (Wen et al. 2005). This is extremely important not only for the binders in particular (Chowhan et al. 1992), but also for their application as crystallization inhibitors (Jain and Banga 2010; Van den Mooter et al. 2010). They are insoluble only in very lipophilic solvents such as the aliphatic or aromatic hydrocarbons.

3.2.2 *Copovidone*

Copovidone is the pharmacopoeia designation for a polymer comprising the monomers vinylpyrrolidone and vinyl acetate in the ratio of 6:4, whereby the distribution of the monomers within the polymer is purely statistical. By introducing the "soft" monomer vinyl acetate, the resulting polymer is given a higher degree

Properties

- Vinylpyrrolidone / vinyl acetate
  - 60/40
- Appearance
  - White or slightly yellowish free flowing powder
- Molecular weight
  - ~ 55 000 g/mol
- K-value (1% in water)
  - 25 – 31
- Glass transition temperature
  - 101°C

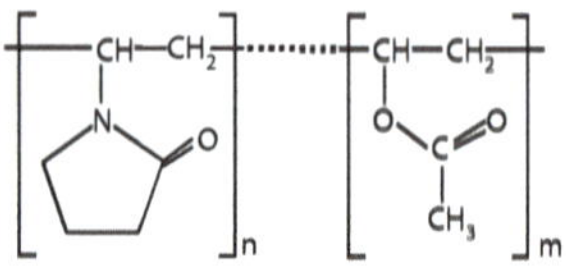

Fig. 3.3 Structure and properties of copovidone

of plasticity, recognizable by its strongly reduced glass transition temperature of 101 °C compared to that of a polyvinylpyrrolidone (149 °C) of the same molecular weight. The K-value of 28 corresponds approximately to that of Kollidon® 30; thus, its molecular weight is also within a similar range (approx. 50,000 Da) (Fig. 3.3).

Just like povidone, copovidone is soluble in water and many organic solvents. The application areas of copovidone are many and various. For example, copovidone can be used as a wet (Ritala et al. 1986) as well as a dry binder (Kolter and Flick 2006); for the latter application especially, its high degree of plasticity ensures that tablets produced have a very high degree of hardness (Maschke et al. 2007). Comprising small particles with a hollow spherical structure, it is the most effective dry binder for roller compaction (Herting et al. 2007) and tableting (Ali and Langley 2010). Further applications are as a film former in oral and dermal dosage forms (Gil et al. 2008; Repka et al. 2012) and, especially, as a base polymer for the production of solid solutions by melt extrusion (Ranzani et al. 2011) or spray drying (Patterson et al. 2008). For melt extrusion in particular, it ensures easy processing and excellent properties of the solid solutions produced.

3.2.3 Polyvinylcaprolactam-Polyvinyl Acetate-Polyethylene Glycol Graft Copolymer (Soluplus®)

Soluplus® was developed by BASF in 2009 and launched in the same year (Kolter et al. 2010). It is characterized by a unique polymer structure as it is produced by grafting vinylcaprolactam and vinyl acetate onto polyethylene glycol in a copolymerization reaction. As a result of this particular method of manufacture, it has a backbone of polyethylene glycol and side-chains comprising the two vinyl monomers. This gives the product a pronounced amphiphilic character—it could

Properties

- **PEG 6000 / vinylcaprolactam / vinyl acetate**
  (13/57/30)
- **Appearance**
  White to yellowish free flowing granules
- **Molecular weight**
  ~ 118 000 g/mol
- **K-value (1% in ethanol)**
  31 – 41
- **Glass transition temperature**
  ~ 70°C
- **Critical micelle concentration (CMC)**
  7.6 mg/L

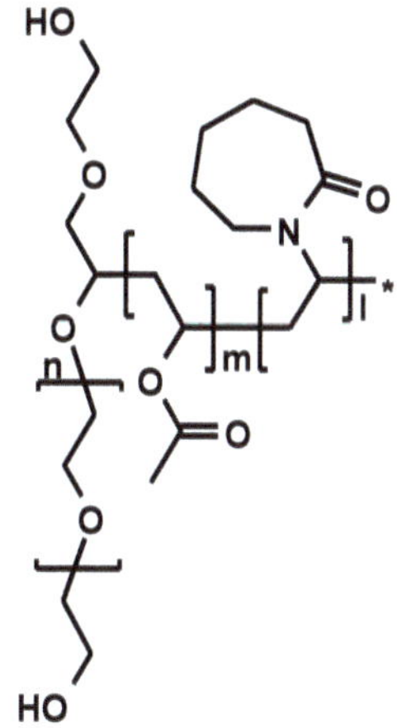

Fig. 3.4 Structure and properties of polyvinylcaprolactam-polyvinyl acetate-polyethylene glycol graft copolymer (Soluplus®)

well be designated as a polymeric surfactant as it reduces interfacial tension and forms micelles in water. The vinyl acetate and polyethylene glycol give rise to a relatively low glass transition temperature that is ideal for melt extrusion applications. It can thus be better extruded at low temperatures than many other polymers. Temperature-sensitive active ingredients can thus be processed using melt extrusion with this unique polymer (Fig. 3.4).

In spite of its relatively high molecular weight of 118,000 Da, it has a very low melt viscosity, probably facilitated by the polyethylene glycol moiety in the polymer. This polyethylene glycol component acts as a plasticizer that is covalently bound to the molecular structure and hence cannot escape.

Soluplus® is soluble in numerous solvents (Kolter et al. 2010) and is hence suitable for the manufacture of solid solutions from organic solutions by means of spray drying.

The dense particle structure of Soluplus® makes for excellent flow properties and high bulk density. It enables melt extrusion to be carried out at extremely high-throughput rates as the polymer can be fed into the extruder rapidly, dust-free, and without tackiness.

3.2.4 *Polyvinyl Acetate/Povidone (Kollidon® SR)*

Kollidon® SR is not a copolymer but a formulated mixture of the two polymers polyvinyl acetate and povidone in a ratio of 8:2 (Bühler 2008a). The polymers are manufactured separately and then combined using special formulation technology.

Properties

- Polyvinyl acetate / polyvinylpyrrolidone
  - 80/ 20
- Appearance
  - White or slightly yellowish, free flowing powder
- Molecular weight
  - Polyvinyl acetate: ~ 450 000 g/mol
  - Povidone: ~ 50 000 g/mol
- K-value (1 % in tetrahydrofuran)
  - 60 – 65

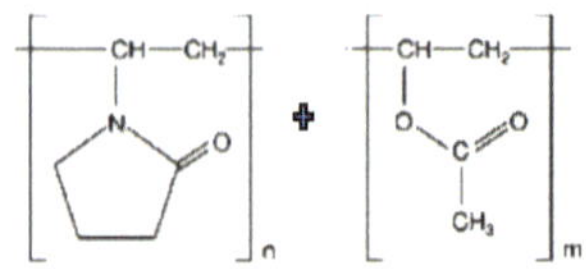

Fig. 3.5 Structure and properties of polyvinyl acetate/povidone (Kollidon®)

In this way, Kollidon® SR contains a water-insoluble polymer, polyvinyl acetate, and a water-soluble polymer, povidone. The molecular weight of polyvinyl acetate is approximately 450,000 Da and that of povidone is 50,000 Da (Fig. 3.5).

Combination of these two very different polymers results in properties that are decisive for the main application—sustained release from matrix tablets. The insoluble polyvinyl acetate renders an extremely high degree of plasticity and hence compressibility but also presents a diffusion barrier that slows down the release of the active ingredient from the drug. The water-soluble povidone on the other hand creates micropores in the framework of the polyvinyl acetate through which water can penetrate the entire system, this in turn dissolves the active ingredient, which can then allow diffusion of the drug. The ratio of the two components has been optimized so that the sustained release process can be applied to most active ingredients. Only in the case of very water-soluble actives is it necessary to include additional retardation agents to slow the release further. Kollidon SR, based on its excellent compressibility and flow properties, is highly suitable for the manufacture of sustained release tablets by direct compression (Strübing et al. 2008; Bühler and Fussnegger 2005). However, it can also be used in roller compaction (Bühler 2008a, b), wet granulation (Kranz and Wagner 2006), and melt granulation (Flick et al. 2000). Recent studies have shown that it can also be meaningfully employed in melt extrusion (Özgüney et al. 2009).

One of the major advantages of sustained release dosage forms based on Kollidon SR is the fact that the release properties can be accurately profiled and forecast by computer modeling (Siepmann et al. 2010).

3.3 Suitability and Applications in Hot-Melt Extrusion

When intended for use in hot-melt extrusion, the polymers must fulfill certain requirements. As an extrusion process is only possible above the glass transition temperature and the extrusion temperature should not be too high, a low glass transition temperature is certainly an advantage (Chokshi et al. 2006). Normally, melt extrusion is carried out at temperatures of 20–40 °C above that of the glass transition temperature (Kolter et al. 2010). Polymers of high molecular weight and showing a strong interaction between the functional groups tend to generate a high degree of melt viscosity (Karl et al. 2011; Schillinga et al. 2008) and are therefore difficult to extrudate.

Further relevant properties are a high degree of solubility for active ingredients in solid solution (Leuner and Dressman 2000), low hygroscopicity, and low toxicity (Ghebre-Sellassie and Martin 2007). The toxicity factor is important as the base polymer is used in higher concentrations and hence substantial amounts are subsequently administered to patients with the drug.

Water-soluble polymers are principally used for formulating poorly soluble active ingredients that exhibit lower bioavailability (Forster et al. 2001a). They can also be used to mask unpleasant taste (Maniruzzamana et al. 2012) or to protect the active ingredient from external influences by embedding it (Crowley et al. 2007). Water-insoluble polymers with a certain degree of plasticity are mainly used to achieve sustained release of the active ingredient (Zhang and McGinity 2000).

3.3.1 Glass Transition and Degradation Temperatures

The glass transition temperature of polyvinyllactams varies quite considerably depending mainly on their molecular weights and on the comonomers and their proportions in the molecule. Vinylpyrrolidone is a "hard" monomer and consequently produces very hard polymers with a high glass transition temperature (Buera et al. 1992). As the glass transition temperature increases with molecular weight, Kollidon® 12 PF (90 °C), has a substantially lower value than Kollidon® 90 F (156 °C). The two polymers also have very different molecular weights (2,500 and 1,250,000 Da, respectively). By introducing the "soft" vinyl acetate, the glass transition temperature sinks significantly. Thus, Kollidon® VA 64 (copovidone) has a glass transition temperature of only 101 °C compared to PVP homopolymer of the same molecular weight (149 °C) (Fig. 3.6).

The glass transition temperature can be lowered by adding plasticizers (Ghebremeskel et al. 2007). Obviously, this is of particular interest for those polymers with a high glass transition temperature. Solid plasticizers are recommended, in such cases as these, exhibit a lower degree of mobility within the polymer framework and provide greater stability (Karl et al. 2010) (Fig. 3.7).

The miscibility of the plasticizer with the base polymer should also be ascertained (Ghebremeskel et al. 2007). In the ideal case, the two components should mix perfectly at the molecular level; this is indicated if there is only one glass transition temperature. Should two glass transition temperatures/melting points occur, the components are not completely miscible but form clusters. Furthermore, the scientific literature has demonstrated that the miscibility of polymers decreases with increasing molecular weight (Kolter et al. 2010). This illustrates that polymeric plasticizers such as polyethylene glycol or poloxamers with a molecular weight of several thousand

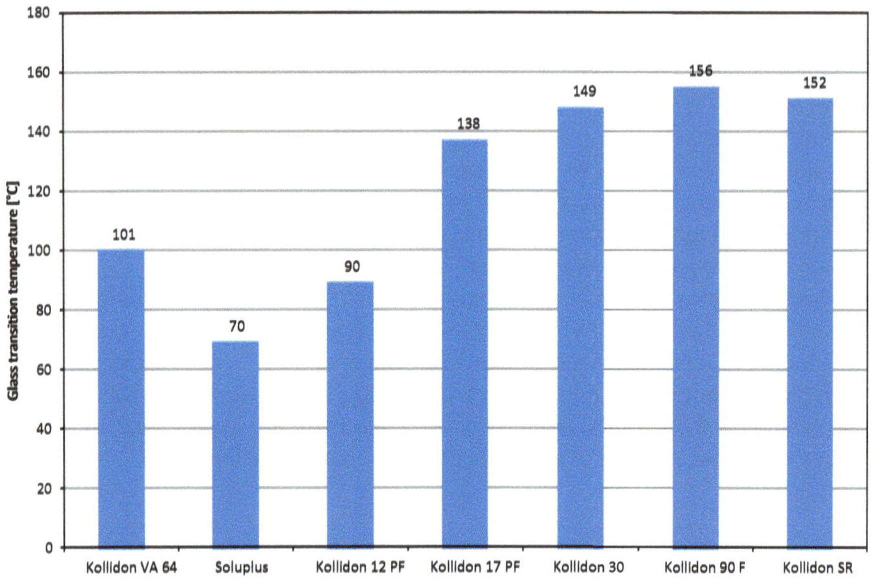

Fig. 3.6 Glass transition temperature of polyvinyllactams

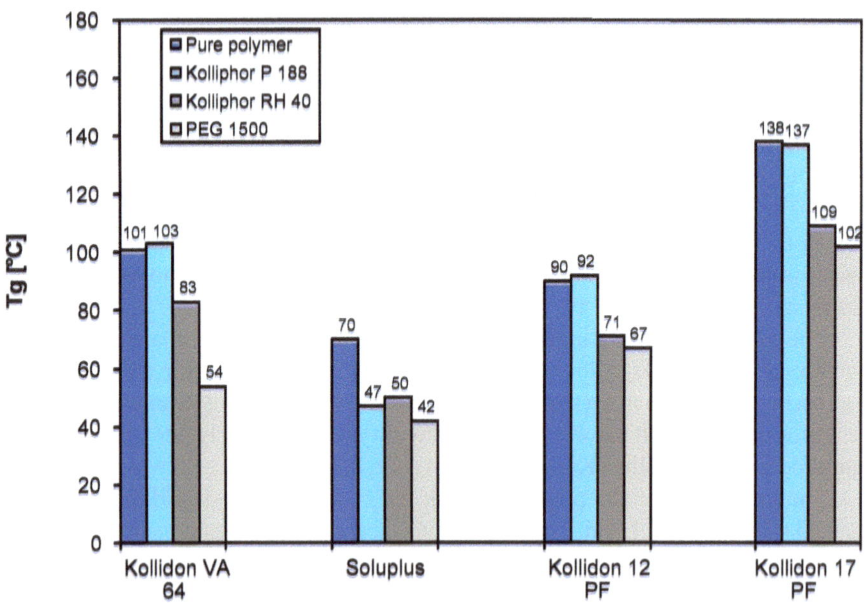

Fig. 3.7 Impact of plasticizers on glass transition temperature

Daltons can only be incorporated into a molecularly dispersed system with moderate difficulty (Fig. 3.8).

As is the case with all organic materials, polymers are also degraded at high temperatures (Pielichowski 2008). The thermostability of polymers can be established

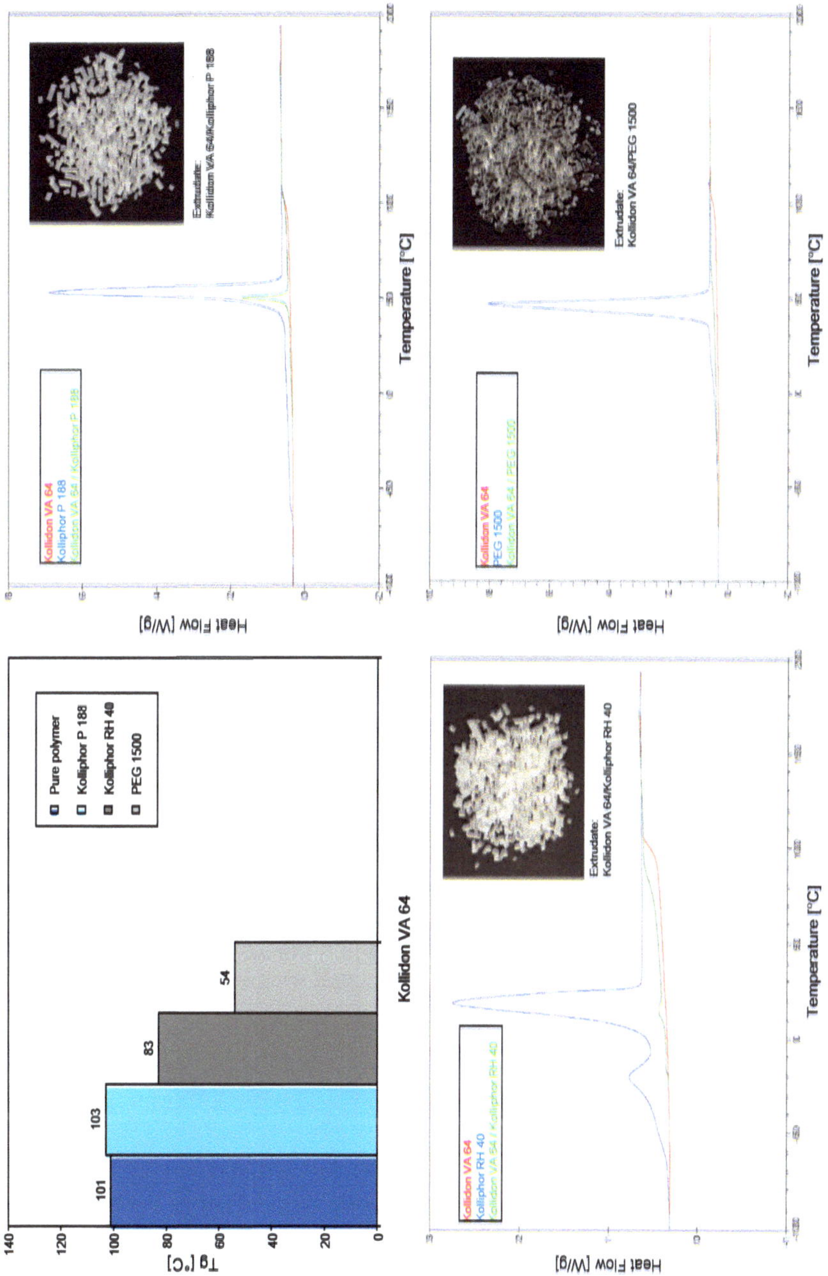

Fig. 3.8 Miscibility of plasticizers with base polymers

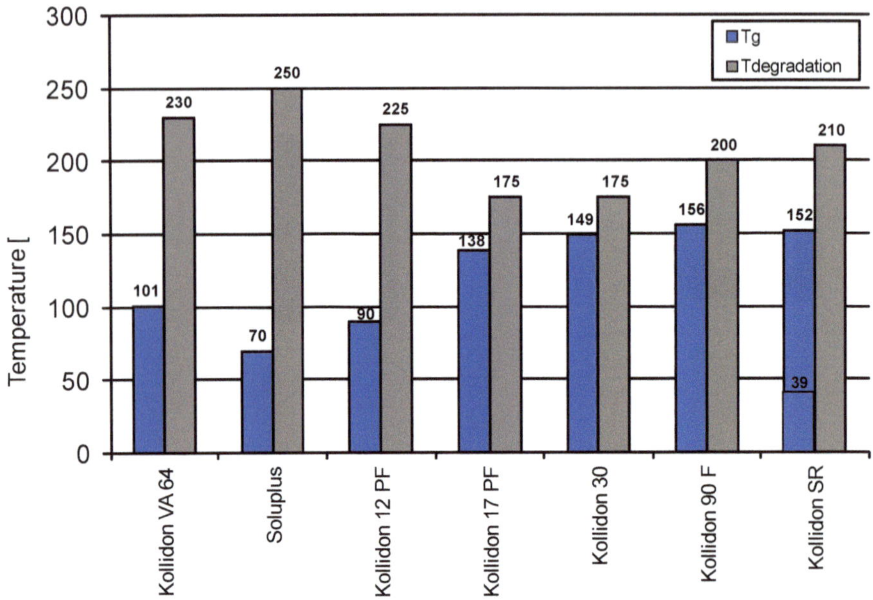

Fig. 3.9 Degradation and glass transition temperatures of polyvinyllactam polymers

by thermogravimetry in a first step analysis (Doyle 1961). With thermogravimetric analysis, the temperature at which loss of mass begins is used. Polyvinyllactams are very stable thermally as the lactam structure does not split easily. Frequently, the degradation temperatures of these polymers are around 200 °C, whereby Soluplus® at 250 °C, uniquely sets itself apart (Fig. 3.9).

The difference between glass transition temperature and degradation temperature serves as an indicator of the extrusion range, which is defined as the temperature range within which extrusion can be performed from a process and stability point of view. The broadest range by far was found with Soluplus®, followed by Kollidon® VA 64, and Kollidon® 12 PF (Kolter et al. 2010). Materials with a higher glass transition temperature such as Kollidon® 30 can—without plasticizer—only be extruded at a relatively high temperature range with a very small window.

3.3.2 *Melt Viscosity*

For a polymer to be processed in an extruder, the melt viscosity must not be too high, otherwise the torque will increase substantially, overloading the motor and the screws. The melt viscosity of polymers increases substantially with the molecular weight and is of course also dependent upon temperature and the shear gradient. It decreases with increasing temperature (Painter and Coleman 2009). However, the extrusion temperature has an upper limit; this is determined by the degradation of

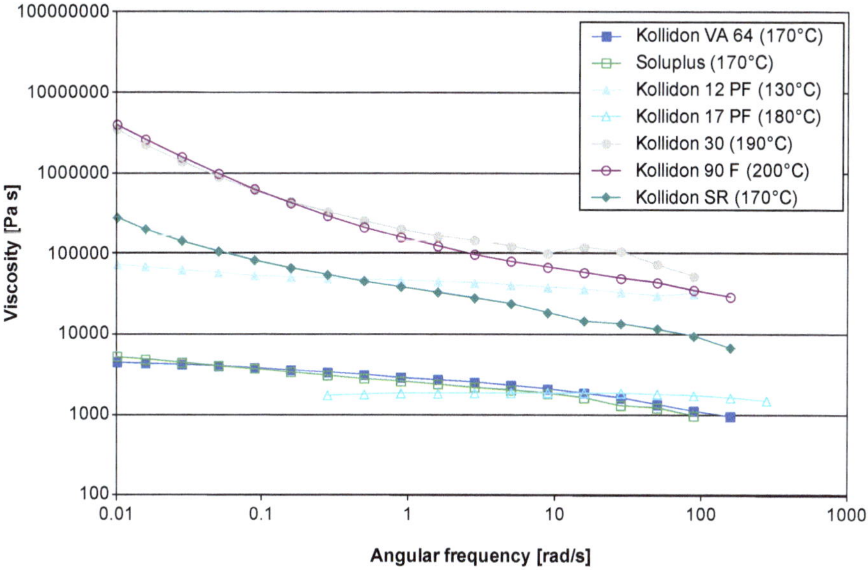

Fig. 3.10 Melt viscosity of polymers as a function of shear stress

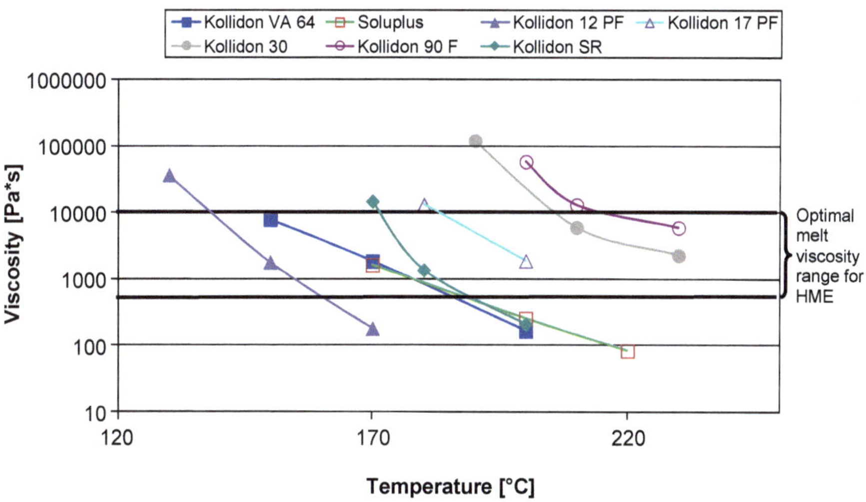

Fig. 3.11 Melt viscosity as a function of temperature

the polymer or the active ingredient incorporated. The melt viscosity should not be too low, however, as this will prevent reasonable extrudate formation through the extrusion nozzle. In such cases, stable strands or bands are thus not possible so that further processing, e.g., cutting and milling, becomes problematic. If possible, the melt viscosity should be within the range of 800–10,000 Pa s. This particular range has been shown to be optimal for melt extrusion and subsequent downstream processing (Kolter et al. 2010) (Figs. 3.10 and 3.11).

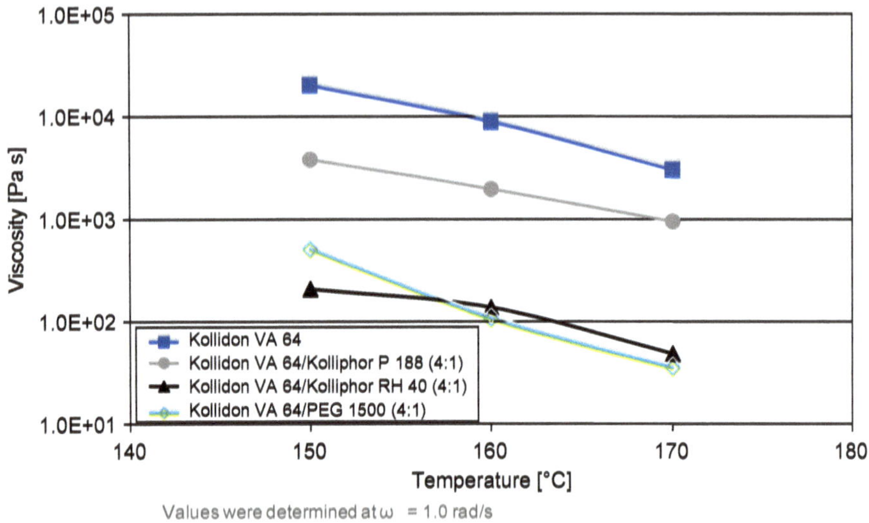

Fig. 3.12 Influence of plasticizers on the melt viscosity of Kollidon® VA 64

Just as with glass transition temperature, the melt viscosity can also be lowered by adding plasticizers. In this way, the extrusion of higher molecular weight polymers becomes possible and the extrusion temperatures can generally be lowered. In general, the more plasticizer added the lower the melt viscosity; however, the effect of other properties such as solubilization capacity and tackiness must be considered (Fig. 3.12).

3.3.3 Solid Solution Capacity

Solid solution capacity is defined as the maximum concentration of an active ingredient in a polymer where it is still completely dissolved (Breitenbach 2002). It thus describes the dissolving capacity of the polymer. This is mainly influenced by the lipophilicity (Leuner and Dressman 2000), solubility parameter (Forster et al. 2001a, b; Greenhalgh et al. 1999; Breitkreutz 1998; Van Krevelen and Hoftyzer 1976), and possible interactions between the functional groups of the polymer and the active ingredient (Douroumis 2012). In the latter case, the hydrogen bonds in particular play an important role. A very special influence is exerted by the amide structure of the solvent, an influence that soon becomes apparent if the solubility of several solvents is compared. All solvents with an amide structure, e.g., dimethyl formamide, dimethyl acetamide, pyrrolidone, or methyl pyrrolidone are able to dissolve much more active ingredient than those with keto- or hydroxy groups such as acetone or alcohols. This principle can also be applied to polymers; thus, polyvinyllactams possess significantly more solid solution capacity than other polymers (Fig. 3.13).

Several methods can be used for determining solid solution capacity, e.g., film casting or hot-melt extrusion. In both cases, the active ingredient concentration is

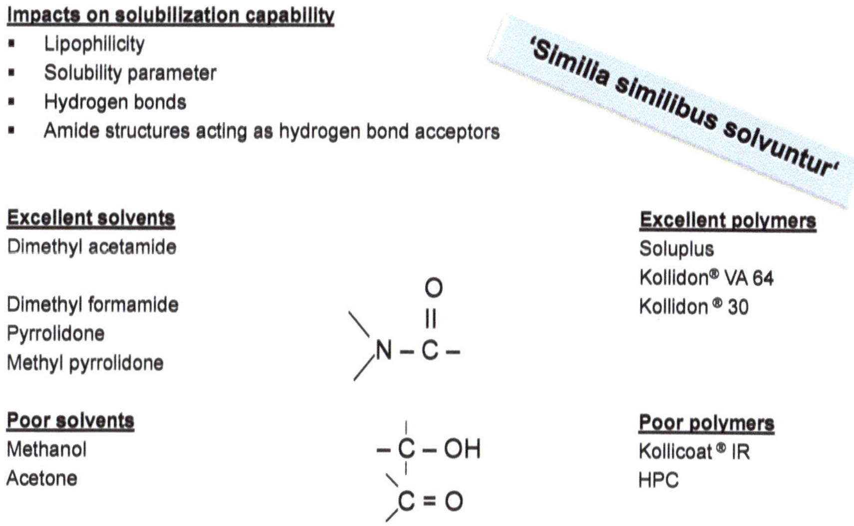

Fig. 3.13 Impact on solid solution capacity

Polymer	Drug content [% dissolved in polymer]		
	Fenofibrate	Carbamazepine	Itraconazole
Kollidon® VA 64	25	40	> 50
Soluplus®	35	30	> 50
Kollidon® 12 PF	20	20	35
Kollidon® 17 PF	15	25	45
Kollidon® 30*	20	15	> 50
Kollidon® 90 F**	< 5	15	> 50
Kollidon® SR	30	35	> 50

* With 10 % PEG 1500 ** With 20 % PEG 1500

Fig. 3.14 Solid solution capacity of polyvinyllactams for various actives (film casting method)

successively increased and analyzed by differential scanning calorimetry (DSC) and X-ray diffraction (XRD) (Liu et al. 2012; Lim et al. 2011) as to whether the substance is still in solution or whether it is at least partly in crystalline form. Prior to analysis, the samples should be stored for a certain length of time so that the crystallization process can be assessed (Figs. 3.14 and 3.15).

It has been shown that both methods produce similar results in many cases, whereby it can be stated that hot-melt extrusion generally produces somewhat lower solution capacity results (Kolter et al. 2010). Film casting, however, can thus be used as a screening method.

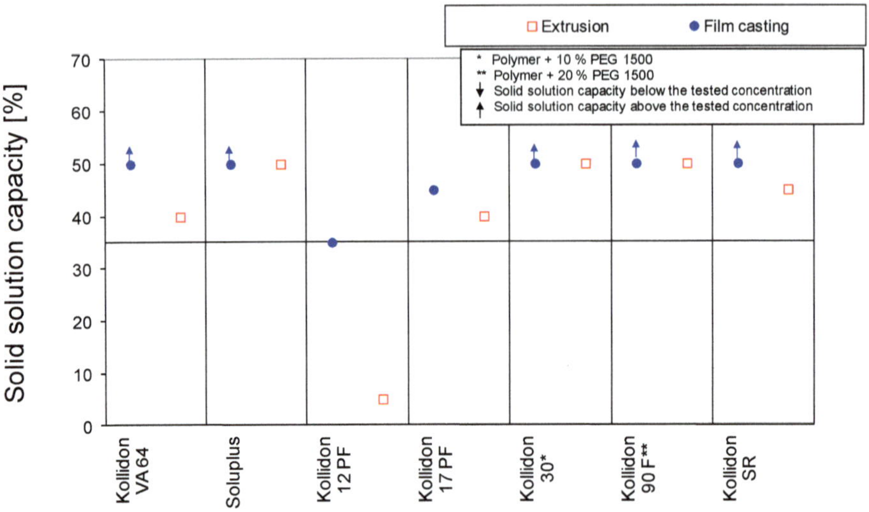

Fig. 3.15 Comparison of solid solution capacities for itraconazole as determined by film casting and hot-melt extrusion

3.3.4 Solubilization Capacity

The solubilization capacity describes the solubilization effect of polymers on active ingredients in aqueous solution and also plays a role in the release of the active. If the polymer is able to retain the active ingredient dissolved in gastric and intestinal fluids, this is an excellent precondition for very good resorption and a high degree of bioavailability.

Of course, it is primarily the amphiphilic polymers that are good active ingredient solubilizers as they have a high tendency to bind within the hydrophobic spaces of the micelles (Qiu and Bae 2003). In the case of nonamphiphilic polymers, solubilization can only take place by attachment to the polymer chains with a subsequent complexing effect (Koltzenburg 2011; Tantishaiyakul et al. 1999).

Soluplus®, based on its structure, is able to solubilize a wide spectrum of active ingredients in high concentrations (Fig. 3.16).

3.3.5 Stability

Thermogravimetric analysis provides a good indication of the thermal stability of polymers. However, much more meaningful results can be obtained from extrusion studies at increasing temperatures whereby the extrudate is analyzed for stability characteristics. The major undesirable characteristics are hydrolysis and pyrolysis products, changes in pH, and splitting of the polymer chains. Reversion to the original monomer units is also an indication and should be assessed for potential formulation

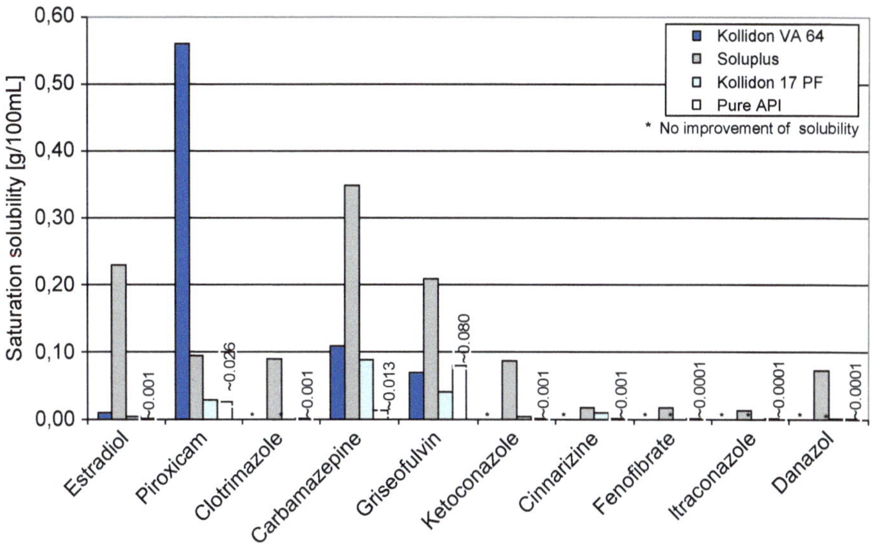

Fig. 3.16 Solubility enhancement of various actives by polyvinyllactams

issues. Polymers containing hydroxyl groups are especially sensitive to thermal stress as the hydroxyl groups can react with each other with the release of water. This particular reaction is especially enhanced by the high temperatures employed, the release of water from the extruder by evaporation and an acid pH. In this way, the polymer chains can become linked; the result is the formation of insoluble or at least sparingly soluble polymers or the polymers themselves can react with the active ingredients. Polymers containing no hydroxyl groups should therefore be preferred if stability is a major concern.

In the case of polymers with an amide structure, the potential for stability is much higher. This has been shown in studies with Kollidon® VA 64 and with Soluplus® at temperatures between 140–220 °C (Fig. 3.17).

3.3.6 *Dissolution and Bioavailability Studies*

When developing formulations for poorly soluble drugs, the general goal is to increase bioavailability. This cannot be tested utilizing in vitro experiments but must be assessed in human studies. Of course, the in vitro dissolution profile provides some indication of the efficacy expected in humans but often there is no exact in vitro–in vivo correlation; the dissolution characteristics in a glass vessel and the behavior in the intestine might well be very different. However, almost complete in vitro dissolution combined with a high degree of supersaturation maintained over hours are good indicators of enhanced bioavailability (Linn et al. 2012).

As an example, itraconazole was formulated with Soluplus® by hot-melt extrusion at 150 °C in a ThermoFisher 16 mm extruder into a solid solution of 20 % active

Test parameter		Powder	Extrudate [extruded at 140° C]	Extrudate [extruded at 180° C]
Identification (IR)		conf.	conforms	conforms
pH-value		4.1	3.9	3.9
Ester value	[mg KOH/g]	197	196	196
Vinyl acetate	[ppm]	<2	<2	<2
Vinyl caprolactam	[ppm]	<10	<10	<10
Caprolactam	[g/100g]	0.3	0.3	0.3
Ethylene glycol	[ppm]	<100	60	63
Acetic acid/Acetate	[ppm]	365	399	396
Peroxide	[ppm]	<20	<20	<20
Molecular weight Mw	[g/Mol]	118,000	119,000	116,000

➡ No change in chemical and physicochemical parameters up to 180° C

Fig. 3.17 Stability of Soluplus® during extrusion at various temperatures

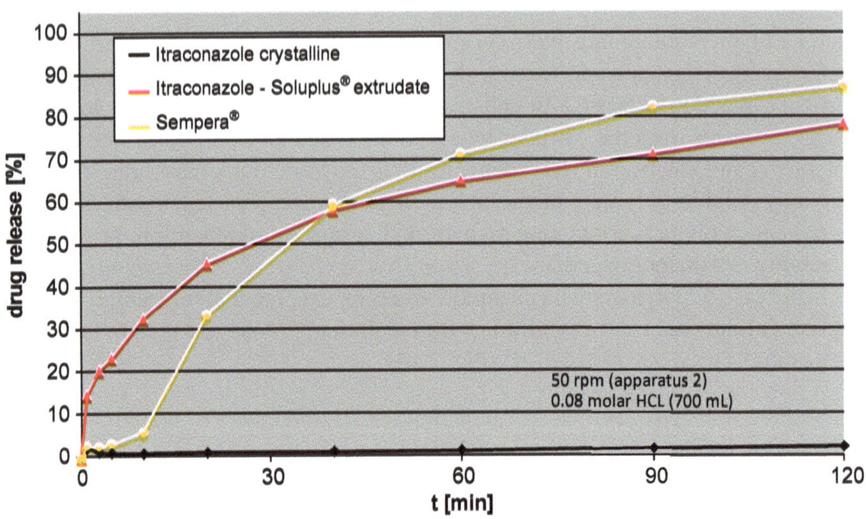

Fig. 3.18 Dissolution of itraconazole formulations

ingredient content, milled and filled into hard gelatin capsules. In order to prevent lumping of the solid solution particles, MCC (15 %) and Kollidon® CL (15 %) as inert spacers, had been blended into the mix. This formulation was tested for dissolution and bioavailability in comparison with the fine active ingredient crystals and the marketed drug Sempera®, which also contains itraconazole (Figs. 3.18 and 3.19).

The plasma curves reveal that itraconazole is an active ingredient that is poorly absorbed without special formulation considerations. The crystalline ingredient did not result in significant plasma levels. In contrast, the solid solution of itraconazole

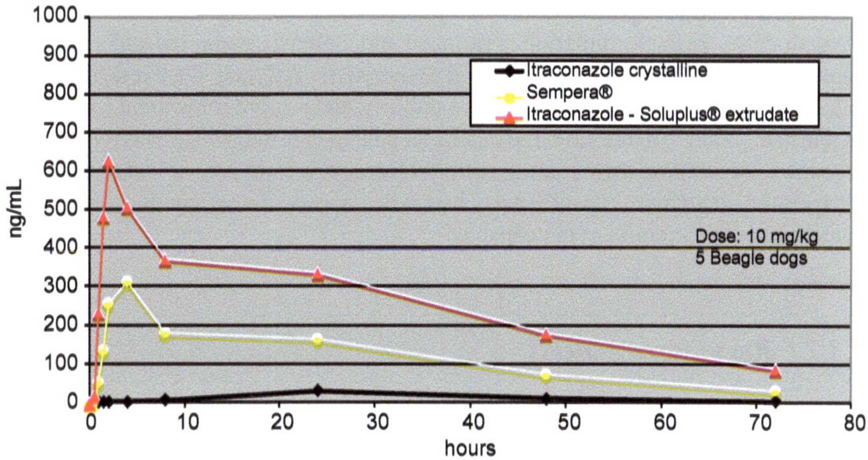

Fig. 3.19 Bioavailability of itraconazole formulations

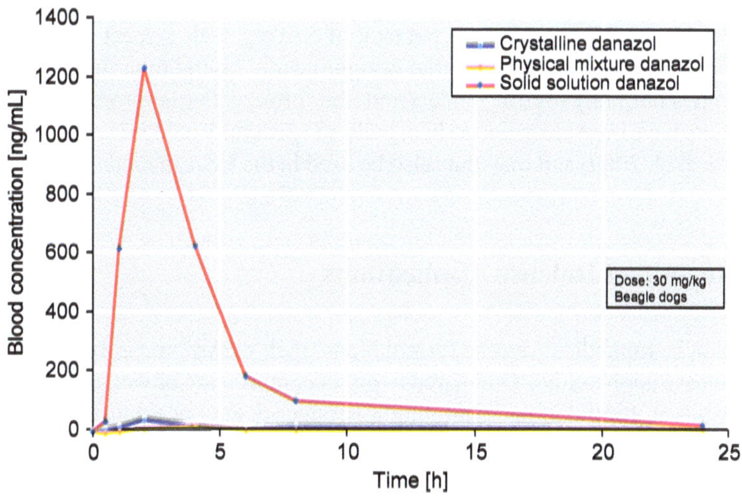

Fig. 3.20 Bioavailability of danazol formulations

in Soluplus® significantly enhanced the absorption and in addition outperformed Sempera® by a factor of 2.3. Compared to the crystalline form, the enhancement of bioavailability was 26-fold.

Similar results were also obtained with other very sparingly soluble drugs. It was shown for example that danazol, in a crystalline form, was sparingly absorbed, even in a physical mixture with Soluplus®. If, however, a solid solution containing 15 % danazol was prepared, extruded at 140 °C and filled into capsules, the bioavailability was significantly enhanced to approximately 15-fold of the crystalline form (Fig. 3.20).

Anti-HIV active ingredients also frequently suffer from low solubility and poor bioavailability. In the formulation of ritonavir and lopinavir, marketed under the trade name Kaletra®, copovidone was utilized as the matrix polymer since it significantly enhances intestinal absorption, improves stability, and is more convenient for patients (Klein et al. 2007). In this case, the older soft gelatin capsule, which was replaced by the new solid solution formulation, suffered from poor stability—it had to be stored refrigerated—and more capsules a day had to be taken by the patient due to the lower content of active ingredient per capsule.

3.3.7 Regulatory Aspects

Povidone and copovidone are well-established pharmaceutical excipients and are regularly used in numerous dosage forms and various administration routes. They have been approved by all major authorities (FDA, EMA and PMDA) and monographs have been published in the relevant pharmacopoeias (USP, Ph. Eur., JP).

The use of copovidone in approved drugs as solid solutions of high polymer content renders a high maximum potency of 854 mg with respect to the Inactive Ingredient List of FDA—a figure that is significantly higher than that produced by the polymers normally used as binders and film formers. Furthermore, povidone and copovidone have been classified as being GRAS (generally recognized as safe) status (Maschke et al. 2009) and may thus also be used in the USA in dietary supplements.

3.4 Sustained Release Applications

In addition to immediate release formulations, melt extrusion can be used for sustained release applications. Compared to direct compression or wet granulation, this often results in formulations exhibiting a stronger degree of retardation (Maschke et al. 2009). Of course, such formulations cannot be based on water-soluble polymers only but also require water-insoluble polymers. In this respect, Kollidon® SR, a combination of a water-soluble and a water-insoluble polymer, is particularly suitable. Because of the plasticity of polyvinyl acetate, the main component, it is relatively easy to extrude and subsequently to be compressed into tablets. In the example shown in Fig. 3.21, a simple formulation consisting of 60 % theophylline and Kollidon® SR was extruded using a 25-mm extruder (Coperion) at 150 °C, milled, blended with 0.5 % magnesium stearate and compressed into 12-mm tablets of 400 mg at 18 kN. This formulation resulted in stronger retardation of the theophylline than the direct compression of the same composition.

It would appear to be quite difficult to develop reliable sustained release formulations of poorly soluble active ingredients because the retardation should not be based on the dissolution of the crystals—which is often insufficient—but on the mode of action of the sustained release polymer. Thus, it is a prerequisite to have the active

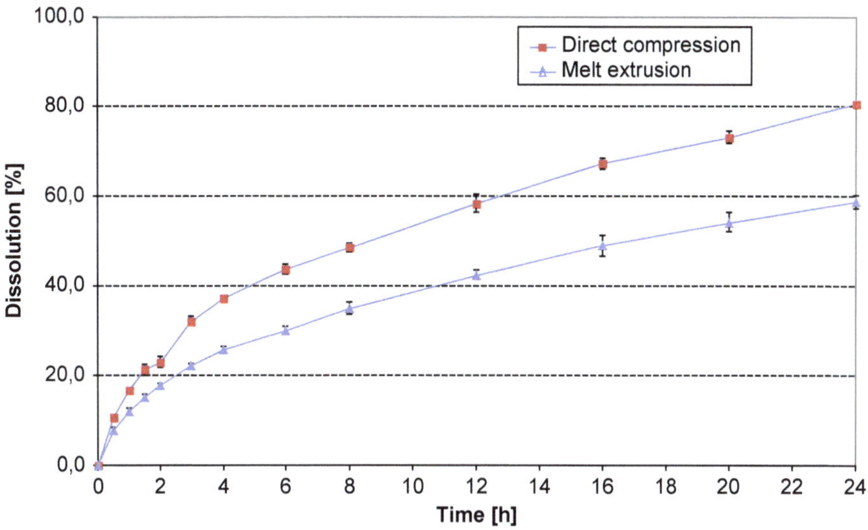

Fig. 3.21 Dissolution of melt-extruded and directly compressed Kollidon® SR formulations

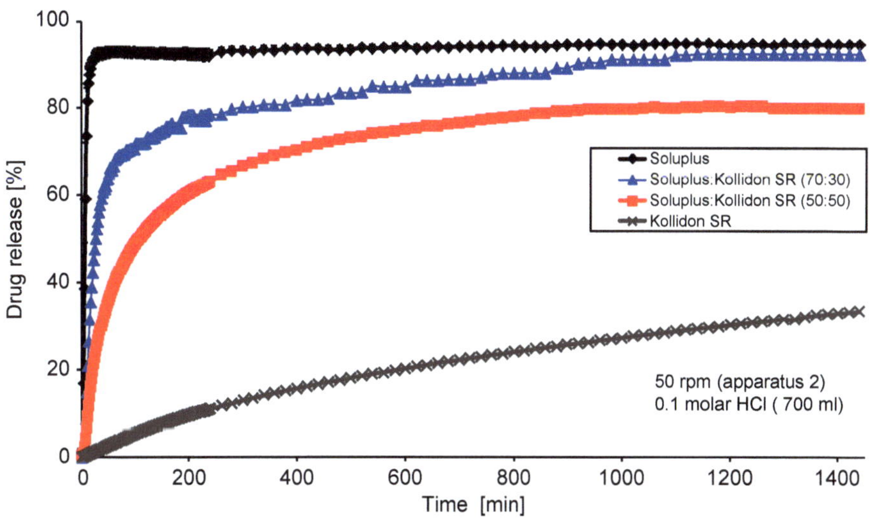

Fig. 3.22 Dissolution of melt-extruded itraconazole—Kollidon® SR formulations

ingredient in the form of a solid dispersion, which should by itself, or produced by another polymer, result in slow dissolution. Kollidon® SR is capable of forming solid solutions with active ingredients. In addition, since it contains the water-insoluble polyvinyl acetate, it is very likely to retard dissolution. Also, due to its excellent flowability and spherical particle structure, it allows high feed and high-throughput rates in the extruder.

As an example (Djuric et al. 2010), itraconazole (15 % in the formulation) was extruded together with Kollidon® SR with increasing amounts of Soluplus® on a 16 mm ThermoFisher extruder at 150 °C. All formulations obtained were amorphous as proven by XRD. Dissolution tests were conducted with milled extrudates containing 100 mg itraconazole using a sieve cut from 315–500 μm to minimize the impact of particle size on dissolution (Fig. 3.22).

This study reveals that Kollidon® SR strongly retards the dissolution of itraconazole and that the release rates can be adjusted by the addition of the water-soluble polymer Soluplus®. Stability data indicated no sign of crystallization since the dissolution profiles were similar at various time points.

3.5 Conclusions

It is quite evident that the polyvinyllactam polymers, including a number of Kollidons®, have tremendous applicability in melt extrusion applications. In addition, one vinylcaprolactam structure (Soluplus®) was designed and developed with melt extrusion applications in mind. From immediate release to sustained release, these classes of polymers can provide the formulation scientist a solution for water-insoluble drugs, as well as soluble actives.

References

Agrawal Y, Prakasam K (1988) Effect of binders on sulfamethoxazole tablets. J Pharm Sci 77: 885–888

Akers M (2002) Excipient—drug interactions in parenteral formulations. J Pharm Sci 91: 2283–2300

Ali S, Langley N (2010) Dry granulation simplifies tableting process. Pharm Form Quality 12:26–29

Amnuaikit C, Ikeuchi I, Ogawara K, Higaki K, Kimura T (2005) Skin permeation of propranolol from polymeric film containing terpene enhancers for transdermal use. Int J Pharm 289:167–178

Breitenbach J (2002) Melt extrusion: From process to drug delivery technology. Eur J Pharm Biopharm 54:107–117

Breitkreutz J (1998) Prediction of intestinal drug absorption properties by three-dimensional solubility parameters. Pharm Res 15:1370–1375

Brouwers J, Brewster M, Augustijns P (2009) Supersaturating drug delivery systems: the answer to solubility-limited oral bioavailability. J Pharm Sci 98:2549–2572

Buera M, Levi G, Karel M (1992) Glass transition in poly(vinylpyrrolidone): Effect of molecular weight and diluents. Biotechnol Prog 8:144–148

Bühler V (2008a) Pharmaceutical technology of BASF excipients, 9th edn. BASF Pharma Ingredients & Service, Ludwigshafen

Bühler V (2008b) Kollidon—polyvinylpyrrolidone excipients for the pharmaceutical industry, 9th edn. BASF Pharma Ingredients & Service, Ludwigshafen

Bühler V, Fussnegger B (2005) Generic Drug Formulations, 5th edn. BASF Pharma Solutions, Ludwigshafen

Chokshi R, Sandhu H, Iyer R, Shah N, Malick A, Hossein Z (2006) Characterization of physicomechanical properties of indomethacin and polymers to assess their suitability for hot-melt

extrusion process as a means to manufacture solid dispersion/solution. J Pharm Sci 94: 2463–2474
Chowhan Z, Amaro A, Ong JTH (1992) Punch geometry and formulation considerations in reducing tablet friability and their effect on in vitro dissolution. J Pharm Sci 81(3):290–294
Crowley M, Zhang F, Repka M, Thumma S, Upadhye S, Battu S, McGinity J, Martin C (2007) Pharmaceutical applications of hot-melt extrusion: part I. Drug Dev Ind Pharm 33:909–929
Damian F, Blaton N, Kinget R, Van den Mooter G (2002) Physical stability of solid dispersions of the antiviral agent UC-781 with PEG 6000, gelucire 44/14 and PVP K30. Int J Pharm 244:87–98
Djuric D, Maschke A, Kolter K (2010) Modified release of poorly soluble itraconazole by means of polymer combinations for hot-melt extrusion. Poster AAPS Annual Meeting and Exposition, Atlanta
Douroumis D (2012) Hot-melt extrusion: pharmaceutical applications. Wiley, Hoboken
Doyle C (1961) Estimating thermal stability of experimental polymers by empirical thermogravimetric analysis. Anal Chem 33:77–79
Flick D, Fraunhofer W, Rock C, Kolter K (2000) Melt granulation with Kollidon® SR for the manufacture of sustained release tablets. Poster 27th International Symposium on Controlled Release of Bioactive Materials, Paris
Forster A, Hempenstall J, Rades T (2001a) Characterization of glass solutions of poorly water-soluble drugs produced by melt extrusion with hydrophilic amorphous polymers. J Pharm Pharmacol 53:303–315
Forster A, Hempenstall J, Tucker I, Rades T (2001b) Selection of excipients for melt extrusion with two poorly water-soluble drugs by solubility parameter calculation and thermal analysis. Int J Pharm 226:147–161
Garekani H, Sadeghi F, Ghazi A (2003) Increasing the aqueous solubility of acetaminophen in the presence of polyvinylpyrrolidone and investigation of the mechanisms involved. Drug Dev Ind Pharm 29:173–179
Ghebremeskel A, Vemavarapu C, Lodaya M (2007) Use of surfactants as plasticizers in preparing solid dispersions of poorly soluble API. Int J Pharm 328:119–129
Ghebre-Sellassie I, Martin C (2007) Pharmaceutical extrusion technology. Drugs and the pharmaceutical sciences 133: 245–260
Gil E, Colarte A, Sampedro J, Bataille B (2008) Subcoating with Kollidon VA 64 as water barrier in a new combined native dextran/HPMC–cetyl alcohol controlled release tablet. Eur J Pharm Biopharm 69:303–311
Greenhalgh D, William A, Timmins P, York P (1999) Solubility parameters as predictors of miscibility in solid dispersions. J Pharm Sci 88:1182–1190
He X, Barone M, Marsac P, Sperry D (2008) Development of a rapidly dispersing tablet of a poorly wettable compound-formulation DOE and mechanistic study of effect of formulation excipients on wetting of celecoxib. Int J Pharm 353:176–186
Herting M, Klose K, Kleinebudde P (2007) Comparison of different dry binders for roll compaction/dry granulation. Pharm Dev Technol 12:525–532
Ishida H, Wu T, Yu L (2007) Sudden rise of crystal growth rate of nifedipine near Tg without and with polyvinylpyrrolidone. J Pharm Sci 96:1131–1138
Jain P, Banga AK (2010) Inhibition of crystallization in drug-in-adhesive-type transdermal patches. Int J Pharm 394(1–2):68–74
Karl M, Djuric D, Kolter K (2011) Pharmaceutical excipients for hot-melt extrusion. Pharm Tech 35(5):74–82
Karl M, Nalawade S, Maschke A, Djuric D, Kolter K (2010) Suitability of pure and plasticized polymers for hot-melt extrusion. Poster AAPS Annual Meeting and Exposition, Atlanta
Klein E, Chiu L, Awni W, Zhu T, Heuser R, Doan T, Breitenbach J, Morris J, Brun S, Hanna G (2007) The tablet formulation of Lopinavir/Ritonavir provides similar bioavailability to the soft-gelatin capsule formulation with less pharmacokinetic variability and diminished food effect. J Acquir Immune Defic Syndr 44:401–410
Kolter K, Flick D (2006) Structure and dry binding activity of different polymers, including Kollidon VA 64. Drug Dev Ind 26:1159–1165

Kolter K, Karl M, Nalawade N, Rottmann N (2010) Hot-melt extrusion with BASF Pharma polymers— extrusion compendium. BASF Pharma Ingredients & Service, Ludwigshafen

Koltzenburg S (2011) Formulation of problem drugs. In: Reintjes T (ed) Solubility enhancement with BASF Pharma Polymers Solubilizer. BASF Pharma Ingredients & Service, Ludwigshafen.

Kranz H, Wagner T (2006) Effects of formulation and process variables on the release of a weakly basic drug from single unit extended release formulations. Euro J Pharm Biopharm 62(1):70–76

Kuentz M, Röthlisberger D (2002) Determination of the optimal amount of water in liquid-fill masses for hard gelatin capsules by means of texture analysis and experimental design. Int J Pharm 236:145–152

Lee Y, Chien Y (1995) Oral mucosa controlled delivery of LHRH by bilayer mucoadhesive polymer systems. J Control Release 37:251–261

Leuner C, Dressman J (2000) Improving drug solubility for oral delivery using solid dispersions. Eur J Pharm Biopharm 50:47–60

Lim H, Lan Y, Yao Y, Fahmy R, Hoag S (2011) Enhancing dissolution rate of carbamazepine via solvent casting and hot-melt extrusion with a novel polymeric solubilizer (Soluplus).

Linn M, Collnot E, Djuric D, Hempel K, Fabian E, Kolter K, Lehr C (2012) Soluplus® as an effective absorption enhancer of poorly soluble drugs in vitro and in vivo. Eur J pharm Sci 45:336–343

Liu X, Lu M, Guo Z, Huang L, Feng X, Wu C (2012) Improving the chemical stability of amorphous solid dispersion with cocrystal technique by hot-melt extrusion. Pharm Res 29:806–817

Ma X, Taw J, Chiang C (1996) Control of drug crystallization in transdermal matrix system. Int J Pharm 142:115–119

Maniruzzamana M, Boatenga J, Bonnefilleb M, Aranyosb A, Mitchella J, Douroumis D (2012) Taste masking of paracetamol by hot-melt extrusion: an in vitro and in vivo evaluation. Eur J Pharm Biopharm 80:433–442

Maschke A, Meyer-Böhm K, Kolter K (2007) Application of Kollidon VA64 fine in roller compaction. Poster AAPS Annual Meeting, San Diego

Maschke A, Kumpp U, Güntherberg N, Kolter K (2009) Effect of preparation method on release behavior of Kollidon SR tablets, hot-melt extrusion versus direct compression. 36th Annual Meeting and Exposition of the Control Release Society, Copenhagen

Nair R, Gonen S, Hoag S (2002) Influence of polyethylene glycol and povidone on the polymorphic transformation and solubility of carbamazepine. Int J Pharm 240:11–22

Özgüney I, Shuwisitkul D, Bodmeier R (2009) Development and characterization of extended release Kollidon SR mini-matrices prepared by hot-melt extrusion. Euro J Pharm Biopharm 73:140–145

Painter P, Coleman M (2009) Essentials of polymer science and engineering. Destech, Lancaster

Patterson J, James M, Forster A, Rades T (2008) Melt extrusion and spray drying of carbamazepine and dipyridamole with polyvinylpyrrolidone/vinyl acetate copolymers. Drug Dev Ind Pharm 34:95–106

Pielichowski K (2008) Thermal degradation of polymeric materials. Smithers Rapra, Shropshire

Plaizier-Vercammen J (1983) Interaction of povidone with aromatic compounds IV: Effects of macromolecule molecular weight, solvent dielectric constant, and ligand solubility on complex formation. J Pharm Sci 72:1042–1044

Plaizier-Vercammen J, Neve R (1983) Interaction of povidone with aromatic compounds III: thermodynamics of the binding equilibria and interaction forces in buffer solutions at varying pH values and varying dielectric constant. J Pharm Sci 71:552–556

Qiu L, Bae Y (2003) Polymer architecture and drug delivery. Pharm Res 23:1–30

Ranzani L, Font J, Galimany F, Santanach A, Gomez-Gomar A, Casadevall G, Gryczke A (2011) Enhanced in vivo absorption of CB-1 antagonist in rats via solid solutions prepared by hot-melt extrusion. Drug Dev Ind Pharm 37:694–701

Repka M, Shah S, Lu J, Maddineni S, Morott J, Patwardhan K, Mohammed N (2012) Melt extrusion: process to product. Expert Opin Drug Deliv 9:105–125

Ritala M, Jungersen O, Holm P, Schæfer T, Kristensen H (1986) A comparison between binders in the wet phase of granulation in a high shear mixer. Drug Dev Ind Pharm 12:1685–1700

Schillinga SU, Brucea CD, Shah NH, Malickb AW, McGinity JW (2008) Citric acid monohydrate as a release-modifying agent in melt extruded matrix tablets. Int J Pharm 361:158–168

Siepmann F, Eckart K, Maschke A, Kolter K, Siepmann J (2010) Modeling drug release from PVAc/PVP matrix tablets. J Control Release 141:216–222
Simonelli A, Mehta S, Higuchi W (1970) Inhibition of sulfathiazole crystal growth by polyvinylpyrrolidone. J Pharm Sci 59:633–638
Strickley R (2004) Solubilizing excipients in oral and injectable formulations. Pharm Res 21: 201–230
Strübing S, Metz H, Mäder K (2008) Characterization of poly(vinyl acetate) based floating matrix tablets. J Control Release 126:149–155
Tantishaiyakul V, Kaewnopparat N, Ingkatawornwong S (1999) Properties of solid dispersions of piroxicam in polyvinylpyrrolidone. Int J Pharm Sci 181:143–151
Tantry J, Tank J, Suryanarayana R (2007) Processing-induced phase transitions of theophylline-implications on the dissolution of theophylline tablets. J Pharm Sci 96:1434–1444
Van den Mooter G, Wuyts M, Blaton N, Busson R, Grobet P, Augustijns P, Kinget R (2010) Physical stabilisation of amorphous ketoconazole in solid dispersions with polyvinylpyrrolidone K25. Eur J Pharm Sci 12:261–269
Van Krevelen D, Hoftyzer P (1976) Properties of polymers—their correlation with chemical structures. Elsevier, Amsterdam
Wen H, Morris K, Park K (2005) Study on the interactions between polyvinylpyrrolidone (PVP) and acetaminophen crystals: partial dissolution pattern change. J Pharm Sci 94(10):2166–2174
Zhang F, McGinity J (2000) Properties of hot-melt extruded theophylline tablets containing poly(vinyl acetate). Drug Dev Ind Pharm 26:931–942

Chapter 4
Properties and Applications of Hypromellose Acetate Succinate (HPMCAS) for Solubility Enhancement Using Melt Extrusion

Sakaé Obara, Fumié K. Tanno and Ashish Sarode

Abstract The development of bioavailable solid dispersions requires the use of secondary materials for stabilizing the amorphous dispersion while also enhancing dissolution and solubility of the new chemical entity. With the increased use of amorphous dispersions, hypromellose acetate succinate (HPMCAS) has emerged as a commonly used excipient for formulation design. This unique material, originally designed to enhance thermoplastic properties of cellulosics, can be used to enhance the solubility of many poorly soluble amorphous products. This chapter details the properties of HPMCAS that make it a beneficial material for use in solid dispersion formulation, with a specific focus on relevant properties for melt extrusion.

4.1 Introduction

Cellulose ethers such as methylcellulose, hypromellose (commonly referred to as hydroxypropyl methylcellulose (HPMC)), and hydroxypropylcellulose (HPC) have been widely used in many industries including construction, food, cosmetic, and pharmaceutical applications. Film coating of pharmaceutical dosage forms using HPMC first appeared in a patent by Singiser of Abbott Laboratories in 1962 (Nagai et al. 1998). At that time organic solvents were typically used, but later, aqueous film coatings became popular due to environmental preference and new developments in coating machine technology which allowed for higher drying efficiency. HPC has also been used in pharmaceutical applications for long time, where it has served as a binding agent for wet-granulation processes. Cellulose esters, such as cellacefate (commonly referred to as cellulose acetate phthalate (CAP)) and hypromellose phthalate (commonly referred to as hydroxypropyl methylcellulose phthalate (HPMCP)) are commonly used as enteric coating materials. These cellulose esters have free carboxylic groups in the polymeric backbone, and therefore dissolve in buffers of higher pH.

S. Obara (✉) · F. K. Tanno
Specialty Chemicals Research Center, Shin-Etsu Chemical Co., Ltd., Niigata, Japan
e-mail: obaras@shinetsu.jp

A. Sarode
Biomedical and Pharmaceutical Sciences, University of Rhode Island, Kingston, RI, USA

M. A. Repka et al. (eds.), *Melt Extrusion*,
AAPS Advances in the Pharmaceutical Sciences Series 9,
DOI 10.1007/978-1-4614-8432-5_4, © American Association of Pharmaceutical Scientists 2013

Of these materials, hypromellose acetate succinate (HPMCAS) has emerged as the most commonly used cellulose ether for solid dispersion production. In general, other materials have shown deficiencies for thermal processing or bioavailability enhancement. For example, HPMC has been used extensively for solid dispersion production but shows substantial limitations for production using extrusion due to the nonthermoplastic nature of the material. Other excipients, including HPMCP and CAP show limitations due to the thermal stability under extrusion conditions. Although applicable for the production of solid dispersions, these systems must often be produced using solvent-based methods.

4.2 Development History of HPMCAS

HPMCAS was first commercialized by Shin-Etsu Chemical Co., Ltd. in Japan in 1986 as an enteric coating agent under the trade name Shin-Etsu AQOAT. The original project to develop this polymer was aimed at creating a material for the production of enteric hard capsule shells that could be manufactured by nonsolvent methods such as melt extrusion or injection molding. As part of the development, polymer chemists screened HPMC-based polymers seeking to identify chemical variants with optimum properties. While modifying this material, which had been marketed for film-coating application for many years, the scientists attempted to attach other functional groups which could introduce thermoplastic behavior to the polymer. Acetic and succinic ester derivatives of HPMC came up in the screening study, but the enteric capsule project was ultimately discontinued. At the same time, however, it was found that micronized powder of this polymer could be dispersed into water and applied as aqueous enteric coating to pharmaceutical drug products. Noting significant demand for enteric materials, HPMCAS was first put into the market as an enteric coating agent for aqueous dispersion coating (Nagai et al. 1998). It was only later that scientists demonstrated the use of the material for bioavailability enhancement. As of 2012, more than 30 pharmaceutical products marketed throughout the world use this polymer as an enteric coating agent and solid dispersion carrier. HPMCAS is currently listed in the Japanese Pharmacopoeia, in which its monograph was transferred from another compendium "Japanese Pharmaceutical Excipients (JPE)" in 2012. It has also been listed in the National Formulary (NF) in the USA since 2005. In the mid 1990s and early 2000s, applications of solid dispersions using HPMCAS were studied and published by several authors (Tamabuchi et al. 1993; Tanno et al. 2004; Curatolo et al. 2009). The first commercial solid dispersion product using HPMCAS was brought into the Japanese market in 1994 by Zeria Pharmaceuticals and Nissan Chemical Co., Ltd., using efonidipine hydrochloride ethanolate as the active pharmaceutical ingredient (API). The solid dispersion product was produced by a method using organic solvents (Miyajima et al. 1991). As of August 2012, there is no official record indicating any commercial melt-extruded solid dispersion products using HPMCAS in the market. However, since there are several ongoing melt extrusion projects using HPMCAS, the properties of this polymer with regard to this application will be discussed in this chapter.

4.3 Manufacturing Process

Figure 4.1 shows the structure of HPMCAS and Fig. 4.2 shows the scheme of the manufacturing process (Onda et al. 1980). The base polymer is a low viscosity grade of HPMC. Using acetic acid as solvent and sodium acetate as catalyst, two additional ester groups, succinoyl and acetyl groups are introduced by succinic anhydride and acetic anhydride. After reaction, the reactant paste is precipitated with water followed by washing, then drying. Impurities related to the process include residual acetic acid (solvent) and sodium acetate (catalyst). Two free acids, acetic acid and succinic acid are also impurities that gradually form by hydrolysis of the polymer after production. Storage in a high humidity condition accelerates the hydrolysis. Care should be taken because some of these impurities may interact with API in the solid dispersion formulations.

4.4 Coating Applications

Although this book is about melt extrusion process, knowledge on the characteristics of HPMCAS in the coating applications may be useful to understand more about this polymer for applications to melt extrusion. In aqueous dispersion coatings, the fine particle grade of HPMCAS is suspended into water in which plasticizer is dissolved in advance. The recommended plasticizer is triethyl citrate (TEC). TEC reduces glass transition temperature of HPMCAS (Klar and Urbanez 2009). TEC is mandatory for aqueous dispersion coating of HPMCAS. Without it, a film does not form. However, coating with an organic solvent solution does not require plasticizer for film formation upon drying. One issue in the aqueous dispersion coating with HPMCAS is that it is recommended to chill the coating dispersion and maintain the temperature at less than 20 °C. At high temperature, the polymer tends to coagulate and nozzle blocking may occur during spraying. This phenomenon is related to the fact that this polymer tends to be thermoplastic to a certain extent. One unique coating method using HPMCAS is the dry powder coating (Obara et al. 1999; Kablitz et al. 2006; Cerea et al. 2008). In this method, the polymer is directly applied as dry powder that is supplied by blowing onto tablet or granule cores using a quantitative powder feeder. TEC is again necessary to spray simultaneously, but organic solvent or water is not required. First, the powder deposits on the surface of core along with droplets of plasticizer. After the curing process, which is carried out by simple heating at 60 °C for 30–60 min, the powder layer turns into a continuous film layer. This is also related to the thermoplastic nature of this polymer.

4.5 Material Properties Critical for Extrusion

For successful extrusion, materials are required to exhibit appropriate melt viscosity, glass transition, and thermal stability characteristics. Additionally, for successful use in bioavailability enhancement applications it is necessary that the material exhibits

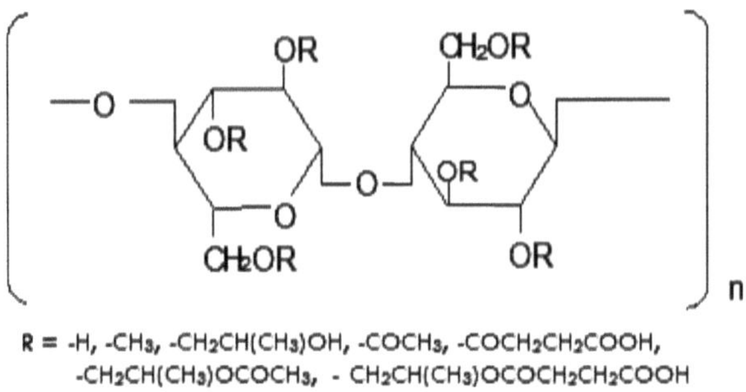

Fig. 4.1 Structure of HPMCAS

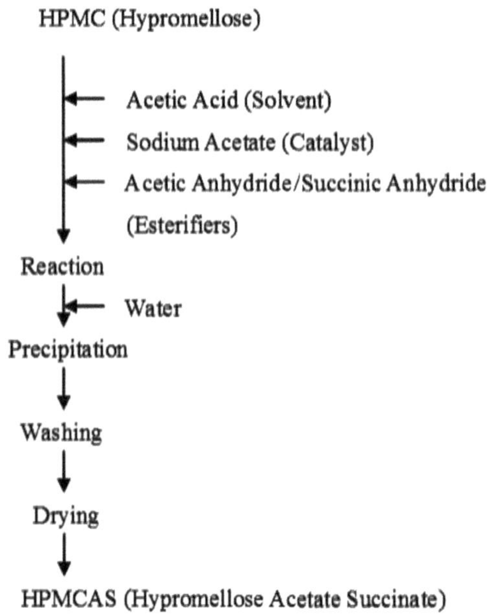

Fig. 4.2 Manufacturing process of HPMCAS

chemical structures that facilitate miscibility with poorly soluble pharmaceuticals as well as interactions which can lead to superior observed drug concentrations in aqueous environments. This section discusses the critical material properties of HPMCAS as they relate to the ability to produce amorphous dispersions via hot melt extrusion.

4.5.1 Grades

Currently six grades of HPMCAS are commercially available (Shin-Etsu Chemical Co., Ltd. 2010). These are listed in Table 4.1. There are two physical grades having different particle size, and for each physical grade, there are three chemical grades. The two physical grades include granular grade ("G"-grade,

Table 4.1 Commercially available grades of HPMCAS

Grade	Particle size[a]	Typical value (specification range)[b]					Soluble pH[c]
		Acetyl (%)	Succinoyl (%)	Methoxy (%)	Hydroxy-propoxy (%)	Viscosity (mPa·s)	
AS-LF	Fine	8 (5.0–9.0)	15 (14.0–18.0)	23 (20.0–24.0)	7 (5.0–9.0)	3 (2.4–3.6)	≥ 5.5
AS-LG	Granular						
AS-MF	Fine	9 (7.0–11.0)	11 (10.0–14.0)	23 (21.0–25.0)	7 (5.0–9.0)	3 (2.4–3.6)	≥ 6.0
AS-MG	Granular						
AS-HF	Fine	12 (10.0–14.0)	7 (4.0–8.0)	23 (22.0–24.0)	7 (6.0–10.0)	3 (2.4–3.6)	≥ 6.5
AS-HG	Granular						

[a] Approximate mean particle size: Fine = 5 μm, Granular = 0.5 mm
[b] Test methods are according to the monograph in the National Formulary
[c] Based on dissolving time of cast film in McIlvaine's buffer (citric acid–Na_2HPO_4). The solubility depends on buffer formulation

which is the grade code ending with G. Mean particle size is approximately 0.5 mm) and fine grade ("F"-grade, which is the grade code ending with F. Mean particle size is approximately 5 μm). The F-grade was originally developed for aqueous dispersion coating and G-grade was designed for dissolving the polymer into organic solvent. The two grades are chemically the same, as F-grade is manufactured by further mechanical milling of G-grade. The F-grade is not recommended for dissolving in organic solvents because the powder tends to create lumps that take a long time to dissolve completely. Especially on large scale, F-grade easily forms a dust cloud, and if this appears in the vapor of organic solvent, it is called a hybrid mixture which has a high risk of dust explosion. Consideration of grade selection related to particle size is also an important factor for extrusion operations. Starve-fed extrusion operations typically require good material flowability to prevent sticking and bridging in the intake area. Fine grade HPMCAS has a significant surface area for interaction, which can often lead to sticking and bridging. For operations requiring higher throughputs, utilization of the G-grade material can improve intake efficiency; however, care must be taken to ensure blend uniformity and appropriate internal mixing within the extruder due to the larger particle size of the polymer.

The three chemical grades are distinguished by different ratios of acetyl and succinoyl groups. These grades show different polymer dissolution profiles, and therefore different drug-release behavior, at various environmental pH. Since this polymer is an enteric coating agent, it does not dissolve in acidic media or purified water, although it dissolves in buffer of pH higher than 5.5–6.5. The dissolution onset and extent depends on the grade of the material. This variation of ester groups was originally designed to control drug release at targeted positions in small intestine, but for application of solid dispersions, it may affect the drug-release level and inhibitory effect of recrystallization, likely due to interaction between API and the functional groups of this polymer. In fact, several recent studies have suggested

Table 4.2 T_g values for various cellulosic derivatives

Polymer	Grade	T_g (°C)
HPMCAS	AS-LG	115
	AS-MG	115
	AS-HG	115
HPMCP	HP-55	129
	HP-55S	136
	HP-50	145
HPMC[a]	Pharmacoat® 606	155
	Metolose® 60SH-50	163

Differential scanning calorimeter: Model 3200SA (Bruker-AXS, WI, USA). Sample: 10 mg in aluminum pan. Reference: Al_2O_3. Atmospheric gas: N_2. Heat rate: 10 °C/min. Program: 1st run (RT to 180 °C) then cool down (to 25 °C), followed by 2nd run (25 to 230 °C)

T_g glass transition temperature

[a] Pharmacoat and Metolose are trade names of Shin-Etsu Chemical Co., Ltd., Japan

that the performance difference between the grades is attributed to hydrophobic interactions between the drug and polymer, leading to inhibition of recrystallization from solution.

4.5.2 Glass Transition Temperature (T_g)

The T_g of HPMCAS measured with differential scanning calorimeter (DSC) is shown in Table 4.2. It was lower compared with other cellulose derivatives that have been used for pharmaceutical applications. This allows for more effective use of the material for extrusion applications than some other cellulosics having relatively high T_g values. Noting the moderate T_g of 115 °C, HPMCAS also balances well requirements for retaining a suitable T_g after production of a solid dispersion formulation. Even with moderate drug loadings of 20–30 % w/w drug substance, the solid dispersion would be projected to maintain a T_g greater than 90 °C in many cases based on the Gordon–Taylor relationship. Chemical composition of HPMCAS had little impact on the T_g of the polymer, whereas chemical variants of HPMC and HPMCP yielded significant changes in the T_g. Noting that elevated T_g values can impact the ability to process via extrusion, some of the chemical variants for the non-AS materials would present challenges to use in extrusion operations.

4.5.3 Melt Viscosity

Melt viscosity of materials is also an important factor that determines utility of an excipient for extrusion operations. Using a capillary rheometer (Capilograph® Model E3B, Toyo-Seiki, Japan), the melting behavior of HPMCAS without API was measured under various conditions. The melt viscosity data, shown in Table 4.3, indicates

Table 4.3 Melt viscosity of HPMCAS and HPMCP at different operating conditions

	Viscosity (Pa·s)				
	HPMCAS			HPMCP	
Temperature (°C)	AS-LF	AS-MF	AS-HF	HP-50	HP-55
150	Overload	1,099	2,101	Overload	Overload
160	961	1,388	1,355	Overload	Overload
170	641	842	721	Overload	Overload
175	–	–	–	22,266	37,101
180	469	604	532	14,133	17,066

The samples contained moisture between 1 and 3 %

that when compared with HPMCP, which is another enteric polymer that has been used commercially for solid dispersion production, HPMCAS provides significantly lower viscosities which can facilitate its use for extrusion operations. Beyond processing, viscosity also impacts diffusivity of the molten material and can facilitate the formation of an amorphous dispersion. This suggested that HPMCAS was a more effective material for the preparation of amorphous dispersions than HPMCP when using melt extrusion. Different HPMCAS chemistries showed subtle differences between the types whereas all three grades showed a temperature dependence, with higher temperatures resulting in a reduced viscosity that could improve processing. This behavior, however, only continues to a certain range beyond which chemical decomposition of the polymer results in an inability to process. Overall, the generally moderate melt viscosity values of HPMCAS signal good applicability of the material for the production of melt-extruded solid dispersions.

4.6 Visual Appearance

On extrusion, HPMCAS forms continuous rods that have some elasticity. At 170 °C and higher, the rods were transparent, but at lower temperature, they were translucent. Images of HPMCAS extrudates can be seen in Fig. 4.3. For preparation of amorphous dispersions, drug substances molecularly dispersed within the material will result in extrudates that share a similar transparent nature; however, the color of the extrudate may change depending on the properties of the drug substance. Downstream processing of the extrudate into powders, pellets, or directly shaped tablets can be achieved using conventional technologies such as die face cutting or pelletization.

4.7 Material Stability After Extrusion

Since the polymer is exposed to heat in the melt extrusion process, it is fundamental to investigate how the physicochemical properties of the polymer are affected by extrusion. Therefore, HPMCAS was extruded using a small-scale extruder and its stability was assessed.

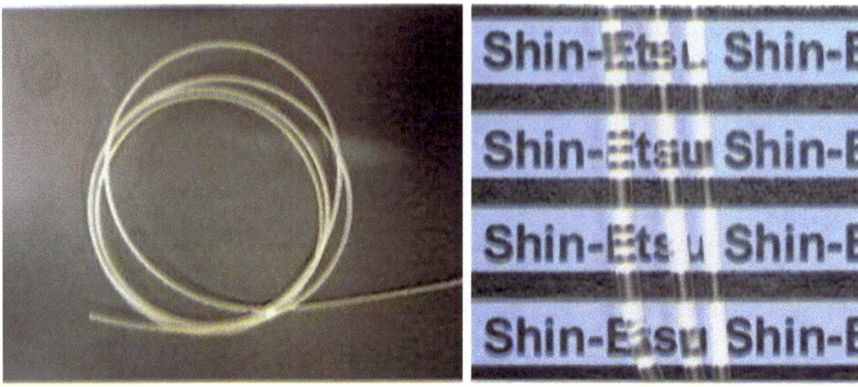

Fig. 4.3 Appearance of HPMCAS extrudates

Samples of HPMCAS (Grades: AS-LF, AS-MF, and AS-HF) were fed manually using a funnel into the Haake MiniLab® Micro Compounder (Thermo Fisher Scientific, MA, USA). The melt extrudates were milled and screened through US mesh #45 for further analysis. Glassy yellowish extrudates were obtained for all three grades.

The results of stability tests are summarized in Table 4.4. Free succinic acid and acetic acid values were increased and the succinoyl content was reduced accordingly with the increase in extrusion temperature and speed. The most significant change was the increase in free succinic acid that raised the total free acid content of HPMCAS. This result suggests that cleavage of succinoyl and acetyl groups occurs at high temperature and succinoyl groups have more cleavage tendency compared to acetyl groups. The ether, methoxy, and hydroxypropoxy groups were stable even after extrusion.

Overall, AS-LF was the most stable grade after melt extrusion with the lowest increase in total free acid (≤ 1 %) even at higher extrusion temperature and speed. Although the changes in substituents and free acid content after melt extrusion did not affect the solubility of AS-LF and AS-MF grades, the AS-HF grade displayed an increase in film-dissolution time with the increase in extrusion temperature and speed. Other observations were a slight reduction in viscosity which may be attributed to cleavage of polymeric chains, and color change at high temperature. The acceptance levels of free acids should be considered for each case to optimize the processing conditions for melt extrusion.

4.8 Case Studies with API

Similar studies of the physicochemical properties of HPMCAS in the presence of APIs have also been conducted. Using the Capilograph® to process solid dispersion formulations, Fig. 4.4 shows the resulting viscosity–shear relationship when incorporating nifedipine into the dispersion at 33 % drug loading. Clearly, when API was

Table 4.4 Stability of HPMCAS after melt extrusion (Haake MiniLab®)

Extruded at		Viscosity (mPa·s)	Loss on drying (%)	Substituents (%)				Free acid (%)		YI	Dissolving time at pH 6.8 (min)
Temperature (°C)	Rotor (rpm)			MeO	HPO	Ac	Suc	AA	SA		
Grade: AS-LF											
Before extrusion		2.79	1.4	22.6	7.2	7.9	14.7	0.03	0.02	6	7
160	100	2.67	1.4	22.6	7.2	7.9	14.6	0.10	0.24	22	7
	200	2.60	1.5	22.7	7.1	7.8	14.4	0.14	0.35	27	8
	300	2.56	1.6	22.4	7.1	7.7	14.3	0.17	0.51	29	7
180	100	2.70	1.6	22.5	7.1	7.7	14.3	0.13	0.38	26	8
	200	2.62	1.5	22.6	7.2	7.9	14.3	0.14	0.44	29	8
	300	2.56	1.6	22.5	7.1	7.7	14.2	0.17	0.50	32	7
200	100	2.53	1.7	22.6	7.1	7.7	14.1	0.14	0.62	40	8
	200	2.49	1.5	22.6	7.1	7.7	14.1	0.18	0.70	40	8
	300	2.50	1.4	22.6	7.0	8.1	14.1	0.23	0.70	39	8
Grade: AS-MF											
Before extrusion		2.76	1.3	23.0	7.2	9.3	11.4	0.04	0.03	12	7
160	100	2.66	1.3	22.9	7.1	9.4	11.1	0.10	0.44	31	8
	200	2.60	1.1	23.1	7.2	9.3	10.8	0.12	0.68	38	8
	300	2.60	1.1	23.0	7.1	9.4	10.7	0.14	0.85	48	8
180	100	2.62	1.2	23.0	7.3	9.2	10.8	0.11	0.72	33	7
	200	2.59	1.1	23.0	7.2	9.3	10.8	0.12	0.77	36	7
	300	2.59	1.1	23.1	7.2	9.3	10.9	0.12	0.88	47	8
200	100	2.50	1.1	23.0	7.2	9.2	10.4	0.16	1.19	38	8
	200	2.46	1.0	23.0	7.2	9.3	10.5	0.15	1.09	36	8
	300	2.50	1.2	23.0	7.2	9.1	10.1	0.16	1.13	44	8
Grade: AS-HF											
Before extrusion		2.69	1.5	23.6	7.5	11.8	7.7	0.04	0.03	8	27
160	100	2.76	1.4	23.4	7.4	11.8	7.3	0.08	0.47	29	34
	200	2.65	1.3	23.6	7.5	11.7	7.1	0.10	0.63	33	38
	300	2.62	1.4	23.6	7.5	11.7	7.1	0.11	0.71	37	43
180	100	2.70	1.4	23.6	7.5	11.7	7.0	0.10	0.72	35	45
	200	2.70	1.5	23.4	7.5	11.7	7.0	0.10	0.78	38	49
	300	2.65	1.6	23.3	7.4	11.7	7.0	0.12	0.88	41	47
200	100	2.65	1.4	23.4	7.4	11.6	6.8	0.12	1.03	41	60
	200	2.65	1.6	23.3	7.4	11.7	6.8	0.11	0.94	41	60
	300	2.64	2.0	23.4	7.4	11.7	6.8	0.13	0.97	42	59

Viscosity, loss on drying, substituents, and free acids were analyzed according to the compendial (National Formulary) methods. Yellowness index was measured using a color computer (Model SM-4, Suga, Japan). Dissolving time at pH 6.8: Films were cast from organic solvent solution of sample. The film specimens (1 cm^2) were immersed in pH 6.8 buffer (simulated intestinal fluid without enzyme) and their disintegration time was measured using a disintegration test apparatus
MeO methoxy, *HPO* hydroxypropoxy, *Ac* acetyl, *Suc* succinoyl, *AA* acetic acid, *SA* succinic acid, *YI* yellowness index

included in the formulation, the melt viscosity during extrusion was lower because the drug substance functions as a solid state plasticizer for the polymer. Furthermore, increasing shear rate resulted in a decrease in melt viscosity due to the thixotropic behavior of the polymer. Notably, the chemistry differences between the grades did

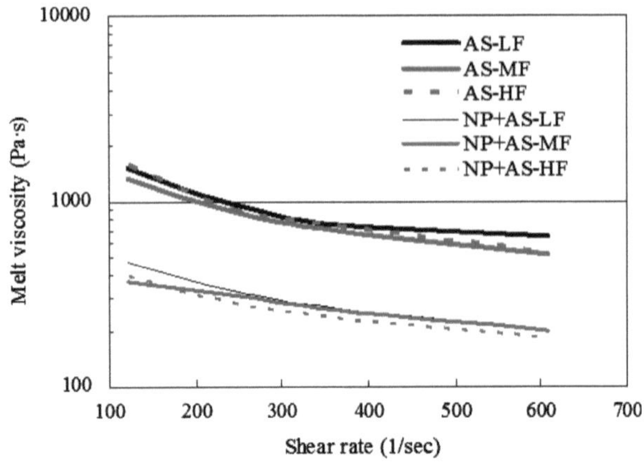

Fig. 4.4 Shear rate and viscosity during melt extrusion (170 °C) for HPMCAS and nifedipine (NP) + HPMCAS (1:2) solid dispersions

not translate into significant differences in melt viscosity. This suggests that formulation and process conditions can be used to control melt viscosity of the dispersion formulations.

On extrusion, clear extrudate rods were formed at temperatures above 160 °C, indicating the successful molecular dispersion of drug in the polymeric carrier. To confirm amorphous nature of the dispersion, x-ray diffraction was conducted. The x-ray diffraction patterns revealed that at temperatures higher than 160 °C, the drug was incorporated in the formulation as an amorphous form. The ability to render nifedipine amorphous was also related to the drug to polymer ratio. As shown in Fig. 4.5,

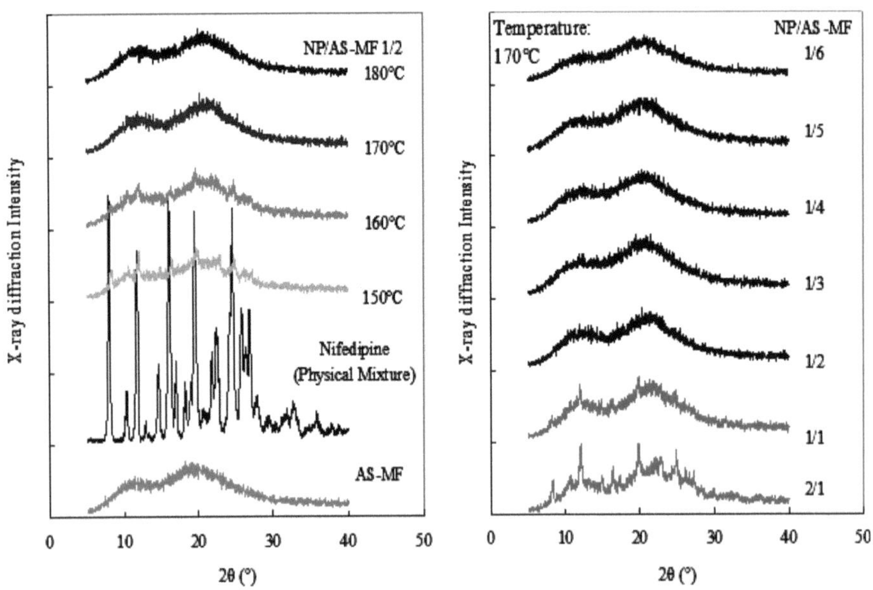

Fig. 4.5 x-ray diffraction pattern of extrudates

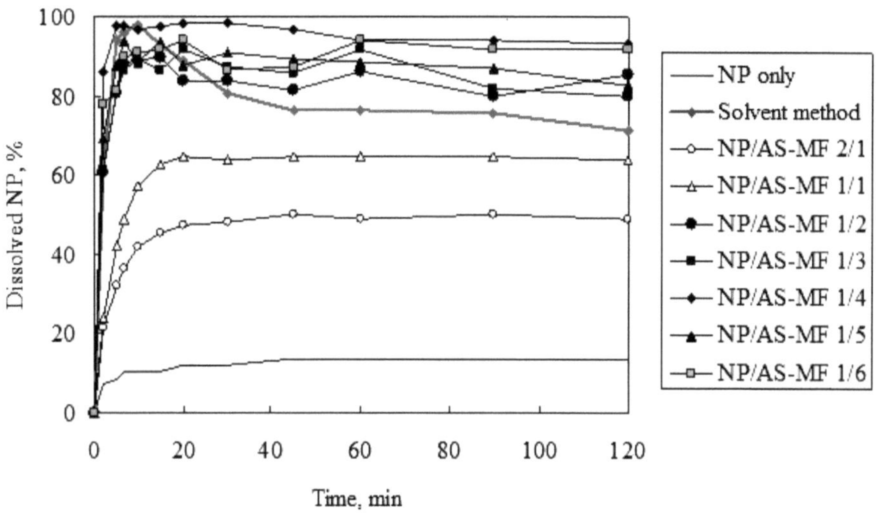

Fig. 4.6 Dissolution test results (extrudates with Capilograph®)

amorphous dispersions could be formed up to the 1:2 ratio; however, for systems at and beyond the 1:1 drug to polymer ratio residual crystallinity was observed.

The extrudates were further milled and a dissolution test was carried out using a pH 6.8 buffer, with release profiles shown in Fig. 4.6. The melt extrudates with HPMCAS showed significant increase in the dissolution level compared to the original API crystalline. When the API to polymer ratio was 1:2 or with higher polymer level, the dissolution was more than 80 % and supersaturated concentration was maintained at least for 2 h. With these samples, there was no clear effect of drug loading on the ability to influence the magnitude or extent of supersaturation. Further comparison to a comparable spray-dried dispersion exhibited no substantial difference and suggested that for nifedipine solid dispersions, manufacturing technology did not strongly influence dissolution behavior. However, dissolution profiles for dispersions having drug loadings greater than 33 % showed decreased behavior, which was most likely due to the residual crystallinity present in the dispersion. This highlights the well known fact that dispersions should be free of residual crystallinity to maximize dissolution performance.

Moving beyond the simple Capilograph® machine, dispersions were also prepared using a Haake Minilab® conical twin-screw extruder. Dissolution testing of nifedipine dispersions prepared using different grades of HPMCAS was also conducted, with the resulting profiles shown in Fig. 4.7. This comparison reveals how different polymer grades affect dissolution behavior, with L-chemistry material providing the lowest stabilizing performance of the three systems tested. In general, all three grades showed a significant increase in dissolution of API compared to the original API. AS-LF showed gradual decrease in the dissolution level due to recrystallization of API. AS-HF showed slightly slower release in the beginning, which

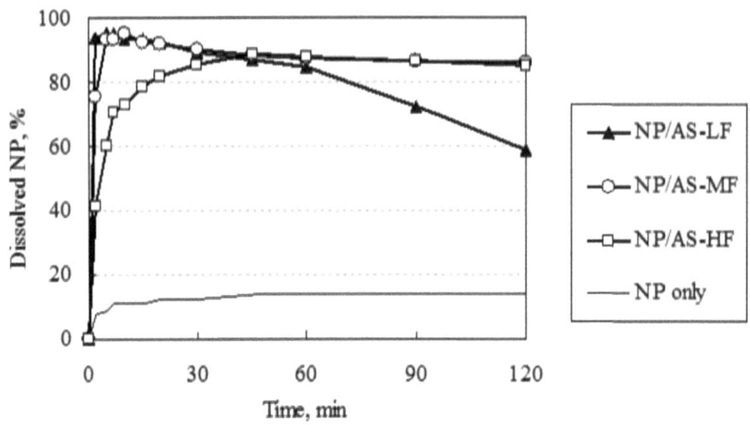

Fig. 4.7 Dissolution test results (extrudates from Haake MiniLab®)

may be due to the higher pH for dissolution onset. In this case, AS-MF seems to be the best grade to choose. This grade dependency would be due to some interaction between the API and polymer, although the nature of this interaction and the impact on dissolution is often material specific. Further consideration should also be given to regional absorption limitations when selecting the grade of HPMCAS. By varying the chemistry, the pH of polymer dissolution will change. For materials with narrow windows of absorption or poor colonic absorption this may be a critical point. It is therefore recommended to previously test all three grades, check dissolution levels, and confirm as necessary in vivo to ensure selection of the appropriate grade.

Dispersion formulations can also be produced on conventional parallel orientation twin-screw extruders. Continuing to scale-up the process to the Pharma 11 twin-screw extruder, a test formulation of 1:2 nifedipine to HPMCAS-M ratio was manufactured using processing conditions shown in Fig. 4.8. Under these operating conditions, it was possible to produce an amorphous dispersion. Dissolution testing of the dispersion was also conducted and compared to a physical mixture of drug and polymer. As shown in Fig. 4.9, a substantial improvement in dissolution rate and

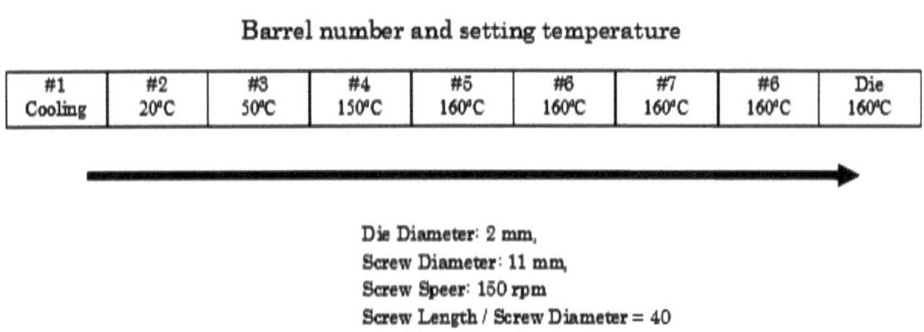

Fig. 4.8 Extrusion settings and temperatures of barrels in a twin-screw extruder (Pharma 11). Nifedipine:HPMCAS (AS-MF) = 1:2

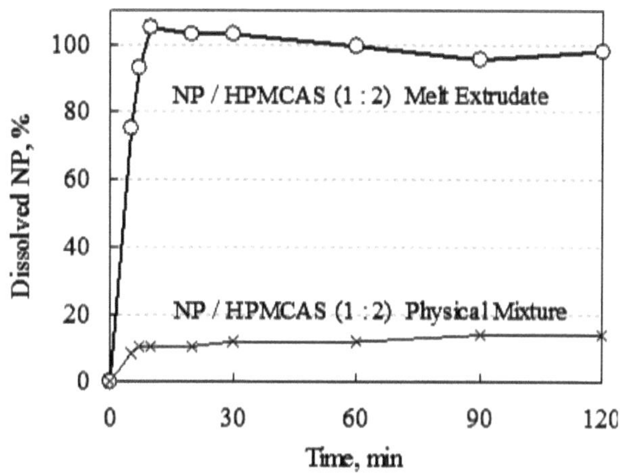

Fig. 4.9 Dissolution test results (extrudates from Pharma 11 twin-screw extruder)

extent of supersaturation is realized when compared to the crystalline mixture. Additionally, similarity of release behavior to previously tested systems is also exhibited, suggesting scale-independent behavior of the HPMCAS dispersions.

4.9 Recommended Extrusion Conditions for HPMCAS

Extrusion is an energy intensive process which, if operated with poor controls, can lead to decomposition of materials. Under sufficient energy input conditions, degradation of HPMCAS is possible, resulting from both temperature and shear. As already discussed, the acetyl and succinoyl groups exhibit the greatest tendency of cleavage which can result in polymer impurities that impact chemical and physical stability of the other components in the product. In order to use HPMCAS effectively, it is important to define an operating range for the material and discuss this range in relation to other cellulosic excipients.

For extruded materials, the processing window is bounded by the T_g of the excipient which establishes the lower processing threshold without incorporation of a plasticizer. The upper bound can be defined by the threshold for decomposition of the excipient, in this case measured through the use of thermal gravimetric analysis (TGA). As shown in Fig. 4.10, the window for HPMCAS is larger than the window for the other commonly used cellulosic materials. This does not exactly mean that the operating window for melt extrusion can be set solely on this information. Other methods for assessment of molecular weight stability and interactions with the drug substance, as well as physical and chemical stability of the API must be considered and confirmed in order to establish the true processing range. However, as a general guidance during early development, it is recommended to maintain extrusion temperatures below 180 °C to minimize ester cleavage and polymer decomposition during processing.

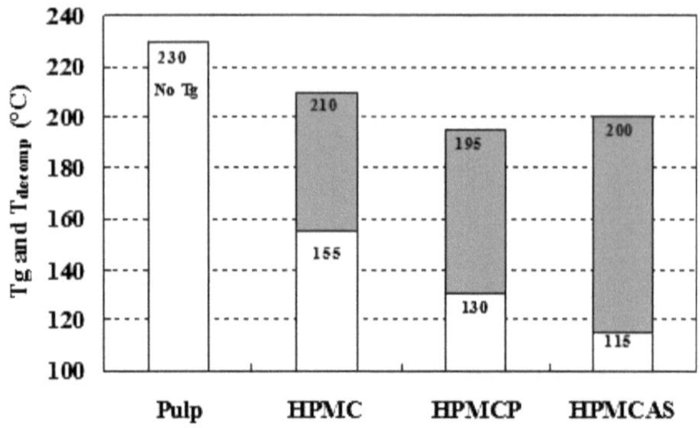

Fig. 4.10 Comparison of temperature "window" (i.e., T_g and decomposition temperature (T_{decomp})) of HPMCAS and other related cellulose materials

4.10 Conclusions

Originally developed for the production of enteric capsules, HPMCAS has shown tremendous versatility as a coating material and is now one of the preferred excipients for use in the development of amorphous solid dispersions. Due to its intentionally engineered thermoplastic properties, HPMCAS is well positioned for use in the production of dispersions by melt extrusion. Further, the moderate T_g and high decomposition temperature (T_{decomp}) provide an excellent window for processing on the extrusion platform. Coupling this with the well-documented concentration-enhancing nature of the material results in an excipient capable of enhancing oral exposure of many poorly soluble compounds. While bioavailability benefits vary on a case-by-case basis for a number of excipients, the broad applicability seen with HPMCAS means that this material will play a pivotal role in enabling formulations for many years to come.

References

Cerea M, Foppoli A et al (2008) Dry coating of soft gelatin capsules with HPMCAS. Drug Dev Ind Pharm 34:1196–1200

Curatolo W, Nightingale JA et al (2009) Utility of hydroxypropylmethylcellulose acetate succinate (HPMCAS) for initiation and maintenance of drug supersaturation in the GI milieu. Pharm Res 26:1419–1431

Kablitz CD, Harder K et al (2006) Dry coating in a rotary fluid bed. Eur J Pharm Sci 27:212–219

Klar F, Urbanez NA (2009) The role of capillary force promoters in dry coating procedures—evaluation of acetylated monoglyceride, isopropyl myristate and palmitate. Eur J Pharm Biopharm 71:124–129

Miyajima M, Yamaguchi Y et al (1991) Pharmaceutical composition of dihydropyridine compound. US Patent 4983593

Nagai T, Sekigawa F, Hoshi N (1998) Application of HPMC and HPMCAS to aqueous film coating of pharmaceutical dosage forms. In: McGinity JW (ed) Aqueous polymeric coatings for pharmaceutical dosage forms. Marcel Dekker, New York

Obara S, Maruyama N et al (1999) Dry coating—an innovative enteric coating method using a cellulose derivative. Eur J Pharm Biopharm 47:51–59
Onda Y, Muto H et al (1980) Ether-ester derivatives of cellulose and their applications. US Patent 4226981
Shin-Etsu Chemical Co., Ltd. (2010) Shin-Etsu AQOAT product brochure
Tamabuchi T, Nishimura M et al (1993) Improving the pharmaceutical properties of 4″-O-(4-Methoxyphenyl) acetyltylosin by preparing solid dispersions with carboxymethyl ethyl cellulose. Yakuzaigaku 53:221–228
Tanno F, Nishiyama Y et al (2004) Evaluation of hypromellose acetate succinate (HPMCAS) as a carrier in solid dispersions. Drug Dev Ind Pharm 30:9–17

Chapter 5
Cellulose Ethers for Extrusion Applications

Elanor Pinto and Thomas Dürig

Abstract Cellulose ethers have been used for years in developing oral dosage forms via batch processes such as direct compression and wet granulation. Recent advances in continuous manufacturing have brought about innovative technologies such as hot melt extrusion, melt extrusion granulation, and continuous wet granulation. Hydroxypropylcellulose and ethylcellulose have advantageous thermal viscoelastic properties making them great polymers to use in extrusion applications for developing immediate and controlled release dosage forms. Hypromellose and its derivatives can be utilized to prevent recrystallization and for solubilization.

5.1 Introduction

Continuous manufacturing leads well into Food and Drug Administration's (FDA) request for better-quality products using lean manufacturing principles (FDA Guidance for Industry 2004). Applications utilizing a twin-screw extruder have been gaining a lot of interest over the past decade in the pharmaceutical industry. The technology can be used for various applications: for developing films (Repka et al. 2000), implant devices (Gosau and Muller 2010), or tablets (Crowley et al. 2004); and for controlled release applications or for immediate release applications. Twin-screw extrusion offers several advantages over conventional processing techniques such as spray drying (Patterson et al. 2007), solvent evaporation, direct compression, and wet granulation. In addition to being more environmentally friendly and cost effective, better content uniformity can be observed with the twin-screw extruder due to intense mixing and agitation inherent in the process.

Several research groups have demonstrated the viability of hot-melt extrusion (HME) in manufacturing amorphous solid dispersions (Forster et al. 2001; Breitenbach and Magerlein 2003; Chokshi et al. 2008). Molecular solid dispersions of nifedipine (Li et al. 2006), nimodipine (Zheng et al. 2007), and itraconazole (Verreck et al. 2003) have been successfully produced using HME. HME is a solvent-free technique, with processing times, often less than 2 min, making this technology

E. Pinto (✉) · T. Dürig
Ashland Specialty Chemical Co, Wilmington, DE, USA
e-mail: epinto@ashland.com

M. A. Repka et al. (eds.), *Melt Extrusion*,
AAPS Advances in the Pharmaceutical Sciences Series 9,
DOI 10.1007/978-1-4614-8432-5_5,

more beneficial than other traditional methods used to prepare solid dispersions (Breitenbach 2002; Crowley et al. 2007). The careful selection of suitable polymeric carriers can be used to make stable amorphous solid dispersions through formation of secondary drug/polymer interactions (Taylor and Zografi 1997; Huang et al. 2008; Andrews et al. 2010) or/and by their antiplasticization effects, which can result in reducing their molecular mobility (Van den Mooter et al. 2001). To achieve the full benefit of amorphous solid dispersions, it is necessary not only to stabilize the drug during storage but also to maintain supersaturated concentration of drug during dissolution. Polymers used to manufacture amorphous solid dispersions have previously been shown to function as stabilizing agents and thus maintain supersaturated concentrations achieved during dissolution of amorphous drugs (Tanno et al. 2004). In addition, amorphous solid dispersions may generate higher solution concentrations than those achieved with the pure amorphous drug (Gupta et al. 2004), suggesting that certain polymers may act as solubilizing agents through complex formation with the drug during dissolution (Loftsson et al. 1996; Usui et al. 1997). The mechanisms driving stabilizing action of polymers remain poorly understood. Previously published articles suggest drug/polymer interactions as possible contributing factors (Repka et al. 2000). Furthermore, alteration of crystal growth and, in particular, the influence of dispersed polymers on crystal habit has been suggested as a possible reason. In this respect, changes in crystal habit were attributed to adhesion of polymer to the growing plane of the drug crystal (Repka et al. 2000). HME is an approach utilized in delivery of poorly water-soluble, class II compounds because of the increased dissolution achievable, and improved absorption and therapeutic efficacy (Chiou and Riegelman 1971; Ford 1986). Extrudates can have greater thermodynamic stability due to a more intimate mixing and better intermolecular interactions.

Polymers play a crucial role in providing a sturdy matrix for incorporating active pharmaceutical ingredients (APIs) into a stable form. Thermoplastic polymers are typically preferred as they can be processed with the extruder at suitable temperatures without affecting the stability of volatile or heat-sensitive APIs. However, plasticizers are used if the processing temperature required is not suitable for the API. In some cases, the API can be an effective plasticizer (Repka et al. 2000; Brabander et al. 2002). Commonly used, pharmaceutically approved polymers include cellulosic derivatives (hydroxypropylcellulose [HPC], hypromellose [HPMC], HPMC acetate succinate [HPMCAS], cellulose acetate [CA], and CA phthalate [CAP]), vinyl polymers (polyvinylpyrrolidone [PVP] and copovidone [PVP-VA]), polyethylene oxide (PEO), polyethylene glycol (PEG), and Eudragit (acrylates).

The potential for twin-screw extrusion to produce a sustained release device was conducted by Mehuys et al. (2004). A fourfold increase in the bioavailability of propranolol was observed when comparing an HME formulation with commercially available Inderal® (Astrazeneca, Belgium). Sustained-release pellets have been successfully developed (Young et al. 2002). Floating tablets developed by twin-screw extrusion have also been produced (Fukuda et al. 2006). Targeted drug delivery systems, including enteric matrix tablets and capsules systems, have been extruded (Mehuys et al. 2005; Andrews et al. 2008). The well-known commercial product utilizing HME is Kaletra®, developed by SOLIQS, which significantly enhanced the bioavailability of the APIs (Klein et al. 2007). SOLIQS also developed a fast-onset ibuprofen and sustained-release verapamil (Isoptin® SRE).

5.2 Hydroxypropylcellulose

5.2.1 Physical, Mechanical, and Thermal Properties

Cellulosic ethers are naturally occurring polysaccharides. They consist of polysaccharide units linked by β-1,4-glycoside bonds. Each glucose unit has three hydroxyl groups that can be derivatized and the average substitution grade cannot exceed three. By alkylation, the units can be derivatized to have hydroxypropyl, hydroxypropylmethyl, and many other semisynthetic cellulosics. Esterification can be used to further derivatize the cellulose ethers to form compounds such as CAP.

The thermal and mechanical properties of Klucel™ HPC, shown in Table 5.1 and Fig. 5.1, gives the polymer its thermoplastic properties making it pliable and easy to use for extrusion. HPC has a glass transition temperature, T_g, of approximately − 4.5 °C. There is debate about HPC having a biphasic glass transition temperature with the second T_g at above 100 °C due to complexities of the molecular mobility of a polymer structure (M. Vasanthavada, Y. Wang, T. Haefele, J. Lakshman, M. Mone, W. Tong, Y. Joshi, A. Serajuddin. Application of Melt Granulation Technology Using Twin-Screw Extruder in Development of High-dose Modified-release Tablet Formulation. *Journal of Pharmaceutical Sciences*. 100(5), 1923). The low T_g provides the polymer its low-melt viscosity and fast-melt flow properties depending on the molecular weight of polymer (Fig. 5.2). Hence, the molecular weight of the polymer can dictate the wide processing window suitable with low molecular weight grade (ELF) processable at temperatures as low as 120 °C and the high molecular weight grade (HF) processable at 200 °C without the use of a plasticizer (Table 5.2). In addition, extruding at different temperatures and molecular weight grades also affects the toughness and flexibility of HPC (Table 5.3).

Table 5.4 lists various plasticizers recommended for Klucel™ HPC. The lipophilicity of the plasticizer can affect the drug release profile so caution should be taken in choosing an appropriate one for an application. For example, a highly soluble plasticizer such as PEG 6000 would be better suited than palmitic acid for immediate release. Solid plasticizers are easier to process since they can be blended easily with the polymer and drug. As shown in Fig. 5.3, solid plasticizers such as stearic acid are effective in lowering the melt viscosity of HPC and would be recommended to use for many volatile and heat-sensitive compounds. The level of plasticizer in the polymer blend can range from 0.5–25 %. Glycerin and triacetin, Fig. 5.4, are also effective plasticizers with HPC. However, liquid plasticizers form tacky blends with HPC and are more difficult to feed into the extruder. Typically, liquid plasticizers often require an additional step where they are injected into the extruder barrel while the dry polymer/compound blend is fed in separately.

Table 5.1 Thermal, physical, and mechanical properties of Klucel™ hydroxypropylcellulose (HPC)

Property	HPC EF	HPC EXF	HPC ELF	Reference
Solid state—particle size (mm), shape	Mean diameter = 250–300 μm	Mean diameter = 50–80 μm	Mean diameter = 250–300 μm	Sympatec helos laser diffraction sensor particle sizer
Grass transition temperature (°C), T_g	– 4.0	– 4.5	– 4.5	DSC: TA instruments DSC Q2000 software: universal V4.7A TA instruments
Grass transition temperature (°C)	– 25–15	– 25–15	– 25–15	
Melting temperature (°C), T_m	191	189	182	
Melting temperature range (°C)	150–210	150–210	150–210	
Effective heat of fusion (J/g)	3.5	3.1	4.1	
Processing temperature: normal[a] (°C)	130	130	100	Leistritz ZSE 18HP
Processing temperature: maximum[b] (°C)	276	285	270	DSC: TA instruments DSC Q2000 software: Universal V4.7A TA instruments
Amorphous density (g/cm^3)	1.088	1.088	1.088	Instron capillary rheometer
Crystalline density (g/cm^3)	2.054	2.054	2.054	X-ray diffraction
Solid density (g/cm^3) (true density)	1.200	1.214	1.208	Micromeritics AccuPyc 1300 pycnometer
Crystallinity (%)	14.9	14.9	14.9	Water-cast film/intron capillary rheometer/X-ray diffraction
Bulk density (g/ml)	0.39	0.28	0.38	
Bulk compressibility (10^{-3} MPa^{-1})	7.12	7.12	7.12	Instron capillary rheometer

[a] Temperature can vary depending on extruder setup
[b] Degradation temperature $T_{96\%}$

The versatile thermal–mechanical properties of HPC make it suitable for many applications. Repka et al. (2000) found bioadhesive benefits with extruded HPC for film applications. With a 5 % polycarbophil additive, these researchers were able to produce HPC films with high tensile strength that performed well in vivo. Repka et al. (1999) dwell even further into the effect of various plasticizers and drugs on the mechanical behavior of extruded HPC films. The addition of the plasticizers PEG 8000, PEG 400, triethylcitrate (TEC), and acetyltributyl citrate (ATBC) and

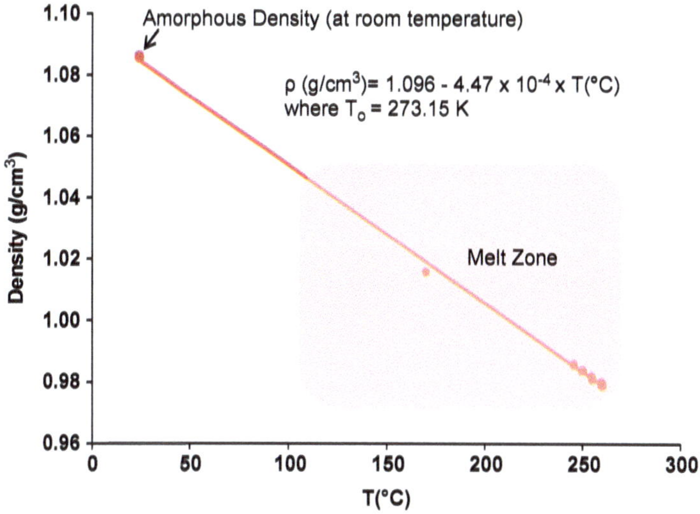

Fig. 5.1 Effect of temperature on Klucel™ hydroxypropylcellulose (HPC) melt density

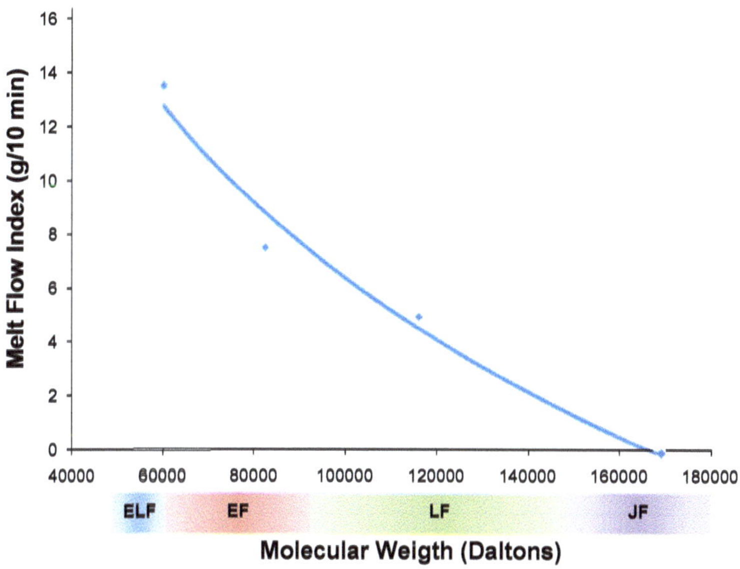

Fig. 5.2 Effect of molecular weight on the melt flow of Klucel™ hydroxypropylcellulose (HPC) at 150 °C using ASTM D1238

drugs (hydrocortisone [HC] and chlorpheniramine maleate [CPM]) lowered the melt viscosity of the polymer making it easier to extrude into films. However, PEG 400 contributed to poor physical stability of HC in HPC films suggesting a negative plasticizer interaction with the drug.

Table 5.2 Processing temperatures of different grades of pure Klucel™ hydroxypropylcellulose (HPC) polymer processed with the Leistritz ZSE 18HP twin-screw extruder

Klucel grade	MW (Daltons)	Recommended extrusion temperature setting (°C)				
		Zone 1	Zone 2	Zone 3	Zone 4–6	Zone 7–8/9
HF	1,150,000	120	150	170	190	200
MF	850,000	120	150	170	180	185
GF	370,000	100	120	150	180	180
JF	140,000	100	120	150	160	160
LF	95,000	80	100	120	140	150
EF	80,000	80	100	120	135	140
ELF	50,000	60	80	90	100	120

Table 5.3 Processability and toughness of extruded strands of Klucel™ hydroxypropylcellulose (HPC) of different molecular weight and processing temperatures using the Leistritz ZSE 18HP at an extruder speed rate of 150 rpm

Grade	Max. processing temperature (°C)	% load	Melt temperature (°C)	Melt pressure (psi)	Three-point bending strength (psi)	Young's modulus (psi)
ELF	140	29	113	210	6,800	450,000
EF	140	27	112	220	5,500	290,000
	160	24	122	150	9,500	630,000
LF	140	30	101	620	7,500	330,000
	160	25	109	210	6,900	320,000
GF	160	24	123	520	6,100	220,000
	180	22	130	310	8,400	520,000
MF	220	27	160	410	3,800	380,000
HF	180	20	141	400	5,000	200,000

Table 5.4 Plasticizers recommended for Klucel™ hydroxypropylcellulose (HPC)

Plasticizer	Form	Aqueous solubility (mg/L)
Glycerin	Liquid	65,000
Triacetin	Liquid	70,000
Triethyl citrate	Liquid	55,000
PEG 400	Liquid	50,000
PEG 6000	Solid	50,000
Palmitic acid	Solid	Insoluble
Stearic acid	Solid	3

PEG polyethylene glycol

5.2.2 *Immediate Release Applications*

Ozeki et al. (1997) investigated the use of HPC for use as a solid dispersion agent. Solid dispersions were made with a slightly water-soluble drug flurbiprofen (FP) and HPC or PEO. The spectrum obtained by powder X-ray diffraction (PXRD), Fig. 5.5, shows that the physical mixture of the drug and HPC (Aa) has the distinct

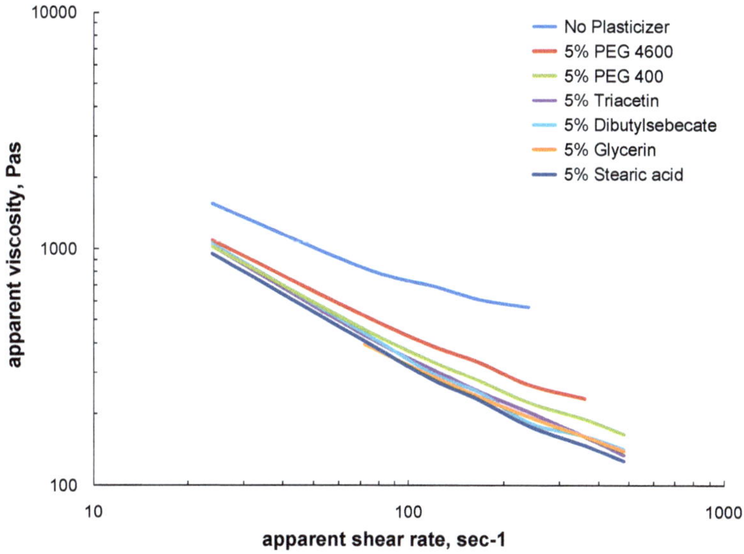

Fig. 5.3 Effect of plasticizer on melt viscosity of Klucel™ hydroxypropylcellulose (HPC) MF (MW ~ 850,000 Daltons) at 180 °C

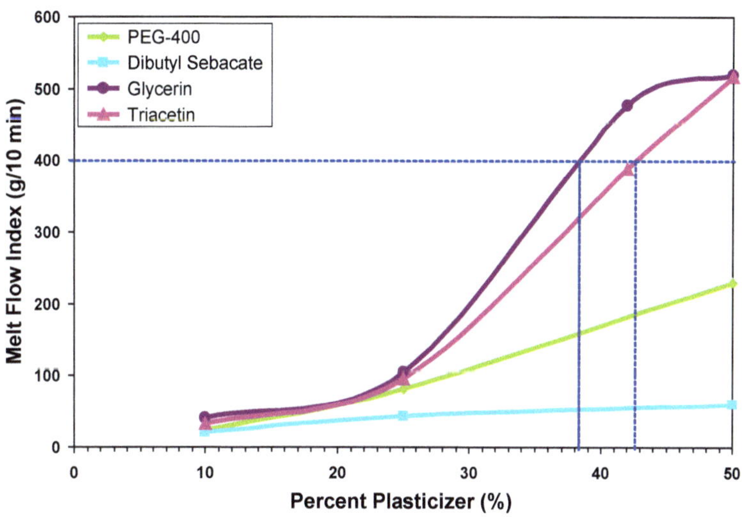

Fig. 5.4 Melt flow index of Klucel™ hydroxypropylcellulose (HPC) ELF blends consisting of different levels of plasticizer at 190 °C

crystalline peaks of FP. However, the crystalline peaks have disappeared with the solid dispersion formulation (Ba). Yuasa et al. (1994) conducted a study into the effect of the chain length and loading of HPC in the solid dispersion on the release profile of FP. With higher HPC loading and lower molecular weight grades of HPC, the FP release rates were improved.

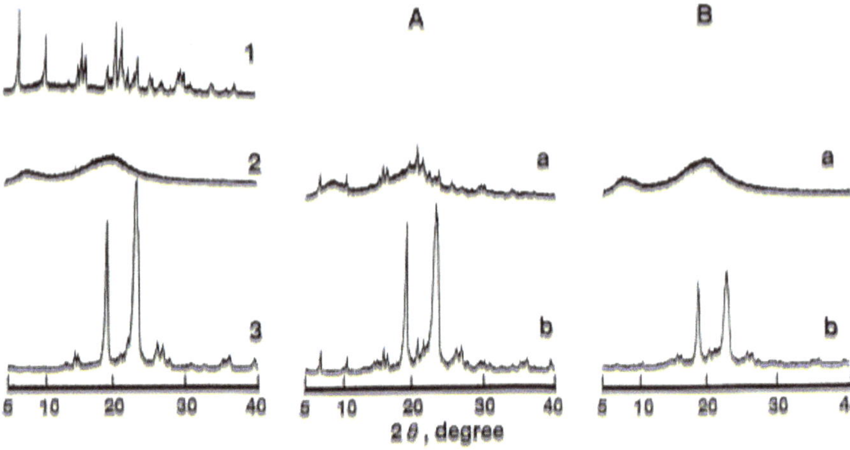

Fig. 5.5 Powder X-ray diffraction (PXRD) patterns of flurbiprofen (FP) (**a**), hydroxypropylcellulose (HPC) (**b**), polyethylene oxide (PEO) (**c**), and physical mixtures (**d**) and solid dispersions (**e**) containing 20 % FP (**1**) FP-HPC system, (**2**) FP-PEO system. (Ozeki et al. 1997)

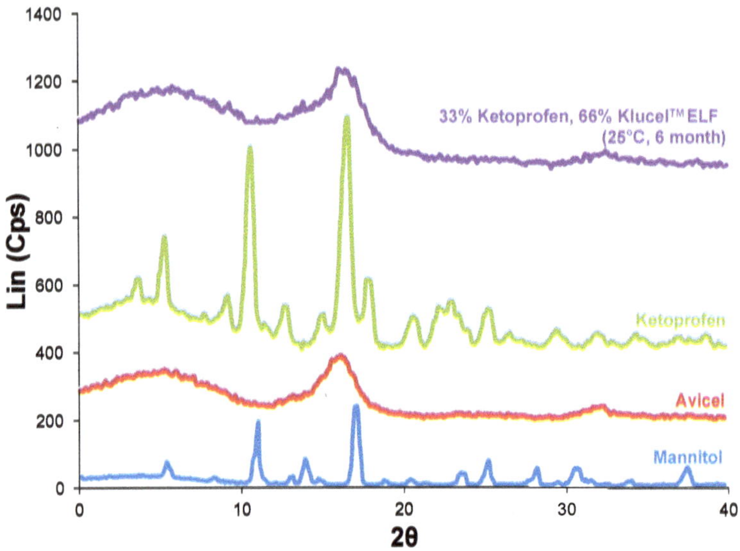

Fig. 5.6 Powder X-ray diffraction (PXRD) patterns of Ketoprofen and extruded Ketoprofen/hydroxypropylcellulose (HPC) tablets at 6 months accelerated stability testing. The extrudate was stored at 40 °C/75 %RH. H4–25T was stored at 25 °C/60 %RH. (Mohammed et al. 2012)

Mohammed et al. (2012) investigated the use of lower molecular weight HPC for solubilizing ketoprofen. The lower molecular weight grade Klucel™ HPC ELF (45,000 Da) was more efficient in increasing the drug release rate than Klucel™ HPC EF (80,000 Da). HPC was able to stabilize ketoprofen in its amorphous form (Fig. 5.6) even after 6 months under accelerated stability storage conditions. However, HPC

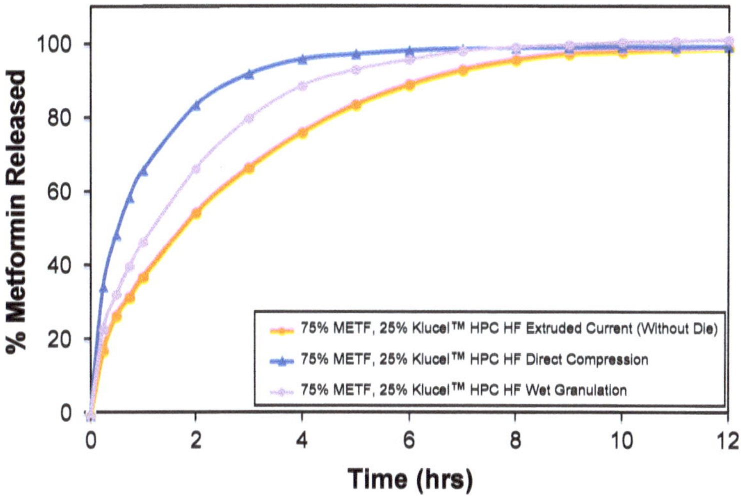

Fig. 5.7 Dissolution profile of Metformin HCl tablets made by different unit operations: direct compression, wet granulation, and hot melt extrusion (HME)

was not able to prevent recrystallization of all drugs as was observed with fenofibrate (Deng et al. 2013). In this case, PVP 17PF and amino methacrylate copolymer worked well as a recrystallization inhibitor of fenofibrate while HPC worked well as a processing aid and as a plasticizing polymer.

5.2.3 *Controlled Release Applications*

Higher molecular weight grades of HPC, Klucel^{TM} HPC HF, are typically recommended for controlled release applications as was observed with the highly soluble drug metformin HCl (METF). It was possible to produce controlled release METF tablets with very high drug loading (up to 75 %; Fig. 5.7). The matrix tablets made by melt extrusion granulation had a slower METF release rate in comparison to ones made by direct compression and wet granulation. Further investigation into the slower release rate showed a significantly lower porosity with tablets made by melt extrusion granulation (Table 5.5). The lower porosity of the hot melt-extruded tablets resulted in slower ingress of media into the tablet (Fig. 5.8) and thus slower diffusion of dissolved drug out of the tablet in the early time phase (first 30 min). After this initial period, a sufficiently strong gel layer envelops the tablet to control the further ingress of water into the system. As expected, the higher the molecular weight of HPC, the denser and stronger the semi-permeable gel layer formed. Vasanthavada et al. (2011) were also able to develop controlled release formulations using the high molecular weight grade Klucel^{TM} HPC HF. Also, they were able to demonstrate the *in vivo* performance of the formulation having a t_{max} of 4–8 h in a clinical study.

In addition, the METF tablets made by melt extrusion granulation were more than two times stronger, smaller, and consequently less porous as compared with

Table 5.5 Tablet physicals. Extrusion results in marked reduction of tablet porosity, increasing strength, and decreased drug release (burst) in early time phase

Unit process	Granule density (g/ml)	Tablet volume (ml)	Porosity (%)	Tablet strength (Kp) 3 kN precompression 15 kN main compression
Extrusion	1.30	0.8	3.4	14.2
Wet granulation	1.35	0.9	12.7	4.0
Direct compression	1.35	0.9	15.3	5.0

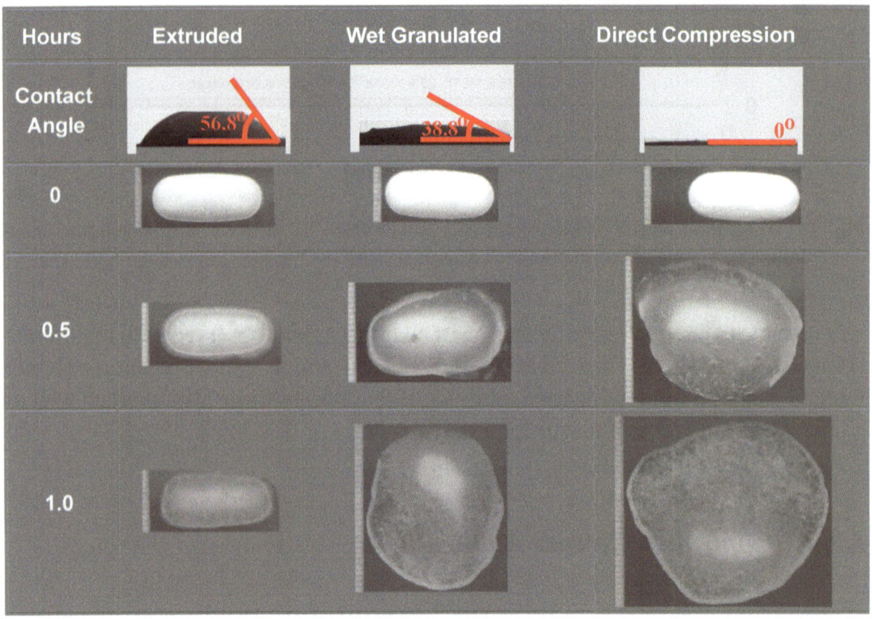

Fig. 5.8 Media uptake of Metformin HCl Tablets. Tablets made by hot melt extrusion (HME) were the least porous of the three methods and hence demonstrated slower dissolution media penetration rates. The porosity of the porous surfaces was further illustrated by the rapidly changing and much lower contact angle for the wet granulated and direct compression tablets

the analogous tablets made by wet granulation and direct compression (Table 5.5). The improved mechanical properties and smaller tablet size for the same weight of unit dose can be attributed to the intimate mixing of drug with polymer in the molten state and the substantial elimination of air in the extrudate and final tablet. In addition, the extrusion process also resulted in improved compactibility and reduced elastic recovery as evidenced by the enhanced tablet strength and reduced friability (Vasanthavada et al. 2011). Lakshman et al. (2011) were able to obtain drug loads as high as 90 % metformin with melt extrusion granulation (Vasanthavada et al. 2010). In addition, these researchers found that melt extrusion granulation improved the hardness and friability of the tablets in comparison to unit processes, wet granulation, and solvent granulation.

Table 5.6 Ethoxyl grades of Aqualon™ ethylcellulose based on the degree of substitution

Grade (type)	Ethoxyl content (%)	Degree of substitution Ethoxyl groups per anhydroglucose unit
K	45.5–46.8	2.28–2.38
N	47.5–49.0	2.42–2.53
T	49.0+	2.53+

5.3 Ethylcellulose

5.3.1 Physical, Mechanical, and Thermal Properties

Ethylcellulose has been commonly used in various hot melt applications due to its unusual combination of properties. It has excellent thermoplasticity, is miscible with various plasticizers, oils, and waxes, and imparts good durability and hardness. A big economical and environmental benefit is that there is no need for volatile organic solvents during the one-step manufacturing process. Ethylcellulose is a cellulose ether derived from ethyl chloride with alkali cellulose. The β-anhydroglucose unit of the cellulose chain contains three replaceable OH groups that are replaced with an ethyl group during the synthesis of the cellulose ether. Complete substitution of the OH group of all three OH groups would result in a triethyl ether possessing a degree of substitution of 3 and ethoxyl content of 54.88. FDA-approved commercial grades of ethylcellulose can be broken down into three types based on their degree of substitution (Table 5.6).

The degree of substitution has a definitive effect on the softening point, hardness, and moisture uptake of ethylcellulose. The more hydrophilic grade of ethylcellulose, the K-type, has the greatest moisture uptake at 3–4 % compared with other EC grades, the highest melt viscosity per a given molecular weight, and good hardness of 95–105 on the R-Scale. Due to its softening temperature and good hardness, the K-type is typically preferred for injection molding applications. It can be injected without surface lamination or mold defects and has better resistance to distortion under heat. The more popular intermediate lipophilic type, the N-type, has moderate moisture uptake of 2–3 %, good softening temperature at 150–156 °C, and the least hardness of 90–95 on the R-Scale. Similarly, the most lipophilic grade, the T-type, has the least moisture uptake, greatest softening temperature of 165 °C and above, and greatest hardness of 95 and above. The T-type is useful where maximum water resistance is important.

The N-type is the most versatile of the ethylcellulose grades and is recommended for twin-screw extrusion applications. A summary of the physical properties has been listed in Table 5.7. Ethylcellulose is prone to oxidation above its softening point of 156 °C and can lead to a brittle product unless stabilized with antioxidants (such as ascorbic acid). However, below the softening point, ethylcellulose is less prone to oxidation under high pressures for extended periods of time making it beneficial in

Table 5.7 Properties of N-type of Aqualon™ ethylcellulose. (Hercules 2002)

Properties	Characteristics
Physical	
Color (Hazen) in solution	2–5
Discoloration (by sunlight)	Very slight
Refractive index	1.47
Odor	None
Taste	None
Moisture absorption, by film in 24 h at 80 % RH, %	2
Water vapor transmission, $g/m^2/24$ h, 3-mil film, ASTM E 96–66	890
Thermal	
Softening point (°C)	152–162
Thermal conductivity ($cal/s/cm^2/°C/cm \times 10^{-4}$)	5.6
Specific heat (cal/°C/g)	0.32–0.46
Coefficient of thermal expansion (10^{-5}/°C)	10–14
Mechanical	
Elongation at rupture (%, 3-mil film, conditioned at 77°F and 50 % RH)	7–30
Elastic modulus ($lb/in^2 \times 10^5$)	1–5
Flexibility (folding endurance, M.I.T. double folds, 3-mil film)	160–2000
Flexural strength (lb/in^2)	4000–12000
Tensile strength (lb/in^2, 3-mil film, dry)	6800–10500
Hardness index, Sward (3-mil film)	52–61
Specific gravity	1.14
Specific volume (cu. In/lb in solution)	23.9

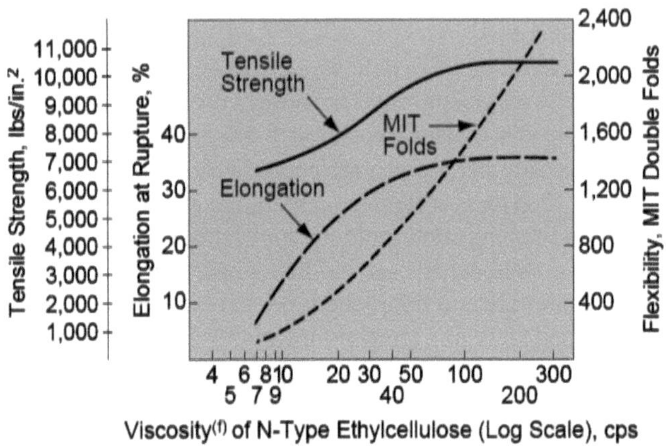

Fig. 5.9 Effect of viscosity on the mechanical properties of Aqualon™ ethylcellulose N-type films. (Hercules 2002)

extrusion. Ethylcellulose has no taste and can be used for taste masking via HME or when used in a coating application.

The mechanical properties of the N-type ethylcelluloses are affected by the molecular weight or the viscosity of the grade as shown in Fig. 5.9. As viscosity of the

Table 5.8 Plasticizers recommended for Aqualon™ ethylcellulose

Plasticizer	Form	Aqueous solubility
Dibutyl sebacate	Liquid	40 mg/L
Glyceryl behenate	Solid	Insoluble
Lt. Mineral oil	Liquid	Insoluble
Castor oil	Liquid	Insoluble
Soybean oil	Liquid	Insoluble
PEG 6000	Solid	50,000 mg/L
Palmitic acid	Solid	Insoluble
Stearic acid	Solid	3 mg/L

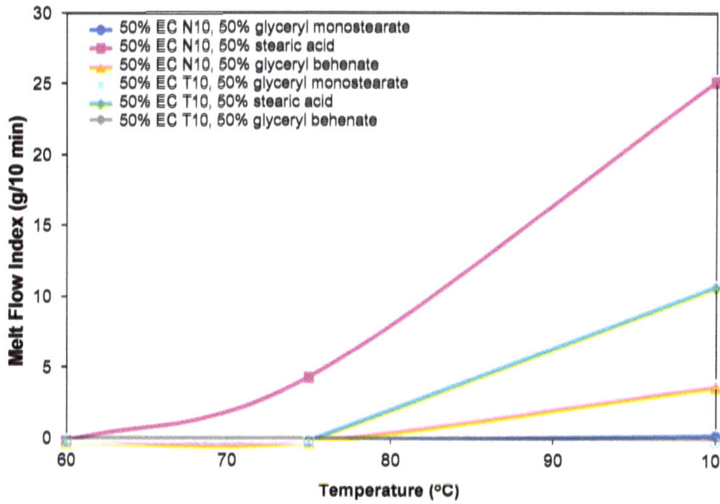

Fig. 5.10 Effect of different plasticizers on the melt flow, using ASTM D1238, of the N and T grade of Aqualon™ ethylcellulose

ethylcellulose grade increases, the tensile strength, percent elongation at rupture, and the flexibility (M.I.T. double folds) of the N-type casted film increases. The mechanical strength of the films correlate with the mechanical performance of the ethylcellulose extrudates. Wherein, the greater the viscosity grade of extruded ethylcellulose, the more durable and harder the extrudates.

5.3.1.1 Plasticizers

As with HPC, various types of plasticizers can be used to lower the softening point of ethylcellulose (Table 5.8). In addition, plasticizers provide easier processing at lower temperatures and can also enhance sustained release profiles utilizing more hydrophobic ones. Stearic acid is an efficient and easy-to-use plasticizer for HME using ethylcellulose. Figure 5.10 shows the melt flow of blends of the N- and T-type ethylcellulose with different plasticizers or lubricants. As expected, since the T-type

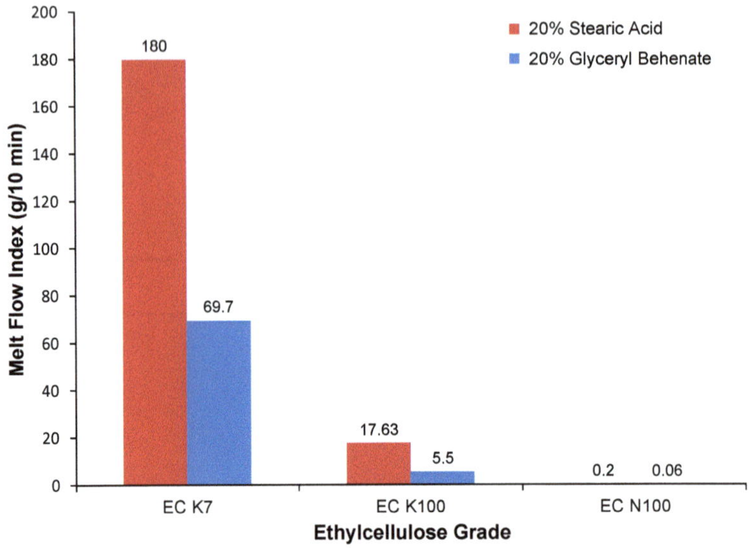

Fig. 5.11 Effect of different plasticizers on the melt flow, using ASTM D1238, of the N and T grade of Aqualon^{TM} ethylcellulose

has a lowest melt flow indicating a very viscous polymer blend at 190 °C suggesting that the T-type blend would need to be processed at a higher temperature. Similarly, the greater the molecular weight of the N-type, the lower the melt viscosity indicating that lower molecular weight grades would be easier to process at lower temperatures. Figure 5.11 shows the melt flow of blends of the N- and K-type ethylcellulose with stearic acid and glyceryl behenate. As expected, stearic acid worked better than glyceryl behenate in increasing the melt viscosity of the ethylcellulose. The K-type of ethylcellulose demonstrated better thermoplasticity than the N-type with much higher melt viscosity. A much lower viscosity grade, EC K7, had even better thermoplasticity than the higher molecular weight EC K100 due to easier alignment and less steric hindrance with the lower molecular weight EC polymer chains.

In addition to reducing the processing temperature and improving extrudability of ethylcellulose (Table 5.9), plasticizers can also improve the toughness or brittleness of ethylcellulose. Triacetin, palmitic acid, stearic acid, and castor oil were more efficient at plasticizing ethylcellulose than glycerin and PEG 6000, making the films easier to process and less plastic. Increasing the temperature increased the melt viscosity of the polymer blend and in some cases the flexibility of the extruded films. The toughness, average peak load, is the maximum force that can be applied to the film before it will crack or break. Palmitic acid, stearic acid, and castor oil worked best at improving the toughness of ethylcellulose. Lower processing temperatures also resulted in tougher films.

Table 5.9 Processability and toughness of extruded films of Aqualon^{TM} EC N10 with various levels and types of plasticizers/lubricants and processing temperatures using the Leistritz ZSE 18HP at an extruder speed rate of 100 rpm

Plasticizer/lubricant	%	Max. barrel zone temperature (°C)	% load	Melt temperature (°C)	Melt pressure (psi)	Toughness (lbf)
Glycerin	10	135	21	97	1320	43
PEG 6000	10	130	20	109	840	24
	10	150	20	117	400	11
Triacetin	10	125	24	84	720	121
	10	135	24	94	650	217
	30	115	18	78	450	82
	30	125	16	81	270	94
Palmitic acid	10	115	32	71	410	251
	10	125	28	78	480	229
	10	140	27	91	390	389
Stearic acid	10	135	31	92	480	38
	10	145	28	95	260	373
	30	115	14	70	80	145
	30	125	15	80	60	150
Castor oil	10	125	37	92	910	290
	10	135	35	94	900	196
	10	145	29	102	650	199
	30	115	18	83	580	89
	30	125	16	83	380	92
	30	135	15	90	260	114

5.3.2 Controlled Release Applications

As shown in Fig. 5.12, Crowley et al. (2004) observed the benefits of using ethyl cellulose for HME applications. Tablets made by HME using ethylcellulose (48.0–49.5 % ethoxyl substitution, 9–11 cP) had much slower guaifenesin release rates than those made by direct compression. In addition, using finer fractions of ethylcellulose was able to further retard drug release. Brabander et al. also observed that ethylcellulose was an effective sustained release agent in HME applications (De Brabander et al. 2003). However, utilizing a 40 % ethylcellulose composition with ibuprofen revealed that the release profile was too slow. By using more hydrophilic polymers as channel formers, such as HPMC and xanthan gum, it was possible to modulate the drug release profile over a wide range. The molecular weight (Fig. 5.13) affected the release profile achieved wherein the higher molecular weight HPMC gave faster release profiles due to higher swelling capability with higher molecular weight grades. The extrudate demonstrated a robust sustained release performance *in vivo* (De Brabander et al. 2004). Vasanthavada et al. were also able to demonstrate the *in vivo* performance of an ethylcellulose-based Metformin-extruded formulation with a t_{max} of 4–12 h during a clinical study (Vasanthavada et al. 2011).

The ethoxyl content and viscosity of the ethylcellulose can modulate the release profiles of extruded matrix tablets. As shown in Fig. 5.14 with a highly soluble drug metformin, using a molecular weight or viscosity grade EC N4, exhibited a faster

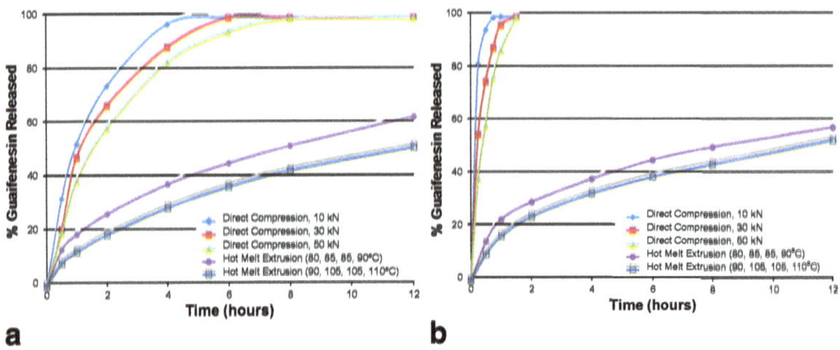

Fig. 5.12 Influence of ethylcellulose particle size, compaction force, and extrusion temperature on guaifenesin release from matrix tablets made by direct compression and HME. **a** Matrix tablets made with the "fine" ethyl cellulose (325–80 mesh) and **b** matrix tablets made with the "coarse" ethyl cellulose (80–30 mesh). (♦) Direct compression, 10 kN; (■) direct compression, 30 kN; (▲) direct compression, 50 kN; (•) HME, 80, 85, 85, 90 °C; (□) HME, 90, 105, 105, 110 °C. (Crowley et al. 2004)

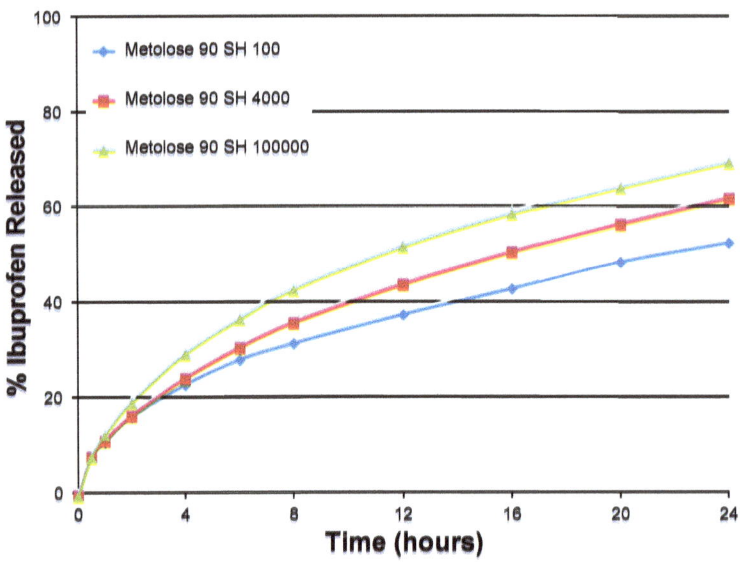

Fig. 5.13 Mean dissolution profiles (± S.D) of the minimatrices containing 60 % ibuprofen, 30 % ethylcellulose, and 10 % HPMC (•) Metolose® 90 SH 100, (■) Metolose® 90 SH 4000, and (▼) Metolose® 90 SH 100 000. (Brabander et al. 2003)

drug release than a higher molecular weight grade EC N10. In addition, using the lower percent ethoxyl grade EC K7 provided a much faster release profile than a higher % ethoxyl grade with molecular weight EC N7 due to the lower lipophilicity of the K-type. As observed by Brabender, incorporating a hydrophilic polymer as a channel former is another alternative to modulating release profiles with EC-based systems. Figure 5.15 shows the effect of using HEC and HPC as channel formers.

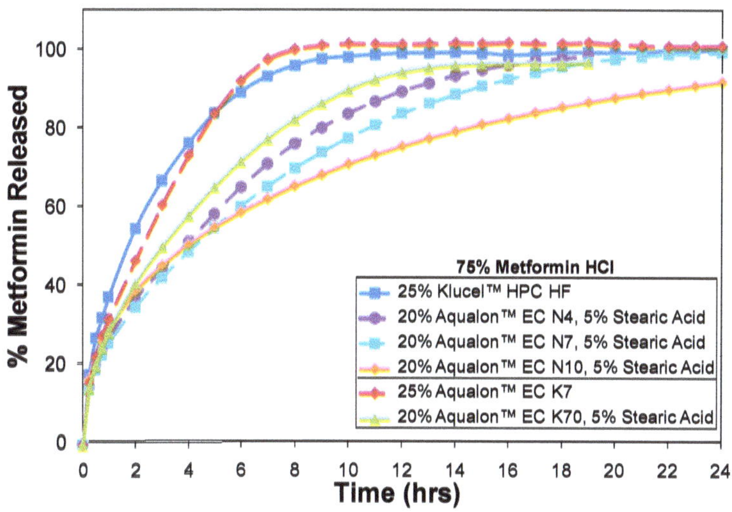

Fig. 5.14 Effect of different ethoxyl % and viscosity grades of Aqualon™ ethylcellulose on the dissolution profiles of melt extrusion granulated metformin matrix tablets

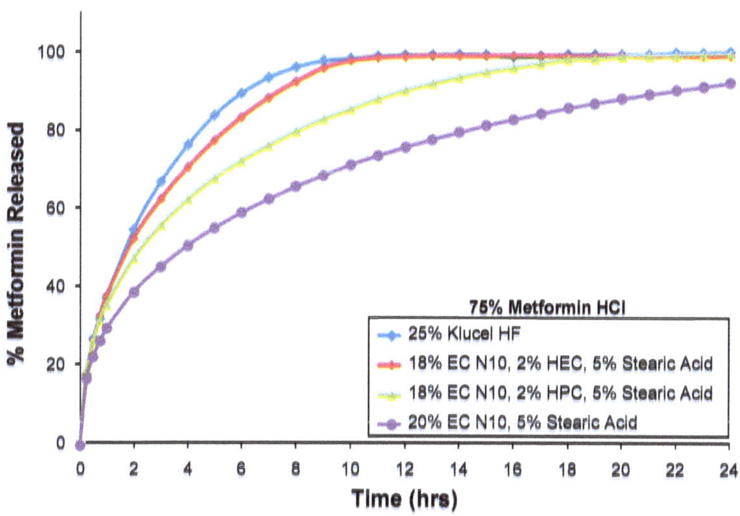

Fig. 5.15 Effect of using a hydrophillic polymer Natrosol™ hydoxyethylcellulose (HEC), and Klucel™ hydroxypropylcellulose (HPC) on modulating the release profile of metformin HCl from hot melt-extruded Aqualon™ ethylcellulose (EC) and HPC type on dissolution profiles of hot melt-extruded acetaminophen matrix tablets

HEC being the more hydrophilic polymer resulted in EC matrix tablets with faster release profiles in comparison to HPC. In addition, increasing the composition of a hydrophilic polymer Klucel™ HPC HF can also enhance and modulate the drug release as observed in Fig. 5.16 with a moderately soluble drug acetaminophen.

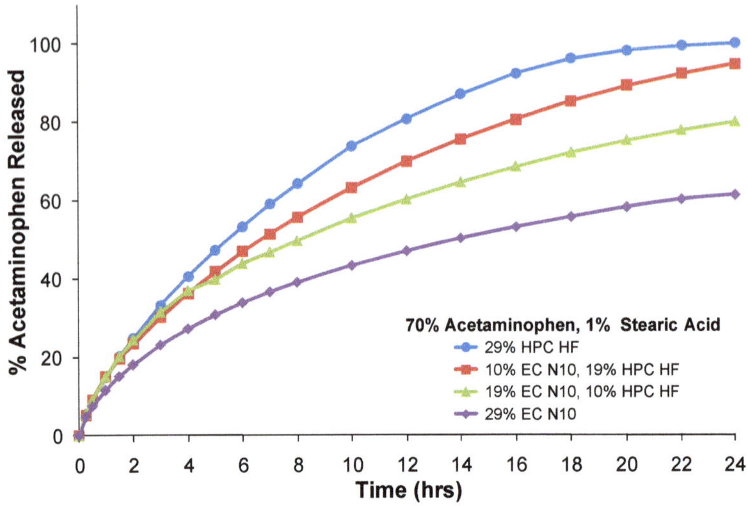

Fig. 5.16 Effect of polymer blend of Aqualon™ EC N10 and Klucel™ HPC HF type on dissolution profiles of hot melt-extruded acetaminophen matrix tablets

5.4 Hydroxypropyl Methylcellulose and Hypromellose Acetate Succinate

HPMC has been shown to be a recrystallization inhibitor in stabilizing amorphous drugs and hence enhance bioavailability of poorly soluble drugs. The supersaturated levels generated by dissolution of the amorphous solid dispersions can arise from the stabilizing effects of the polymers (Perissutti et al. 2002; Chokshi et al. 2005) or complexation of the crystalline drug in the polymer matrix, hence reducing the degree of supersaturation and lower thermodynamic tendency toward recrystallization (Mehuys et al. 2004; Li et al. 2006). One can compare the performance of itraconazole melt extrudates prepared using HPMC and PVP-VA/Eudragit E polymeric platforms (Six et al. 2005). HPMC was shown to significantly increase the bioavailability in comparison to the other extrudates despite having a slower dissolution rate. DiNunzio et al. (2010) utilized HME to produce solid dispersions of HC using HPMC E3. The extrusion was performed with a twin-screw extruder at a maximum processing temperature of 180 °C. HPMC E3 was effective in solubilizing HC in its amorphous form as shown in Fig. 5.17. The HC distinct crystalline peaks at 15 and 18 2θ were observed with the HPMC E3 physical blend but disappeared in the HPMC E3 solid dispersion.

HPMCAS has become a very popular polymer used with HME techniques since it is well known as a recrystallization inhibitor. HPMCAS MF generated greater supersaturated levels of felodipine at pH 6.8 and maintained these concentrations for longer periods of times compared with PVP and hydroxypropylmethylcellulose (HPMC);

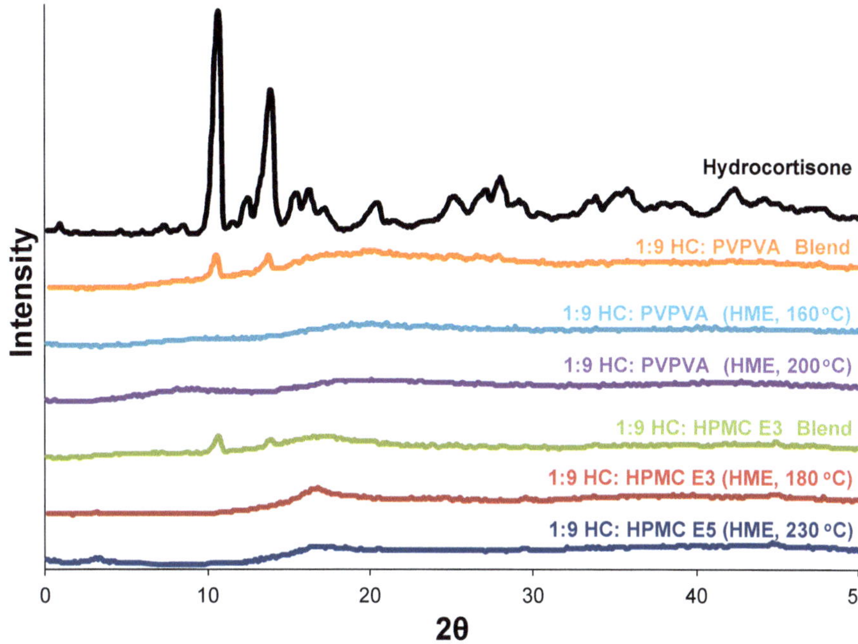

Fig. 5.17 Powder X-ray diffraction (PXRD) spectrum of hydrocortisone (HC), HC: polymer mixture blends, and HC: polymer solid dispersions. (DiNunzio et al. 2010)

Mehuys et al. 2004). The acetyl and succinate groups present within HPMCAS provide sufficient capacity for API/polymer hydrogen bonding. Secondary forces have been shown to be highly significant in the formation and stabilization of solid dispersions (Konno and Taylor 2006). HPMCAS has also been shown to be of particular benefit in the preparation of solid dispersions and for the production of enteric dosage forms (Mehuys et al. 2005; Tanno et al. 2004).

5.5 Conclusion

The use of cellulose ethers has grown over the years in various industries and applications. The wide variability of options in derivatization, degree of substitution, and molecular weight play a role in this class of polymers' versatility. A more hydrophobic derivation of the cellulosic backbone, i.e., substitution with ethoxyl, can result in a polymer for controlled release applications. A higher molecular weight grade of HPC or HPMC can promote a denser gel barrier film formation upon dissolution, promoting retardation of drug release. A lower degree of substitution of ethylcellulose impacts the thermoplasticity of the polymer making it possible to extrude the polymer at lower extrusion processing temperatures. In addition, derivation of the cellulosic backbone with ionic substitutes, i.e., HPMCAS, promotes stabilization of amorphous APIs through molecular ionic interactions with the polymer and API. It will be interesting to see the advances of cellulosics in pharmaceutical extrusion applications over the next 10 years.

Acknowledgments While there is a long list of brilliant people at Ashland that contributed work and ideas to make this book chapter possible, I would specifically like to thank Jim Brady for all his guidance in helping me understand cellulosic chemistry and functionality. In addition, I would like to thank Hong Yang for all his guidance in helping me understand extrusion and polymer viscoelastic behavior.

References

Andrews GP et al (2008) The manufacture and characterization of hot-melt extruded enteric tablets. Eur J Pharm Biopharm 69:264–273

Andrews GP, Abu-Diak O, Jones DS (2010) Physicochemical characterization of hot-melt extruded bicalutamide-polyvinylpyrrolidone solid dispersions. J Pharm Sci 99(3):1322–1335

Breitenbach J (2002) Melt extrusion: from process to drug delivery technology. Eur J Pharm Biopharm 54:107–117

Breitenbach J, Magerlein M (2003) Melt extruded molecular dispersions. In: Ghebre-Sellassie L Martin C (eds) Pharmaceutical extrusion technology. Drugs and the pharmaceutical sciences. Marcel Dekker Inc, New York

Chiou WL, Riegelman S (1971) Pharmaceutical applications of solid dispersion systems. J Pharm Sci 60(9):1281–1302

Chokshi RJ et al (2005) Characterization of physico-mechanical properties of indomethacin and polymers to assess their suitability for hot-melt extrusion process as a means to manufacture solid dispersion/solution. J Pharm Sci 94:2463–74

Chokshi RJ et al (2008) Stabilization of low glass transition temperature indomethacin formulations: impact of polymer-type and its concentration. J Pharm Sci 97(6):2286–2298

Crowley MM et al (2004) Physicochemical properties and mechanism of drug release from ethylcellulose matrix tablets prepared by direct compression and hot-melt extrusion Int J Pharm 269:509–522

Crowley MM et al (2007) Pharmaceutical applications of hot-melt extrusion: Part I. Drug Dev Ind Pharm 33(9):909–926

De Brabander C et al (2002) Characterization of ibuprofen as a nontraditional plasticizer of ethyl cellulose. J Pharm Sci 91(7):1678–1685

De Brabander C, Vervaet C, Remon JP (2003) Development and evaluation of sustained release mini-matrices prepared via hot melt extrusion. J Control Release 89(2):235–247

De Brabander C et al (2004) Bioavailability of Ibuprofen from Hot-Melt Extruded Mini-Matrices. Int J Pharm 271(1–2):77–84

Deng W et al (2013) Stabilization of fenofibrate in low molecular weight hydroxypropylcellulose matrices produced by hot-melt extrusion. Drug Dev Ind Pharm 39(2):290–298

DiNunzio JC et al (2010) Fusion production of solid dispersions containing a heat-sensitive active ingredient by hot melt extrusion and Kinetisol dispersing. Eur J Pharm Biopharm 74:340–351

FDA Guidance for Industry (2004) Pharmaceutical cGMPs for the 21st Century: a Risk-Based Approach. [Online] September. www.fda.gov

Ford JL (1986) The current status of solid dispersions. Pharm Acta Helv 61:69–88

Forster A, Hempenstall J, Rades T (2001) Characterization of glass solutions of poorly water-soluble drugs produced by melt extrusion with hydrophilic amorphous polymers J Pharm Pharmacol 53:303–315

Fukuda M, Peppas NA, McGinity JW (2006) Floating hot-melt extruded tablets for gastroretentive controlled drug release system. J Control Release 115 121–129
Gosau M, Muller BW (2010) Release of gertamicin sulphate from biodegradable PLGA-implants produced by hot melt extrusion. Pharmazie 65:487–492
Gupta P, Kakumanu KV, Bansal AK (2004) Stability and solubility of celecoxib-PVP amorphous dispersions: a molecular perspective. Pharm Res 21(10):1762–1769
Hercules (2002) Aqualon ethylcellulose physical and chemical properties. Hercules Inc., Wilmington
Huang J, Wigent RJ, Schwartz JB (2008) Drug-polymer interaction and its significance on the physical stability of nifedipine amorphous dispersion in microparticles of an ammonio methacrylate copolymer and ethylcellulose binary blend. J Pharm Sci 97(1):251–262
Klein CE, Chiu YL, Awni W (2007) The tablet formulation of lopinavir/ritonavir provides similar bioavailability to the soft-gelatin capsule formulation with less pharmacokinetic variability and diminished food effect. J Acquir Immune Defic Syndr 44:401–410
Konno H, Taylor LS (2006) Influence of different polymers on the crystallization tendency of molecularly dispersed amorphous felodipine. J Pharm Sci 95(12):2692–705
Lakshman J et al (2011) Application of melt granulation technology to enhance tabletting properties of poorly compactable high-dose drugs. J Pharm Sci 100(4):1553–1565
Li L, AbuBaker O, Shao Zezhi J (2006) Characterization of poly(ethylene oxide) as a drug carrier in hot-melt extrusion. Drug Dev Ind Pharm 32(8):991–1002
Loftsson T, Fridriksdottir F, Gudmundsdottir TK (1996) The effect of water-soluble polymers on aqueous solubility of drugs. Int J Pharm 127:293–296
Mehuys E et al (2004) In vitro and in vivo evaluation of a matrix-in-cylinder system for sustained drug delivery. 2. J Control Release 96:261–271
Mehuys E, Remon JP, Vervaet C (2005) Production of enteric capsules by means of hot-melt extrusion. Eur J Pharm Sci 24:207–12
Mohammed NN et al (2012) Klucel EF and ELF polymers for immediate-release oral dosage forms prepared by melt extrusion technology. AAPS Pharm Sci Tech 13(4):1158–1169
Ozeki T, Yuasa H, Kanaya Y (1997) Application of the solid dispersion method to the controlled release of medicine. IX. Difference in the release of flurbiprofen from solid dispersions with poly(ethylene oxide) and hydroxypropylcellulose and the interaction between medicine and polymers. Int J Pharm 155(2):209–217
Patterson JE et al (2007) Preparation of glass solutions of three poorly water soluble drugs by spray drying, melt extrusion and ball milling. Int J Pharm 336(1):22–34
Perissutti B et al (2002) Preparation of extruded carbamazepine and PEG 4000 as a potential rapid release dosage form. Eur J Pharm Biopharm 53:125–132
Repka MA et al (1999) Influence of plasticizers and drugs on the physical-mechanical properties of hydroxypropylcellulose films prepared by hot melt extrusion. Drug Dev Ind Pharm 25:625–633
Repka MA, McGinity JW (2000) Physical-mechanical, moisture absorption and bioadhesive properties of hydroxypropylcellulose hot-melt extruded films. Biomaterials 21(14):1509–1517
Six K et al (2005) Clinical study of solid dispersions of itraconazole prepared by hot-stage extrusion. Eur J Pharm Sci 24:179–86
Tanno F et al (2004) Evaluation of hypromellose acetate succinate (HPMCAS) as a carrier in solid dispersions. Drug Dev Ind Pharm 30:9–17
Taylor LS, Zografi G (1997) Spectroscopic characterization of interactions between PVP and indomethacin in amorphous molecular dispersions. Pharm Res 14:1691–1698
Usui F et al (1997) Inhibitory effects of water-soluble polymers on precipitation of RS-8359. Int J Pharm 154:59–66
Van den MG et al (2001) Physical stabilization of amorphous ketoconazole in solid dispersions with polyvinylpyrrolidone K25. Eur J Pharm Sci 12:261–269
Vasanthavada M et al (2010) Process for making compositions with poorly compressible therapeutic compounds. US 2010/0152299 A1 United States, June 17, 2010

Vasanthavada M et al (2011) Application of melt granulation technology using twin-screw extruder in development of high-dose modified release tablet formulation. J Pharm Sci 100(5):1923–1934
Verreck G et al (2003) Characterization of solid dispersions of itraconazole and hydroxypropylmethycellulose prepared by melt extrusion—part 1. Int J Pharm 251:165–174
Young CR, Koleng JJ, McGinity JW (2002) Production of spherical pellets by a hot-melt extrusion and spheronization process. Int J Pharm 242(1–2):87–92
Yuasa H et al (1994) Application of the solid dispersion method to the controlled release of medicine. VII. Release mechanism of a slightly water soluble medicine and interaction between flurbiprofen and hydroxypropyl cellulose in solid dispersion. Chem Pharm Bull 42:354–358
Zheng X et al (2007) Part 1: characterization of solid dispersions of nimodipine prepared by hot-melt extrusion. Drug Dev Ind Pharm 33(7):791–802

Chapter 6
Properties and Applications of Polyethylene Oxide and Ethylcellulose for Tamper Resistance and Controlled Drug Delivery

Sampada B. Upadhye and Ali R. Rajabi-Siahboomi

Abstract Polyethylene oxide, by virtue of its thermoplastic behavior, may be a suitable polymer for pharmaceutical dosage form using hot-melt extrusion technology. Applications of high molecular weight POLYOX allows formulation of extended-release melt-extruded matrices. Combination of different grades of POLYOX and other polymers enables formulators to tailor the release profile of drugs, as well as enhance melt-extrusion processing. POLYOX has been widely studied and used in pharmaceutical applications such as controlled release of drugs in osmotic pumps, hydrophilic matrices, mucoadhesion, oral films, gastroretentive systems, and abuse-resistance technologies. Ethylcellulose is well known for its barrier membrane coatings for extended-release applications. Due to its thermal properties, it has been successfully used in hot-melt extrusion and injection molding. In this chapter, specific pharmaceutical applications, where thermal behavior of POLYOX and ethyl cellulose has been utilized in the design and manufacture of the melt-extruded dosage form, will be discussed.

6.1 Polyethylene Oxide—Chemistry and Properties

Polyethylene oxides (PEOs) are nonionic, high molecular weight hydrophilic (water-soluble) polymers (CAS Registry Number 25322-68-3) available in the form of white, free-flowing powders. They are manufactured by Dow Chemical Company under the trade name of POLYOX™ Water Soluble Resins (WSR; http://dowwolff.custhelp.com). The pharmaceutical grades of POLYOX are available in molecular weight ranges of 100,000–7,000,000 Da as shown in Table 6.1 (http://dow.com). PEO polymers are safe and are not absorbed through the gastrointestinal tract. On hydration, PEO forms a gel and have high swelling capacity. All of the above physical and chemical properties have attracted significant attention to the potential uses of PEO in pharmaceutical dosage forms. PEO has been used in oral solid drug delivery technologies such as osmotic pumps, hydrophilic matrices, gastroretentive dosage forms, hot-melt extrusion (HME) and abuse-deterrent

A. R. Rajabi-Siahboomi (✉) · S. B. Upadhye
Colorcon Inc., Harleysville,
PA, USA
e-mail: asiahboomi@colorcon.com

M. A. Repka et al. (eds.), *Melt Extrusion*,
AAPS Advances in the Pharmaceutical Sciences Series 9,
DOI 10.1007/978-1-4614-8432-5_6,

Fig. 6.1 Illustration of POLYOX synthesis

$$(C_2H_4)O \xrightarrow{\text{Catalyst}} RO-[CH_2CH_2O]_n\text{-}H \xrightarrow{\text{Controlled Scission}} RO-[CH_2CH_2O]_n\text{-}H$$

n ~ 45,000 to 180,000 — POLYOX 301, Coagulant, 303

n ~ 2,000 to 45,000 — POLYOX N10 to N60K

Table 6.1 Commercial grades of POLYOX used in the pharmaceutical industry

POLYOX resins	Molecular weight (Da)	Aqueous viscosity range at 25 °C (mPa s)
WSR N-10 NF	100,000	12–50 (at 5 % w/v)
WSR N-80 NF	200,000	65–115 (at 5 % w/v)
WSR N-750 NF	300,000	600–1,000 (at 5 % w/v)
WSR 205 NF	600,000	4,500–8,800 (at 5 % w/v)
WSR 1105 NF	900,000	8,800–17,600 (at 5 % w/v)
WSR N-12K NF	1,000,000	400–800 (at 2 % w/v)
WSR N-60K NF	2,000,000	2,000–4,000 (at 2 % w/v)
WSR 301 NF	4,000,000	1,650–5,500 (at 1 % w/v)
WSR coagulant NF	5,000,000	5,500–7,500 (at 1 % w/v)
WSR 303 NF	7,000,000	7,500–10,000 (at 1 % w/v)

formulations (Coppens et al. 2005; Bartholomaeus et al. 2012). POLYOX has precedence of use in pharmaceutical products with an IID (inactive ingredient database) limit of 543 mg per tablet for oral applications.

The general structure of PEO is $[-CH_2-CH_2-O-]_n$ where n is average number of oxyethylene groups (Braun 1980). PEO structure is similar to that of polyethylene glycol (PEG), but has longer molecular chains, i.e., higher molecular weights. The ethylene oxide monomer is an epoxide ring and in the presence of a catalyst it forms a chain having the repeating unit $-CH_2-CH_2-O-$ (Fig. 6.1).

The end group reactivity of these resins is almost negligible, due to their high molecular weights. They do possess some reactivity associated with the polyether backbone. The PEO resins are available from Dow Chemical Company in different molecular weight grades as shown in Table 6.1 along with their solution viscosities. Their typical physical and chemical properties are summarized in (Table 6.2).

As a polyether, POLYOX may undergo auto-oxidation and thus chain scission during storage, resulting in reduction of molecular weight and therefore, lower solution viscosity of the polymer. The degradation occurs in the presence of oxygen, and is catalyzed by UV light, metal ions, and heat. Butylated hydroxytoluene (BHT) is added (100–500 ppm) to all POLYOX WSR grades to reduce oxidative degradation. Table 6.3 shows suggested concentrations of common antioxidants used in formulations containing POLYOX. Oxidative degradation of high molecular weight grades of PEO on long-term storage can lead to formation of potentially reactive groups such as formate ions and therefore, the compatibility with sensitive drugs during the preformulation phase should be evaluated (Shamblin 2010).

Table 6.2 Typical physical and chemical properties of POLYOX WSRs

Physical and chemical properties	Values and comments
Appearance	Off-white powder, with specified limits for off-color particles
Bulk and tap densities (g/cm^3)	~0.47 and 0.51 Good for blending other excipients/materials
Typical particle size (μm)	~150 Similar to common direct compression excipients
Crystallinity (%)	~95 This gives a melting point for PEO and a low inherent moisture content (< 1 % at ambient conditions)
Melting point (°C)	~63–68 Low melting temperature makes PEO suited for hot-melt extrusion applications
Swellability in water	Polyethylene oxide swells ~7 times its initial dry state, independent of pH of solution
Solution pH	~8

6.2 Pharmaceutical Applications of POLYOX

POLYOX has been widely studied and used in pharmaceutical applications such as controlled release of drugs in osmotic pumps, hydrophilic matrices, mucoadhesion, gastroretentive systems, and abuse-resistance technologies. The use of PEO in such a wide range of applications is attributed to its physical, chemical as well as thermal stability, compressibility, hydrophilic nature, and high capacity to swell. In this chapter, specific pharmaceutical applications, where thermal behavior of POLYOX has been utilized in the design and manufacture of the dosage form, will be discussed.

6.3 Applications of POLYOX in HME

The choice of polymer used during formulation development for HME is critical, to facilitate processing in the extruder as well as to obtain the desired drug-release performance. Torque, melt pressure, and drive-motor amperage are used as indirect measures of melt viscosity. Torque is a measure of the mechanical work needed to move material through an extruder. Melt pressure is the force generated within the extruder as materials are compacted, melted, and forced through an aperture at the end of the extrusion system such as a die. Melt pressure measurements are typically taken at the end of the extruder barrel. Under a given set of processing conditions, higher viscosity materials result in higher values of torque, melt pressure, and drive-motor

Table 6.3 Commonly used antioxidants suitable for formulations containing POLYOX. (Rowe et al. 2013)

Antioxidant	Physical properties	Suggested use level (% w/w)	Known sensitivities	Comments
BHT	Crystalline powder with melting point of 70–73 °C	0.005–0.020	Light, heat, humidity; turns yellow	Micronized grades of BHT are most commonly used
BHA	Waxy solid with melting point of 48–55 °C	0.005–0.020	Light, some metals	Low melting point and thus may sublime
Alpha tocopherol (Vitamin E)	Available in liquid form	0.050–0.075	Light, heat, some metals	May be dissolved in ethanol and sprayed on to formulation

amperage (Ghebre-Selassie et al. 2003). These are important considerations to ensure that extrusion equipment do not exceed the maximum values of these attributes. Improper conditions may lead to degradation of the drug, excipient, or additives. The melt flow index of the polymers in pharmaceutical melt extrusion is fairly low thus having no significant effect on the die temperature itself (Coppens et al. 2005).

Despite its high molecular weight, POLYOX is highly crystalline and has a melting point around 65 °C, above which, the polymer becomes thermoplastic. PEO by virtue of having low melting point and good melt flow index may be considered as a suitable polymer for use in HME formulations. Addition of an appropriate plasticizer to higher molecular weight PEO resins may be recommended. PEO can be extruded at processing temperatures 20–30 °C higher than its melting point without significant degradation. Thermogravimetric analysis (TGA) data have indicated that PEO does not exhibit significant weight loss (degradation) until 350 °C (Coppens et al. 2005). At temperatures far above the crystalline melting point, high molecular weight polymers of POLYOX still retain a very high degree of crystalline character (http://dow.com). For this reason, low molecular weight PEO grades are more suitable for HME applications whereas high molecular weight grades provide a better sustained release dissolution profile for oral solid dosage forms. Crowley et al. (2002) reported that lower molecular weight PEO degraded more rapidly than higher molecular weight PEO. This study used a single-screw microtruder (Model RC 0750, Randcastle) equipped with a rod-shaped die. PEO degradation was determined in terms of average molecular weight using gel-permeation chromatography (GPC). The study also investigated the addition of low molecular weight PEO (100,000 Da) at concentrations of 10, 20, and 40 % to a high molecular weight PEO (1,000,000 Da) matrix of chlorpheniramine maleate formulations. The low molecular weight PEO acted as a processing aid for the higher molecular weight polymer, resulting in decreased extruder-drive amperage with the increase in its content in the blend. There was no significant difference in the release profile of chlorpheniramine maleate from the combination matrix as compared to the matrix containing only PEO 1,000,000 Da.

Table 6.4 Melt-flow properties of POLYOX resins

POLYOX resins	Melt flow at 190 °C (g/10 min)
WSR N-10	> 100
WSR N-80	25–35
WSR N-750	3–4
WSR N-3000	1–1.5
WSR N-301	< 0.5
WSR coagulant	< 0.5
WSR 303	< 0.5

Table 6.4 shows the melt-flow properties of POLYOX resins (http://dow.com). These data were collected at 190 °C, which represent a modification to the American Society for Testing and Materials (ASTM) method. The high molecular weight grades of PEO require plasticizer addition in order to enable melt extrusion at moderate temperatures.

It has also been noted that the melt viscosity of PEO resins decreased considerably with increasing shear rate which indicate their pseudoplastic character above their melting point (http://dow.com).

Since HME exposes polymers to elevated temperatures, thermal stability of the polymers is crucial. Critical variables such as melt viscosity in HME is a valuable preformulation tool in determining the HME processing temperatures to achieve thermally stable formulations. Meyer (2011) has reported the use of dynamic mechanical thermal analysis (DMTA) to determine the melt viscosity of PEO in the absence and presence of plasticizers and active pharmaceutical ingredient (API) in order to find the ideal processing conditions. DMTA records the temperature-dependent viscoelastic properties and determines the modulus of elasticity and the damping values by applying an oscillating force to the sample. It was reported that PEO reached the critical melt viscosity value of approximately 10 kPa s during HME at around 62 °C. The addition of a plasticizer, PEG at 10 and 20 % levels, decreased the temperature to achieve the critical viscosity to 49 and 45 °C, respectively. However, the addition of a model drug, theophylline led to an increase in temperature to 100 °C to achieve the critical viscosity for HME. Therefore, the theophylline seemed to have an antiplasticization effect during the HME of PEO.

6.4 Applications of POLYOX in Melt-Extruded Films

Thin films for transmuosal drug delivery are generally fabricated using organic or aqueous film casting techniques. The main disadvantages of these techniques are long processing times, use of organic solvents, and structural changes in the polymer leading to altered mechanical stability of the films (Ghebre-Selassie et al. 2003). The use of POLYOX in melt-extrusion applications has provided significant benefits in the area of thin films research due to the elimination of solvents and one step processing, leading to physically as well as mechanically stable films for oral mucosal use (Repka et al. 2002). A bioadhesive film has the benefit of simplifying dosage form design and reducing preparation costs, due to the elimination of the adhesive layer in the system. It is desirable for the film to have adequate adhesion strength so that retention of the film at the application site can be achieved.

Bioadhesive hot-melt-extruded films for oral drug delivery have been reported using a combination of HPC, POLYOX, and a plasticizer (Mooney 2000). The films adhere to the wet mucosal surfaces, providing a protective barrier for the tissue and deliver controlled release of drug (e.g., antibiotic) to the local area. In another study, a novel transmucosal hot-melt-extruded film comprising HPC, PEO, and drug in the absence of a plasticizer was studied (Repka et al. 2002). The plasticizer-free hot-melt-extruded films demonstrated better stability and increased efficiency of drug delivery to the patient. They characterized HME films of HPC or PEO, containing clotrimazole, a drug used for oral candidiasis therapy (Repka et al. 2003; Prodduturi et al. 2004, 2005). The film was processed at a temperature range of 125–130 °C utilizing a Killon extruder (ModelKLB-100) equipped with a 6-in. flex-lip die. The extruded films demonstrated good content uniformity with a postprocessing drug content of 93.3 %, and exhibited desirable and consistent release properties.

In another study, a topical film formulation of ketoconazole, an antifungal drug, was prepared by HME technology using PEO (Repka et al. 2004). The films demonstrated good postprocessing drug content and efficacy. Differential scanning calorimetry studies indicated that ketoconazole was in solid solution within hot-melt-extruded films and this phenomenon was postulated to be responsible for the increase in permeability and efficacy of the drug.

HME of oral film formulations of antiemetic drug; delta 9-tetrahydro cannabinol (THC) has been studied in order to overcome its poor solubility and avoid its first-pass metabolism. The buccal delivery of these types of drugs using oral films could enhance the performance of the drug; easy access, rapid action, and controlled release and therefore improved patient compliance. THC has been incorporated into transmucosal films using PEO and HPC by HME, utilizing various processing aids (Munjal et al. 2006a, b; Repka et al. 2006). The chemical stability of the drug in the polymeric films was investigated with respect to processing temperature, processing time, formulation additives, and storage conditions. Oral films comprising modified synthetic cyclodextrins to enhance the stability and bioavailability of a hemisuccinate ester prodrug of THC (THC-HS) in PEO hot-melt-extruded matrices were further investigated (Upadhye et al. 2010). Earlier preformulation studies have revealed that THC-HS is hydrolyzed to its parent compound, THC which then oxidized to cannabinol (oxidative degradation product of THC). Hence the effect of various classes of antioxidants on the stability of THC-HS was investigated. Hot-melt cast polymeric matrices of POLYOX-N80 were fabricated with various antioxidants. Two classes of antioxidants were studied alone or in combinations: (i) free radical scavengers such as BHT, (ii) reducing agents or oxygen scavengers such as ascorbic acid. The antioxidants incorporated into the melt-extruded matrices reduced the degradation of THC-HS. The combination of BHT + propyl gallate showed significant advantage ($p < 0.05$) with 76.35 and 71 % postprocessing content of THC-HS over the other antioxidants.

Crowley et al. (2004) investigated the physicochemical and mechanical properties of hot-melt-extruded POLYOX films containing either guaifenesin (30 %w/w) or ketoprofen (15 % w/w) as model drugs. Drug films were successfully prepared by a HME process and the results demonstrated no degradation of the drugs during the process. Crystallization of guaifenesin on the surface of the extruded films was reported at all concentrations; however, no crystallization of ketoprofen was observed at concentration up to 15 % w/w loading levels. Melting points corresponding to the crystalline drugs were not observed in the films, suggesting the miscibility of the drugs in the molten polymer. Both drugs decreased the drive load and plasticized the polymer during extrusion. The Hansen solubility parameters predicted miscibility between PEO and ketoprofen and poor miscibility between PEO and guaifenesin. Increasing concentrations of both drugs decreased the tensile strength of the extruded films (Crowley et al. 2004).

6.5 Melt Extrusion of POLYOX for Extended-Release Matrix Applications

Although there have been much interest in the development of extended-release (ER) formulations using melt extrusion of POLYOX in hydrophilic matrices, many such studies have not been published and are not in the public domain. Here, some examples of typical studies in the design and evaluation of the performance of these systems are shown. Zhang and McGinity 1999 described a novel method to prepare POLYOX sustained-release matrix tablets using HME directly from a single screw extruder. In this study, chlorpheniramine maleate was used as a model drug and the influence of PEO properties on drug release was investigated. PEG 3350 was included as a plasticizer to facilitate the extrusion processing and 4.5 mm diameter rods were extruded and cut into blocks as tablets. The stability of PEO as a function of processing temperature was determined using GPC. It was reported that polymer type, temperature, and residence time in the extruder impacted the PEO stability. It was also shown that additional mixing of the components occurred in the barrel of the extruder, since the content uniformity of the extruded tablets was within 99.0–101.0 % of the theoretical content. As the plasticizer (PEG 3350) concentration increased, drug release from the melt-extruded matrices increased. On the other hand, the rate of drug release was only slightly affected by changes in drug contents until the drug loading reached around 20 % w/w. Drug loading levels greater than 20 %w/w were not investigated because a solid cylindrical extrudate could not be obtained. It was postulated that the rate of hydration and dissolution of the entire matrix system accelerated due to the presence of the plasticizer and the soluble drug.

Similarly, Crowley et al. (2002) studied the thermal stability of POLYOX in sustained-release tablets prepared by HME, using GPC. The chemical stability of PEO was found to be dependent on both the storage and processing temperature, and the molecular weight of the polymer. Storage of the polymer above its melting point significantly increased polymer degradation. The thermal stability of high molecular weight PEO (1,000,000) in sustained-release chlorpheniramine maleate tablets prepared by HME was found to depend on the processing temperature and screw speed. Lower molecular weight PEO (100,000) was demonstrated to be a suitable processing aid for higher molecular weight PEO (1,000,000). It was described that incorporation of PEO 100,000 reduced the melt viscosity, friction, and chain entanglements between the larger PEO molecules thereby lowering its degradation. As the percentage of PEO 100,000 in the powder blend increased, the drive amperage decreased and the stability of PEO increased. Incorporation of PEO 100,000 at the levels of 10, 20, and 40 % did not alter the drug-release rate. Vitamin E, Vitamin E Succinate, and Vitamin E TPGS (d-alpha tocopheryl polyethylene glycol 1000 succinate) were found as suitable stabilizers for PEO, however, ascorbic acid was shown to degrade the polymer in solution. Thermal analysis demonstrated that Vitamin E Succinate and Vitamin E TPGS were dispersed at the molecular level in hot-melt-extruded tablets (Crowley et al. 2002), and suppressed the melting point of the POLYOX.

6.6 Applications of POLYOX in Abuse-Resistant Matrix Formulations

Research from patients on substance abuse treatment programs, has shown the progression of abuse from oral route of administration to inhalation and finally to intravenous use of drugs (Butler et al. 2010). Significant efforts have been made in the pharmaceutical industry to develop abuse-resistant and abuse-deterrent technologies to prevent the misuse of substances meant for medicinal applications. Abuse-resistant formulations decrease the likelihood of the abuser being able to extract and manipulate the drug within a formulation (Bartholomaeus et al. 2012).

ER opioid tablets have played an important role in chronic pain management therapy by enhancing the patient care and compliance through timed dosing of opioid analgesics. Unfortunately, opioids have been extracted and abused because of their positive subjective effects such as euphoria (Substance Abuse and Mental Health Services 2011). ER opioids seem to be particularly popular among the abusers as they contain higher amounts of the drug than their immediate-release dosage formulations.

Various strategies have been used as abuse deterrence/resistance for opioids (Bartholomaeus et al. 2012). For example, Embeda® is a morphine sulfate extended-release formulation containing naltrexone, which is a sequestered opioid antagonist. If the product is crushed, the naltrexone is released, neutralizing the opioid effects of morphine sulfate. In another strategy, an immediate-release formulation of oxycodone (Oxecta®), contains nonsequestered aversive excipients that cause mucosal irritation if abused via inhalation and exhibit gelling properties that deter injection. In other technologies such as INTAC®, the softening and melting properties of POLYOX, coupled with its high molecular weight have been used in formulation of abuse-resistant oral dosage forms. The high molecular weight of PEO controls the release of drug and imparts superior crush resistance to the formulation (Fig. 6.2). Due to its low glass-transition temperature (Tg), PEO melts with the drug and extraction of drug from the molten phase becomes difficult. A formulation containing PEO due to its gel-forming properties, when subjected to solvent extraction by abusers leads to a viscous solution, impedes its syringeabilty and injectability as shown in Fig. 6.2 (Bartholomaeus et al. 2012).

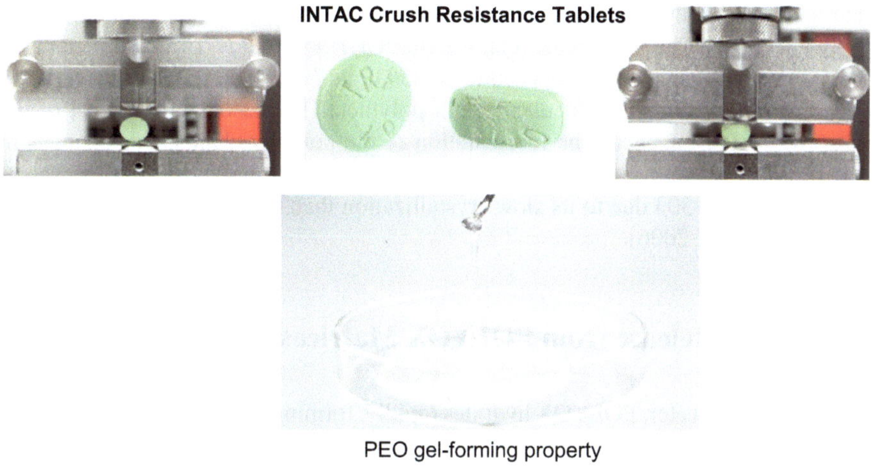

Fig. 6.2 Illustration of INTAC® technology's resistance to tablet crushing and high viscosity gels due to POLYOX properties (Used with Permission, Grünenthal GmbH, copyright 2013)

Abuse-resistant extended-release formulations of tramadol hydrochloride have been developed using POLYOX as the primary matrix former (Bartholomaeus et al. 2012). Tablets were manufactured by compressing mixtures of drug with PEO in a molten state followed by solidification on cooling. A novel molten technique was used to enhance the interactions between PEO and the drug with improved mechanical strength of the tablets. In this technology, the upper and lower punch toolings as well as the die of the tablet press are heated. The most feasible and fastest approach to manufacture abuse-resistance tablets may involve HME combined with downstream processing equipment. Therefore, a manufacturing process may consist of: blending, melt extrusion, cooling, cutting, forming, and film coating. Consistent drug release with increasing concentrations of 5–40 mg of oxymorphone hydrochloride has been demonstrated with excellent stability data (Bartholomaeus et al. 2012).

The choice of an appropriate grade of POLYOX for abuse-resistant formulations is important as the release of the active as well as crush resistance properties might be affected due to the plasticization of the polymer during its shelf-life of the formulation. This physical aging may result in volume and enthalpy relaxation causing structural changes in the polymeric backbone (Kiss et al. 2006). Techniques such as positron annihilation lifetime spectroscopy (PALS), differential scanning calorimetry, and scanning electron microscopy have been used to study the physical aging of POLYOX (Kiss et al. 2006). PALS is a technique used to determine the size distribution of the free volume in polymer. The measurement is based on the interaction of the free volume with the orthopositronium atoms. These atoms are present in the free volume of the polymer and the size of the free volume around them determines their lifetime. Two grades of POLYOX; WSR N-12K (MW 1,000,000) and WSR 303 (MW 7,000,000) were investigated at 40 °C and 75 % Relative humidity (RH) for a period of 4 weeks and a decrease in the orthopositronium lifetime values and an increase in the melting enthalpies as a function of storage time indicated a reorientation of the polymer chains. The lower molecular weight POLYOX WSR N-12K exhibited faster crystallization, which was proposed to be due to its faster water-absorbing capacity as compared to higher molecular weight polymer. The plasticization effect of the absorbed water enhanced the reorientation of the polymer chains. The changes in the orthopositronium lifetime were more pronounced for higher molecular weight POLYOX WSR 303 due to its slow crystallization than the lower molecular weight PEO (Kiss et al. 2006).

6.7 Drug Release from POLYOX Matrices

In contact with water, POLYOX hydrates rapidly, forming a gel-like structure around the dosage form. The polyether chains of POLYOX form strong hydrogen bonds with water, structuring the water around the polymer and facilitating polymer–polymer interactions. The strength of the gel structure depends on the polymer molecular weight and the presence of other additives. Higher molecular weight and high concentration of polymer produce stronger and more viscous hydrated gel structures. Therefore, drug release from matrices (irrespective of manufacturing methods) may be controlled by polymer swelling, diffusion in the hydrated gel, erosion, or by all these processes together (Verhoeven et al. 2009; Tiwari et al. 2011). A variety of release profiles may be obtained, depending on the POLYOX molecular weight and the physicochemical properties of the drug. The pH of the dissolution medium generally does not affect the drug release regardless of the molecular weight of the PEO, but may be affected by solubility of the drug in different pH media.

Mucoadhesion properties of POLYOX have been explored for the design of various drug delivery devices. For example, POLYOX has been used in the design of buccal delivery, ensuring intimate contact with the mucosa for an adequate time interval, providing desired release rates. Prior to the drug passing through the mucosal barrier and reaching blood stream, it needs to dissolve in the medium. This is critical for lipophilic drugs that have good absorption, but exhibit a slow dissolution rate in aqueous media. Cappello et al. (2006) employed hydroxypropyl-β-cyclodextrin in POLYOX matrices for the buccal delivery of the poorly soluble drug carvedilol. It was shown that enhancement of drug dissolution in the hydrated tablet was more important than higher solubility in solution for increasing the delivery rate. Cyclodextrins caused an increase in the erosion rate of the tablet and also improved the dissolution of the drug inside the hydrated polymeric matrix, leading to faster release and higher permeability through the mucosal membrane.

6.8 Ethylcellulose (EC)—Properties and Applications in HME

Ethylcellulose (EC) is a cellulose ether prepared by treating cellulose with an alkaline solution to produce alkali-cellulose which is subsequently reacted with ethyl chloride. Ethylcelulose (Fig. 6.3) is commercially available from various manufacturers and ETHOCEL^{TM} is the brand name used by Dow Chemical Company. It is a

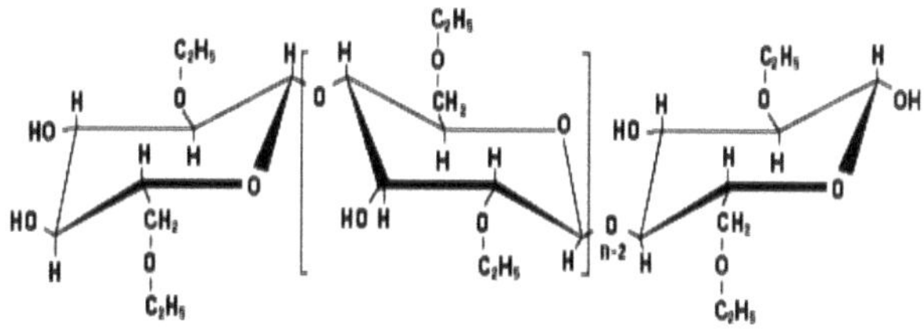

Fig. 6.3 Chemical structure of ethylcellulose

Table 6.5 Common ETHOCEL grades used in pharmaceutical applications (ethoxyl content 48.0–49.5 %)

Product ranges	Viscosity range[a] mPa s (cP)
ETHOCEL Std. 4	3–5.5
ETHOCEL Std. 7	6–8
ETHOCEL Std. 10	9–11
ETHOCEL Std. 20	18–22
ETHOCEL Std. 45	41–49
ETHOCEL Std. 100	90–110

[a] Viscosities are for a 5 % solution measured at 25 °C in an Ubbelohde viscometer and the solvent is toluene:ethanol at 6:4 or 8:2 ratios

partially substituted ethyl ether of cellulose, which is insoluble in water, but remains hydrophilic in nature due to the residual free hydroxyl groups on the cellulose backbone with a degree of substitution from 2.2 to 2.6 out of a maximum of 3 as shown in Fig. 6.3 (ETHOCEL® Ethylcellulose Polymers Technical Handbook 1998; Rekhi and Jambhekar 1995)

ETHOCEL is used in pharmaceutical applications for microencapsulation of drugs or coating for multiparticulates for extended release and taste masking, inert matrices and tablet binding. ETHOCEL is available in various viscosity grades as shown in Table 6.5, has a Tg of 129–133 °C and a crystalline melting point of 180 °C. EC is a good candidate for HME because it exhibits thermoplastic behavior at temperatures above its Tg and below the temperature at which it exhibits degradation (250 °C) (Coppens et al. 2005). Typically, plasticizers are added to reduce Tg of EC and permit processing at lower temperatures. Few drugs may act as plasticizers to polymers and eliminate the need for addition of plasticizers for melt-extrusion. For example, the extruded films of ibuprofen with EC had a lower Tg than films containing the polymer alone (De Brabander et al. 2002). The ibuprofen films were made on a laboratory-scale corotating twin screw extruder. An increase in the amount of ibuprofen from 5 to 20 %w/w lowered the Tg of the blend from a range of 106–109 to 53–66 °C. A complete miscibility appeared between ibuprofen and EC as was indicated by a single Tg of the extruded films. Ibuprofen release from EC-extruded materials was modified by adding hydrophilic polymers like hypromellose or xanthan gum (De Brabander et al. 2004).

Crowley et al. (2004) investigated the physicochemical properties and drug-release mechanism from EC matrix tablets prepared by direct compression or HME of binary mixtures of a water-soluble drug, guaifenesin and the polymer. EC was separated into fine or coarse particle size fractions directly compressed with 30 %

w/w guaifenesin at 10, 30 or 50 KN compaction forces. The extruded tablets were processed at temperatures of 80–90 and 90–110 °C. The results of this study demonstrated that the guaifenesin release rate was slower when fine particles were used. Tablets prepared by HME exhibited considerably slower drug release relative to those prepared by direct compression. The surface morphology of the hot-melt-extruded tablets was found to depend upon processing temperature. The release profiles of hot-melt-extruded tablets were found to be in good agreement with the Higuchi diffusion model while those prepared by direct compression using coarse particles were found to release guaifenesin by both diffusion and erosion mechanisms.

The use of EC in injection molding technology has also been studied, where the core contained drug, hypromellose, and glycerides and was surrounded by a hot-melt extruded EC membrane (Mehuys et al. 2004a, b). The length and not the diameter of the melt-extruded device had a profound effect on the drug-release rate. It was postulated that drug release was through the erosion of the core from the open ends of the device and the EC melt-extruded layer provided a barrier membrane for water ingress and drug release.

The rheological and thermal characteristics of EC intended for HME formulations have been investigated by Maru et al. (2011). The influence of zidovudine, lamivudine (first-line antiretroviral drugs), and plasticizers on the properties of EC and their suitability for HME to produce extended-release matrices were evaluated. Plasticizers such as triethylcitrate and PEG-6000 lowered the melt viscosity of the formulation. In addition, the drugs were reported to have plasticizing effects too. In another study, the utilization of EC as a sustained-release agent in metoprolol tartrate extended-release mini-tab matrices using HME has been investigated (Verhoeven et al. 2009). PEO was added to the formulation to increase drug-release rate. Increasing the hydrophilic polymer concentration in the matrix yielded faster drug release irrespective of polymer molecular weight. The extrudates were smooth at 70 °C for PEO formulations with metoprolol tartrate homogeneously distributed in the mini-matrices. This study illustrated that mini-matrices of EC and POLYOX may be used for extended-release formulations with a wide range of drug-release profiles.

6.9 Conclusions

PEO, by virtue of its thermoplastic behavior, may be a suitable polymer for pharmaceutical dosage form using HME technology. Applications of high molecular weight POLYOX allows formulation of extended-release melt-extruded matrices. Combination of different grades of POLYOX and other polymers enables formulators to tailor the release profile of drugs, as well as enhance melt-extrusion processing. The use of PEO in melt-extruded bioadhesive thin film applications has provided the benefits of processing ease, solvent-free manufacturing operations, and unique applications such as buccal dosage forms. One of the most significant applications of PEO is for abuse-resistant dosage forms for drugs such as opioid analgesic. PEO affords unique benefits of being thermoplastic with high swelling capacity and crush resistance, preventing drug abuse through inhalation or injection.

EC is well known for its barrier membrane coatings for extended-release applications. Due to its thermal properties, it has been successfully used in HME and injection molding.

The recent advancements in extrusion engineering and fundamental understanding of materials properties are paving the way for HME technology to have a significant contribution and benefit in the development of solid dosage forms. This technology may enable formulation and manufacturing of one or more challenging drugs for extended release and/or abuse-resistance dosage forms.

References

Bartholomaeus JH, Arkenau-Marić E, Galia E (2012) Opioid extended-release tablets with improved tamper-resistant properties. Expert Opin Drug Deliv 9(8):879–891

Braun DB (1980) Poly (ethylene oxide). In: Davidson RL (ed) Handbook of water-soluble gums and resins. McGraw-Hill, New York

Butler SF, Fernandez KC, Chang A, Benoit C, Morey LC, Black R, Katz N (2010) Measuring attractiveness for abuse of prescription opioids. Pain Med 11(1):67–80

Cappello B, De Rosa G, Giannini L, Rotonda M, Mensitieri G, Miro A, Quaglia F, Russo R (2006) Cyclodextrin-containing poly(ethyleneoxide) tablets for the delivery of poorly soluble drugs: potential as buccal delivery system. Int J Pharm 319:63–70

Coppens KA, Hall MJ, Mitchell SA, Read MD (2005) Hypromellose, ethylcellulose, and polyethylene oxide use in hot melt extrusion. Pharm Technol ogy pp 1—4

Crowley MM, Zhang F, Koleng JJ, McGinity JW (2002) Stability of polyethylene oxide in matrix tablets prepared by hot-melt extrusion. Biomaterials 23(21):4241–4248

Crowley MM, Fredersdorf A, Schroeder B, Kucera S, Prodduturi S, Repka MA, McGinity JW (2004) The influence of guaifenesin and ketoprofen on the properties of hot-melt extruded polyethylene oxide films. Eur J Pharm Sci 22(5):409–418

De Brabander C, Van Den Mooter G, Vervaet C, Remon JP (2002) Characterization of ibuprofen as a nontraditional plasticizer of ethyl cellulose. J Pharm Sci 91(7):1678–1685

De Brabander C, Vervaet C, Görtz JP, Remon JP, Berlo JA (2004) Bioavailability of ibuprofen from hot-melt extruded mini-matrices. Int J Pharm 271(1–2):77–84

ETHOCEL® Ethylcellulose Polymers Technical Handbook (1998), Dow Chemical Company. http://dow.custhelp.com

Ghebre-Selassie I, Charles M (eds) (2003) Pharmaceutical extrusion technology. Marcel Dekker, New York

Kiss D, Süvegh K, Marek T, Dévényi L, Novák C, Zelkó R (2006) Tracking the physical aging of Poly(ethylene oxide): a technical note. AAPS PharmSciTech 7(4):E1–E4

Maru SM, de Matas M, Kelly A, Paradkar A (2011) Characterization of thermal and rheological properties of zidovudine, lamivudine and plasticizer blends with ethyl cellulose to assess their suitability for hot melt extrusion. Eur J Pharm Sci 44(4):471–8

Mehuys E, Vervaet C, Gielen I, Van Bree H, Remon JP (2004a) In-vitro and in-vivo evaluation of a matrix-in-cylinder system for sustained drug delivery. J Control Release 96(2):261–71

Mehuys E, Vervaet C, Remon JP (2004b) Hot-melt extruded ethylcellulose cylinders containing a HPMC-Gelucire core for sustained drug delivery. J Control Release 94(2–3):273–80

Meyer F (2011) Dynamic Mechanical Thermal Analysis (DMTA) on Polyethylene Oxide (PEO) for evaluation of ideal processing conditions for pharmaceutical hot melt extrusion (HME) applications. APV experts workshop on hot melt extrusion and its use in the manufacturing of pharmaceutical dosage forms. Tarrytown, New York

Mooney MT, Schiraldi MT (2000) Extrudable compositions for topical or transdermal drug delivery. US Patent No. 6,072,100

Munjal M, Elsohly MA, Repka MA (2006a) Polymeric systems for Amorphous D9-tetrahydrocannabinol produced by a hot-melt method. Part I: chemical and thermal stability during processing. J Pharm Sci 95(8):1841–1853
Munjal M, Elsohly MA, Repka MA (2006b) Polymeric systems for amorphous D9-tetrahydrocannabinol produced by a hot-melt method. part II: effect of oxidation mechanisms and chemical interactions on stability. J Pharm Sci 95(11):2473–2485
Prodduturi S, Manek RV, Kolling WM, Stodghill SP, Repka MA (2004) Water vapor sorption isotherms of hot-melt extruded hydroxypropyl cellulose films: effect on physico-mechanical properties, release characteristics and stability. J Pharm Sci 93(12):3047–3056
Prodduturi S, Manek RV, Kolling WM, Stodghill SP, Repka MA (2005) Solid-state stability and characterization of hot-melt extruded poly(ethylene oxide) films. J Pharm Sci 94(10):2232–45
Rekhi G, Jambhekar S (1995) Ethylcellulose—a polymer review. Drug Dev Ind Pharm 21(1):61–77
Repka MA (2002) Bioadhesive hot-melt extruded film for topical and mucosal adhesion applications and drug delivery and process for preparation thereof US 6,375,963B1
Repka MA, Prodduturi S, Stodghill SP (2003) Production and characterization of hot-melt extruded films containing clotrimazole. Drug Dev Ind Pharm 29(7):757–765
Repka MA, Mididoddi PK, Stodghill SP (2004) Influence of 'Etching' on human nails for the assessment of treatment modalities for onychomycosis. Int J Pharm 283(1–2):95–106
Repka MA, ElSohly MA, Munjal M, Ross SA (2006) Temperature stability and bioadhesive properties of D9-tetrahydrocannabinol incorporated hydroxypropyl cellulose polymer matrix systems. Drug Dev Ind Pharm 32(1):21–32
Rowe RC, Sheskey PJ, Quinn ME (Eds) (2013) Handbook of pharmaceutical excipients. Jobson Publication, New York
Shamblin S (2010) Controlled release using bilayer osmotic tablet technology: reducing theory to practice. In: Wen HKP (ed) Oral controlled release formulation design and drug delivery: theory to practice. Wiley, Hoboken
Substance Abuse and Mental Health Services Administration (2011) Results from the 2010 National Survey on Drug Use and Health: summary of national findings. NSDUH Series H-41, HHS Publication No. SMA 11-4658. Rockville, MD: Substance Abuse and Mental Health Services Administration, 2011. http://www.samhsa.gov/data/NSDUH/2k10NSDUH/2k10Results.htm
Tiwari SB, DiNunzio J, Rajabi-Siahboomi AR (2011) Drug–polymer matrices for extended release. In: Wilson CG, Crowley PJ (eds) Controlled release in oral drug delivery. Springer, New York
Upadhye SB, Kulkarni SJ, Majumdar S, Avery MA, Gul W, ElSohly MA, Repka MA (2010) Preparation and characterization of inclusion complexes of a hemisuccinate ester prodrug of delta 9-tetrahydrocannabinol with modified beta-cyclodextrins. AAPS PharmSciTech 11(2):509–517
Verhoeven E, De Beer TR, Schacht E, Van den Mooter G, Remon JP, Vervaet C (2009) Influence of polyethylene glycol/polyethylene oxide on the release characteristics of sustained-release ethylcellulose mini-matrices produced by hot-melt extrusion: in-vitro and in-vivo evaluations. Eur J Pharm Biopharm 72(2):463–70
Zhang F, McGinity JW (1999) Properties of sustained-release tablets prepared by hot-melt extrusion. Pharm Dev Technol 4(2):241–50

Part III
Formulation Design

Chapter 7
Formulation Development of Amorphous Solid Dispersions Prepared by Melt Extrusion

James C. DiNunzio and Dave A. Miller

Abstract Amorphous systems have been applied effectively in the pharmaceutical industry for a number of commercial and developmental products, although they are still considered a choice of last resort to enable therapy because of the metastable nature of the drug product. Of the technologies for preparing amorphous dispersions, melt extrusion is considered a highly effective and cost-efficient platform that is the primary technology for many major pharmaceutical companies. Successful development of melt-extruded amorphous dispersions requires strong understanding of formulation and process to produce a system having the necessary product attributes. As a result of the complexity associated with formulation research, a structured approach for amorphous formulation design is necessary to ensure that major development criteria are satisfied. This chapter discusses the fundamental aspects for formulation development of melt-extruded systems, the interplay of formulation with manufacturing process, and a structured design approach to turn molecules into medicines using melt extrusion.

7.1 Introduction

Over the past two decades, since the advent of high-throughput screening (HTS) methodologies in drug discovery, the percentage of poorly water-soluble (PWS) drug candidates in development has continuously increased (Macarron 2006). Furthermore, molecules emerging from discovery groups of late are deviating even further from conventional drug-like properties as the boundaries of chemical space are continually expanded in search of new and better drugs. On the basis of literature reports and the authors' experience, the percentage of PWS candidates in contemporary pipelines can range from 40 to 90 %, depending on the therapeutic area (Hauss 2007). This phenomenon is attributable to HTS methods that elegantly optimize candidates toward receptor binding, yet are essentially nondiscriminating with regard to

J. C. DiNunzio (✉)
Merck & Co., Inc., Summit, NJ, USA
e-mail: james.dinunzio@merck.com

Hoffmann-La Roche, Inc., Nutley, NJ, USA

D. A. Miller
DisperSol Technologies, LLC., Austin, TX, USA

M. A. Repka et al. (eds.), *Melt Extrusion*,
AAPS Advances in the Pharmaceutical Sciences Series 9,
DOI 10.1007/978-1-4614-8432-5_7, © American Association of Pharmaceutical Scientists 2013

the molecular properties crucial to drug delivery (Curatolo 1998). Consequently, a trend toward greater molecular weight and lipophilicity, two properties significantly influencing solubility, has been identified over the last few decades (Lipinski 2000).

Dissolution of a drug substance is necessary to achieve systemic exposure and binding to the biological target. Poor water solubility retards dissolution, thereby limiting the concentration of the active compound at the target site, often to the extent that the therapeutic effect cannot be elicited. Increasing dose can be an effective strategy for improving exposures; however, increasing dose with PWS drugs leads to nonlinear increases or plateauing exposures. Ultimately, absorption can be highly erratic, compromising the safety and efficacy profile. Furthermore, in the case of high-dose compounds, further increase in dose may not be practical from a patient convenience perspective. In many cases, solubility enhancement technologies must be employed to improve or enable therapies with PWS drugs.

There are several techniques available for enhancing the solubility of PWS drugs, and the appropriate formulation strategy is selected primarily according to the properties of the drug molecule. These techniques principally include solid-state approaches, such as the formation of salts, metastable polymorphs, or cocrystals (Hughey and Williams 2012); particle size reduction (Morales et al. 2012); solubilization in partially or nonaqueous vehicles (Zhang and DiNunzio 2012); and formulation as crystalline or amorphous solid dispersion (ASD) systems (Leuner and Dressman 2000). Solid-state modifications are often not possible owing to molecular properties that are unfavorable for salt or cocrystal formation. Metastable polymorphs are rarely viable because of inherent physical instability. As is more commonly the case nowadays, the aqueous solubility of crystalline drug forms is so limited that target exposures cannot be achieved with even the most advanced particle size reduction or crystalline solid dispersion techniques. In the same direction, modern drug candidates often exhibit poor solubility in suitable nonaqueous vehicles, limiting the dose that can be loaded into solubilized formulations such as soft gelatin capsules.

It is estimated that 10–30 % of leads identified by HTS have solubility issues related to crystal lattice stability (Horspool and Lipinski 2003). In early in vitro screening, leads are typically solubilized in solvents such as dimethyl sulfoxide (DMSO), and hence solubility issues associated with crystalline forms are not initially realized. Such a problem may not be identified until later in development when the synthesis process is scaled up and the bulk crystalline form is isolated (Horspool and Lipinski 2003).

Following from this discussion, ASD technology is increasingly utilized to enable the delivery of today's challenging molecules. Owing to increased free energy and specific surface area, ASD systems offer the capacity for supersaturation; providing dissolved drug concentrations greatly exceeding that of the crystalline form. Further, through judicious carrier selection, stability can be imparted to the high energy system in both the solid and liquid solution states. Hence, by proper design, ASD systems can be made physically stable for relevant time periods, and supersaturation generated by these systems can be maintained for the duration of intestinal transit. The result is a stable drug product that provides vast improvements in systemic exposure, thereby enabling or greatly enhancing a drug therapy.

Chiou and Riegelman (1971) have defined solid dispersions to be: "the dispersion of one or more active ingredients in an inert carrier or matrix at solid state prepared by melting (fusion), solvent, or melting-solvent method." Following from this definition and ignoring the morphology of the carrier, an ASD is one in which the active ingredient is dispersed in the carrier matrix in a substantially noncrystalline form. This could include single-phase systems (solid solutions) in which the drug is mixed at a molecular level with the carrier, two-phase systems in which the drug and polymer exist as separate amorphous domains, or a mixture of the two. There are inherent stability and performance benefits to a single-phase ASD system, and hence it should be the formulation scientist's goal to design such a system.

Initially, solvent-based methods were most preferred for the production of ASD systems owing to the thermal stresses and viscosity limitations of the batch melt processes. However, with the advent of pharmaceutical melt extrusion, a renaissance occurred with the use of the fusion method for the production of ASDs (Leuner and Dressman 2000). Screw extrusion provided, for the first time, a means of briefly processing pharmaceutical materials at elevated temperatures to limit thermal stress, and a process by which viscous molten materials could be thoroughly mixed. These attributes coupled with the numerous benefits of eliminating solvents have made hot-melt extrusion (HME) the preferred method of manufacture for ASD systems.

In this chapter, the application of HME to the design and production of ASD systems is discussed in depth. First, a structured approach for the design of amorphous systems using melt extrusion will be discussed. The chapter then delves into the key formulation considerations one must take into account when developing melt-extruded ASD formulations. Finally, process development and product manufacturing at each stage of drug development is discussed in detail. It is the goal of the authors to provide the reader with the in-depth knowledge needed to develop melt-extruded ASD systems and develop the process at each stage of manufacturing.

7.2 Structured Design for Amorphous Systems Using Melt Extrusion

Although amorphous formulations have been increasingly relied on by pharmaceutical companies to provide the required exposure, this approach still presents unique challenges due to the metastable nature of the drug product produced (Serajuddin 1999; Brouwers et al. 2009). Conventional approaches, which rely on particle size reduction (Morales et al. 2012) or solubilization (Davis and Brewster 2004; Gao 2012), that provide physically stable dosage forms still remain the preferred choice by both early-stage and commercial groups within many organizations because of the lower risk associated with these approaches as well as the preexisting infrastructure to support drug product manufacture. Although the view and capabilities of many commercial organizations are changing within Large Pharma to more positively position melt extrusion, many mid- and small-size companies still lack the infrastructure and technical development expertise. Such preexisting biases for the adoption of amorphous formulations and melt extrusion require sound justification for the technological approach and a comprehensive development strategy to ensure success.

Drug delivery limitations associated with solubility of a compound will be detected early in development, oftentimes being flagged at the discovery stage on the basis of low solubility in biorelevant media or limited oral exposure in preclinical models. Similarly, physical and chemical liabilities associated with the molecule, such as limited solubility in volatile organic solvents or insufficient thermal stability, may also be characterized, aiding in the choice of manufacturing technology. The strategy leading to the selection of an amorphous formulation to address solubility limitations and the choice of melt extrusion as the preferred manufacturing process are made within the space of a structured design approach. This approach is based on the input from multiple functional groups within the organization and an extensive R & D plan to appropriately characterize the critical product attributes of the drug product being developed. The road map for the structured design of ASDs prepared by melt extrusion focuses on the development of systems capable of providing appropriate manufacturability, oral bioavailability, and stability. Shown in Fig. 7.1, this modular approach utilizes the basic properties of drug and candidate excipients to identify lead systems and initial processing conditions leading to a systematic design of prototype formulations to assess key product attributes. By applying this approach, it is possible to quickly identify robust formulations to support first-in-man (FIM) studies, structuring experiments to provide data on the key performance attributes of the system. Information gained during this approach is also critical for future development, providing a baseline understanding of compositional interactions and the impact of critical process parameters.

Prior to the production of prototype dispersions, key data will have been generated to clearly establish the need for solubility enhancement and justification of the use of an amorphous concept. These data may be provided from early chemical analysis or preclinical exposures in relevant animal models. Similarly, multiple conventional formulations may have already been tested and shown limited improvement, thus helping to justify the use of an amorphous formulation. Further supporting this will be the characterization of specific molecular properties of the new chemical entity. The attributes characterized will provide insight into the viability of amorphous formulation, as well as necessary information regarding the stability of the system under proposed processing conditions. As needed, thermodynamic modeling may also be applied to assess the performance of prototype systems in silico to aid in the selection of systems for prototype manufacturing.

Application of accurate early characterization aids in the selection of target prototype formulations. After these systems have been identified and initial processing conditions selected, prototype manufacturing will be conducted using laboratory-scale equipment. This equipment is typically designed to support batch sizes less than 20 g (preferably ~5 g) and is capable of simulating temperature and shear aspects observed during melt extrusion on larger pilot- and production-scale units (DiNunzio et al. 2012). These systems also provide critical information regarding melt temperature, motor load, melt pressure, and feed rate during the operation

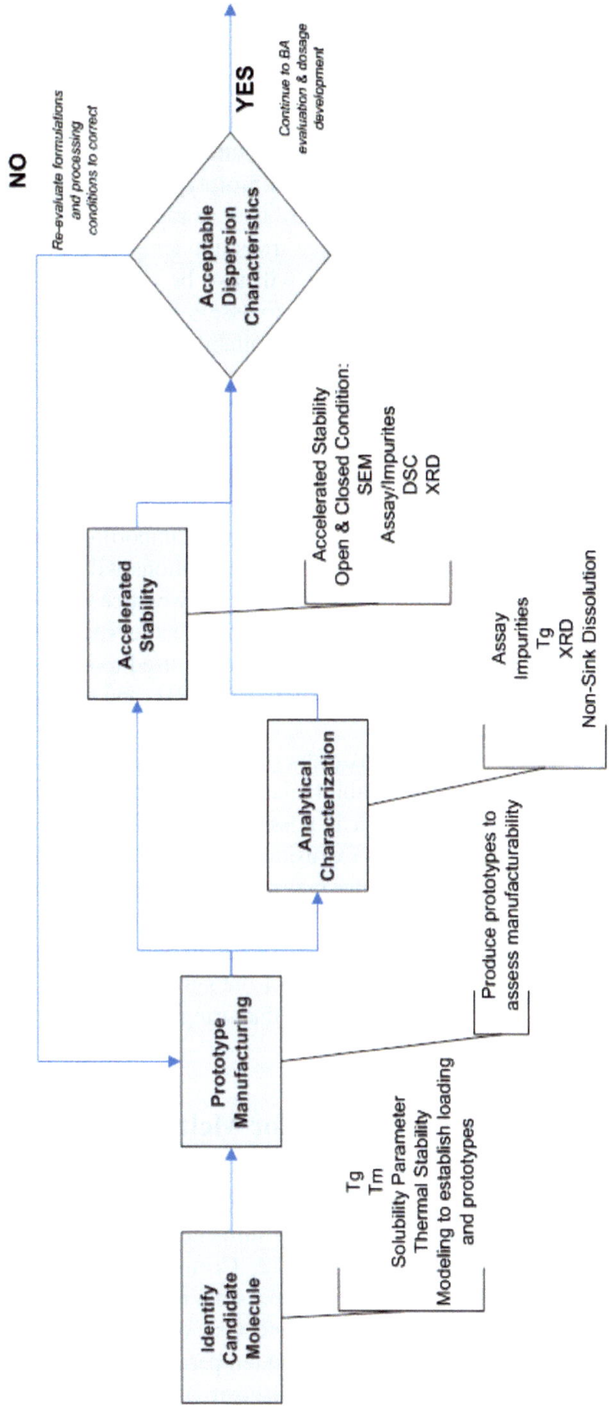

Fig. 7.1 Road map for structured design of melt-extruded formulations in early development. (Reproduced with permission from DiNunzio et al. 2012)

allowing for simultaneous evaluation of manufacturability. At-line assessment can easily be made on the basis of the appearance of the extrudate, noting that homogeneous ASDs are generally transparent glasses, to modify processing conditions such as screw speed, feed rate, and barrel temperature to achieve an amorphous form. In many cases, in-line or at-line monitoring of product attributes can also be conducted to support process modification during manufacture of prototypes.

Following the manufacture of amorphous prototypes, the formulations will be milled and tested for physicochemical attributes to rank order performance. The attributes include potency, impurities, glass transition temperature, amorphous content, and moisture content. Formulations will also be rank ordered for solubility improvement using in vitro methodologies to assess free drug in solution or total drug in solution. Early evaluation of oral bioavailability in preclinical models is also recommended at this stage to provide a more accurate assessment of formulation performance and validation of the in vitro testing approach.

Stability studies will be coupled to the initial prototype manufacturing, with samples set down on stability immediately after manufacturing using a range of conditions. Storage conditions will be selected in such a way that they provide a rapid rank order of stability performance while also supporting evaluation of conventional conditions. In general, these storage conditions will span a range from 40 °C/75 %RH open container storage to simulate the worst-case regulatory requirements to room temperature and refrigerated conditions as specified in regulatory guidances. Other more aggressive conditions may be used to stress samples, evaluating the impact of elevated moisture (40 °C/100 %RH) and elevated temperature (50, 60, 80 °C) on product stability.

From the cumulative results, a rank order of test formulations based on manufacturability, stability, and oral bioavailability can be established. Using this structured approach to develop formulations, it is possible to rapidly identify target formulations and their corresponding melt extrusion processing conditions to achieve the desired product attributes. The sequence may also be repeated as necessary to address deficiencies in preferred formulations through the incorporation of other functional additives. Selection of the final formulation can then be made based on a body of data covering the key performance metrics of the system, ensuring robust product performance prior to scale-up and FIM manufacturing.

7.3 Formulation Considerations for Melt Extrusion

During melt extrusion, a strong interplay between the formulation and process is observed, such that both factors significantly influence the final product attributes, as well as one another (DiNunzio et al. 2012; Crowley et al. 2007; Repka et al. 2007; Schenck et al. 2011). Thus, it is not possible to truly decouple formulation considerations from process requirements and vice versa. For example, melt viscosity is a strong function of composition and temperature for most pharmaceutical systems. Increasing temperature to reduce viscosity is bounded by the upper limit of the stability for the composition. Changes to the composition to reduce viscosity are possible, for example, the addition of a plasticizer; however, such a change impacts the molecular mobility of the system and may compromise key attributes of the dispersion. Proper design of composition and process is predicated on the development of a fundamental understanding of the individual materials and their related interactions. This understanding is achieved through a combination of basic material characterization, small-scale experimentation, and in silico modeling, which can lead to rational design selection supporting a structured development approach.

7.3.1 Drug Substance

Detailed characterization of the pharmaceutical drug substance is conducted to support Chemistry, Manufacturing and Controls (CMC) development and regulatory filings for FIM studies and commercial approval. This information includes a large number of experimentally determined and derived values, encompassing a range of physicochemical attributes. Many of these properties identify the need for solubility enhancement and the potential applicability of melt extrusion in comparison with other amorphous manufacturing technologies. A list of major properties related to amorphous formulation development is provided in Table 7.1. Many of the experimentally determined values can be assessed using compendial methodologies. Derived values are calculated from information based on molecular structure assessment or calculation using experimentally determined values. These properties are not an "all-encompassing" list but rather those that are important for quickly assessing the viability of amorphous formulation design and suitability of melt extrusion for production.

Table 7.1 Relevant properties of drug substance for amorphous formulation development, with key properties related to amorphous-form viability and production technology selection presented in italics

Molecular	Physical	Biopharmaceutical
Molecular formula	Number of polymorphs	BCS classification
Molecular weight	*Melting temperature of polymorphs*	*Biorelevant solubility*
H-bond donor groups	Glass transition temperature	pH solubility
H-bond acceptor groups	T_m/T_g	*Log P*
Solubility parameter	*Degradation onset temperature*	Permeability
Chiral centers	Hygroscopicity	CYP450 behavior
Acid/base	Morphology	PGP behavior
pKa	Particle size	
Solubility in organic solvents	Density	

BCS biopharmaceutics classification system, T_g glass transition temperature, T_m melting temperature, *CYP450* cytochrome P450

Solubility of a solute in a solvent is described by the enthalpy of solution and the entropy of mixing, where the enthalpy of solution is related to the melting temperature (T_m) of the drug substance and lattice energy associated with the melting event (Brouwers et al. 2009). As the melting temperature increases and lattice energy becomes more significant, the benefit of an amorphous form becomes greater. Melting temperature is easily measured by standard differential scanning calorimetry (DSC) methods. Rapid cooling and testing of the second heating cycle using modulated DSC (mDSC) can also be used for assessment of the glass transition temperature (T_g) of the compound. By evaluating the melting behavior of the drug substance, it is possible to quickly assess the potential benefits of an amorphous form and provide critical information for design of the solid dispersion formulation. For example, the low-melting point drug darunavir ($T_m = 105\,°C$) has been developed as a crystalline material because of the limited benefit of an amorphous form (Thommes et al. 2011; Van Gyseghem et al. 2009), whereas itraconazole ($T_m = 168\,°C$) (Six et al. 2002) is a commercially marketed ASD.

If an amorphous form may provide the needed solubility advantage, or if all other conventional means have been exhausted, the appropriate initial manufacturing technique may quickly be determined by a comparison of thermal stability and solubility in volatile organic solvents. Thermal stability will be measured by the behavior of the compound under heating using thermal gravimetric analysis (TGA), with the evaluation generally conducted from 25 to 300 °C to cover a range of potential processing temperatures while also providing some insight into melt stability (Hughey et al. 2010). Isothermal methods may also be performed at anticipated processing temperatures and/or above the melting temperature of the compound, followed by purity analysis. Compounds should generally provide acceptable stability in the anticipated temperature range, with impurity levels not more than fivefold the anticipated specification limit. This fivefold rule of thumb has been established based on historical experience with extrusion process optimization to reduce impurities below the levels observed during initial screening. It should be noted that the level of allowable impurity will vary on the basis of the properties of the parent compound and the nature of the impurity (i.e., genotoxic vs. nongenotoxic). Similarly, solubility evaluation in organic solvents such as acetone, methanol, ethanol, isopropanol, and methyl ethyl ketone should be performed using standard methodologies. Compounds suitable for spray drying will exhibit solubility of 10 mg/ml and preferably 50 mg/ml or greater to support commercial applications. Additional solvents such as dichloromethane, tetrahydrofuran, and chloroform may also be considered; however, their application must be considered in relation to the toxicity of material. By weighing the benefits of each manufacturing technology, it may be possible to select the most appropriate process early in development. It should also be noted, however, that differences in solid dispersion density, surface area, and degree of homogeneity because of manufacturing technology selected have also been reported (Patterson et al. 2007). Following proof-of-concept formulation development, additional manufacturing process consideration should be given to ensure selection of the appropriate technology.

Stability of the amorphous form will become a function of the drug substance and stabilizing polymers used in the solid dispersion. When examining the properties of the drug substance, molecular structure and T_m/T_g ratio have been shown to be key predictors of stability (Angell 1995, 2002). Generally, compounds exhibiting planar and easily "stack-able" structures tend to recrystallize rapidly, whereas those existing as racemic mixtures with greater structure diversity are generally more stable in the amorphous form. The ability to form intermolecular interactions with stabilizing polymers has also been shown to improve physical stability of the formulation. The glass behavior of the drug substance is also related to the physical stability of the system. Evaluating the T_m/T_g ratio provides an indication of glass strength, with strong glasses having values less than ~ 1.3. For these systems, higher drug loadings will be possible and greater physical stability is anticipated.

Miscibility of the drug substance with common pharmaceutical carriers must also be considered. The most basic method for determining miscibility is performed by evaluating the solubility parameters of the proposed materials (Greenhalgh et al. 1999). Solubility parameters are a measure of cohesive energy density calculated by group contribution theory, related to the dispersive (δ_d), polar (δ_p), and hydrogen-bonding (δ_h) components of the material, as shown in Eq. 7.1. In general, materials having solubility parameter differences less than 7 $MPa^{1/2}$ are considered miscible, whereas those having differences greater than 10 $MPa^{1/2}$ are considered immiscible. More detailed analysis comparing the partial solubility parameters via a Bagley diagram can provide even greater predictability of the miscibility by evaluating dispersive and polar components as a combined factor (Albers et al. 2011).

$$\delta_T^2 = \delta_d^2 + \delta_p^2 + \delta_h^2 \qquad (7.1)$$

Fundamental properties of the drug substance also influence the nature of the melt extrusion process. By mapping critical properties of the drug substance, it is possible to understand the nature of the extrusion process that will be used in manufacture. Melt extrusion manufacturing is conducted in one of two regimes to yield an amorphous formulation based on the processing temperature in relation to the melting temperature of the drug substance. Miscibility regime processing occurs when the processing temperature is greater than the melting temperature of the drug substance. Manufacturing in the solubilization regime occurs when the processing temperature is less than the melting temperature of the drug substance, relying on melt solubilization to generate an ASD. The T_m/T_g ratio for a pharmaceutical compound describes the nature of the glass and the likelihood for amorphous stability (Friesen et al. 2008). It is also related to the maximum attainable drug loading. In general, as the T_m/T_g ratio increases, the degree of difficulty for generating an ASD of sufficient drug loading increases as well. Similarly, as the melting temperature of the drug substance increases, the processing temperatures required become higher, until processing via melt solubilization is required to prevent degradation of the materials. Described graphically and shown in Fig. 7.2, it is possible to understand how the properties of the drug substance will influence the processing conditions in melt extrusion.

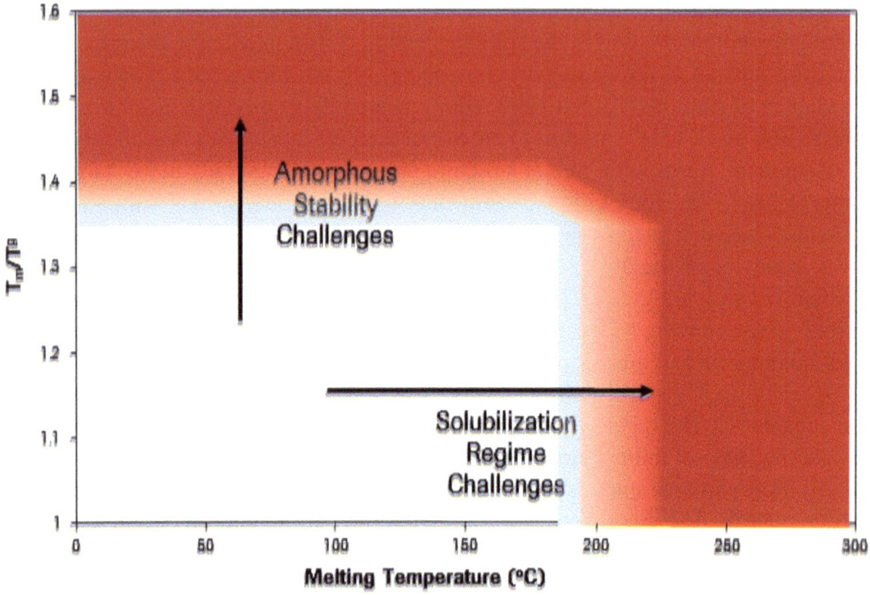

Fig. 7.2 Process mapping for melt extrusion based on drug substance properties

7.3.2 *Excipients*

Excipients used in amorphous melt-extruded formulations are classified into two general categories: stabilizing polymers and functional additives. Stabilizing polymers, as the name suggests, serve as the carrier for the amorphous drug and are the material in which the active pharmaceutical ingredient (API) is homogeneously dispersed to create the final solid dispersion. These materials are generally higher molecular weight amorphous polymers that provide appropriate melt viscosities for manufacturing and sufficient glass transition temperatures for stabilization of the amorphous form. Functional additives include solubilizers, plasticizers, and pore formers. These materials are included into the formulation to influence the processability or critical product attributes of the solid dispersion. Material selection, both of stabilizing polymer and optional functional additives, will strongly influence many aspects of the manufacturing and drug product performance. Therefore, a detailed understanding of the material properties is necessary to ensure appropriate excipient selection.

Looking at the available materials for preparing amorphous dispersions, presented in Table 7.2, one notes only a limited number of pharmaceutically acceptable polymers for use with melt extrusion. Of these materials, the vast majority were originally developed for other applications, and it is only within the last several years that materials specifically designed for melt extrusion, such as Soluplus®, have reached the market. Given this, it is a pleasant surprise to see that these materials have been

Table 7.2 Common polymeric pharmaceutical excipients for use in melt extrusion

Polymer	T_g (°C)	T_{deg} (°C)	Grades
Vinylpyrrolidone	90–170[a]	175–250[a]	Povidone® K12, K17, K30, K90; Plasdone® K25, K29/32, K90
Vinylpyrrolidone–vinylacetate copolymer	101	230	Kollidon® VA 64; Plasdone® S630
Polyethylene glycol, vinyl acetate, vinyl caprolactam graft copolymer	70	250	Soluplus®
Polymethacrylates	130	155	Eudragit® L100, L100–55
Hypromellose acetate succinate	120–135[a]	185	AQOAT®-L, M, H
Amino methacrylate copolymer	56	> 200	Eudragit® E PO

T_g glass transition temperature; T_{deg} degradation temperature
[a]Dependent on the grade of the material selected

successfully used to produce a range of ASDs. Polymer selection for melt extrusion will be driven by considerations related to manufacturability, oral bioavailability, and stability of the dispersion. These product attributes are directly related to the fundamental properties of the materials and careful evaluation can aid in polymer selection a priori.

Stability considerations refer to both the initial stability and long-term stability of the dispersion, and are impacted by miscibility and glass transition temperature of the final solid dispersion. Evaluation of solubility parameters provides an indication of miscibility for drug and polymer, allowing for initial screening of those materials exhibiting a parameter difference greater than 10 $MPa^{1/2}$. Polymeric materials having higher glass transition temperatures will also generally result in solid dispersions with higher glass transition temperatures and lower molecular mobility, although it must be noted that stability will also be strongly influenced by specific intermolecular interactions and moisture absorption during storage. Therefore, consideration should also be given to the chemistry of the polymer in relation to the drug substance, as well as the hygroscopic nature of the carrier polymer. Chemical stability of the polymer will also factor into the decision, noting that performance is a function of temperature and processing conditions.

Manufacturability is governed by the melt viscosity, glass transition temperature, and thermal stability of the materials. Preferred materials will have a wide range between degradation onset and glass transition temperature. As the window for processing is narrowed, it does not eliminate the ability to process with the material but may require the drug to function as a plasticizer or require the use of functional excipients to improve the processing range. Melt viscosity will also strongly influence manufacturability, impacting motor load during production as well as melt diffusivity according to the Stokes–Einstein equation (Eq. 7.2). Careful consideration should be given for the selection of high melt viscosity materials when processing in the solubilization regime due to the impact of diffusion rates on process performance.

$$D = \frac{k_B T}{6\pi \eta r} \tag{7.2}$$

Oral bioavailability of the solid dispersion is perhaps the most critical of the three product attributes, since exposure limitations will typically be the rate-limiting step in developing the compound and the driver for application of ASD technology. Selection of the polymer will be largely determined by specific interactions between the drug and candidate excipients, to which the performance cannot be predicted a priori. The ability of the system to supersaturate free drug levels and maintain these levels is necessary to increase the driving force for absorption in the gut. Both ionic and nonionic polymers have been shown to improve the level of supersaturation, although generally the ionic enteric polymers tend to show the greatest stabilization and best performance as concentration-enhancing polymers. While a number of mechanisms have been proposed, it is currently believed that the stabilization potential is owing to hydrophobic interactions between the drug and polymer with a contribution from colloidal charge due to ionization of ionic materials (Friesen et al. 2008; Alonzo et al. 2010; DiNunzio et al. 2010). The reader is referred to Chap. 8 for additional discussion of the mechanisms for supersaturation. Utilization of early-stage screening technologies and prototype manufacturing will be able to provide sufficient insight into performance to allow for selection of the appropriate drug–polymer systems based on free drug concentration. Beyond the ability to supersaturate and maintain free drug concentrations in vitro, a detailed understanding of the biopharmaceutical properties of the drug substance is necessary for selection. For example, a compound not absorbed in the later stages of the gastrointestinal tract may not benefit much from the application of high pH dissolving enteric polymers, whereas a compound well absorbed throughout may exhibit excellent performance. In other cases, where exposure is limited by solubility and metabolism, alternate strategies such as boosters may need to be evaluated to ensure sufficient exposure. Coupling performance evaluation in relevant preclinical animal models will be necessary to ensure accurate assessment of performance in relation to biopharmaceutical constraints.

The incorporation of functional additives is generally undertaken to address deficiencies of binary drug–polymer systems and include a range of materials that serve as boosters, plasticizers, solubilizers, pore formers, and antioxidants. Although it may be possible to identify the need for certain materials prior to prototype manufacturing, many of these materials will be incorporated after the initial evaluation. Consideration of the type of system being designed and mode of incorporation are also necessary, as these materials span a range of physical states at room temperature. While the incorporation of these materials is intended to address the deficiency in the system, it must also be noted that the presence of these additives may also

Table 7.3 Common pharmaceutical functional additives for use in melt extrusion

Plasticizers	Surfactant/solubilizer	Antioxidant
Triethyl citrate	Pluronic	Butylated hydroxytoluene
Dibutyl sebecate	Span	Butylated hydroxyanisole
Citric acid	Tween	Edetate disodium
Butyl stearate	Vitamin E TPGS	
Glycerol monostearate	Docusate sodium	
Diethyl phthalate	Cremophor	

influence every other critical product attribute. For example, a solubilizer intended to improve dissolution rate may also reduce the glass transition temperature and/or inhibit intermolecular interactions between the drug and polymer, resulting in physical instability during storage. A range of commonly used functional additives are presented in Table 7.3.

In addition to a range of biopharmaceutics classification system (BCS) II compounds currently in development, a number of new chemical entities are BCS IV compounds that are permeability-limited because of p-glycoprotein (Pgp) efflux or cytochrome P450 (CYP450) metabolism. For compounds where these limitations significantly reduce exposure, formulations may be developed to increase solubility as an amorphous form while also incorporating another material to saturate or inhibit the permeability barrier. These materials are referred to as boosters and include excipients as well as active moieties. A commonly cited example using this strategy is Kaletra®, where ritonavir is used to boost the systemic concentrations of lopinavir (Breitenbach 2006). Excipients can also be used for this purpose, with several common materials such as Tween 80, Cremophor® EL, and vitamin E TPGS having been used for this purpose.

For many drug–polymer formulations, particularly those relying on high molecular weight materials, a high melt viscosity will be observed during manufacturing. This may be particularly difficult to address using modification of process parameters if the drug does not provide significant plasticizing effect. This may also become a rate-limiting step for systems processed in the solubilization regime. The incorporation of a plasticizer can reduce melt viscosity. These materials increase free volume between polymer chains, reducing viscosity and glass transition temperature (Aharoni 1998). As a result, the incorporation of these additives can have a marked impact on the physical properties of the systems. Addition of plasticizers has been shown to alter tensile strength and elastic modulus. Physical stability also decreases because of the greater molecular mobility caused by the reduction of glass transition temperature. Even with these changes to mechanical attributes, a number of materials have been successfully used to plasticize melt-extruded formulations, including polyethylene glycol, triethyl citrate, and dibutyl sebecate. Commercially, plasticizers have also been incorporated into Rezulin®, a melt-extruded solid dispersion that was approved by the FDA but withdrawn from the market because of safety concerns not related to extrusion.

The solubility limitations faced by systems cannot often be addressed in the confines of binary drug–polymer amorphous formulations. The incorporation of solubilizers, such as surfactants-like poloxamer and polysorbate, to increase solubility and dissolution rate has also been reported for the development of melt-extruded systems (Ghebremeskel et al. 2007). These materials enhance wettability of the solid dispersion and can be very effective at low concentrations to improve the dissolution rate of drug products. It is also a convenient aspect that many of these materials also provide functionality as boosters and plasticizers, allowing these materials to serve multiple functions within melt-extruded formulations.

Particularly applicable for controlled-release systems, pore formers are incorporated as functional materials to increase the surface area for mass transport from the system. For extruded dispersions, these can be incorporated to achieve a homogeneous dispersion or as a secondary phase. Within the design of the system, these materials are intended to dissolve at a faster rate than the other components of the matrix, yielding a more porous structure with greater surface area. Given the nature of incorporation, a range of amorphous, crystalline, and semicrystalline materials can be used for this function. Having been shown to improve drug release from a number of extruded matrices intended for controlled-release application, little utility to date has been shown for these materials for improving oral bioavailability from ASDs.

Chemical stability of melt extrusion systems is also critical to the development of a viable drug product. During processing, temperatures up to 185 °C are routinely used, exposing materials to conditions at which reactions occur at velocities approximately 50,000 times faster than at room temperature.[1] Minor incompatibilities observed during preformulation testing can become significant limitations in developing melt-extruded formulations. Of the common degradation pathways, oxidation is a major concern for many pharmaceuticals. If process modification is not possible to alter the microenvironmental conditions, additives may be incorporated to reduce impurity formation (Wu and McGinity 2003; Crowley et al. 2002). Oxygen-scavenging antioxidants such as ascorbic acid can be incorporated into the system to provide protection by preferential oxidation, preserving the drug substance. Chelating agents can also reduce degradation rates by complexing with metal ions that catalyze oxidation (Crowley et al. 2007).

7.3.3 Early Evaluation of Binary and Multicomponent Systems

Prior to manufacturing prototype formulations, a number of methodologies are available for characterizing the performance of binary and multicomponent systems using material sparing approaches. These range from simple glass transition temperature calculations to phase diagram mapping and can provide a wealth of information

[1] Assuming reaction rate doubles for every 10 °C temperature increase.

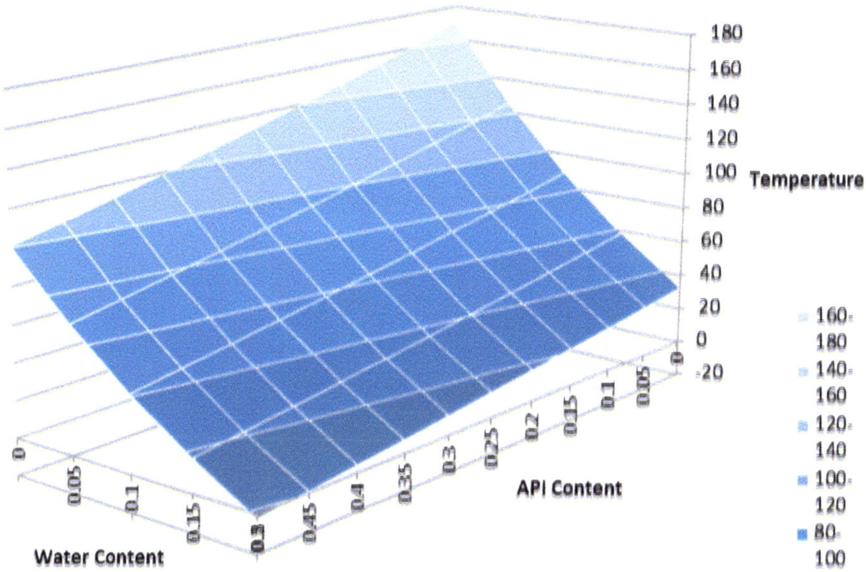

Fig. 7.3 Three-dimensional plot of glass transition temperature using Fox equation for a solid dispersion where the drug substance T_g is 50 °C and polymer T_g is 170 °C

regarding the composition and required processing conditions. Based on the fundamental solid-state aspects of ASDs, the primary aim of early evaluation is to identify unsuccessful formulations using a minimum amount of drug substance so that prototype production is conducted on systems most likely to provide success.

Some of the earliest characterization of amorphous materials was undertaken by Fox et al., proposing the Fox equation to describe the composite glass transition temperature of a homogeneous system. Expressed in Eq. 7.3, it is possible to rearrange for calculation of glass transition temperature as a function of composition. Applying the rule of 50, it becomes easy to identify maximum theoretical drug loading for stability for compositions having T_g values greater than 90 °C (Yoshioka et al. 1995). This approach can provide a simple method for identifying appropriate drug loadings in prototype dispersions. Further derivation can also allow for evaluation of the impact of moisture on the basis of the known hygroscopicity profile for the stabilizing polymer. Shown in Fig. 7.3, a three-dimensional plot can be generated to understand the impact of drug loading and moisture content on dispersion glass transition temperature.

Equation 7.3: Fox equation for glass transition temperature of amorphous dispersions

$$\frac{1}{T_g} = \sum \frac{w_i}{T_{gi}} \tag{7.3}$$

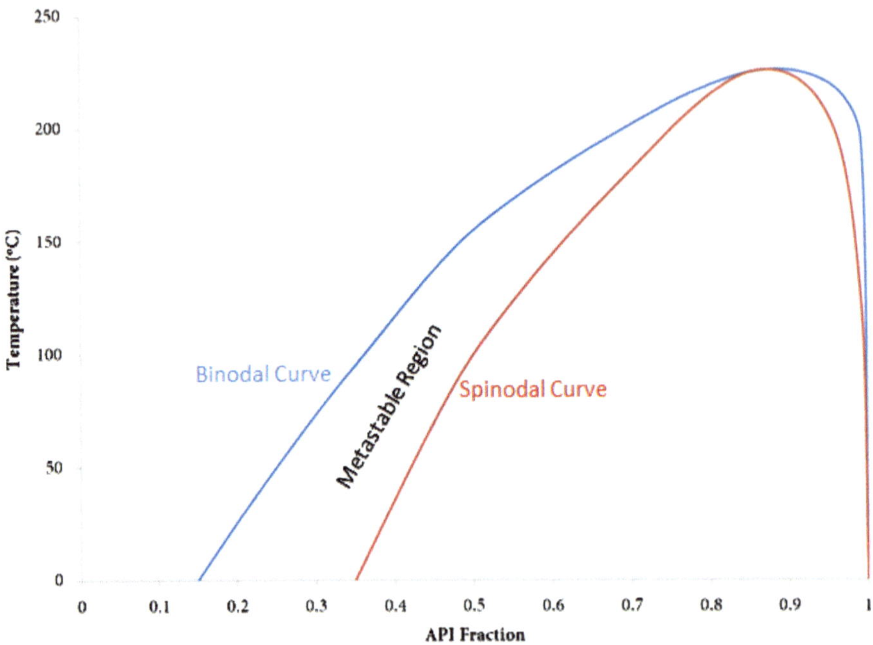

Fig. 7.4 Phase diagram showing stable, metastable, and unstable regions as a function of temperature and composition

More detailed modeling can also be performed on the basis of Flory–Huggins theory and using limited experimental testing of melting point depression to understand the drug–polymer interaction parameter (Marsac et al. 2006; Zhao et al. 2011). In these models, it is assumed that the drug–polymer pair is analogous to the solute–solvent pair in the Flory–Huggins lattice model, allowing for description of miscibility based on Gibbs free energy change before and after mixing. Application of melting point depression trials through the incorporation of low levels of polymer into the drug substance allow for determination of the Flory–Huggins interaction parameter (χ). Evaluation of the free energy–drug loading relationship at different temperatures, as well as determination of the binodal and spinodal curves via evaluation of the first and second derivatives, allows for the development of a phase diagram depicting stability, metastability, and instability as a function of temperature and composition. A representative diagram is presented in Fig. 7.4. By utilizing such diagrams, it is possible to estimate drug loadings attainable under different processing conditions, as well as estimate the risks associated for long-term stability at different conditions. Care should be taken to avoid blindly applying this information in place of actual processing or stability data, as the model does not account for lattice dissociation energies or kinetic storage phenomena such as moisture uptake during storage. Additionally, this approach may be limited when working with compounds having very high melting points.

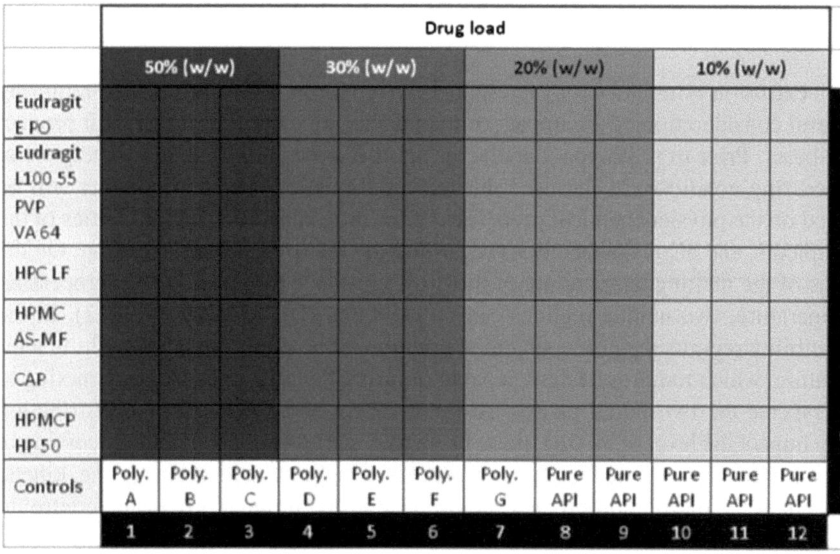

Fig. 7.5 High-throughput amorphous formulation screening design. (Reproduced with permission from Shah et al. 2012)

Material sparing high-throughput methodologies also exist to simulate the behavior of ASDs, although the majority of these involve the preparation of dispersions via solvent or precipitation methodologies (Vandecruys et al. 2007; Lauer et al. 2010; Van Eerdenbrugh and Taylor 2010). In most of these cases, a large range of formulations can be tested at varying drug loadings and compositions, as shown in Fig. 7.5. The design of these systems allows for rapid sequential evaluation of amorphous nature, homogeneity, kinetic solubility, and physical stability. Several of these systems have been reported in the literature and have been shown to provide an excellent platform for screening a large number of potential compositions. While these systems can provide important insight into drug–excipient interactions and the subsequent impact on stability and kinetic solubility, there are inherent differences in the performance of solid dispersions prepared by different methodologies. It has been demonstrated that based on the processing technology, uniformity differences can exist between compositionally identical dispersions (Dong et al. 2008). It has also been shown that differences in manufacturing method can impact the biopharmaceutical performance of the solid dispersion. Given this, a need still exists for high-throughput systems to simulate melt extrusion. Although a number of methodologies have been tested to simulate extrusion in material sparing modes, such as repeated thermal cycling of physical blends and hot-stage microscopy of physical mixtures, these techniques lack the high shear critical to melt extrusion performance. The absence of shear in melt extrusion simulating systems can often lead to false negatives, and care must be taken when applying the results of solvent processing high-throughput methods to the selection of formulations for melt extrusion prototype manufacturing.

7.4 Process Design for Early-Stage Melt Extrusion

Melt extrusion exhibits a strong interplay between formulation and process, requiring careful consideration of the impact of manufacturing conditions on critical product attributes. Prior to prototype manufacturing, it is necessary to define some general processing conditions to be used during initial studies. These conditions will be based on the physicochemical properties of the drug substance, the properties of the excipients, and also considerations regarding the interplay during extrusion. On the basis of the melting temperature of the drug substance in relation to the processing temperature, two unique regimes can be identified (DiNunzio et al. 2012). In the miscibility regime, the processing temperature will be greater than the melting temperature, which requires designs capable of providing greater distributive mixing to intersperse the two liquid phases. Although shear and residence time distribution may impact the level of mixing, these properties are less critical than if processing in the solubilization regime. For systems processed in the solubilization regime, kinetic aspects of the dissolution process should be adjusted through processing control to maximize mass transfer rates. Therefore, greater specific energy (SE) input should be applied, along with more aggressive screw designs to increase shear and extend residence time. This will aid in the formation of an amorphous form, although excessive designs may result in unnecessary impurities being formed during production. Anticipated polymer melt viscosities can also influence the choice of screw design, noting that high viscosity systems may not be able to support a large number of kneading elements owing to torque constraints. Further preliminary consideration of processing temperatures and drug loading of the systems can be gained by examination of the phase diagrams. For compositions that require high temperature to achieve the target drug loading within the stable range, it is advisable to consider reducing the drug loading in order to demonstrate proof of concept. Additional consideration can also be given when dealing with heat-sensitive materials, specifically focusing on reducing the residence time of the formulation and minimizing the temperatures required. Through judicious application of the initial characterization information, it is possible to select processing conditions most likely to yield initial success; however, it is important to understand that process development will be a comprehensive activity that only begins at the prototype manufacturing stage.

7.5 Prototype Manufacturing

Application of structured early-stage characterization can provide critical insight into the anticipated performance of the solid dispersion formulations while also helping to design the manufacturing to be used for prototype manufacturing. Through this process, it is possible to identify target excipients, drug loadings, processing temperatures, and even process section designs. Once the systems have been rank ordered on the basis of desired product attributes, the next stage is a campaign-style series

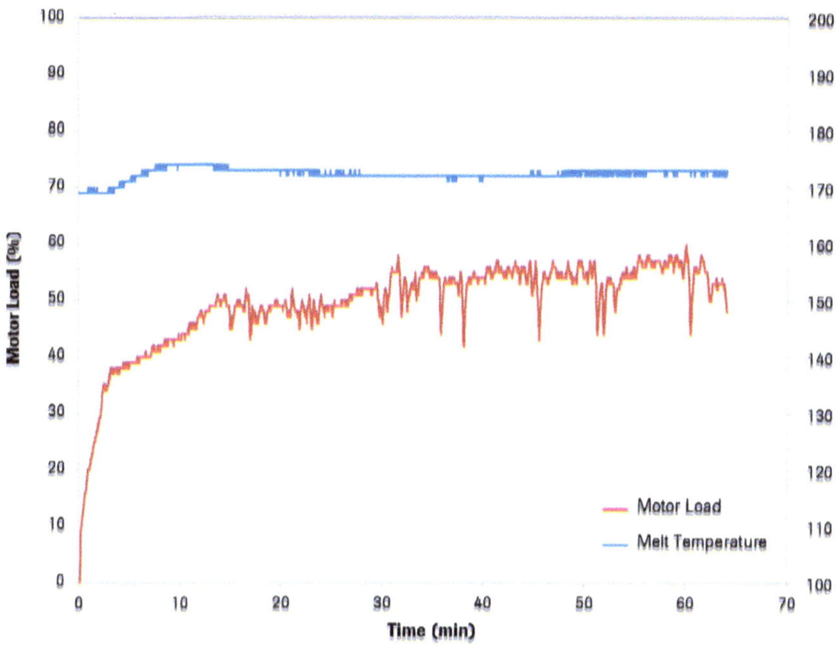

Fig. 7.6 Melt pressure and motor load plots measured in-line during twin-screw extrusion

of manufacturing trials designed to assess the manufacturability, bioavailability, and stability of ASDs prepared by melt extrusion.

Depending on drug substance availability, prototype manufacturing of initial formulations is generally performed on laboratory-scale extruders ranging in size from 7 to 16 mm diameters and capable of supporting batch sizes of approximately 20 g. If constrained by drug substance availability, even smaller units may be used; however, care should be given to ensure similar energy inputs for representative evaluation of process performance. During this stage, key information regarding process temperature, motor load, melt pressure, and screw speed will be evaluated. Most systems are capable of providing real-time displays for these critical attributes, as shown in Fig. 7.6, which can then be used to assign a rank order performance. Processing conditions of systems showing opaque materials should be modified to achieve an amorphous form. If processing conditions are unable to yield a clear glass, the system can be eliminated owing to the presence of residual crystallinity, although it should be noted that heterogeneity of the dispersion may lead to a semitransparent nature. This type of at-line characterization should also be supported by the implementation of polarized light microscopy and can be further supported by the implementation of process analytical technology (PAT). Secondary issues such as high melt pressure and motor load may also be addressed at this stage through regulation of screw speed, feed rate, and zone temperatures; however, it is advisable to conduct a more thorough evaluation of process parameters for lead formulations identified from the

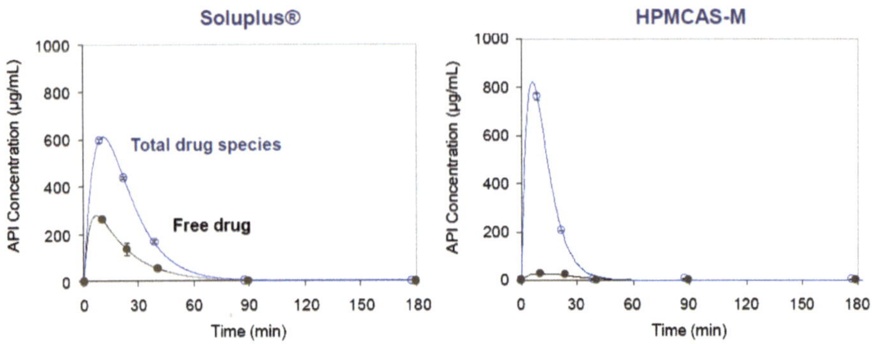

Fig. 7.7 Free drug concentration–time profiles of two different formulations compared to the crystalline drug substance. (Reproduced with permission from Agere Pharmaceuticals)

prototype manufacturing after completion of the initial screening. Following extrusion, acceptable product will be milled as needed using conventional technologies to support further evaluation of the drug product.

The oral bioavailability for BCS II compounds is related to the solubility of the drug substance. Careful measurement of kinetic solubility can provide a rank order for dispersion performance by accurate measurement of free drug concentration. Similarly, formulations that show a relationship between free drug and total drug can use total drug measurements as a surrogate free drug concentration. Free drug and total drug dissolution profiles, shown in Fig. 7.7, can then be used to provide a performance rank order for the different formulations. Some insight can also be gained by comparison of this rank order to that projected in high-throughput solvent evaporation-based screening to identify differences associated with the manufacturing process. In cases where the compound is a BCS IV molecule, more detailed studies may be required to assess the performance of the dispersions. In all cases, it is recommended to support the decision of lead dispersions with testing in relevant animals models. Retrospective comparisons should also be made to identify accuracy of preclinical models in relation to in vitro methodologies.

Physical and chemical stability of the dispersion after manufacturing and on long-term storage are prerequisites in the development of an acceptable drug product. Chemical stability is evaluated using drug-specific methodologies capable of identifying impurities generated during manufacture and storage. Preformulation evaluation will help to identify potential interactions, and systems can be rank ordered on the basis of impurity formation. Similarly, physical stability is evaluated using a series of conventional techniques, including mDSC, x-ray diffraction (XRD), Raman spectroscopy, polarized light microscopy, and scanning electron microscopy. Qualitative techniques will provide an early indication of stability, while more sensitive techniques may be able to identify systems with levels at or below 2 % residual crystallinity (Newman et al. 2008). Preferred compositions will show no recrystallization during storage, particularly under conditions required to support regulatory approval of the drug product. Failure of products observed in more aggressive conditions outside regulatory requirements will not preclude the use of the dispersion formulation for subsequent development; however, care should be given to understand the mode of failure and how the system may be modified in subsequent development to reduce instability. For example, systems exhibiting recrystallization as a result of excessive moisture uptake during prototype manufacturing can be developed in moisture-protective drug products or in moisture-resistant packaging to reduce the influence of water on storage stability.

Key aspects from each of the major phases (manufacturability, bioavailability, and stability) will contribute to the overall ranking of the system. The extent of characterization is fully dependent on the approach of the individual scientist. Systems showing inappropriate performance may be dropped from future studies or carried through to better understand the specific modes of failure. A cumulative evaluation of all formulations will be necessary to provide final rank order. It is also highly recommended that evaluation of bioavailability be conducted in relevant preclinical models, and if necessary, even in man to screen performance. Stability in both standard and aggressive storage conditions is also highly recommended to provide a rapid identification of system stability. In general, priority ranking will be given to oral bioavailability as this property will be the most critical metric. Physical stability is the next most important metric for evaluation of the prototype formulation, noting that options exist for modifying the drug product formulation and restricting storage conditions to enhance stability. Finally, manufacturability will be considered. Systems with high motor loads or excessive impurities may be further improved during optimization to minimize the impact of these issues. Ultimately, by thoroughly rank ordering the solid dispersion formulations, it is possible to identify the lead composition for good laboratory practicc (GLP) and FIM studies, although additional optimization may be required prior to initiation.

7.6 Optimization for GLP and Good Manufacturing Practice (GMP) Manufacturing

7.6.1 Considerations for Selection of Prototypes for GLP and Clinical Stages

By following the structured design approach to developing melt-extruded ASDs and establishing a rank order as detailed earlier, typically the formulation scientist will arrive at a small group of viable formulations. When prototype screening is conducted very early in the development program, prior to entry into GLP studies, the formulation scientist must judiciously select the most appropriate formulations for preclinical and clinical studies from this group. In many cases, the optimal formulation will be utilized for both phases of development. In other instances, the system utilized for GLP studies will be different from the formulation employed in clinical studies. It simply depends on the product requirements at both stages and how the formulation fulfills those requirements. When the GLP system does not optimally satisfy the clinical need, and vice versa, different formulations should be employed. In this section, the formulation requirements for GLP studies and clinical studies will be discussed in the context of selecting the appropriate melt-extruded ASD prototype for each phase.

7.6.1.1 Formulation Selection for GLP Studies

ASD systems are becoming increasingly more utilized for GLP toxicology studies as new molecular entities are becoming increasingly more insoluble, not only in aqueous media but also in common nonaqueous vehicles. Inadequate exposures during toxicity studies can lead to misleading safety evaluations that can delay entry into the clinic and jeopardize the development program (Zheng et al. 2012). Substantial solubility enhancement of NMEs through formulation intervention is often required for these studies where steep ascending dose designs up to very high doses (hundreds to thousands of mg/kg of body weight) are required to achieve target safety margins for entry into humans (Kwong et al. 2011). Supersaturating ASDs have become compound-enabling strategies in these cases as they typically provide vast improvements in dose linearity and overall exposure, thus permitting thorough toxicity profiling of the compound (Moser et al. 2008a, b).

The properties of an optimal GLP formulation may not necessarily be the same as those of the optimal clinical or commercial formulation owing to the different demands of a GLP formulation. First, an ASD formulation for GLP studies is typically dosed as a suspension in an aqueous vehicle. A large quantity of the suspension formulation is prepared and group dosing is performed from a single lot. Dosing can require as much as several hours to complete, and therefore, the ASD formulation must remain physically stable in suspension for the duration of constitution and dosing. One must consider this as a critical feature of the GLP formulation when selecting from melt-extruded prototypes. Systems largely comprising a carrier that is soluble in the aqueous vehicle can lead to dissolution of the API followed by recrystallization prior to dosing. Recrystallization of the amorphous form will negatively impact the performance of the formulation and diminish the results of the toxicity assessment.

Perhaps the best strategy to employ for GLP ASD formulations is to utilize a primary carrier that is insoluble in an aqueous vehicle. It is strongly advised to consider melt-extruded ASDs containing anionic polymeric carriers, such as hypromellose acetate succinate or methacrylic acid copolymers type A and C. Amorphous systems with carriers such as these can typically be dosed in aqueous vehicles that have been titrated to a pH of 4.0–5.0 with acceptable physical stability in suspension. Stability evaluation of such systems in suspension can be accomplished according to the method described by Moser et al. (Moser et al. 2008a). Several authors have demonstrated these systems to achieve target performance with respect to dose linearity and exposure (Zheng et al. 2012; Kwong et al. 2011; Moser et al. 2008a, b).

Table 7.4 Attributes of an ideal melt-extruded ASD formulation for preclinical studies

Remains amorphous in aqueous suspension for suitable duration for group dosing (2–4 h)	Provides linear increases in exposures with ascending dose
Enables achievement of exposures suitable for toxicity profiling and establishment of safety margins	Easily dispersed in vehicle and results in a stable homogenous suspension
Sufficient drug loading to facilitate high doses	Amorphous powder is stable on storage for duration of study or for sufficient time for resupply
Forms a homogenous suspension that is easily dosed through gavage needle and/or feeding tube	

The formulation scientist must be aware of the solids limitations for dosing ASD formulations in suspension and balance this with the appropriate drug loading. Generally, total solids loading in suspension is limited to approximately 200 mg/ml, beyond which the suspension becomes prohibitively viscous to dose via gavage. Also, dosing volume for aqueous vehicles is limited to about 10 ml/kg; hence, the approximate maximum amount of deliverable solids per dose administration is 2,000 mg/kg. The maximum API dose deliverable depends on drug loading in the ASD formulation, i.e., at 10 % drug loading, the maximum API dose is 200 mg/kg, whereas at 50 % drug loading, the top dose is 1,000 mg/kg. Drug loading must be optimized to achieve the target high dose while balancing the demands of physical stability and bioavailability. The optimization of drug loading with respect to bioavailability and suspension stability should be conducted as enabling development to ensure the suitability of an ASD formulation for GLP studies.

Particle size is a key factor to consider in the design of a melt-extruded ASD product for GLP studies. Therefore, downstream processing of the amorphous extrudate is critical to the success of the formulation. Particle size must be optimized and sufficiently controlled to ensure rapid dispersal and the formation of a homogenous suspension when the ASD powder is constituted in the vehicle. Particle size must also be controlled to ensure that the formulation can be easily administered through a gavage needle and/or feeding tube. Suspendability studies and gavageability studies should be conducted on formulations milled under varying conditions to identify the optimum particle size distribution.

Shelf-life stability for a melt-extruded ASD system intended for GLP use is less critical than a clinical formulation because study durations are typically much shorter, frequent product resupply is often not an issue, and refrigerated storage is usually a viable option. Therefore, systems that may have not performed optimally with regard to physical stability during prototype evaluation could be considered for GLP use, assuming acceptable stability in suspension. This allows for utilization of ASDs with higher drug loads, which is often necessary to achieve the target high dose. Physical stability of an ASD for GLP studies is important; however, there are options to mitigate instability, such as refrigerated storage, shorter expiration dates, and intermittent resupply, for longer studies (Table 7.4).

7.6.1.2 Formulation Selection for Clinical Studies

Selecting the appropriate formulation for clinical studies should directly follow from the rank order established during prototype screening. The composition should be optimal with respect to bioavailability, stability, and manufacturability. Further, the absorption characteristics of the active molecule as well as the desired pharmacokinetic (PK) profile for the therapy must be considered when selecting the clinical ASD formulation. Careful consideration must also be paid to the impact that the ASD composition will have on the performance, convenience, and manufacturability of the final clinical dosage form. In this section, the critical attributes of melt-extruded ASD systems as they relate to the performance and manufacturability of clinical dosage forms will be discussed.

Selection of the appropriate formulation for clinical studies must involve consideration of the absorption properties of the active and the PK/pharmacodynamic (PD) profile of the therapy. For drugs that exhibit site-specific absorption, an ASD system that is designed to release just prior to, or at the site of, absorption in the gastrointestinal (GI) tract should be selected. When a drug is predominantly absorbed in the proximal small intestine, a nonionic system may be preferred as drug release can be initiated and supersaturation maintained in the stomach, increasing the free drug concentration as the stomach contents enter the duodenum. In contrast, a drug that is well absorbed over the entire intestinal tract may benefit from an anionic system, which can sustain supersaturated drug concentrations for the duration of intestinal transit (DiNunzio et al. 2010; DiNunzio et al. 2008; Miller et al. 2008). Further, therapies that benefit from release of supersaturated drug concentrations in the distal regions of the GI tract are best formulated as anionic systems designed to release at pH levels of 6.5 or greater. For drugs that have toxicity associated with elevated C_{max}, the appropriate ASD system should be designed to blunt peak concentrations while keeping total exposure constant. Enteric carriers can be useful in this case along with higher molecular weight nonionic polymers.

Drug loading is a critical aspect to consider when selecting and optimizing an ASD formulation for clinical studies. For low-to-moderate dose therapies, drug loading should be optimized entirely with respect to bioavailability and physical stability as described previously. For high-dose therapies, maximizing ASD drug loading is critical to minimizing dosage form size and/or number of units per administration toward optimizing patient comfort, convenience, and compliance. Drug loading must be maximized while balancing performance and stability aspects to arrive at the optimal drug product. An appropriate experimental campaign must be conducted to identify the highest acceptable ASD drug loading that provides acceptable biopharmaceutical performance and stability.

For moderate-to-high dose capsules, the scientist must optimize drug loading in the ASD to maximize the API quantity that can be loaded into the desired capsule shell. Bulk density is also important in this regard as will be discussed in a subsequent section. For high-dose tablets, it is also critical to optimize drug content in the ASD formulation to maximize API load within the geometric limitation of the tablet. However, solids fractions for tablets are significantly greater than capsules, allowing for superior API loads. Hence, for most high-dose ASD applications, tablets are preferred to capsules, as dose administration can be achieved with smaller and fewer units.

Selecting and optimizing a melt-extruded ASD prototype for clinical use requires forethought as to how the ASD system will impact performance of the final dosage form and how this can be managed to accommodate the clinical requirements. Specifically, the ASD composition chiefly influences capsule dispersion, tablet disintegration, and ultimately API dissolution. For low-to-moderate dose therapies, this influence can be mitigated by external excipients to ensure desired performance. However, for high-dose therapies where there is less capacity for external excipients in the formulation, the ASD composition will largely dictate critical performance attributes.

For capsules, formulation optimization must be conducted to achieve dispersion of the capsule's powder contents following dissolution of the shell. It is often the case that ASD systems tend to form an insoluble plug when improperly formulated as a capsule dosage form, leading to a sustained release of the active rather than the desired immediate release. Serajuddin et al. (1988) explained the mechanism of plug formation as the faster dissolution of the hydrophilic carrier at the surface leading to the formation of a hydrophobic drug-rich layer that becomes a retardant to release. Some ASD carrier polymers, specifically nonionic polymers, tend to form plugs because of more rapid surface dissolution as well as coalescence on hydration leading to viscous hydrogel formation at the plug surface. The ancillary excipients both internal and external to the ASD system must be optimized to ensure adequate dispersion of capsule contents. Such excipients include surfactants, wetting agents, soluble fillers, insoluble fillers, disintegrants, and the like. These excipients function to promote wetting, sterically interfere with hydrophobic interactions and gel formation, and physically forcing adjacent particles apart, thereby acting synergistically to promote dispersion of capsule powder contents. Anionic carriers tend to have less issue with respect to plug formation, and thus external capsule formulation optimization is often more straightforward.

For tablets, nonionic polymers commonly used for melt-extruded ASD systems, such as hypromellose, povidone, and copovidone, typically act as binders on compression and hydration. ASD formulations designed with nonionic carriers typically form nondisintegrating matrices when compressed into tablets at high solid dispersion fractions. Additionally, these polymers can act to retard disintegration by a fusion mechanism on hydration. By either mechanism, the end result is an eroding tablet with slow drug release. Often, these systems cannot be formulated into disintegrating tablets, and thus effort must be spent optimizing the internal ASD composition as well as the external composition to achieve a rapidly eroding system. The Kaletra and Norvir tablets are two commercial examples of nondisintegrating tablets that have been elegantly designed for rapid erosion (Kanzer et al. 2010). Enteric polymers tend to be poor binders; therefore, high-dose tablets designed with anionic ASD systems tend to be faster disintegrating. This makes optimizing for dissolution performance more straightforward; however, achieving acceptable compression profiles and tablet hardness can be more challenging with ionic ASDs.

Almost all melt-extruded ASD systems are milled prior to incorporation into a final dosage form. The purpose of milling is primarily to generate a granular or powder intermediate that can be further processed into a final dosage from. Milling also increases surface area to enhance dissolution properties and ultimately maximize the biopharmaceutical performance of the dosage form. Melt-extruded ASDs typically exhibit high bulk densities, making them ideal for capsule filling and tablet compression, particularly in high-dose applications. Milling will impact bulk density. Particularly in fine milling applications, powder density of finely milled product tends to be lower. An optimum must be established with respect to particle size and density to achieve the desired performance attributes while not adversely impacting fill volumes for capsule and table manufacturing. Milling can also play an important role in tablet compression for high-dose tablets where external excipient content is constrained. Particularly for polymer carriers that are poor dry binders, i.e., anionic polymers, increasing surface area provides greater surface for bonding interactions, thus providing improved compression profiles and tablet hardness. For nonionic ASD carriers that act as efficient dry binders, particle size distribution is less critical. For capsules or disintegrating tablet systems, particle size will also impact the dissolution rate of the ASD following powder dispersion or tablet disintegration with fine milling typically leading to rapid dissolution. For eroding systems, milling enhances water permeation into the monolith to accelerate erosion. In summary, the optimum particle size distribution of an ASD system must be identified during development and milling conditions must be optimized such that the final ASD powder facilitates performance and manufacturability of the final dosage form.

7.6.2 *Process Development and Optimization for GLP and GMP Manufacturing*

Manufacturing of prototype formulations and drug product for preclinical non-GLP studies is typically conducted in R & D laboratories on small-scale equipment. As a drug development program transitions into GLP and clinical development phases, the manufacturing operations transition into a GMP environment, and typically to larger-scale equipment. Particularly in the case of chronic toxicology studies and large clinical studies, batch sizes increase substantially. By this point, the melt extrusion process for the ASD system must be scaled up and optimized to ensure robust process performance and the avoidance of deviations during manufacturing campaigns. In this section, process development considerations for transitions into GLP and GMP manufacturing of melt-extruded ASD formulations will be discussed in detail. For a more in-depth discussion of twin-screw extrusion and extrusion systems in GMP environments, the reader is referred to Chap. 2.

7.6.2.1 Process Length

Process length is the length of the extrusion system and is described in dimensionless units as the ratio of length to the diameter of the screw (L/D). For modular systems, process length can be adjusted by multiples of the barrel segment length, typically four times the screw diameter (4D) (Dreiblatt 2003). For clamshell systems, the process length is dictated by the furthest upstream location. Typically, process lengths are in the range of 20:1 to 40:1, or longer up to a limit of about 50 to 60 times the screw diameter at which point torque and maintenance of clearances become issues (Dreiblatt 2003; Crowley et al. 2007). For detailed information on process length and barrel design, the reader is referred to Chap. 2.

Process length is dictated by the number of different unit operations that are required to consistently achieve the desired quality attributes for the ASD system. When melt processing to achieve ASD systems, the critical unit operations are dictated by the processing regime: miscibility, or solubilization. The miscibility regime is the more straightforward of the two, and melting and dispersive mixing operations are most critical. Because of the heat and mass transfer efficiency of twin-screw extrusion systems, homogenous melt mixing can occur quite rapidly. Therefore, when processing above the melting point of the API, melting of the drug substance with simultaneous mixing with the carrier system can occur over a very short process length. By the simplest scenario, the required unit operations will be (1) feeding the preblend, (2) melting the API while simultaneously rendering the polymer molten, (3) distributively mixing the API into the polymer, and (4) pressurizing the homogenous molten system through the die. Generally, a length of 4D can be assigned to each operation, and hence the total process length would be 16D. If additional unit operations are required for venting, liquid injection, extended metering for greater die pressurization, etc., barrel modules would be added to this basic process, increasing the length by about 4D per operation. Although adding additional process length may be required for additional unit operations, consideration must be given to the increased thermal stress resulting from a greater residence time distribution RTD. In the case of extended metering zones for greater die pressurization as in the case for direct product shaping, consideration must be paid to increased thermal stress resulting from a greater RTD and melt pressure.

When processing in the solubilization regime, it is likely that an extended process length will be required to accommodate several dispersive mixing operations and extended residence times required to solubilize the drug substance. Solubilization is a kinetic process, and when formulation components and temperature profile are fixed, the key process variables are shear and time. The extent to which greater shear and time are required for solubilization will depend on the difference between the maximum processing temperature and the melting temperature of the API ($\Delta T_{\mathrm{MP-Process}}$). As this value increases, so too will the shear stress required to reduce the drug's domain size to molecular levels and the time required to complete the operations homogenously throughout the melt. In the simplest scenario, the required unit operations will be (1) feeding the preblend, (2) rendering the polymer molten, (3) dispersively mixing the API into the molten polymer, and (4) pressurizing the homogenous molten system through the die. Additional dispersive mixing operations with appropriate shear intensities should be added to this basic design until the desired ASD system is achieved. Additional process length will also be required when secondary feed streams are needed or additional unit operations are required.

7.6.2.2 Temperature Profile

Extrusion systems are designed to allow for temperature control in the various zones of the extruder. The temperature at each zone depends on the composition being processed and the unit operations being performed in each zone. Each zone is equipped with heating and cooling mediums controlled by a programmable logic controller (PLC) that balances heat and cooling inputs to maintain the set point. During an extrusion process, the system imparts a substantial amount of mechanical energy at the interface between the screw, the material, and the barrel wall leading to a significant heat generation (Kapp and Palmer 2003). However, external heating is often required to bring the extruder segments and die up to the operating temperature and to maintain temperature during the process (Kapp and Palmer 2003). When mechanical forces generate excessive thermal energy, the PLC triggers the cooling mechanism to reduce the temperature in that segment to the set point. It is important during process development phases to identify the optimum temperature for each zone according to the unit operations being performed in the respective segment.

When processing in the miscibility regime, it is critical that the API be completely melted to ensure molecular mixing with the polymer. Hence, temperature set points in the melt zones should be at or above the melting point of the drug. A plateauing temperature profile will be typical in this case in which the temperature ramps up from the feed zone to at or above the API melting point in the melt, mixing, and metering zones.

When processing an ASD system in the solubilization regime, it may be necessary for a significant portion of the process length to be near the maximum temperature (typically dictated by the thermal stability of excipients) to facilitate complete solubilization of the API and the achievement of a single-phase amorphous system. The key determinant of temperature profile in the solubilization regime will be the $\Delta T_{\mathrm{MP-Process}}$. As this difference increases, API dissolution rate in the viscous polymer decreases, and a maximum number of barrel segments must be set at the peak temperature to accelerate dissolution. The temperature in the feed and conveying zones must be sufficiently low to enable efficient conveying into the melt zones. Once in the melt zone, the temperature profile may remain constant through the die.

7.6.2.3 Screw Design

As described in Chap. 2, twin-screw extrusion systems have segmented screw designs allowing for unique configuration of the screw to meet specific product requirements. Screw elements essentially consist of conveying and mixing elements. Mixing elements can be further broken down to distributive elements, e.g., combing elements and dispersive mixing elements, e.g., kneading elements. Distributive mixing elements are less shear-intense and support mixing of low viscosity phases into high-viscosity phases. Dispersive mixing elements are more shear-intense and are designed for reducing domain sizes of a minor phase into the primary phase of a melt. This section will briefly discuss screw design as it specifically pertains to the manufacture of ASD formulations. For a detailed discussion of screw design, the reader is referred to Chap. 2.

As one progresses from prototype manufacturing into GLP/GMP production, particular attention must be paid to screw design to achieve the desired and consistent product properties. Again, for ASD systems, a key consideration for screw design will be whether processing is occurring in the miscibility regime or the solubilization regime. As discussed earlier, judicious polymer selection in this case should include in silico and empirical data supporting complete miscibility of the drug substance and the carrier at relevant ratios. Melt mixing miscible components should be a somewhat low shear operation when compared with melt solubilization. Therefore, distributive mixing elements should be employed to perform mixing operations during the process. Mixing should essentially consist of dividing and subdividing the melt until homogenous. Depending on the affinity of the different molten components, homogenous mixing could be achieved in a short distance of the process length. Process development from prototype manufacturing to GLP/GMP production should include iterative titration of the number of mixing operations performed along the screw length as well as the length of the individual mixing operations. In general, a common screw design for processing an ASD composition in the miscibility regime should include the following operations: (1) feed conveying, (2) melt pumping, (3) mixing (predominately distributive), and (4) metering of the homogenous melt product at pressure. For venting operations, a low-pressure conveying operation can be incorporated to provide a low-pressure and high-surface area section for removal of volatiles. Note that these are very simplified descriptions of screw operations intended only to provide the reader with general guidance toward designing the appropriate screw. Again, the reader is referred to Chap. 2 for more detailed discussion.

As discussed previously, when processing in the solubilization regime, shear is a key process variable dictating the amount of drug that can be solubilized within residence time distribution of the process. An adequate screw for this regime will include sufficient dispersive mixing operations to ensure the entire drug content is consistently solubilized prior to exiting the barrel. The aggressiveness of the screw will depend on the magnitude of the $\Delta T_{\mathrm{MP-Process}}$ parameter. The larger this parameter, the more aggressive the screw design must be and the greater the SE input. However, the aggressiveness of the screw must be balanced against the shear sensitivity of the excipient materials, as many polymers can degrade under high shear. Titration of shear, meaning total process length consisting of shear elements and the degree of intensity of the shear components, should be conducted between prototype manufacturing stages and GLP/GMP production stages. The optimum shear intensity and SE input will be identified when a homogenous ASD product is consistently produced without excessive degradation of the excipient materials. Such optimization should be conducted moving from low-to-high shear screw systems, making at-line observations regarding transparency of the extrudate. Additional kneading elements, or more aggressive kneading elements, can be added iteratively to the screw until the extrudate appears consistently transparent at line. Thorough analytical characterization of the product should be conducted to confirm a single-phase system with minimal degradation of the excipient materials before identifying the screw design as optimum. Representative screw designs for early development are presented in Fig. 7.8.

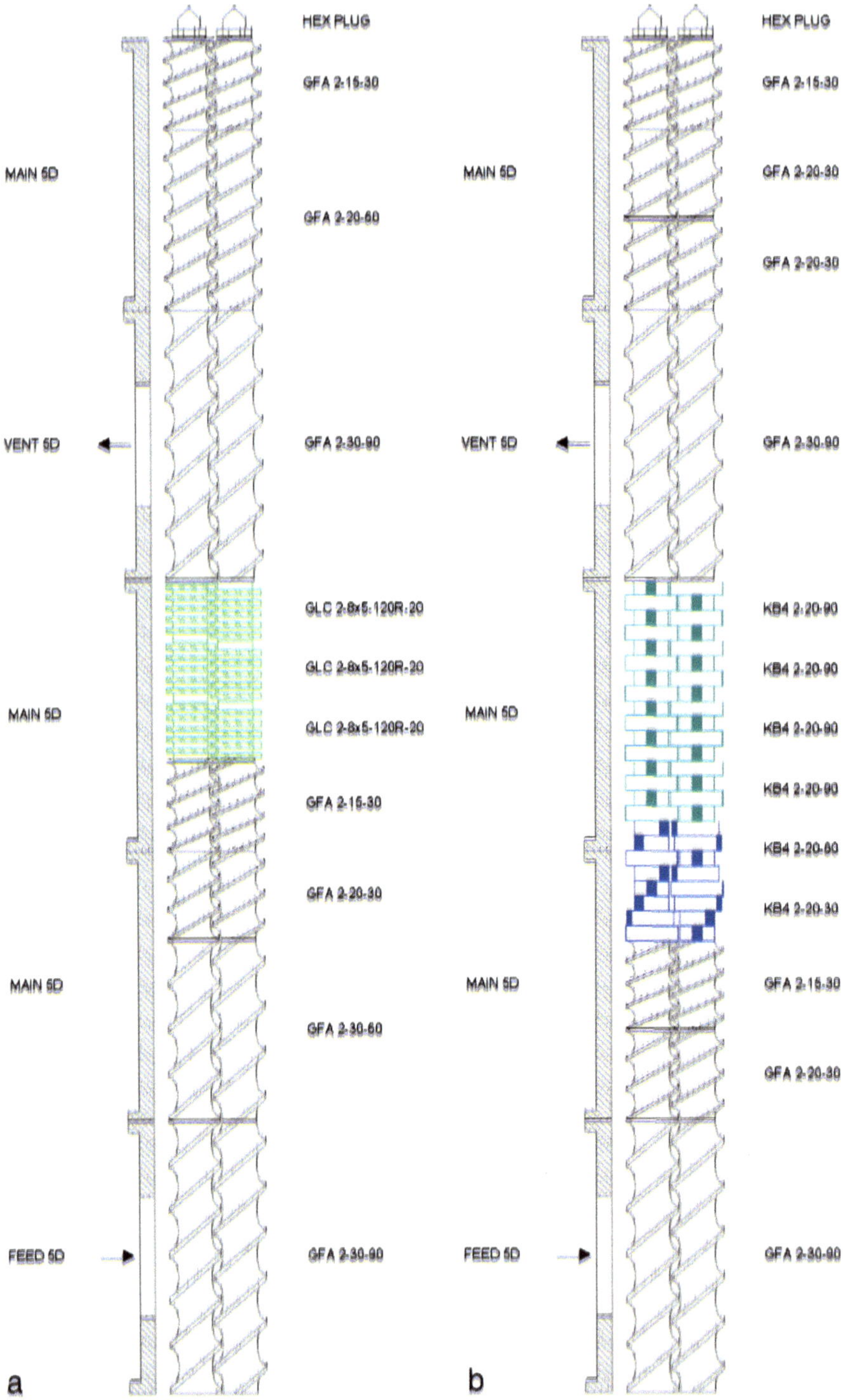

Fig. 7.8 Representative screw designs for distributive and dispersive mixing. **a** High distributive mixing design. **b** High dispersive mixing design

Screw design directly impacts process residence time. In general, as the percentage of mixing operations in the total process length increases, so does the residence time distribution for a given feed rate and screw speed. This should be considered in the context of minimizing thermal stress for processing of heat-labile materials. Reducing the number of mixing elements to a minimum is also important to maximize process throughput, as these elements typically have minimal forwarding capacity. When working to minimize mixing elements, care must be taken to ensure that all required product attributes (amorphous, homogeneous) are achieved prior to the die region. Evaluation of these properties can be achieved using in-line monitoring or sampling from the postmixing to predie region. By ensuring that critical product attributes are achieved within the length of the screw and not because of processing in the die region, it is possible to avoid unnecessary issues during scale-up.

Complete optimization of screw design is critical between the prototype manufacturing and GLP/GMP manufacturing stages of development because the screw design developed during this phase will largely determine the screw design that is employed in later stage and even in commercial production. The principles of scale-up of twin-screw extrusion system are well defined for geometrically similar systems. Therefore, in order to ensure seamless transfer between scales, the screw design ideally would not change between scales. Consequently, screw design at early stages of development has a direct impact on the scale-up and high-volume manufacturing. Adequate resources should be applied to this task and proper experimental design should be executed to ensure success of the melt extrusion process and ultimately the product.

7.6.2.4 Specific Energy

As defined earlier in this chapter and elsewhere in this text, SE is the amount of energy input by the extrusion system per unit mass of feed. It is important to define the optimum SE during initial process development in preparation for GLP/GMP manufacturing because SE will be a key scale-up factor and critical process parameter throughout the life of the process. An appropriate experimental design should be executed to identify the optimum SE by varying the ratio of screw speed (rpm) to feed rate and evaluate the resulting effects on key product properties. Once the optimum range of SE is defined for a given process configuration, it should be held constant, e.g., if throughput is increased, the screw speed should be correspondingly increased to maintain SE.

The criticality of SE for the production of ASD systems will depend on the processing regime. In the miscibility regime, the key operations are melting the API and distributing the liquid active into the molten carrier. These operations require efficient heat transfer with the barrel wall and relatively low shear input. Additionally, product properties are typically more robust with respect to variances in SE.

In the solubilization regime, achieving a homogenous amorphous system will be highly dependent on shear input from the system per unit mass. Therefore, optimization of SE is critical to consistently producing product with the desired attributes. A

Table 7.5 Attributes of a melt extrusion process for production of ASD systems in the miscibility and solubilization regimes

Miscibility regime	Solubilization regime
Higher feed rates	Lower feed rates
Lower shear	Higher shear focused on dispersive mixing
Shorter process lengths	Longer process length
Lower specific energy	Higher specific energy
Bulk API properties less critical	Bulk API properties are critical
Bulk excipient properties less critical	Bulk excipient properties can be critical
Primarily distributive mixing	Primarily dispersive mixing

API active pharmaceutical ingredient

higher amount of shear will be required per unit mass to solubilize the drug vs. melt mixing; hence, SE will be greater relative to processing in the miscibility regime. The SE ratio will also increase with the difference in $\Delta T_{\text{MP-Process}}$, i.e., the further the processes deviates from the API melting point, the more shear energy is required to solubilize the drug. The degree of fill will be lower and screw speed higher with respect to processing in the miscibility regime, and hence throughput is typically lower for solubilization processes. Considering the criticality of SE to the properties of an ASD product produced in the solubilization regime, it is essential to establish an acceptable processing space with respect to SE prior to transitioning into GLP/GMP manufacturing. Moreover, SE is a critical scale-up parameter, and therefore it is vital to establish the optima early in development as it will propagate into each stage of manufacturing (Table 7.5).

7.6.2.5 Feed Material Properties

As with any pharmaceutical unit operation, establishing specifications for feed materials to ensure product consistency and quality is critical. For production of ASD systems by melt extrusion, the feed materials will be the API, polymer(s), ancillary excipients (surfactants, pore formers, glidants, etc.), and processing aids. Since the API and polymer carrier are the primary components of a melt-extruded ASD system, this section will focus on defining feed properties for these materials. The principles of establishing the key feed properties for these materials will also apply to additional formulation components.

In terms of API, it is of the utmost importance to ensure that the bulk drug substance meets specifications with respect to purity, morphology, melting point, and particle size. Therefore, it is the role of the formulation scientist to define the critical API properties for achieving the desired product properties prior to GLP/GMP manufacturing and to see the maintenance of these properties for all manufacturing campaigns. Particularly in the GLP stages, purity of the API can vary. It is important for the formulation scientist to identify if and how these impurities can influence product performance. Moreover, GLP toxicology studies can provide an opportunity to increase limits for potential process-related degradation products by intentionally incorporating API into these studies with elevated levels of the target impurities. Identifying negative influence of API impurities on the properties of melt-extruded ASD will require dialogue between the formulation development and API synthesis groups to ensure that the purity specifications are met. By the time a program has moved into the clinic, the API synthesis process should be well established and purity should no longer be an issue.

The same is true for the interrelated properties of morphology, crystal habit, and melting point. These API properties can vary at early stages of development and can impact the quality of melt-extruded ASD systems. Obviously, changes in crystal structure that influence the thermal properties of the drug substance can significantly impact melting or solubilization of the API during extrusion. The formulation scientist should ensure that the specifications for drug substance are met prior to ASD manufacture for preclinical studies. Again, as the synthesis process is established moving into FIM studies, these properties should not vary and should not pose issues to ASD manufacturing.

Particle size is a property that can vary at all stages of development and can have a substantial impact on an HME process and ASD product. It is therefore essential for the formulation scientist to conduct thorough studies regarding the influence of API particle size on ASD product properties and establish an acceptable particle size distribution range for the HME process. In the miscibility regime, it is important to ensure that the drug particles are below an established limit to avoid incomplete or variable melting during the process.

In the solubilization regime, API particle size is often a critical parameter. As described previously, in this regime, solubilization of the API in the molten carrier is analogous to a dissolution process and can be described by the Noyes–Whitney equation. Examination of this equation reveals that increasing the surface area of the solute increases the rate of dissolution. Putting this in terms of a melt extrusion process, increasing the surface area of the API increases the amount of drug that can be solubilized within the residence time of the extrusion process. Once again, the criticality of API particle size will depend on the magnitude of the $\Delta T_{\text{MP-Process}}$ parameter as well as the target drug loading. The greater the ΔT and drug loading, the finer the API particles must be to achieve the desired product. In very challenging cases, micronization of the API may be required to achieve the desired product. The formulation scientist must diligently evaluate a range of particle size distributions in the melt extrusion process to determine the effect on ASD properties and establish particle size specifications.

API particle size and bulk density can also have a significant impact on throughput of an extrusion system. Maximizing the feed volume with a low bulk density API can still result in low mass flow and poor product output. When faced with this issue, the ideal solution is to mitigate the problem on the API synthesis side. If this cannot be accomplished, another option is to preprocess the API to improve bulk density, e.g., by roller compaction. Another option is to utilize a crammer feeder that densifies the API in the process of feeding into the barrel. Yet another option is to implement shovel elements that can convey larger powder volumes than conventional conveying elements. Once again, API density issues should be identified early and a solution implemented prior to drug product manufacturing.

Moisture content (water and/or solvents) can greatly influence melt processing and properties of ASD formulations. Water content in the bulk API is typically not an issue for PWS drugs; however, water content can be detrimental to drug–polymer miscibility and can also influence the process feed. At a minimum, a cursory analysis of API water content should be conducted to identify any potential issues to processing or product performance. More common, particularly at preclinical stages, is residual organic solvent content in the API from synthesis. Residual solvent can have a number of effects on an HME process and resulting ASD composition. The formulation scientist should therefore monitor residual solvent content in the API at all stages and be aware of the potential effects on the process and product. It is recommended that during and following prototype development, the bulk API be dried in a vacuum oven to minimize residual solvent content as this will be more representative of future bulk API properties. Residual organic solvents that are evaporated during processing can impact the process feed in the absence of proper barrel configuration. To ensure appropriate devolatilization during extrusion, it is recommended to place one vent port in the metering zone. If off-gassing in the feed region is observed, baffled feed inserts can be used to improve vapor removal. Processing can be further enhanced by the removal of residual moisture from drug substance and excipients, although this process may not be desirable for long-term production. Again, through open dialogue with the API synthesis group, these issues can be resolved on scale-up and development of the synthesis process; therefore, by FIM studies, moisture in the bulk API should not be an issue.

For excipient materials, the critical properties to monitor and specify are purity, moisture content, and particle size. Purity is almost always ensured by the supplier and verified with a certificate of analysis for each lot. This property is therefore straightforward to monitor and verify. In some rare instances, lot-to-lot variability within the manufacturer's specifications can cause unwanted variability in the product properties of melt-extruded ASD systems. Such sensitivity should be identified early in development and dialogue with the supplier should be initiated to ensure delivery of excipient materials within the product specifications.

Moisture content of excipients can in some instances negatively influence feeding. Off-gassing through the feed port can occur, causing residual moisture and solvents to condense at various locations leading to powder agglomeration, reduced feed flow, and bridging. Options for dealing with these issues include employing special coatings, such as Teflon, on surfaces in the solids feeding region and baffled hopper designs. Over time, this agglomeration becomes substantial, forming large chunks of agglomerated feed that can eventually lead to bridging and choking of the feed to the extruder. One simple solution is to place a vent port between the evaporation point and the feed throat to evacuate vapors before they can impact the feed. Another possible solution is to minimize residual moisture by controlled excipient storage conditions or preprocessing to remove excess moisture. Any of these strategies can be effective and it is the role of the formulation scientist to identify issues with residual moisture and implement the most effective solution prior to GLP/GMP manufacturing.

Perhaps the most important excipient property to optimize and control is particle size distribution. Particle size of the primary carrier components can impact the feed and the mixing efficiency of the process for a given configuration. With regard to the process feed, larger particle sizes tend to increase the feeding efficiency of the extrusion process. Finer excipient particles can lead to poor hopper flow in gravimetric feeder systems leading to erratic mass flow to the extrusion system. Fine excipient particles are also more prone to aeration and agglomeration in the feed throat of the barrel that can lead to bridge formation and feed choking. Although aeration can be partially addressed with feeder height positioning, these effects are largely dependent on the polymer itself, and these aspects must be monitored during process development prior to GLP/GMP manufacturing. If feed issues are identified relating to particle size of the excipient carrier, a coarser grade should be evaluated. Fine polymer grades also tend to have a lower bulk density, which can negatively impact mass flow into the system and consequently product output. If a coarser polymer grade cannot be utilized to improve mass flow, the implementation of shovel screw elements in the feed zone of the extruder are recommended for utilization of a crammer feeder to increase feed density and improve throughput.

Excipient particle size can also impact mixing efficiency and homogeneity of the ASD product. This is the case primarily when processing in the solubilization regime where intimacy of contact between drug and polymer particles can be critical to complete solubilization of the API. When considering the solubilization of an API by melt extrusion from the perspective of the Noyes–Whitney equation, an increase in surface area accelerates solubilization of the solute. In the case of melt processing, increasing the surface area of the solute (drug substance) and the solvent (polymer carrier) are both applicable. Therefore, by reducing the particle size of the polymer, one can effectively increase the surface area for drug–polymer interactions and increase the amount of the drug substance that is able to be dissolved within the residence time distribution of the extrusion process. Proper design of experiments should be conducted to identify the optimum and acceptable limits of excipient particle size distribution with respect to ASD homogeneity. These specifications should be implemented prior to drug product manufacturing.

The bulk properties of the excipient carrier tend to be less critical to product properties when processing in the miscibility regime. Because processing is conducted above the melting point of the API and the T_g of the polymer, mixing occurs between two liquid phases, and hence surface area considerations are not overly critical. However, one should avoid excipient materials that are too coarse, as this could potentially lead to polymer-rich regions within the ASD system. DSC analysis of ASD product can be utilized to detect such issues during prototype development allowing for proper definition of excipient properties prior to manufacturing.

The earlier discussion focused on the primary excipient carrier; however, the same issues apply to ancillary excipients as well, with the magnitude of the impact depending upon the concentration in the formulation. The same attention should be paid to purity, water content, particle size distribution (when relevant), and viscosity (for liquid feeds). Potential issues should be identified prior to manufacturing stages in order to mitigate process disruptions.

7.6.2.6 Feed Configuration

During prototype manufacturing, the ASD formulation components are typically fed to the extrusion system preblended either manually, with a single-screw auger, or a small twin-screw gravimetric feeder system. On bench-scale extrusion systems, there is often only a single feeding port, making feed configuration a moot issue. However, when scaling the process up in preparation for manufacturing, there are several different options to consider with respect to feed configuration. First, the formulation components can be fed together or separately by individual feed systems. Second, when feeding components individually, they can be fed at the same position on the barrel or they can be fed at different ports depending on specific requirements. In this section, feed configuration will be briefly discussed as it pertains to the development of an ASD manufacturing process in preparation for drug product manufacturing.

In early process development and manufacturing, small batch sizes often dictate preblending and metering the formulation from a single feeder. Also, feed system availability, familiarity with single-feed orientations, and the bias of pharmaceutical manufacturing toward batch weighing often lead to utilization of a single-feed orientation in early development. A benefit of a preblend is that, excluding the possibility of segregation in the hopper, the content uniformity of the extruded product is dependent on the blending operation and largely independent of the extrusion and feeding system (Dreiblatt 2003). However, as process development progresses toward larger-scale manufacturing, metering the individual components via multiple feeders should be evaluated as it simplifies the overall process by eliminating dispensing and preblending operations. In this case, the content uniformity of the extruded product depends on the accuracy and precision of the individual feeders (Dreiblatt 2003). However, metering individual components with multiple feed systems is common practice for extrusion operations in a number of industries, including pharmaceuticals. Modern gravimetric feeders provide the required precision and accuracy to generate content uniformity within USP standards, and the extrusion system is typically sufficient to dampen brief feed variations (Schenck et al. 2011). As the program advances through development, additional resources can be implemented to leverage in-line monitoring and real-time quality control. Successful in-line monitoring has been used for late-stage developmental products to ensure appropriate uniformity of multiple feed streams (Bigert and Smith-Goettler 2011).

As discussed previously, utilization of a crammer feeder or side stuffer is a common means managing low-bulk density formulation components. By individually force feeding these components, higher feed rates can be achieved, improving process efficiency and product output.

In the case of thermally labile APIs, downstream feeding of the drug substance into the already molten carrier reduces the residence time distribution and minimizes thermal stress on the active. Downstream feeding can be accomplished by top feeding with a gravimetric feeder or side feeding with an auger system. The same is true for shear-sensitive materials, where downstream feeding avoids shear stresses encountered when the materials are most viscous, i.e., prior to and during the melt transition.

Liquid injection nozzle systems should be employed when the ASD formulation calls for feeding of liquid or molten components. These are typically equipped with mass flow meters or loss-on-weight reservoir systems to monitor and control the mass flow. Examples of liquid components used in ASD formulations include wetting agents (Tween or Span), surfactants (vitamin E TPGS or docusate sodium), and plasticizers (TEC or DBS). In the case of semisolid, they can be fed via a heated system such that the semisolid component is fed as a liquid. Liquid injection systems should be employed early in process development and product manufacturing to avoid the need for a pregranulation operation in GLP/GMP manufacturing. This improves the overall process efficiency as well as product uniformity.

In summary, preblending and feeding ASD formulation components as a single feed is often the preferred option in early GLP/GMP manufacturing owing to relatively small batch sizes. Separate component feeds can be required when formulation components exhibit poor flow, are thermally sensitive, or are nonsolids. In these cases, the optimal feed configuration should be established early in process development in preparation for drug product manufacturing. Finally, component feeding should be evaluated and implemented prior to high-volume manufacturing to improve overall process efficiency and product uniformity.

7.6.2.7 Shaping and Downstream Processing

A vast majority of ASD compositions produced by melt extrusion will be milled prior to incorporation into a final dosage form. Therefore, the shaping operation performed at the die is often not critical. Strand dies are most typically used, but the strand diameter and number of strands is of limited relevance to downstream milling. However, die geometry can be critical to process development as it is a key determinant of die pressurization. Melt temperature increases with increasing pressure due to viscous dissipation. Therefore, one must be mindful of the pressure formed at the die as it relates to the thermal stress imparted on processed material. As a general rule, a die with the largest open area cross section should be utilized to minimize die pressure while providing an appropriate linear velocity. Obviously, this is most critical when one or more of the ASD formulation components is thermally unstable. Strand dies with numerous and large-diameter orifices or sheet dies are recommended in these cases. Mating of gear pumps can also improve performance and mitigate issues due to back pressure and viscous dissipation.

Back pressure at the die leads to product hold-up at the end of the barrel in proportion to the magnitude of the pressure. Stagnant product at the end of the barrel is subjected to prolonged shear and thermal stress, which can adversely affect the impurity profile of the product (Puaux et al. 2000). As mentioned previously, this can present a degradation issue for sensitive materials. However, this is also an issue as it relates to the quality of an ASD formulation. The ASD product exiting the extrusion system should not be dependent on stagnant compounding to achieve the desired properties. That is to say that if the desired product is only achieved if material is allowed to collect at the die and experiences elevated temperatures and prolonged shear, then the process is not robust. In this case, the product properties are dependent on pressure at the die, which can fluctuate significantly with the feed, and hence disruption in the feed can lead to out-of-specification product. The desired ASD product properties should be attained irrespective of die geometry. It is therefore advised to develop the process in the absence of a die initially to be certain that the desired product properties are achieved independent of the die. Alternatively, sampling or in-line monitoring of melt material prior to the die region can also be used to ensure attainment of desired critical properties. In this way, a more robust process is ensured that produces the desired ASD product independent of feed perturbations.

Calendering, in which a molten feed is forced between two calender rollers to produce films, flakes, or sheets of tablet cores or pellets (Breitenbach 2002), is an operation that has been utilized for commercial production of ASD systems by melt extrusion. However, in solubility enhancement applications, direct shaping for the formation of a final dosage form is almost never employed. This is because the high density of the shaped article eliminates the possibility for a disintegrating system and substantially limits surface area and porosity, reducing the rate at which the article can erode. More often, direct shaping is utilized to control the geometry of the melt-extruded product to ensure that a consistent particle size distribution results from the downstream milling operation. Milling of calendared intermediates greatly increases surface area, thus allowing for disintegrating or rapidly eroding systems, which are desired in solubility enhancement applications.

7.6.2.8 Scale-Up

The key to scale-up of a melt extrusion process for ASD formulations is to optimize the process early in development, fix key parameters, maintain geometric similarity between scales, and maintain a constant SE. Optimization of the key parameters has been discussed earlier, but to reiterate, these parameters are feed material properties, feed configuration, process length, screw/barrel design, temperature profile, die design, and feed rate to screw speed ratio. Identifying target specifications and acceptable operating spaces for each of these parameters will significantly increase the likelihood for successful process transfer between scales. In maintaining geometric similarity between scales, it is recommended to adhere to the same design of extruder (modular or clamshell) and ideally remain with the same equipment make. Changing extrusion system type between scales can introduce unforeseen issues on scale-up and prolong transfer between scales. Finally, it is critical that SE remains constant between scales for the manufacture of ASD systems. This is particularly true when processing in the solubilization regime where energy input per unit mass is critical to achieving the desired product attributes. Therefore, if an increase in throughput is desired, screw speed should be increased congruent with the increased feed rate. In summary, when scaling up a melt extrusion process for the manufacture of an ASD drug product, one should adhere to target specifications, ensure geometric similarity between scales, and maintain a fixed SE. For a more detailed discussion regarding scale-up of melt extrusion processes, the reader is referred to Chap. 2.

7.6.2.9 Quality-by-Design and Process Analytical Technology

Melt extrusion is a complex process with multiple independent variables that affect multiple dependent variables in a convoluted manner. Successful process design principally involves understanding the interplay between both sets of variables and decoupling their interactions in the mapping of process space. Process development requires rigorous and in-depth analysis of the influence of each independent variable as well as interactions between independent variables, as they influence the process-dependent variables and ultimately key product attributes. A statistical experimental design approach will aid in structuring process development work, managing experimental data, and decoupling multivariate interactions. By adhering to a design approach, one can arrive at a thorough understanding of the process and identification of the operating space boundaries.

PAT is a powerful tool to employ toward process development and product manufacturing that leverages process modeling and in-line measurements to provide real-time control. PAT is essential to mapping process space as it provides real-time chemical composition data and allows for monitoring of product property responses to input variations/system fluctuations. The incorporation of PAT during process development allows for monitoring of product quality following normal system perturbations to ensure the process is sufficiently robust. Continued development and use of PAT during product manufacturing assist in the identification and isolation of out-of-specification product. At present, several late-stage products have incorporated such technology for real-time product release based on analytical data collected during production (Bigert and Smith-Goettler 2011).

PAT is typically not implemented until high-volume melt extrusion manufacturing has commenced. However, in-line monitoring can be useful at all manufacturing scales to ensure drug product quality. The proper in-line analytical technique and implementation should be investigated prior to entry into a manufacturing setting and applied to the process as early as it is feasible. Implementation of techniques such as near-infrared spectroscopy (NIR) can provide vital in-process data to ensure that a substantially amorphous product is achieved throughout a melt extrusion campaign. At steady state, no changes in product attributes should be observed and product quality is ensured. However, if a disruption in the process occurs, an in-line analytical technique becomes essential for determining when process recovery is complete and product collection should resume. There are a few different analytical techniques and PAT strategies that can be applied to melt extrusion manufacturing of ASD systems. A detailed discussion of these techniques and strategies is beyond the scope of this chapter. The inquiring reader is referred to Chap. 16 for further information regarding PAT strategies and implementation.

7.7 Summary

Driven by the ever changing needs of new compounds, drug delivery limitations are becoming increasingly common, necessitating technologies that can enable therapy. Melt extrusion has emerged as one of the premier technologies, supporting high-volume manufacturing of novel formulations. Seen by many as the first-line technology, the development of amorphous formulations using extrusion requires an intimate knowledge of material science and process engineering. Given these challenges, it is natural to develop these products within the scope of a structured development strategy encompassing early characterization and stepwise process optimization. As summarized within this chapter, key experiments must be developed for each compound to assess the bioavailability enhancement, stability, and manufacturability of the formulations. As these challenges are met through compositional and process modification, key advantages of later-stage development and production can be leveraged using extrusion to yield a highly efficient and cost-effective manufacturing process. The adaptability of the technology to continuous manufacturing and in-line measurement for PAT support, as discussed in later chapters, position melt extrusion as a key industry enabling technology as the production paradigm shifts away from batch manufacturing. Currently supporting 13 approved products, this number is only going to increase as new therapies in development reach the market.

References

Aharoni SM (1998) Increased glass transition temperature in motionally constrained semicrystalline polymers. Polym Adv Technol 9:169–201

Albers J, Matthée K, Knop K, Kleinebudde P (2011) Evaluation of predictive models for stable solid solution formation. J Pharm Sci 100:667–680

Alonzo DE, Zhang GGZ, Zhou D, Gao Y, Taylor LS (2010) Understanding the behavior of amorphous pharmaceutical systems during dissolution. Pharm Res 27:608–618

Angell CA (1995) Formation of glass from liquids and biopolymers. Science 267:1924–1935

Angell CA (2002) Liquid fragility and the glass transition in water and aqueous solutions. Chem Rev 102:2627–2650

Bigert M, Smith-Goettler B (2011) PAT to support the hot melt extrusion platform. In: 6th Annual Leistritz Pharmaceutical Extrusion Seminar. Leistritz, Clinton, NJ. p 41

Breitenbach J (2002) Melt extrusion: from process to drug delivery technology. Eur J Pharm Biopharm 54:107–117

Breitenbach J (2006) Melt extrusion can bring new benefits to HIV therapy: the example of Kaletra tablets. Am J Drug Deliv 4:61–64

Brouwers J, Brewster ME, Augustijns P (2009) Supersaturating drug delivery systems: the answer to solubility-limited oral bioavailability? J Pharm Sci 98:2549–2572

Chiou WL, Riegelman S (1971) Pharmaceutical applications of solid dispersion systems. J Pharm Sci 60:1281–1302

Crowley MM, Zhang F, Koleng JJ, McGinity JW (2002) Stability of polyethylene oxide in matrix tablets prepared by hot-melt extrusion. BioMaterials 23:4241–4248

Crowley MM, Zhang F, Repka MA, Thumma S, Upadhye SB, Battu SK, McGinity JW, Martin C (2007) Pharmaceutical applications of hot-melt extrusion: part I. Drug Dev Ind Pharm 33: 909–926

Curatolo W (1998) Physical chemical properties of oral drug candidates in the discovery and exploratory development settings. Pharm Sci Technol Today 1:387–393

Davis ME, Brewster ME (2004) Cyclodextrin-based pharmaceutics: past, present and future. Nat Rev Drug Discov 3:1023–1035

DiNunzio JC, Miller DA, Yang W, McGinity JW, Williams RO (2008) Amorphous compositions using concentration enhancing polymers for improved bioavailability of itraconazole. Mol Pharm 5:968–980

DiNunzio JC, Hughey JR, Brough C, Miller DA, Williams RO III, McGinity JW (2010) Production of advanced solid dispersions for enhanced bioavailability of itraconazole using KinetiSol dispersing. Drug Dev Ind Pharm 36:1064–1078

DiNunzio JC, Zhang F, Martin C, McGinity JW (2012) Melt extrusion. In: Miller DA, Watts AB, Williams RO III (eds) Formulating poorly water soluble drugs. Springer, New York, NY, pp 311–362

Dong Z, Chatterji A, Sandhu H, Choi DS, Chokshi H, Shah N (2008) Evaluation of solid state properties of solid dispersions prepared by hot-melt extrusion and solvent co-precipitation. Int J Pharm 355:141–149

Dreiblatt A (2003) Process design. In: Ghebre-Sellassie I, Martin C (eds) Pharmaceutical extrusion technology. Marcel Dekker, Inc., New York, pp 153–170

Friesen DT, Shanker R, Crew M, Smithey D, Curatolo WJ, Nightingale JAS (2008) Hydroxypropyl methylcellulose acetate succinate-based spray-dried dispersions: an overview. Mol Pharm 5:1003–1019

Gao P (2012) Design and development of self-emulsifying lipid formulations for improving oral bioavailability of poorly water-soluble and lipophilic drugs. In: Miller DA, Watts AB, Williams RO III (eds) Formulating poorly water soluble drugs. Springer, New York, NY, pp 243–266

Ghebremeskel AN, Vemavarapu C, Lodaya M (2007) Use of surfactants as plasticizers in preparing solid dispersions of poorly soluble API: selection of polymer–surfactant combinations using solubility parameters and testing the processability. Int J Pharm 328:119–129

Greenhalgh DJ, Williams AC, Timmins P, York P (1999) Solubility parameters as predictors of miscibility in solid dispersions. J Pharm Sci 88:1182–1190

Hauss DJ (2007) Oral lipid-based formulations. Adv Drug Deliv Rev 59:667–676

Horspool KR, Lipinski CA (2003) Advancing new drug delivery concepts to gain the lead. Drug Deliv Technol 3:1–9

Hughey JR, Williams RO (2012) Solid-state techniques for improving solubility. In: Miller DA, Watts AB, Williams RO III (eds) Formulating poorly water soluble drugs. Springer, New York, NY, pp 95–131

Hughey JR, DiNunzio JC, Bennett RC, Brough C, Miller DA, Ma H, Williams RO III, McGinity JW (2010) Dissolution enhancement of a drug exhibiting thermal and acidic decomposition characteristics by fusion processing: a comparative study of hot melt extrusion and KinetiSol dispersing. AAPS PharmSciTech 11:760–774

Kanzer J, Hupfeld S, Vasskog T, Tho I, Hölig P, Mägerlein M, Fricker G, Brandl M (2010) In situ formation of nanoparticles upon dispersion of melt extrudate formulations in aqueous medium assessed by asymmetrical flow field-flow fractionation. J Pharm Biomed Anal 53:359–365

Kapp SJ, Palmer PA (2003) Controls and instrumentation. In: Ghebre-Sellassie I, Martin C (eds) Pharmaceutical extrusion technology. M. Dekker, New York, pp 361–382

Kwong E, Higgins J, Templeton AC (2011) Strategies for bringing drug delivery tools into discovery. Int J Pharm 412:1–7

Lauer ME, Grassmann O, Siam M, Tardio J, Jacob L, Page S, Kindt JH, Engel A, Alsenz J (2010) Atomic force microscopy-based screening of drug-excipient miscibility and stability of solid dispersions. Pharm Res 28:572–584

Leuner C, Dressman J (2000) Improving drug solubility for oral delivery using solid dispersions. Eur J Pharm Biopharm 50:47–60

Lipinski CA (2000) Drug-like properties and the causes of poor solubility and poor permeability. J Pharmacol Toxicol Methods 44:235–249

Macarron R (2006) Critical review of the role of HTS in drug discovery. Drug Discov Today 11:277–279

Marsac PJ, Shamblin SL, Taylor LS (2006) Theoretical and pratical approaches for prediction of drug-polymer miscibility and solubility. Pharm Res 23:2417–2426

Miller D, DiNunzio J, Yang W, McGinity J, Williams R (2008) Targeted intestinal delivery of supersaturated itraconazole for improved oral absorption. Pharm Res 25:1450–1459

Morales JO, Watts AB, McConville JT (2012) Mechanical particle-size reduction techniques. In: Miller DA, Watts AB, Williams III RO (eds) Formulating poorly water soluble drugs. Springer, New York, NY, pp 133–170

Moser JD, Broyles J, Liu L, Miller E, Wang M (2008a) Enhancing bioavailability of poorly soluble drugs using spray dried solid dispersions: part II. Am Pharm Rev 60–78

Moser JD, Broyles J, Liu L, Miller E, Wang M (2008b) Enhancing bioavailability of poorly soluble drugs using spray dried solid dispersions: part I. Am Pharm Rev 68–71

Newman A, Engers D, Bates S, Ivanisevic I, Kelly RC, Zografi G (2008) Characterization of amorphous API:polymer mixtures X-ray powder diffraction. J Pharm Sci 97:4840–4856

Patterson JE, James MB, Forster AH, Lancaster RW, Butler JM, Rades T (2007) Preparation of glass solutions of three poorly water soluble drugs by spray drying, melt extrusion and ball milling. Int J Pharm 336:22–34 (2007)

Puaux JP, Bozga G, Ainser A (2000) Residence time distribution in a corotating twin-screw extruder. Chem Eng Sci 55:1641–1651

Repka MA, Battu SK, Upadhye SB, Thumma S, Crowley MM, Zhang F, Martin C, McGinity JW (2007) Pharmaceutical applications of hot-melt extrusion: part II. Drug Dev Ind Pharm 33:1043–1057

Schenck L, Troup GM, Lowinger M, Li L, McKelvey CA (2011) Achieving a hot melt extrusion design space for the production of solid solutions. In: am Ende DJ (ed) Chemical engineering in the pharmaceutical industry: R & D to manufacturing. Wiley, New York, NY

Serajuddin ATM (1999) Solid dispersion of poorly water-soluble drugs: early promises, subsequent problems, and recent breakthroughs. J Pharm Sci 88:1058–1066

Serajuddin ATM, Sheen P-C, Mufson D, Bernstein DF, Augustine MA (1988) Effect of vehicle amphiphilicity on the dissolution and bioavailability of a poorly water-soluble drug from solid dispersions. J Pharm Sci 77:414–417

Shah N, Sandhu H, Choi DS, Kalb O, Page S, Wyttenbach N (2012) Structured development approach for amorphous systems. In: Miller DA, Watts AB, Williams RO III (eds) Formulating poorly water soluble drugs. Springer, New York, NY, pp 267–310

Six K, Leuner C, Dressman J, Verreck G, Peeters J, Blaton N, Augustijns P, Kinget R, Van den Mooter G (2002) Thermal properties of hot-stage extrudates of itraconazole and Eudragit E100. Phase separation and polymorphism. J Therm Anal Calorim 68:591–601

Thommes M, Baert L, Rosier J (2011) 800 mg Darunavir tablets prepared by hot melt extrusion. Pharm Dev Technol 16:645–650

Van Eerdenbrugh B, Taylor LS (2010) Small scale screening to determine the ability of different polymers to inhibit drug crystallization upon rapid solvent evaporation. Mol Pharm 7:1328–1337

Van Gyseghem E, Stokbroekx S, Novoa deAH, Dickens J, Vanstockem M, Baert L, Rosier J, Schueller L, Van den Mooter G (2009) Solid state characterization of the anti-HIV drug TMC114: interconversion of amorphous TMC114, TMC114 ethanolate and hydrate. Eur J Pharm Sci 38:489–497

Vandecruys R, Peeters J, Verreck G, Brewster ME (2007) Use of a screening method to determine excipients which optimize the extent and stability of supersaturated drug solutions and apllications of this system to solid formulation design. Int J Pharm 342:168–175

Wu C, McGinity JW (2003) Influence of methylparaben as a solid-state plasticizer on the physicochemical properties of Eudragit® RS PO hot-melt extrudates. Eur J Pharm Biopharm 56:95–100

Yoshioka M, Hancock BC, Zografi G (1995) Inhibition of indomethacin crystallization in poly(vinylpyrrolidone) coprecipitates. J Pharm Sci 84:983–986

Zhang F, DiNunzio JC (2012) Solubilized formulations. In: Miller DA, Watts AB, Williams RO III (eds) Formulating poorly water soluble drugs. Springer, New York, NY, pp 171–208

Zhao Y, Inbar P, Chokshi H, Malick AW, Choi DS (2011) Prediction of the thermal phase diagram of amorphous solid dispersions by Flory-Huggins theory. J Pharm Sci 100:3197–3207

Zheng W, Jain A, Papoutsakis D, Dannenfelser R-M, Panicucci R, Garad S (2012) Selection of oral bioavailability enhancing formulations during drug discovery. Drug Dev Ind Pharm 38:235–247

Chapter 8
Nonsink In Vitro Dissolution Testing of Amorphous Solid Dispersions

Jeff T. Gautschi

Abstract Solid dispersion technology has been used over the last three decades to improve the dissolution and oral absorption of poorly soluble compounds. While the characterization of dissolution performance of crystalline pharmaceutical systems has long been established, the dynamic nature of the amorphous dissolution processes requires the use of unique methodologies. The in vitro differentiation of the drug and drug-containing species of these systems is crucial to accomplishing the measurement of the critical-to-performance free drug concentrations as a function of time. This chapter describes the theoretical aspects of amorphous dissolution and recent examples applying free drug dissolution testing to the oral bioavailability assessment of solid dispersion formulations.

8.1 Introduction

Amorphous solid dispersions (ASDs)[1] have long been noted for their potential advantages to the pharmaceutical industry (Chiou and Reigelman 1971; Goldberg et al. 1965). More than ever before, however, ASDs are being embraced in the formulation development and manufacturing of oral drug candidates that have solubility-limited absorption. Such insoluble Biopharmaceutics Classification System (BCS) II and IV compounds (Amidon et al. 1995) are increasingly filling drug development pathways[2] as newer and better biological targets demand hydrophobic interactions for

[1] Amorphous solid dispersion (ASD) is used to describe a homogenous dispersion of noncrystalline API and excipient(s) at molecular compositions. Similar systems are often described as solid dispersions, amorphous molecular dispersions, solid solutions, solid liquids, and others.
[2] It is typically stated that between 40 and 70 % of new chemical entities under development in the pharmaceutical industry are insoluble (van de Waterbeemd and Gifford 2003; Benet and Wu 2006).

J. T. Gautschi (✉)
Department of Chemistry, Oregon State University, 153 Gilbert Hall, Corvallis, OR, USA
e-mail: jeff.gautschi@oregonstate.edu

Agere Pharmaceuticals, Inc., Bend, OR, USA

M. A. Repka et al. (eds.), *Melt Extrusion*,
AAPS Advances in the Pharmaceutical Sciences Series 9,
DOI 10.1007/978-1-4614-8432-5_8,

selectivity and therapeutically balanced potency. It is well known that ASDs can enhance dissolution, provide large supersaturation, and maintain that supersaturation in aqueous media for these poorly soluble, but otherwise promising, active pharmaceutical ingredients (APIs; reviewed in Janssens and Van der Mooter 2009; Yu 2001; Leuner and Dressman 2000).

Commercial products recently launched with ASD solubilization technology are shown in Table 8.1,[3] and many more are reported to have entered human clinical trials (Friesem et al. 2008). The most used commercial processing methods to manufacture ASDs are hot-melt extrusion (HME; Repka et al. 2007; Crowley et al. 2007; DiNunzio et al. 2008; Breitenbach 2002) and spray drying (SD; Patel and Suthar 2009; Ronald 1997; Oakley 1994; Shoyele and Cawthorne 2006; Vehring 2008) because these processes are mature (Breitenbach 2002; Patel and Suthar 2009; Ronald 1997), scalable (Breitenbach 2002; Oakley 1994), and controllable (Repka et al. 2007; Crowley et al. 2007; Shoyele and Cawthorne; Vehring 2008). Additional process methodologies to produce ASDs exist, but until recently, only SD and HME were employed on a commercial scale. The current exception is the use of antisolvent controlled precipitation processes to make Roche's Zelboraf® (vemurafenib),[4] the most recently launched ASD product (see Table 8.1). It is anticipated that ASDs will continue to gain momentum in the industry as enabling formulation strategies owing to the recent commercial launches of ASDs on the market, the well-known oral bioavailability enhancements to be gained, and the increasing amount of promising, but poorly soluble, candidates entering drug development pipelines.

From formulation design and development, to cGMP manufacturing and commercialization, amorphous molecular dispersions (AMDs) undergo similar treatment and testing to that of traditional drug formulations. These ASDs, however, require special considerations that extend to the material's critical-to-performance attributes, such as dissolution, physicochemical properties, and stability. For example, Fig. 8.1 depicts the intestinal dissolution events for a low soluble drug formulated as a crystalline drug (Fig. 8.1a), and formulated as an ASD (Fig. 8.1b). As a consequence of the formulation-derived drug species (Fig. 8.1b, box), the dissolution performance of a low soluble drug in an ASD (and any formulations thereof) follow more complex dissolution pathways than crystalline forms.

The dissolution rate to produce freely solvated drug (i.e., "free drug") is often the rate-limiting step to overall oral drug absorption for low soluble drugs (e.g., BCS II compounds). This rate can therefore greatly impact the oral bioavailability. Fundamental factors affecting this rate for crystalline material are described by the well-known modified Noyes–Whitney expression:

$$\frac{dm}{dt} = \frac{D_D \cdot S}{V \cdot h}[C_s - C_t]$$

---

[3] (a) *Zelboraf* (*vemurafenib*) CDER (2011), (b) *Incivek* (*telaprevir*) Bottorf et al. (2007), (c) *Intelence* (*etravirine*) Pomerantz (2007), (d) *Kaletra* (*lopinavir and rotinavir*) Rosenberg et al. (2008), and (e) *Cesamet* (*nabilone*) Dong (2005).

[4] Center for Drug Evaluation and Research Application No. 202429, Clinical Pharmacology and Biopharmaceutics Review(s), www.accessdata.fda.gov.

Table 8.1 Commercial products employing amorphous molecular dispersion (AMD) technology

Product	Manufacturing Process	Polymer (and Estimated API Loading)	Firm
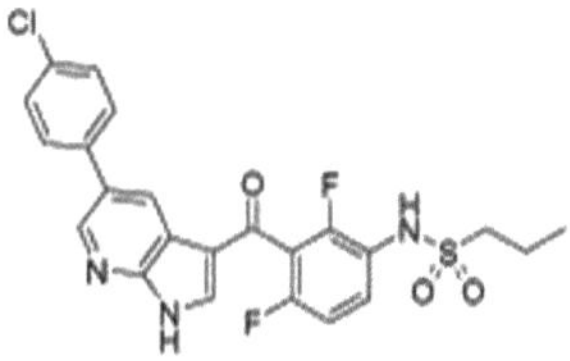 Zelboraf (vemurafenib)	Anti-solvent Precipitation	HPMCAS (20%–40%)	Roche
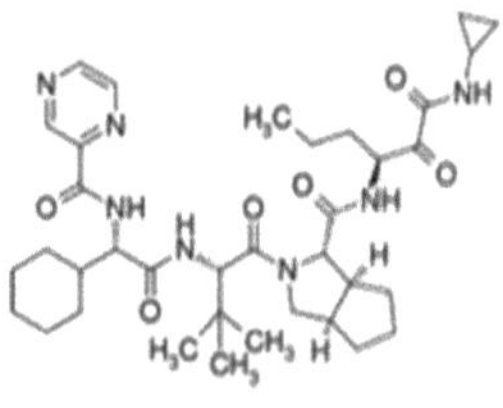 Incivek (telaprevir)	Spray Drying	HPMCAS (25–40%)	Vertex
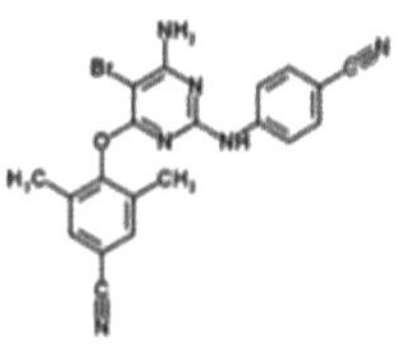 Intelence (etravirine)	Spray Drying	HPMC (20%)	Tibotec/ Johnson & Johnson
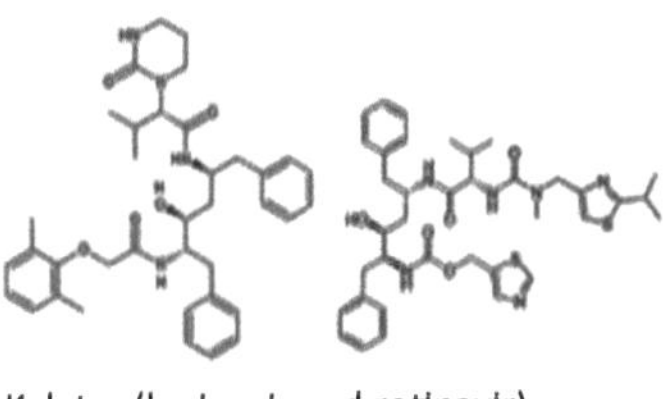 Kaletra (lopinavir and rotinavir)	Melt Extrusion	PVP-VA (40%)	Abbott
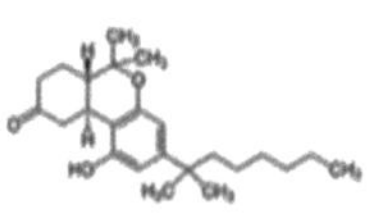 Cesamet (nabilone)	Melt Extrusion	PVP (20%)	Valeant

API active pharmaceutical ingredient, *HPMC* hydroxypropyl methylcellulose, *HPMCAS* hydroxypropyl methylcellulose acetate succinate, *PVP* polyvinylpyrrolidone, *VA* vinyl acetate

where dm/dt is the rate expressed at the change in the amount of freely dissolved drug (m) per unit time (t), D_D is the drug's diffusivity in a uniform boundary layer, S is the surface area, V is the volume of dissolution media, h is the thickness of the uniform boundary layer, and $C_s - C_t$ describes the difference in the concentration

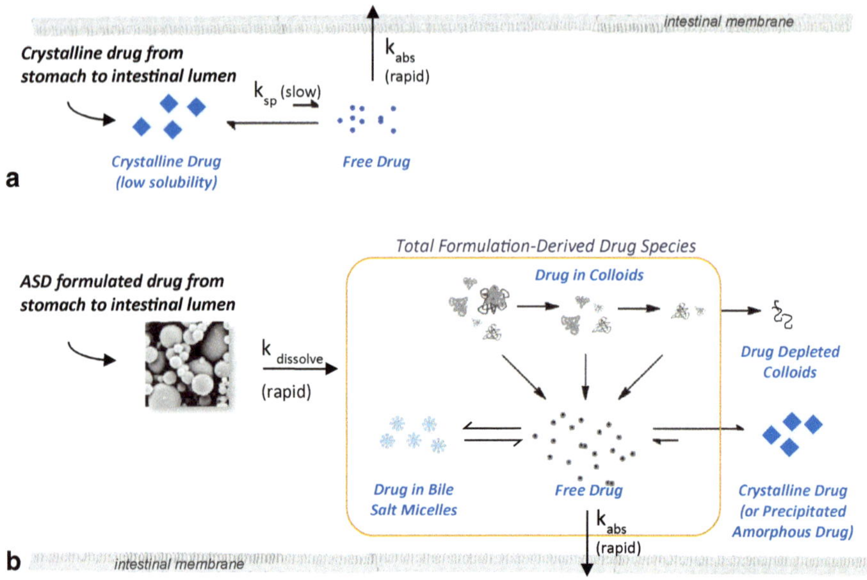

Fig. 8.1 Simple illustration of drug speciation absorption for a low soluble, orally administered drug in **a** crystalline form (unformulated) and **b** amorphous solid dispersion (ASD) formulation. Total formulation-derived drug species upon disintegration are shown in *box* for drug in ASD

of drug and the drug's saturation solubility in the medium. Expressed as the integral, free drug amounts can be written as:

$$m = \frac{D_S}{V}(1 - \exp(-Kt)),\ \text{where } K = S \cdot D_D/h.$$

Limitations of this for describing ASDs are well known (Martinez and Amidon 2002; Balakrishnan et al. 2004). Static surface areas and boundary layers, for example, may not be fixed for ASD-formulated drugs as the primary particles disintegrate into colloidal drug-containing species (see Fig. 8.1b) and these transit down the gastrointestinal (GI) tract. While a crystalline drug deposited in the GI lumen is limited in solubility by the crystalline equilibrium (see Fig. 8.1a), drug presented in the higher energy amorphous state can achieve supersaturation that can be described thermodynamically as:

$$\frac{S_a}{S_c} = \exp\left[\frac{\Delta G_c}{RT}\right],\ \text{where } \Delta G_c = \Delta H_f \times \frac{(T_m - T)T}{T_m^2},\ \text{and}$$

where S_a/S_c is the ratio of the amorphous solubility to crystalline solubility, ΔH_f is the heat of fusion, and T_m is the melting point of the crystalline material.

Successful ASD-formulated drugs are designed to achieve this supersaturation, but this places the free drug in a thermodynamic regime that will promote rapid re-crystallization or amorphous precipitation unless interference occurs. It is therefore critical to understand and manipulate the precipitation rate of amorphous drug at supersaturations achieved by ASDs since the sustainment of any solubility enhancements can greatly impact drug absorption for BCS II compounds. The "spring and parachute" dissolution profile (Guzmán et al. 2007) from ASD formulations (and other insoluble drug formulations) is currently used to describe the events of drug dissolution from a supersaturating formulation. This includes the rapid dissolution ("spring") of the amorphous form followed by the inhibition ("parachute") of the compound's propensity to rapidly return to its most thermodynamically stable form. Proper in vitro dissolution testing of these systems remains the important empirical approach to accurately determine free drug enhancement and differentiation among different ASDs. This chapter will focus on the in vitro dissolution performance testing of ASDs to measure the absorbable free drug species in solution separately from the other drug-containing species in solution.

8.2 Sink In Vitro Dissolution

For oral, immediate-release final drug products, the importance of dissolution testing is exemplified by the regulated use of United States Pharmacopeia (USP) method <711>.[5] Guidelines take into consideration the apparatus, sample preparation, sampling procedure, and other aspects of the methodology. In the context of quality control testing of final drug product per USP methodology and guidelines (US Department of Health and Human Services, Food and Drug Administration, Center for Drug Evaluation and Research (CDER) 1997, 2000), the goal is twofold. First, testing the dissolution-rate consistency of the product provides some insight on the potential in vivo consistency. Second, dissolution testing also provides a performance metric for the manufacturing processes used to produce the ASD. Currently, USP methodology is the gold standard for in vitro dissolution testing of final drug product. Consequently, final drug products containing ASD solubilization technology are tested for complete dissolution over the course of a practical testing duration.

In order to establish this complete release metric, USP dissolution testing for final drug product is developed under sink conditions. The in vitro volumes are based on the accepted values for human GI volumes. For example, in the fasted state, the volumes are considered to be around 250 mL. A soluble compound as determined by the BCS categorization (i.e., BCS I, III; Amidon et al. 1995) is therefore considered one that is soluble in less than 250 mL of test media at its highest dose. Insoluble compounds, on the other hand, are those that are not soluble in these media volumes at their expected doses. For these compounds, sink dissolution conditions cannot be achieved in conventional aqueous media like 0.1 or 0.01 N HCl (gastric mimic pH = 1 or 2, respectively), or phosphate-buffered saline (PBS, intestinal mimic pH = 6.5).

Sink conditions are established during analytical method development to ensure that the drug product completely releases the API as free drug over a brief period

[5] http://www.pharmacopeia.cn/v29240/usp29nf24s0_c711.html.

Table 8.2 Dissolution testing outcomes by drug development stage

Stage	Dissolution testing outcomes
Discovery	Performance screening of leads
Formulation design	Performance screening of formulation prototypes
Formulation development	Performance screening of formulation compositions
Formulation process development	Performance variability as a function of process variables
Formulation process scale-up	Performance variability as a function of process variables
Manufacturing	Performance quality control

of time (e.g., < 3 h). For insoluble compounds, the appropriate use of ionic and nonionic surfactants in the media to achieve sink conditions has been discussed (Gowthamarajan and Singh 2010). Likewise, the merits of the four established apparatuses (i.e., apparatus 1, basket; apparatus 2, paddle; apparatus 3, reciprocating cylinder; and apparatus 4, flow-through cell) to test products containing low soluble compounds under sink conditions have also been reported (Uddin et al. 2011; Brown et al. 2004). Understanding these parameters for the release testing of clinical trial materials (CTMs) that contain AMDs and their insoluble molecules continues to be discussed. Currently, however, this is addressed similarly to release testing of soluble drug products.

8.3 Physiologically Relevant Dissolution

In vitro dissolution testing is not relegated to testing final drug products, though. This valuable performance assessment is utilized extensively at all stages of oral drug development. Table 8.2 lists the rationale for dissolution testing at various stages of drug/formulation development. Even discovery groups are utilizing dissolution testing of formulations for lead compound selection (Padden et al. 2011). By measuring the drug candidate's ability to disintegrate, dissolve, and release free drug into physiologically relevant media, performance can be established and tracked against a number of design, development, and manufacturing variables.

In the best-case scenario, the in vitro dissolution testing predicts the in vivo performance. Toward this end, there has been considerable attention paid to in vitro–in vivo correlation (IVIVC; Cardot et al. 2007). In general, IVIVC efforts continue to focus on dissolution apparatus and methodology (Fotaki and Vertzoni 2010), media (Taupitz et al. 2013; Otsuka et al. 2013; Arndt et al. 2013; Reppas and Vertzoni 2012; Bevernage et al. 2010), and physiologically based pharmacokinetic (PBPK) modeling and statistical analyses (Zhao et al. 2011). In some cases, remarkable correlations have been reported; however, achieving universal methods for accurate IVIVC remains elusive.

Nevertheless, it is reasonable to still strive toward the development of physiologically relevant dissolution procedures. This is especially important during the design and development of performance-enhancing solubilized formulations like ASDs. By incorporating in vitro performance assays during formulation development, comparisons between formulations, and between formulated and unformulated API can be assessed for rank ordering, process-to-performance metrics, and stability sample performance. Rank ordering formulations based on in vitro performance can greatly reduce the number of in vivo studies necessary to screen different formulations. During downstream formulation development and manufacturing scale-up, in vitro dissolution of ASDs can be used to assess performance improvements or detriments of altering the formulation composition or unit operation processes.

8.4 Nonsink In Vitro Dissolution

In order to accomplish in vitro dissolution performance for insoluble drug candidates, nonsink conditions should be utilized since this more closely represents the physiological situation. For example, if a moderate-to-low potency compound is to be examined in a fasted-state simulated intestinal fluid (FaSSIF), the projected highest dose may be ≥ 250 mg, especially for in vivo (nonhuman) PK and toxicology studies. This would require the solubility of the API to be ≥ 1.0 mg/mL in the dissolution media to be considered sink conditions. Compounds requiring solubilization formulation strategies such as those incorporated into ASDs, are typically much lower in solubility and it is often not possible to adjust the media to reach these levels without the concomitant loss of relevance to in vivo physiology. Thus, nonsink dissolution assays are most often applied to test poorly soluble API during the formulation development stages.

Using nonsink dissolution procedures to examine the performance of ASDs provides the dissolution rate, amorphous API solubility supersaturation enhancement, and the potential sustainment inherent to successful ASDs. Importantly, the nonsink dissolution affords insight into the biorelevant API precipitation rates. For example, Fig. 8.2 shows the dissolution profiles for the amorphous insoluble compound fenofibrate presented in a 1:3 API : HPMCAS-M ASD (blue line) made via SD and comparatively as the unformulated API (red line). The profile exhibited by the ASD has been described as a "spring and parachute" dissolution profile (Brouwers et al. 2009). The hallmarks of this profile are that the higher energy amorphous API dissolves rapidly, reaches a supersaturation state that is sustained for some period of time, and eventually precipitates as a lower energy crystalline form, or as an amorphous solid. In contrast, the unformulated API only reaches a very low solubility of about 0.35 μg/mL after 60 min in the dissolution media.

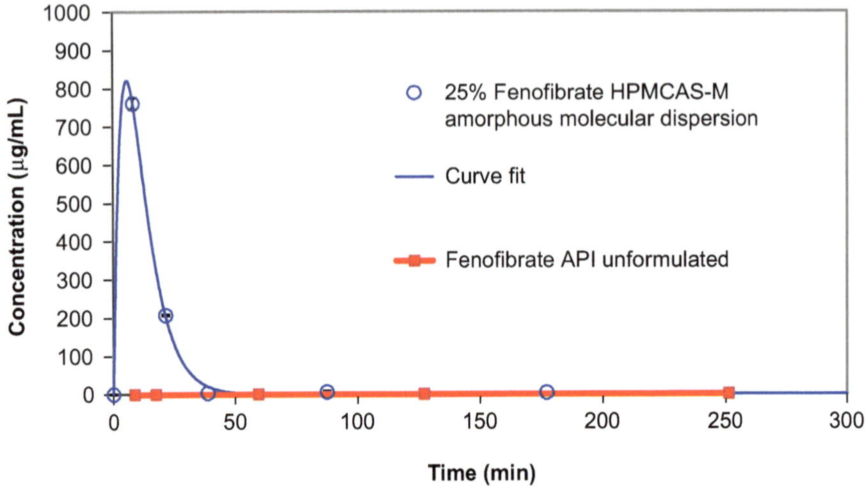

Fig. 8.2 In vitro nonsink dissolution of 1:3 fenofibrate: HPMCAS-M (*blue line*) and unformulated fenofibrate (*red line*)

8.5 Drug Speciation of AMDs and Free Drug Analyses

An important aspect to consider when testing the in vitro dissolution performance of ASDs is the different drug-containing species that can occur in the media. Figure 8.1 differentiates this drug speciation as it occurs for crystalline API during dissolution (Fig. 8.1a), and as it transpires for AMD dissolution (Fig. 8.1b). In the case of the crystalline API, two or three physiologically relevant drug species can be present during testing. These are crystalline drug, freely dissolved drug (or "free drug"), and drug in micelles, if micelles are present in the media. It is critical to separate the crystalline drug from the mixture and to measure the free drug as a function of time in the media since it is only the free drug that is absorbed in vivo. This is relatively straightforward to do and is typically accomplished by filtering and/or centrifugation steps in the sample processing.

In contrast to the dissolution of crystalline API, Fig. 8.1b shows the various drug species that are formed during the dissolution of drug from an ASD. As shown, the ASD produces similar drug species as that of unformulated API (i.e., free drug, crystalline drug, drug in micelles); however, there also exists drug–excipient interactions that can lead to physiologically relevant API–excipient complexes and colloid species during dissolution. These colloidal materials form during the wetting and disintegration of the primary particles of the ASD and vary in size and density according to the physicochemical and solution properties of the particular drug–excipient interaction (Bikiaris 2011). For example, a high degree of API–polymer interaction (e.g., hydrophobic or hydrogen bonding) in one polymeric ASD can lead to a longer duration of API chain disentanglement and therefore a potentially greater amount of colloidal material, as compared to an ASD made from a different polymer (Balata et al. 2010).

Removing the colloidal species during the sample preparation, and separately measuring free drug as a function of time, can be challenging. Traditional in vitro dissolution sample preparation methodologies may not sufficiently remove the resulting submicron and/or nanometer-sized colloidal material that can form. Instrument-intensive methodologies such as nuclear magnetic resonance (NMR) have been employed to examine complex dissolution profiles for API–excipient systems (Kojima et al. 2012; Abhishek and Chandrakumar 2011; Dahlberg 2010; Zhang et al. 2011), but it is desirable to be able to measure free drug using more practical dissolution setups. More recently, several authors have adapted diffusion and partitioning methodologies to the assessment of free drug in solution. In a study by Alonzo et al. (2011), researchers used dialysis bags to conduct free drug dissolution testing by measuring the diffusion across the membrane. While effective at determining the free drug concentration, dialysis methods similar to this can be time and labor intensive thus limiting the throughput of the technology. Other, more resource-efficient methods have also been reported, including the use of immiscible liquids to determine the free drug concentration via partitioning into the water immiscible phase (Shi et al. 2010). Although researchers have shown acceptable correlations, many concerns, immiscible phase, including the incorporation of colloidal material into the water immiscible phase remain. Despite the difficulties in assaying for free drug during ASD dissolution, proprietary methods have been developed that can greatly impact the formulation design of polymeric ASDs.

8.6 Case Study 1

It is often the case that the formulation scientist will utilize an in vitro dissolution assessment to screen different ASDs during formulation development studies. Screening efforts can help rank order a larger set of formulations and reduce the scope of in vivo studies by selecting a smaller set of materials to test. It is therefore important to measure the free drug, in addition to measuring the total drug species, during in vitro performance assessments of ASDs (see also Fig. 8.2). Fenofibrate is currently marketed as an adjunctive therapeutic to treat hypertriglyceridemia or mixed dyslipidemia in adult patients. Fenofibrate is considered a BCS II molecule because of its dose and physicochemical properties (e.g., maximum dose of 120 mg/day; aqueous solubility $<<480$ μg/mL; log $P \approx 5.3$). Five amorphous molecular dispersions of fenofibrate and polymer were produced at a 25 % API loading by weight using laboratory-scale SD techniques (Smithey et al. 2010) and these compositions are shown in Table 8.3.

The 1:4 fenofibrate : polymer ASDs were assessed for in vitro performance in PBS (pH = 6.5) media, and dosed as dry powders at 1 mg active equivalent (4 mg ASD) per milliliter of media (i.e., nonsink conditions) in scintillation vials stirred at approximately 150 rpm. Aliquots of media were removed at set time points

Table 8.3 Compositions of fenofibrate amorphous molecular dispersions (AMDs) and the processes to manufacture them

API loading (%)	Polymer	Manufacturing process
25	Soluplus®	Spray drying
25	HPMCP-H55	Spray drying
25	CAP	Spray drying
25	HPMCAS-M	Spray drying
25	PVP-VA	Spray drying

API active pharmaceutical ingredient, *HPMCP* hydroxypropyl methylcellulose phthalate, *CAP* cellulose acetate phthalate, *HPMCAS* hydroxypropyl methylcellulose acetate succinate, *PVP* polyvinylpyrrolidone, *VA* vinyl acetate

(i.e., T = 10, 20, 40, 90, 120, 180 min), and both the total drug and free drug concentrations (see Fig. 8.1b) were processed and measured separately. Total drug was assessed using a procedure similar to microcentrifuge methodologies previously described (Friesem et al. 2008). Free drug was measured by proprietary means.

Figure 8.3 depicts the C_{max} results of the total drug and free drug analyses for these five ASDs. Although there is some differentiation of the different solubilized formulations when assessing total drug species only, the C_{max} for these dispersions are very similar (see Fig. 8.3a). If a rank order were to be produced from the total drug analysis, it would be HPMCAS-M > CAP > HPMCP-H55 > Soluplus > PVP-VA >>> crystalline API. In contrast, when free drug is assessed (see Fig. 8.3b), there is a much greater distinction among the ASDs. Moreover, the rank ordering results are completely different than that which arise from the total drug analysis. Using the free drug concentrations at C_{max}, the rank ordering is Soluplus > HPMCP-H55 > CAP > HPMCAS-M > PVP-VA > crystalline API.

By obtaining the concentrations of the true free drug portion of the drug speciation that occurs during nonsink dissolution of ASDs, more accurate computational calculations can also be made than if total drug values are used. Using the fenofibrate free drug concentrations of the five compositions, the calculated fraction absorbed can be estimated as a function of free drug concentrations and dose. Figure 8.4 displays these calculations using the physiological parameters for beagle dogs. The fraction absorbed for unformulated, crystalline fenofibrate is also calculated, and an overlay of the range of in vivo performance is estimated from previous reports (Chen et al. 2009).

8.7 Case Study 2

Clearly the in vitro nonsink dissolution free drug analyses can be important during the selection of excipients in the formulation design stages. These measurements can also be used to determine the extent of performance advantages and disadvantages of different API loadings in ASDs. Figure 8.5 shows the in vitro and in vivo dissolution profiles for two ASDs, each produced with the same polymeric excipient, but at two different API loadings (i.e., 25 and 40 % w/w). As in the previous fenofibrate example, the in vitro performance was measured in PBS (pH = 6.5), at a dry powder

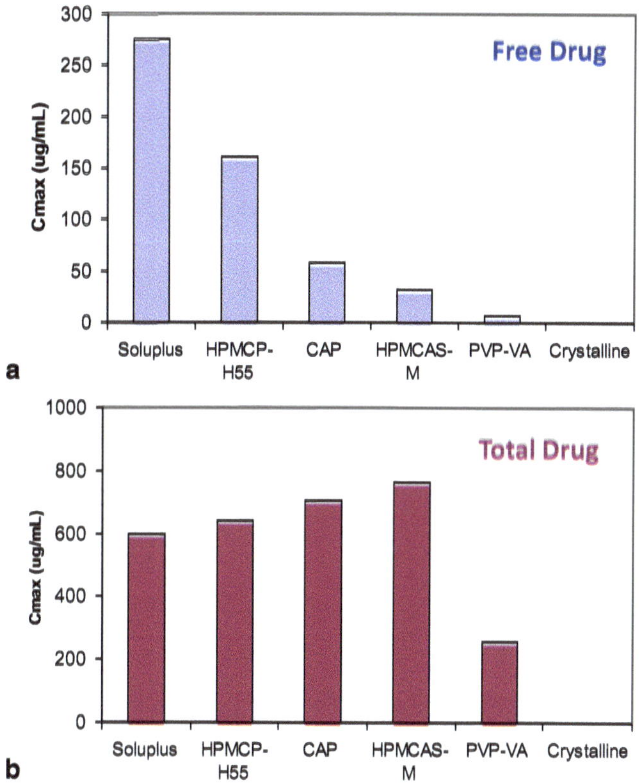

Fig. 8.3 C_{max} values for five amorphous molecular dispersions (AMDs) of 1:3 fenofibrate: polymer as determined for in vitro nonsink dissolution measurements of **a** total drug species and **b** free drug species

dose of 1 mg active equivalent per milliliter of media (i.e., nonsink conditions). The in vivo PK analysis was carried out on Sprague Dawley rats ($n=3$) at a 30 mg/kg dose of SDI suspension.

In Fig. 8.5, the total drug analyses would suggest that little or no performance differences would be expected between the 25 and 40 % API loadings. The free drug analysis, however, forecasts greater performance for the ASD with the lower API loading. This result is not surprising because lower drug loading often leads to a dose-equivalent enhanced performance for ASDs made with the same excipient. This is due, in part, because there is more polymer present in solution to help sustain the API supersaturation state. It is therefore also no surprise that the in vivo study results show a better relationship with the in vitro free drug analysis as compared to the total drug analysis.

The aforementioned case studies demonstrate the importance of proper in vitro dissolution studies for ASDs. When accomplished in nonsink conditions, and analyzed for free drug concentrations as a function of time in the dissolution media, the

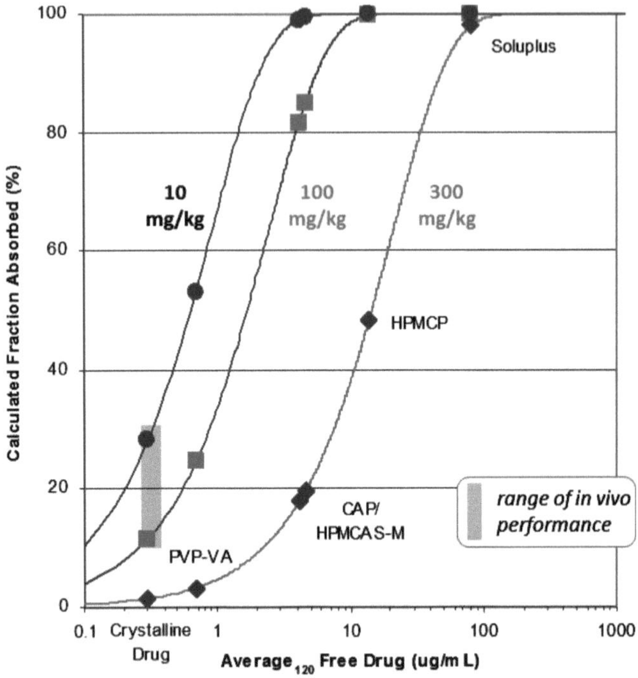

Fig. 8.4 Calculated fraction absorbed (%) of 1:3 fenofibrate: polymer amorphous molecular dispersions (AMDs) and crystalline API as a function of the average in vitro free drug concentrations (μg/mL) and dose (mg/kg) for beagle dog physiology. The *shaded box* is the range of in vivo performance as estimated from literature values

results can be used to make informative decisions regarding excipient choice and drug loading. This can also be extended to monitoring any process changes during formulation development and scale-up.

8.8 Conclusion

The number of poorly soluble compounds in the pharmaceutical development pipelines is increasing. Concomitantly, so too are the drug delivery systems suited for oral delivery of promising, but low soluble, drug candidates. ASDs, however, remain well known for their potential formulation advantages for BCS II and IV compounds. More than ever before, ASD formulations are found in the marketplace as an answer for formulating drugs with solubility-limited absorption. Critical-to-performance attributes of ASDs, such as in vitro dissolution, require specialized knowledge and testing procedures to accurately determine their potential for bioavailability enhancements during formulation design and development, manufacturing, and commercial production.

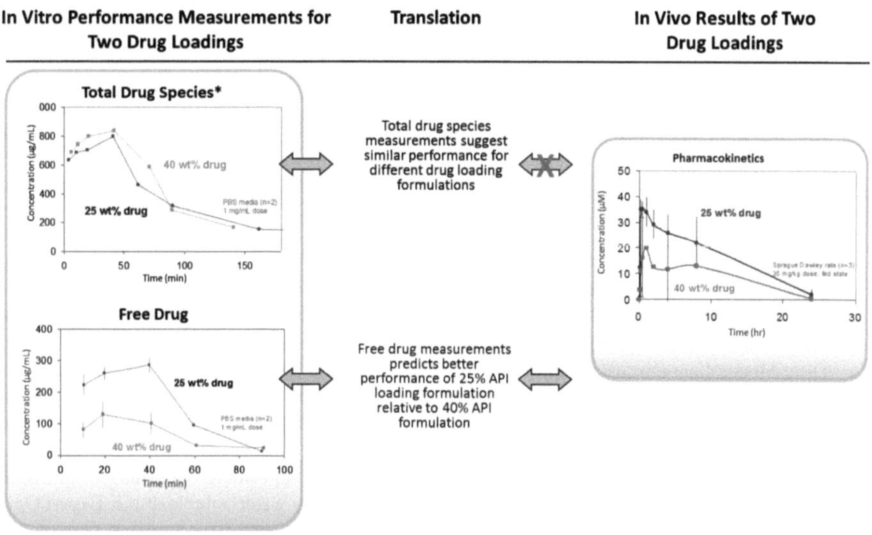

Fig. 8.5 In vitro performance of total drug species and free drug species concentrations and their in vivo relationship for two amorphous molecular dispersions (AMDs) of the same active pharmaceutical ingredient (API) and polymeric excipient, but at two different drug loadings

An important aspect to consider when testing the in vitro dissolution performance of ASDs is the different drug-containing species that can occur in the media. Removing the colloidal species during the sample preparation, and separately measuring free drug as a function of time, can be challenging; however, accurately determining the amount of free drug that is available for absorption remains an important empirical measurement. Case studies discussed in this chapter are examples of how the data obtained from such measurements can lend insight into the potential physiological impacts of ASD formulations. Such test results allow for the proper formulation design and selection, the efficient adjustments during formulation development, and the monitoring of quality attributes during manufacturing.

Acknowledgments The author would like to thank the following individuals and organizations: Marshall Crew, Dan Smithey, and James Fennewald at Agere Pharmaceuticals, Inc. for contributions to the in vitro drug speciation dissolution studies, and Shaukat Ali and Nigel Langley at BASF for the generous samples of Soluplus® polymeric excipient.

References

Abhishek B, Chandrakumar N (2011) Real-time in vitro drug dissolution studies of tablets using volume-localized NMR (MRS). Appl Magn Reson 40:251–259

Alonzo DE, Gao Y, Zhou D, Mo H, Zhang GGZ, Taylor LS (2011) Dissolution and precipitation behavior of amorphous solid dispersions. J Pharm Sci 100:3316–3331

Amidon GL, Lennernäs H, Shah VP, Crison JR (1995) A theoretical basis for a biopharmaceutical drug classification: the correlation of in vitro drug product dissolution and in vivo bioavailability. Pharm Res 12:413–420

Arndt M, Chokshi H, Tang K, Parrott NJ, Reppas C, Dressman JB (2013, Aug) Dissolution media simulating the proximal canine gastrointestinal tract in the fasted state. Eur J Pharm Biopharm 84(3):633–641

Balakrishnan A, Rege BD, Amidon GL, Polli JE (2004) Surfactant-mediated dissolution: contributions of solubility enhancement and relatively low micelle diffusivity. J Pharm Sci 93:2064–2075

Balata G, Mahdi M, Bakera RA (2010) Improvement of solubility and dissolution properties of ketoconazole by solid dispersion and inclusion complexes. Asian J Pharm Sci 5:1–12

Benet LZ, Wu C-Y (2006) Using a biopharmaceutics drug disposition classification system to predict bioavailability and elimination characteristics of new molecular entities. New Jersey Drug Metabolism Discussion Group (NJDMDG), Somerset, NJ. October 5:2006

Bevernage J, Brouwers J, Clarysse S, Vertzon M, Tack J, Annaert P, Augustijns P (2010) Drug supersaturation in simulated and human intestinal fluids representing different nutritional states. J Pharm Sci 99:4525–4534

Bikiaris DN (2011) Solid dispersions, part I: recent evolutions and future opportunities in manufacturing methods for dissolution rate enhancement of poorly water-soluble drugs. Expert Opin Drug Deliv 8:1501–1519

Bottorf KJ, Katstra JP, Gasper F (2007) US Patent 2007/0218138. Sept. 20th, 2007

Breitenbach J (2002) Melt extrusion: from process to drug delivery technology. Eur J Pharm Biopharm 54:107–117

Brouwers J, Brewster ME, Augustijns P (2009) Supersaturating drug delivery systems: the answer to solubility-limited oral bioavailability? J Pharm Sci 98:2549–2572

Brown CK, Chokshi HP, Nickerson B, Reed RA, Rohrs BR, Shah PA (2004) Acceptable analytical practices for dissolution testing of poorly soluble compounds. Pharm Technol 28:56–65

Cardot J-M, Beyssac E, Alric M (2007) In vitro–in vivo correlation: importance of dissolution in IVIVC. Dissolution Technol 14:15–19

CDER (2011) NDA submission application #202429. March 21, 2011

Chen Y, Lu Y, Chen J, Lai J, Sun J, Hu F, Wu W (2009) Enhanced bioavailability of the poorly water-soluble drug fenofibrate by using liposomes containing bile salt. Int J Pharm 376:153–160

Chiou WL, Reigelman S (1971) Pharmaceutical applications of solid dispersion systems. J Pharm Sci 60:1281–1302

Crowley MM, Zhang F, Repka MA, Thumma S, Upadhye SB, Battu SK, McGinity JW, Martin C (2007) Pharmaceutical applications of hot melt extrusion: part I. Drug Dev Ind Pharm 33:909–926

Dahlberg C (2010) Doctoral thesis, Royal Institute of Technology, Stockholm, Sweden

DiNunzio JC, Miller DA, Yang W, McGinity JW, Williams RO (2008) Amorphous compositions using concentrating enhancing polymers for improved bioavailability of itraconazole. Mol Pharm 5:968–980

Dong W (2005) Multiparticulate drug delivery system for lipophilic drugs and macromolecules. PhD dissertation in Chemistry, Freie Universität, Berlin

Fotaki N, Vertzoni M (2010) Biorelevant dissolution methods and their application in in vitro–in vivo correlations for oral formulations. Open Drug Deliv J 4:2–13

Friesem DT, Shanker R, Crew M, Smithey DT, Curatolo WJ, Nightingale JS (2008) Hydroxypropyl methylcellulose acetate succinate-based spray-dried dispersions: an overview. Mol Pharm 5:1003–1019

Goldberg AH, Gibaldi M, Kanig JL (1965) Increasing dissolution rates and gastrointestinal absorption of drugs via solid solutions and eutectic mixtures I. Theoretical consideration and discussion of the literature. J Pharm Sci 54:1145–1148

Gowthamarajan K, Singh SK (2010) Dissolution testing for poorly soluble drugs: a continuing perspective. Dissolution Technol 17:24–32

Guzmán HR, Tawa M, Zhang Z, Ratanabanangkoon P, Shaw P, Gardner CR, Chen H, Moreau J-P, Almarsson Ö, Remenar JF (2007) Combined use of crystalline salt forms and precipitation inhibitors to improve oral absorption of celecoxib from solid oral formulations. J Pharm Sci 96:2686–2702

Janssens S, Van der Mooter G (2009) Review: physical chemistry of solid dispersions. J Pharm Pharmacol 61:1571–1586

Kojima T, Higashi K, Suzuki T, Tomono K, Moribe K, Yamamoto K (2012) Stabilization of a supersaturated solution of mefenamic acid from a solid dispersion with EUDRAGIT® EPO. Pharm Res 29:2777–2791

Leuner C, Dressman J (2000) Improving drug solubility for oral delivery using solid dispersions. Eur J Pharm Biopharm 50:47–60

Martinez MN, Amidon GL (2002) A mechanistic approach to understanding the factors affecting drug absorption: a review of the fundamentals. J Clin Pharmacol 42:620–643

Oakley DE (1994) Scale-up of spray dryers with the aid of computational fluid dynamics. Dry Technol 12:217–233

Otsuka K, Shono Y, Dressman JB (2013) Coupling biorelevant dissolution, methods with physiologically based pharmacokinetic modelling to forecast in-vivo performance of solid oral dosage forms. J Pharm Pharmacol 65:937–952

Padden BE, Miller JM, Robbins T, Zocharski PD, Prasad L, Spence JK, LaFountaine J (2011) Amorphous solid dispersions as enabling formulations for discovery and early development. Am Pharm Rev 14:66–73

Patel RP, Suthar AM (2009) Spray drying technology: an overview. Indian J Sci Technol 2:44–47

Pomerantz RJ (2007) Combining biomedical research within academia and industry in the 21st century (keynote address), AAPS Annual Meeting, 2007, San Diego, CA

Repka MA, Battu SK, Upadhye SB, Thumma S, Crowley MM, Zhang F, Martin C, McGinity JW (2007) Pharmaceutical applications of hot melt extrusion: part II. Drug Dev Ind Pharm 33:1043–1057

Reppas C, Vertzoni M (2012) Biorelevant in-vitro performance testing of orally administered dosage forms. J Pharm Pharmacol 64:919–930

Ronald CD (1997) Spray drying: innovative use of an old process. Des Elem 7:97–113

Rosenberg J, Reinhold U, Liepold B, Breitenbach J, Alani LL, Ghosh S (2008) US Patent 2008/0299203 A1. Dec. 4th, 2008

Shi Y, Gao P, Gong Y, Ping H (2010) Application of a biphasic test for characterization of in vitro drug release of immediate release formulations of Celecoxib and its relevance to in vivo absorption. Mol Pharm 7:1458–1465

Shoyele SA, Cawthorne S (2006) Particle engineering techniques for inhaled biopharmaceuticals. Adv Drug Deliv Rev 58:1009–1029

Smithey D, Fennewald J, Gautschi J, Crew M, Ali S, Lan Y, Langley N (2010) Evaluation of the polymer Soluplus® for spray-dried dispersions of poorly soluble compounds, AAPS 2010, Poster R6081

Taupitz T, Dressman JB, Klein S (2013) In vitro tools for evaluating novel dosage forms of poorly soluble, weakly basic drugs: case example of ketoconazole. J Pharm Sci (Epub ahead of print)

Uddin R, Saffoon N, Bishwajit SK (2011) Dissolution and dissolution apparatus: a review. Int J Cur Biomed Phar Res 1:201–207

US Department of Health and Human Services, Food and Drug Administration, Center for Drug Evaluation and Research (CDER) (1997) Guidance for industry, dissolution testing of immediate release solid oral dosage forms, August 1997

US Department of Health and Human Services, Food and Drug Administration, Center for Drug Evaluation and Research (CDER) (2000) Guidance for industry, waiver of in vivo bioavailability and bioequivalence studies for immediate-release solid oral dosage forms based on a biopharmaceutics classification system, August 2000

Waterbeemd H van de, Gifford E (2003) ADMET in silico modelling: towards prediction paradise? Nat Rev Drug Disc 2:192–204

Vehring R (2008) Pharmaceutical particle engineering via spray drying. Pharm Res 25:999–1022
Yu L (2001) Amorphous pharmaceutical solids: preparation, characterization, and stabilization. Adv Drug Deliv Rev 48:27–42
Zhang Q, Gladden L, Avalle P, Mantle M (2011) In vitro quantitative ^{1}H and ^{19}F nuclear magnetic resonance spectroscopy and imaging studies of fluvastatin™ in Lescol® XL tablets in a USP-IV dissolution cell. J Control Release 156:345–354
Zhao P, Zhang L, Crillo JA, Liu Q, Bullock JM, Moon YJ, Song P, Crar SS, Madabushi R, Wu TC, Booth BP, Rahman NA, Reynolds KS, Gil Berglund E, Lesko LJ, Huang S-M (2011) Applications of physiologically based pharmacokinetic (PBPK) modeling and simulation during regulatory review. Clin Pharmacol Ther 89:259–267

Part IV
Applications

Chapter 9
Manufacture of Pharmaceutically Relevant Materials by Mechanochemistry Using Twin Screw Extrusion

Dominick Daurio, Karthik Nagapudi and Fernando Alvarez-Núñez

Abstract Mechanochemistry broadly refers to the class of chemical reactions that are induced by the application of mechanical force. In the context of pharmaceutical materials, mechanochemistry has been described in the literature for the preparation of cocrystals, salts, and amorphous complexes. In almost all these examples, laboratory-scale mills have been used to demonstrate the production of the aforementioned materials. While laboratory-scale mills demonstrate the utility of the mechanochemical concept, they typically produce small quantities of material and are not considered scalable processes. In this chapter, the application of twin-screw extrusion (TSE) in the production of cocrystals, salts, and amorphous complexes is described. Unlike other mechanical mixing procedures, TSE is a continuous process and lends itself to scalability. TSE can be considered an efficient, scalable, and environmentally friendly process for the consistent manufacture of pharmaceutically relevant systems.

9.1 Introduction

While extrusion technology has been in use since the 1930s in a variety of industries, its application in the pharmaceutical industry spans within the few decades. The applications of twin-screw extrusion (TSE), also known as hot melt extrusion (HME), in the pharmaceutical field ranges from the early development of new and creative drug delivery alternatives at the proof-of-concept level, to the routine manufacture of marketed products. Based on this wide range of applications, it is clear that TSE has been vetted as a manufacturing process and should be considered a mature technology for the pharmaceutical industry.

The application of TSE for the development of oral drug delivery alternatives at an early proof-of-concept stage has been widely published (Andrews et al. 2008; Chokshi et al. 2008; Di Nunzio et al. 2010; Ghosh et al. 2011; Hasa et al. 2011a; Kalivoda et al. 2012; Lyons et al. 2007; Roblegg et al. 2011; Windbergs et al. 2009)

F. Alvarez-Núñez (✉) · D. Daurio · K. Nagapudi
Pharmaceutical R&D, Small Molecule Process and Product
Development, Amgen Inc., Thousand Oaks, CA, USA
e-mail: Faalvare@amgen.com

M. A. Repka et al. (eds.), *Melt Extrusion*,
AAPS Advances in the Pharmaceutical Sciences Series 9,
DOI 10.1007/978-1-4614-8432-5_9,

and was recently captured in two review articles (Crowley et al. 2007; Repka et al. 2007). TSE now seems to be a routine tool that pharmaceutical scientists consider for the development of advantageous oral delivery systems, as well as to overcome other formulation challenges such as taste masking (Maniruzzaman et al. 2012). Keeping pace with the number of publications, the number of patents on the application of TSE in pharmaceutics also shows a linear increase since early 1990s (Crowley et al. 2007).

There are a number of commercial products in the market that are made using extrusion technology. Kaletra® (Abbott), Rezuiln®, which was withdrawn from the market due to toxicological issues (Parke-Davis/Pfizer), Gris-PEG® (Novartis), Cesamet® (Eli Lilly), NuvaRing® (Merck), Implanon® (Merck), Zoladex® (AstraZeneca), and Ozurdex® (Allergan) are some examples of successfully commercialized drug molecules that utilize TSE/HME technology. These successes have served to increase the level of confidence of the pharmaceutical industry in TSE technology and correspondingly the investment in extrusion equipment has also increased in the last decade. In addition, clinical trial material for several active clinical programs is being supplied with formulations that utilize TSE in different ways, further showing the increased acceptance of extrusion processes. This increasing interest in TSE is due to its well-known advantages such as versatility, small foot print, solvent-free or green process, continuous manufacturing capability, and PAT-friendly tool that lends itself to a quality-by-design (QbD) paradigm.

In spite of the increased use of TSE in the pharmaceutical field, the applications so far have been restricted to the manufacture of systems where no clearly defined molecular-level interactions are produced. On the contrary, TSE or HME processes where chemical reactions occur *in situ* have been employed in the polymer industry for decades. This chapter will elaborate on a new opportunity for the pharmaceutical industry, namely the in situ modifications of solid forms of active pharmaceutical ingredients by mechanochemistry using TSE technology.

Mechanochemistry broadly refers to the class of chemical reactions that are induced by the application of mechanical force. While mechanochemistry is an old field, the practical and theoretical aspects of mechanochemistry are less understood when compared to other classes of reactions such as thermochemistry, photochemistry, and electrochemistry (Beyer and Clausen-Schaumann 2005). Nevertheless, applications of mechanochemistry have been reported in a diverse variety of materials such as polymers, metal alloys, inorganic oxides and composites, and organic molecular crystals (James et al. 2012). The primary advantage of mechanochemical reactions is that they use little to no solvent. As such, these reactions can be considered environmentally friendly and, where applicable, can dramatically reduce solvent consumption. In addition, mechanochemistry may also provide a route to induce chemical reactions at reduced temperature and pressure.

In the context of pharmaceutical materials, mechanochemistry has been gaining traction over the last decade primarily in the preparation of cocrystals (Delori et al. 2012). In addition to cocrystallization, examples of salt and amorphous complex formation through mechanochemical means have also been reported in pharmaceutical literature (Friscic et al. 2012; Hasa et al. 2011b; Trask et al. 2006a; Bahl and Bogner 2006; Bahl et al. 2008). In almost all these examples, laboratory-scale mills have been used to demonstrate the production of the aforementioned materials. While laboratory-scale mills are suitable for demonstrating the utility of the mechanochemical concept, they typically produce small quantities of material and are not considered scalable processes.

TSE was recently introduced as a scalable and solventless process that provides a viable route to scale up mechanochemical reactions (Daurio et al. 2011; Medina et al. 2010). This chapter elaborates on the application of TSE in the production of the following classes of pharmaceutically relevant materials: (a) cocrystals, (b) salts, and (c) amorphous complexes.

9.2 Twin-Screw Extrusion—Equipment Operation and Design Considerations

TSE was selected to enable intimate mixing of pharmaceutically relevant materials using little or no solvent with the ultimate goal of modifying the final solid form. A Prism PharmaLab 16 mm corotating twin-screw extruder (25:1 L/D) equipped with a Brabender single-screw volumetric feeder was used for all experiments.

The extruder was equipped with four controllable temperature zones (including the die zone). The temperature zone set points varied according to the model system studied as each system had unique physicochemical properties that needed to be considered. In general, the temperature was set below the melting point of the desired final product to enable collection of a solid powder from the extruder. Although collection of a solid powder was not typically critical to the success of the experiments, it allowed for less postprocessing of the extruded material (e.g., milling or grinding of extrudate). For these experiments, the die zone was not utilized. It was found that when a die was used, it would lead to surplus powder backing up behind the die plate, increasing the extruder torque, and ultimately leading to automatic shutdown of the system.

The extruder screws were designed to have a high mixing capacity and relatively long residence time of powders in the barrel in order to enhance conversion from the initial state to the desired final state (cf. Fig. 9.1). The screw design was set up with alternating 16-mm segments of Zone A and B throughout the barrel. Zone A is a purely conveying zone with minimal mixing capacity and is a single element that is 16 mm in length with a pitch of 7.5 mm. Zone B is a highly mixing zone with neutral conveying capacity and is composed of four distributive elements that are 4 mm in length, each for a total of 16 mm. The elements in Zone B are offset by 90° between each element (accomplished by combining alternating 0° and 90° distributive elements). Monitoring torque during processing is critical as this screw design can require a tremendous amount of power output, especially at lower speeds, potentially leading to automatic shutdown of the extruder.

Appropriate stoichiometric blends of the starting components were prepared in an initial mixing step. Once the materials were uniformly mixed, they were charged

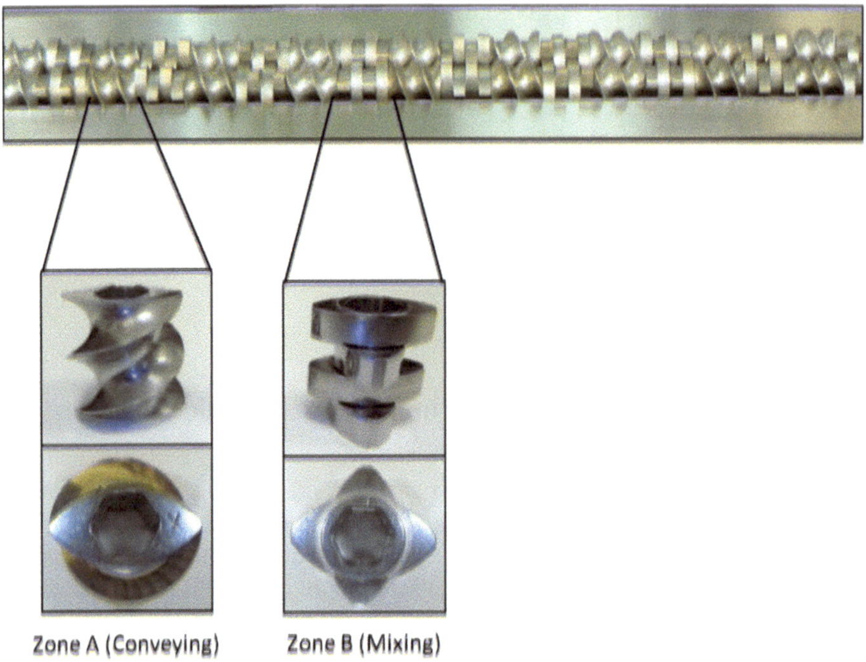

Fig. 9.1 Screw design employed for twin-screw extrusion showing conveying and mixing elements

to extruder manually or through a Brabender single-screw volumetric feeder that fed directly into the extruder hopper at a 5 % feed rate. Hand feeding of the material was utilized when the batch sizes were not suitable for the volumetric feeding system. Liquid-assisted extrusion was performed with appropriate amounts of solvent added to the powder blends prior to extrusion. All remaining details of the extrusion experiments are provided in the discussion section for each of the systems studied. Each experiment processed anywhere between 20 and 100 g of material.

9.3 Examples of Mechanochemistry Using Twin-Screw Extrusion

9.3.1 Production of Cocrystals

Cocrystals have been defined as crystalline materials that comprise two or more components that are solids at room temperature (in order to distinguish them from hydrates and solvates) held together by noncovalent forces (Bond 2007; Desiraju 2003; Dunitz 2003). The distinction between a salt and a cocrystal lies in the fact that there is no proton transfer occurring between the constituents of a cocrystal. There has been an increased interest in cocrystals in the last few years and a number of publications have highlighted the beneficial properties of cocrystals (Chen et al. 2007; Childs et al. 2004; Remenar et al. 2003; Trask et al. 2006b; Variankaval et al. 2006). Traditionally, cocrystals have been prepared by grinding the constituents together. More recently, liquid-assisted grinding has been developed as a more effective method to make cocrystals (Friscic et al. 2006; Trask et al. 2004). In this section, we describe the TSE process to make cocrystals of carbamazepine–saccharin and theophylline–citric acid (Daurio et al. 2011). These cocrystals have been extensively studied, thus the data generated from the TSE process could be compared with literature data to gauge the effectiveness of the TSE process in making cocrystals.

9.3.1.1 Carbamazepine–Saccharin Cocrystal

Cocrystals made from carbamazepine and saccharin have been extensively investigated. Rodriguez-Hornedo et al. have investigated the formation of this cocrystal using grinding methods (Jayasankar et al. 2006). They observed that cocrystal formation was mediated by generation of the amorphous phase. The extent of amorphous formation was dependent on grinding temperature with higher levels of amorphous phase being produced upon cryogrinding. The amorphous phase was found to continuously crystallize on storage into the anhydrous cocrystal Form I. The extent of cocrystal formation from the amorphous phase at room temperature was impacted by relative humidity upon storage. The extent of cocrystal formation was dependent on the hydration state of the starting material. For example, if carbamazepine dihydrate was used in place of anhydrous carbamazepine, the rate of cocrystal formation was accelerated when grinding was conducted under similar conditions. This result was rationalized by the fact that water in starting material serves as a plasticizer aiding in decreasing the glass transition temperature of the amorphous phase, thereby providing a better chance to crystallize during grinding. This system was selected as a model cocrystal to determine if TSE can produce cocrystals where the mechanism of formation is mediated by the amorphous phase.

The X-ray powder diffraction (XRPD) patterns of the material generated using TSE of anhydrous carbamazepine and saccharin are shown in Fig. 9.2a. No conversion to the cocrystal was observed when the extrusion was conducted at 50 °C with a screw speed of 25 RPM (XRPD pattern B). However, cocrystal formation was observed when the anhydrous materials were coground for 30 min in a ball mill. The final bulk temperature after the milling process was found to be 45 °C. Thus, the extrusion result does not match the milling result using similar temperature conditions. These results may be explained considering the milling intensity, milling time, and the amount of material processed. Ball milling is a much more energy-intensive process than TSE. Ball milling can therefore lead to a greater disruption of the crystalline lattice as compared to extrusion. It was also demonstrated by Nair et al. that even with these levels of intensive energy generated by the ball mill, at least 30 min of milling time is required to achieve a cocrystal conversion of about 90 %. In contrast, the processing time (residence time) in the extruder is in the order of about 2–5 min. This value cannot be significantly altered due to practical considerations of

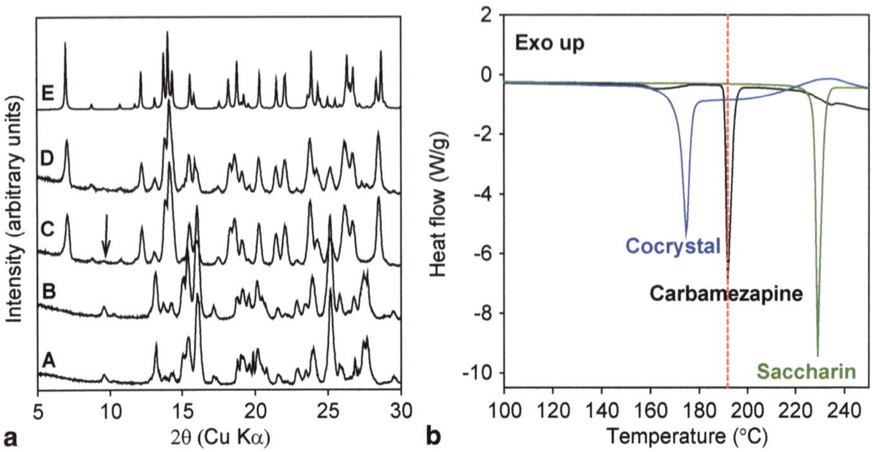

Fig. 9.2 **a** X-ray powder diffraction (XRPD) of materials obtained from the twin-screw extrusion (TSE) of carbamazepine and saccharin: *A* Physical blend prior to extrusion, *B* Extrusion at 50 °C, *C* Extrusion at 190 °C, *D* Extrusion with water at 100 °C, *E*: Reference Form I cocrystal pattern calculated from the published crystal structure obtained from the Cambridge structural database (CSD). **b** Differential scanning calorimetry (DSC) thermogram showing comparison of carbamazepine, saccharin, and the 1:1 carbamazepine–saccharin cocrystal. The *dotted line* in **b** represents the extrusion temperature

extrusion processing. The amount of material processed in the extruder was another point of difference as compared to ball milling. The total amount processed at any given time is about 20 times higher in an extruder than that of a ball mill. Owing to practical considerations, the amount that is processed cannot be reduced below 10 g in the extruder. Based on this discussion, it is clear that cogrinding results using ball milling of anhydrous starting components were not representative of the conditions used in extrusion and should not be expected to yield the cocrystal.

In order to improve cocrystal formation using TSE, the temperature in the extruder was altered. A temperature of 190 °C was selected, which was close to the onset of melting of carbamazepine. XRPD pattern of the material obtained after TSE at 190 °C shows near complete conversion to the cocrystal (XRPD pattern C). Approximately, 95 % conversion to the cocrystal was calculated using ^{13}C solid-state nuclear magnetic resonance (SSNMR) spectroscopy. Consistent with the ^{13}C SSNMR data, the XRPD pattern also showed the presence of unreacted material (marked with an arrow in Fig. 9.2a). In addition to increasing the temperature, further optimization of the extrusion parameters can achieve better conversions to the cocrystal.

An evaluation was also conducted to determine if TSE could replicate the published results, which show the rate of cocrystal formation increased by the presence of water. In previous reported studies, carbamazepine dihydrate was used as the starting material and the water in the dihydrate was found to increase the cocrystal conversion (Jayasankar et al. 2006). This can be considered as an equivalent of liquid-assisted grinding where liquid is intentionally added to the system to facilitate cocrystal formation. Jones et al. have extensively reviewed liquid-assisted grinding (Karki et al. 2007). Liquid-assisted grinding was found to be superior to neat grinding in most cases in effecting cocrystal transformation. The liquid is typically used in small amounts and could end up in the final product resulting in the formation of a solvated phase. In the case of anhydrous carbamazepine–saccharin cocrystal, the liquid is expected to have only a catalytic role as the cocrystal is anhydrous.

In our studies, liquid-assisted extrusion was investigated by processing the anhydrous starting components in the presence of water. About 5 ml of water was added to 10.4 g of a 1:1 molar blend of anhydrous carbamazepine and saccharin and the extrusion was carried out at 100 °C. Water was chosen as the model liquid phase since use of water does not require any additional environmental controls and 100 °C temperature was chosen to obtain a product, free of water. The XRPD pattern of the material obtained from this experiment shows that while conversion to the cocrystal has occurred, the degree of conversion is less than what was obtained with neat extrusion (XRPD pattern D). Approximately, 87 % conversion to the cocrystal was calculated using ^{13}C SSNMR spectroscopy. Since this experiment was only intended to demonstrate proof of concept, it was not further optimized. It is believed that with further optimization, using a gradient temperature in the barrel, better conversions will be observed. The use of hydrated starting material could also be used as a means of improving the extent of cocrystal formation. The differential scanning calorimetry (DSC) thermograms of carbamazepine, saccharin, and the 1:1 carbamazepine–saccharin cocrystal are shown in Fig. 9.2b. The melting temperature onset (171.6 °C) of the cocrystal was found to be lower than that of the starting components and in good agreement with published data.

Matzger et al. have reported the formation of another anhydrous polymorph (Form II) of carbamazepine and saccharin (Porter et al. 2008). This polymorph was prepared from an ethanol solution in the presence of poly(4-methyl 1-pentene) as heteronuclei. TSE was attempted to produce anhydrous Form II by mixing carbamazepine and saccharin with ethanol and poly(4-methyl 1-pentene). The temperature along the barrel for this experiment was increased from 20 to 80 °C. The final temperature was selected to purge all the ethanol from the system. The experiment resulted in the production of Form I cocrystal. While Form II polymorph could not be produced, this experiment showed that either water or ethanol could be used in the liquid-assisted extrusion process to make the Form I cocrystal. Notably, the use of catalytic amounts of solvents led to a lowering of experimental temperatures required to form the cocrystal in the extruder.

9.3.1.2 Theophylline–Citric Acid Cocrystal

Theophylline and citric acid are known to cocrystallize in both the anhydrous and hydrated forms (In the hydrated form, all three components are present in equal stoichiometric proportions). Jones et al. have reported a systematic study of the system and the outcome of their work is summarized in Table 9.1 (Karki et al. 2007). Briefly, the anhydrous cocrystal was formed when neat grinding of the anhydrous

Table 9.1 Outcome of grinding experiments performed with theophylline–citric acid system. The data have been summarized from the work of Jones et al. (Karki et al. 2007)

Component 1	Component 2	Type of grinding	Product
TA	CA	Neat	AC
TA	CM	Neat	HC
TM	CA	Neat	HC
TM	CM	Neat	HC
TA, TM	CA, CM	Water-assisted	HC

AC anhydrous cocrystal, *CA* citric acid anhydrous, *CM* citric acid monohydrate, *HC* hydrated cocrystal, *TA* theophylline anhydrous, *TM* theophylline monohydrate

Table 9.2 Outcome of twin-screw extrusion (TSE) experiments performed with theophylline–citric acid system

Lot number	Component 1	Component 2	Type of extrusion	Temperature	Product
TC1	TA	CA	Neat	153	AC + unreacted[a]
TC2	TA	CM	Neat	153	AC + unreacted[a]
TC3	TM	CA	Neat	153	AC + unreacted[a]
TC4	TA	CM	Neat	50	AC
TC5	TA	CA	Ethanol-assisted	20	AC
TC6	TA	CM	Neat	20	HC
TC7	TA	CM	Water-assisted	20	HC

AC anhydrous cocrystal, *CA* citric acid anhydrous, *CM* citric acid monohydrate, *HC* hydrated cocrystal, *TA* theophylline anhydrous, *TM* theophylline monohydrate

[a] Unreacted starting material was observed by X-ray powder diffraction (XRPD)

reactants was carried out at room temperature. However, the hydrated cocrystal results when either reactant was a hydrate or when both reactants were hydrates. The hydrated cocrystal could also be produced when catalytic amounts of water were added to the system. In the case of liquid-assisted grinding with water, the nature of the starting materials becomes immaterial as the externally added water acted to produce the hydrated cocrystal. In contrast to the previous example, for the theophylline–citric acid case, the liquid component ends up in the final cocrystal product. Thus, theophylline–citric acid was selected as an additional model system to examine if TSE is capable of making both anhydrous and hydrated variety of the cocrystal.

The list of TSE experiments conducted with this system along with the outcome of the experiments is summarized in Table 9.2. The outcome of the TSE experiments mirrored the findings observed with the grinding experiments. However, TSE experiments differed in one important aspect as compared to grinding. The difference was that as opposed to grinding experiments, a wider range of controlled temperatures could be used with TSE experiments. The temperature control is achieved by heating and cooling the extruder barrel as needed to maintain an appropriate heat source or heat sink. As with the grinding experiments, hydrated cocrystal was produced in

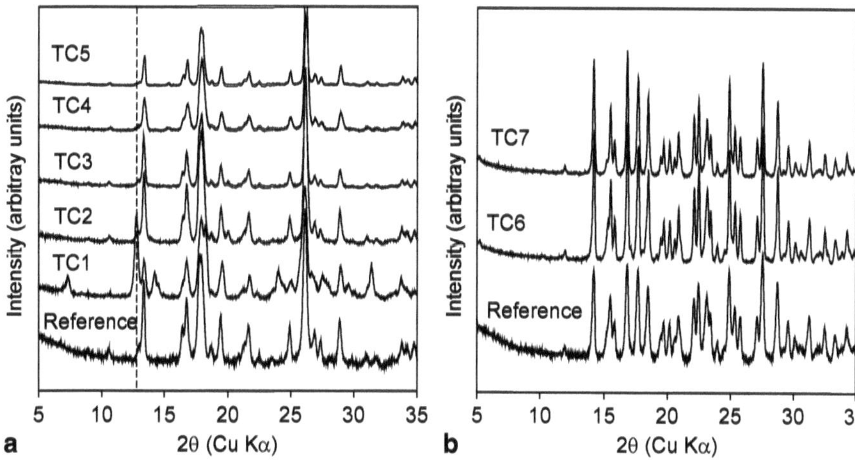

Fig. 9.3 X-ray powder diffraction (XRPD) of materials obtained from the extrusion of theophylline and citric acid. The descriptions of samples TC1 through TC7 are provided in Table 9.2. **a** Samples producing the anhydrous cocrystal phase and **b** Samples producing the hydrated cocrystal. The *dotted line* in **a** tracks the peak at 12.8 ° 2θ and provides information about unreacted starting material in the extruded samples

TSE at room temperature when the starting reactant contained water. As an example, the TSE at room temperature of a blend of anhydrous theophylline and citric acid monohydrate (sample TC6 in Table 9.2) produced the hydrated cocrystal. Hydrated cocrystal was also produced when the sample blend was subjected to TSE with a catalytic amount of water added to it (sample TC7 in Table 9.2). The XRPD patterns of TC6 and TC7 are shown in comparison with a reference hydrated cocrystal pattern in Fig. 9.3b. In contrast to extrusion at 20 °C, extrusion of a blend of anhydrous theophylline and citric acid monohydrate at 50 °C produced the anhydrous cocrystal (sample TC4 in Table 9.2). Thus, by simply modifying the extrusion temperature, the same blend of starting materials could be made to produce the anhydrous cocrystal.

When the extrusion temperature was increased to 153 °C, as expected, the anhydrous cocrystal was formed (sample TC2 in Table 9.2). Regardless of the starting components (samples TC1, TC2, and TC3 in Table 9.2) when the extrusion was carried out at 153 °C the anhydrous cocrystal was produced. The XRPD patterns of the materials TC1 through TC4 are shown in comparison to the reference anhydrous cocrystal in Fig. 9.3a. Following the reflection at 12.8° 2θ (characteristic of anhydrous theophylline), it is clear that samples TC1 through TC3 have unreacted starting components. Only sample TC4 was found to be converted almost entirely to the cocrystal. In addition, the extent of conversion to the anhydrous cocrystal was found to be higher for TC2 and TC3 when compared to TC1.

Thus, when extrusion was carried out at 20 °C (TC6), water resided in the crystal lattice in the form of a hydrated cocrystal. However, when extrusion was carried out at 50 °C (TC4), water served to aid in the cocrystal formation of the anhydrous phase. Even when extrusion was carried out at 153 °C (TC2 and TC3), water aided

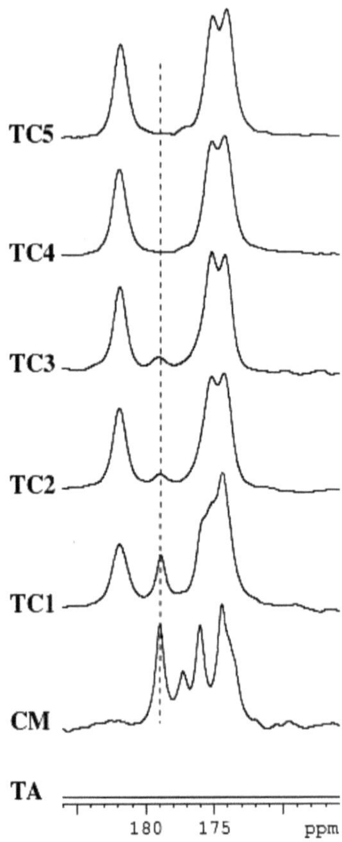

Fig. 9.4 The ^{13}C solid-state nuclear magnetic resonance (SSNMR) spectra of samples *TA* anhydrous theophylline (no peaks in this ppm range), *CM* citric acid monohydrate, TC1, TC2, TC3, TC4, and TC5 (sample descriptions shown in Table 9.2). The extent of conversion to the cocrystal followed the order TC1 < TC3 < TC2 < TC4 = TC5

in conversion to the anhydrous, albeit not completely. However, in the absence of water, when extrusion was carried out at 153 °C (TC1), the extent of conversion to the cocrystal was found to be the lowest. These observations point to the fact that the presence of a liquid phase (water that came through the starting reactants) aided in formation of the anhydrous and hydrated cocrystal. In order to further show the utility of liquid-assisted extrusion, TSE of a blend of anhydrous theophylline and anhydrous citric acid was carried out at 20 °C in the presence of ethanol (sample TC5 in Table 9.2). In this case, complete conversion to the cocrystal was observed. The ^{13}C SSNMR data in the region of the carboxyl group are shown for materials TC1, TC2, TC4, and TC5 in comparison to the starting materials in Fig. 9.4.

The peak at 178.9 ppm corresponding to citric acid monohydrate was followed to determine the extent of conversion to the cocrystal. These data clearly show that conversion, scales as TC1 < TC3 < TC2 < TC4 = TC5. Thus, as seen in the previous case study with carbamazepine and saccharin, liquid-assisted extrusion was able to facilitate cocrystal formation at lower temperatures and resulted in an increased extent of conversion when compared to neat extrusion. These findings using TSE nicely paralleled those observed with grinding experiments.

9.3.2 Production of Salts

The screening of salts of pharmaceutical materials is an essential activity during drug development (Serajuddin 2007; Serajuddin and Pudipeddi 2011). The formation of a salt can help in purification of the drug substance as well as in physical property optimization. For example, careful selection of salts could lead to improved physical and chemical stability, improved pharmacokinetic profiles and improved bulk mechanical properties. Typically, the screening for salts is done via high-throughput solution-based experiments. Recently, a mechanochemical approach, such as grinding (solventless or liquid-assisted), was introduced as a viable method for the screening of crystalline salts (Trask and Haynes 2006). In the industrial setting, however, the large-scale synthesis of salts has always been performed via solution crystallization route (Wermuth and Stahl 2011). In this section, we illustrate for the first time, the production of pharmaceutical salts by mechanochemistry using TSE. Two salt systems were investigated: (a) pyrimethamine maleate and (b) naproxen sodium.

9.3.2.1 Pyrimethamine Maleate

Pyrimethamine is an antimalarial compound that is known to make salts with a wide variety of acidic counter ions. Mechanochemical production of pyrimethamine salts has been reported by Jones et al. (Trask et al. 2006a). In their study, they reported the formation of salts of pyrimethamine with formate, acetate, maleate, fumarate, succinate, glutarate, and salicylate. Of particular interest was the pyrimethamine maleate system. This salt system could only be produced by solvent-assisted grinding and not by neat grinding. Catalytic amounts of methanol were used in production of the salt by solvent-assisted grinding. Due to the relative complexity in forming this salt system, it was chosen to test the utility of TSE in acidic salt formation.

Appropriate quantities of pyrimethamine and maleic acid were blended together prior to extrusion. The powder was then manually fed into the extruder. Two types of extrusion experiments were conducted on the system: (a) Neat extrusion at 35, 100, and 150 °C and (b) solvent assisted extrusion using methanol as the catalytic solvent with progressively increasing temperature in the extruder (First zone 26 $<$ second zone 33 $<$ third zone 75 °C). The final zone was increased to 75 °C in order to dry off the methanol solvent. The XRPD patterns of the extruded materials are shown in comparison to the physical blend and the ball-milled material in Fig. 9.5. Ball milling of the physical blend was performed under published conditions using solvent-assisted grinding with methanol. It is clear from the XRPD in Fig. 9.5 that ball milling produces the pyrimethamine maleate salt. However, extrusion conducted at 35 and 100 °C does not produce the salt. The XRPD of the extrusion conducted under these conditions shows production of amorphous material with no obvious conversion to the salt. However, extrusion conducted at 150 °C clearly shows conversion to the salt. The conversion was found to be incomplete as peaks from original physical blend (marked with an arrow in Fig. 9.5) were also seen in the XRPD of the

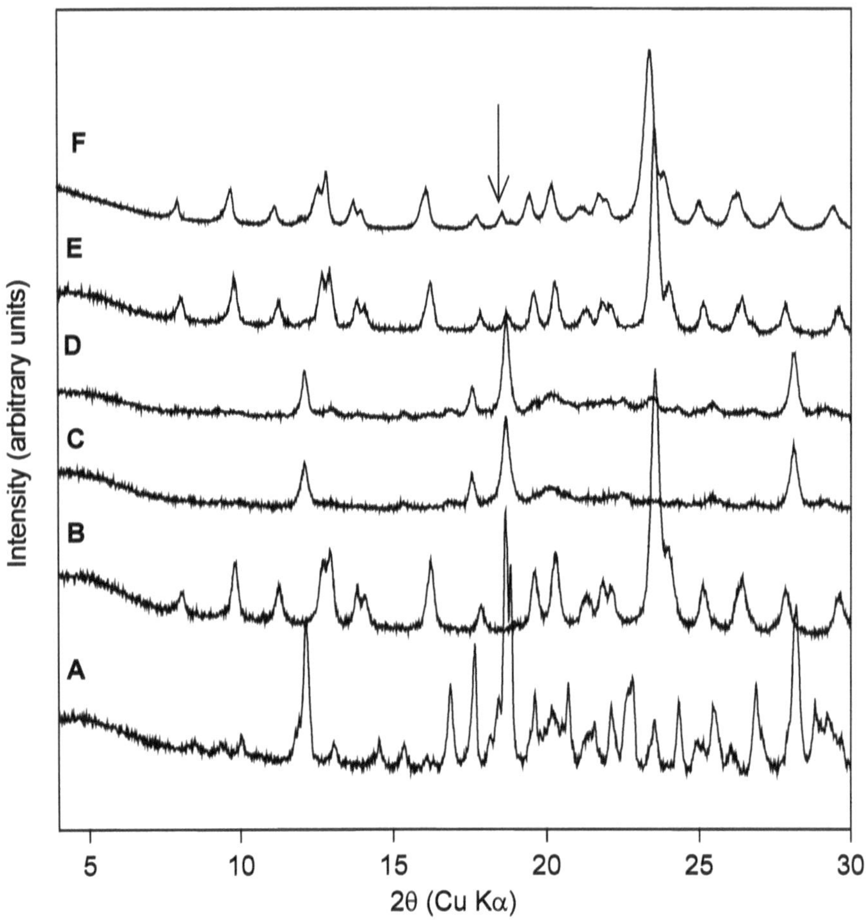

Fig. 9.5 X-ray powder diffraction (XRPD) of materials obtained from the extrusion of pyrimethamine and maleic acid. *A* Physical blend prior to extrusion, *B* Ball-milled sample, *C* Extrusion at 35 °C, *D* Extrusion at 100 °C, *E* Extrusion at 150 °C, and *F* Extrusion with methanol using a temperature gradient (26–75 °C)

sample extruded at 150 °C. However, since this was a proof-of-concept experiment, the extrusion conditions were not further optimized. 150 °C is above the melting point of maleic acid ($T_m = 135$ °C) but below the melting point of pyrimethamine ($T_m = 233.5$ °C). Thus, TSE when conducted above the melting point of the counter ion led to salt formation. This was to be expected as melting of the maleic acid led to more intimate mixing of the counter ion and the drug in the extruder resulting in salt formation. This result was in contrast to what was published with ball milling, where no conversion to the salt was observed under neat grinding conditions. Therefore, TSE, with its access to larger temperature range than ball milling, was able to produce the salt. In addition, solvent-assisted extrusion of the system was conducted with methanol using a temperature gradient from 26 to 75 °C. Conversion to the salt was observed with solvent-assisted extrusion, showing that methanol addition was able to produce the salt at a decreased temperature relative to the 150 °C required for neat extrusion. These results are consistent with what was observed for the cocrystal systems described in the previous section.

9.3.2.2 Naproxen Sodium

A second example, the production of a sodium salt of naproxen was investigated. In contrast to the previous example, which was an acidic salt, this example discusses a basic salt to show the capability of TSE in producing all varieties of salts. Two types of extrusion experiments were conducted on the system: (a) Neat extrusion at 175 °C and (b) solvent-assisted extrusion with sodium hydroxide introduced from a concentrated water solution at 175 °C. For the neat extrusion experiment, appropriate quantities of naproxen and sodium hydroxide were physically blended and hand fed into the extruder. In earlier descriptions of solvent-assisted extrusion, both components needed were introduced as powders and then catalytic amount of the appropriate solvent was added. In contrast, for the solvent-assisted extrusion experiment in this case, sodium hydroxide was dispensed from a 10 N solution into the naproxen powder. This was the first demonstration where one of the constituents was introduced through a solvent medium.

As literature data did not exist for the production of this salt using ball milling, it was decided to proceed directly to the extrusion experiment. Figure 9.6 shows the XRPD patterns of the extruded materials shown in comparison to the physical blend. It was clear from the data presented that both neat extrusion and solvent-assisted extrusion were able to produce the naproxen sodium salt. Nevertheless, the solvent-assisted process is preferable from an operational standpoint as the process is much gentler on the extruder.

9.4 Production of Amorphous Complexes

Use of the amorphous phase to improve the oral bioavailability of poorly soluble drugs is well known. From the standpoint of maximizing exposure, the amorphous phase is of great interest for pharmaceutical scientists. It is at higher energy and offers the promise of higher solubility and faster dissolution rate, and thereby the potential to increase bioavailability (Kennedy et al. 2008; Yu 2001). Along with these advantages, amorphous materials possess the inherent risks of physical and chemical stability issues, as well as processing difficulties (Serajuddin 1999). Physical instability leads to the transformation of the amorphous state to the thermodynamically favored crystalline state. The typical approach to improve the physical stability of amorphous pharmaceuticals is to combine them with inactive ingredients such as

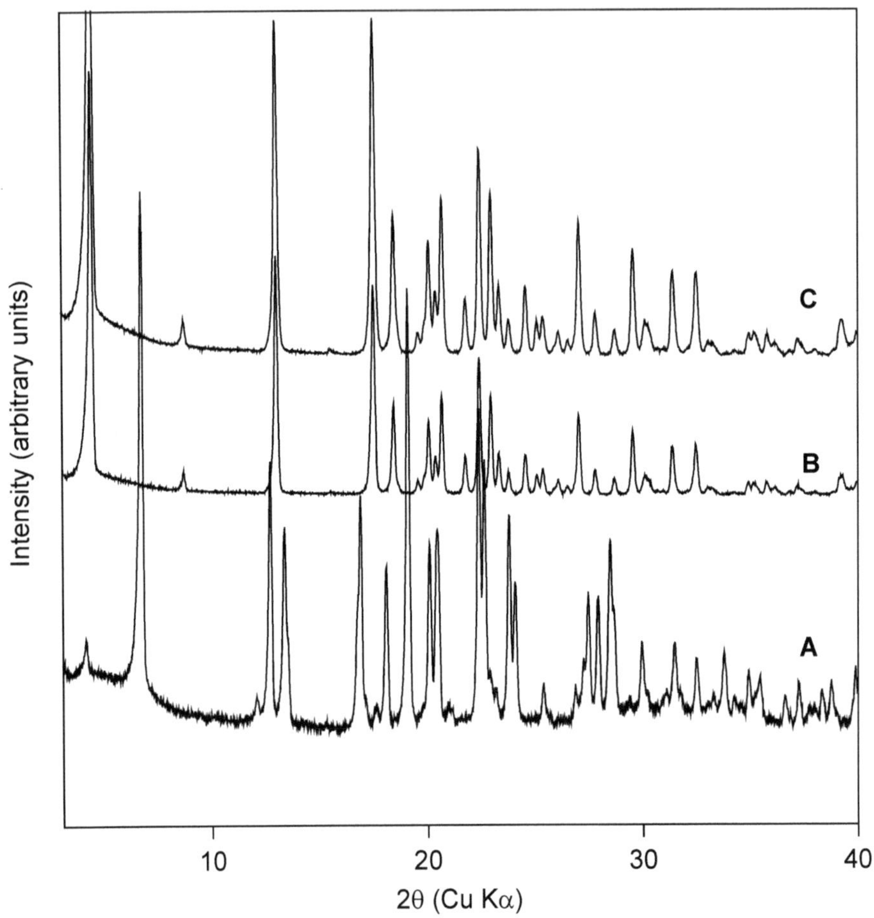

Fig. 9.6 X-ray powder diffraction (XRPD) of materials obtained from the extrusion of naproxen and sodium hydroxide. *A* Physical blend prior to extrusion, *B* neat extrusion at 175 °C, and *C* solvent-assisted extrusion in water at 175 °C

polymers to form amorphous solid dispersions. There are a number of reports in literature describing the use of organic polymers to make amorphous solid dispersions and the solubility and dissolution advantage that can be obtained with these systems both in vitro and in vivo (Janssens et al. 2008; Janssens et al. 2010; Konno et al. 2008; Law et al. 2004; Miyazaki et al. 2004; Rumondor et al. 2009).

In contrast to organic excipients, the application of inorganic materials to improve the physical stability of amorphous API has not been explored in great detail. There have been a few reports of stabilization of acidic drugs with silicates (Bahl and Bogner 2006; Gupta et al. 2003; Kinoshita et al. 2002; Mallick et al. 2008; Watanabe et al. 2002a, b). In almost all these studies, the amorphous complex between the drug and silicate was formed by ball milling. In this section of the chapter, we describe the

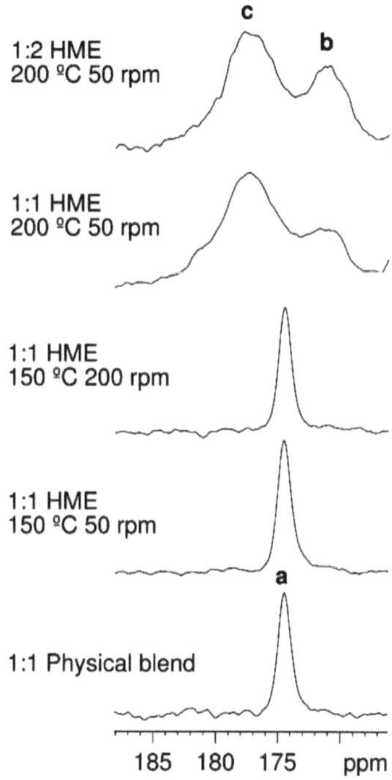

Fig. 9.7 Carboxyl region of the ^{13}C solid-state nuclear magnetic resonance (SSNMR) spectra of HME samples. Peak labels: *a* crystalline sulindac; *b* amorphous sulindac; and *c* amorphous sulindac-Neusilin complex

production of amorphous complex of an acidic drug, sulindac, with Neusilin by both ball milling and HME. The physicochemical stability and dissolution behavior of these amorphous complexes were investigated. While ball milling provided a convenient route to generate small quantities of sulindac-Neusilin amorphous complexes in the laboratory, it is not a scalable process. The full descriptions of the results have been published at length in a separate publication (MacLean et al. 2011).

The ^{13}C SSNMR spectroscopy was used to monitor the progress of the amorphous complex formation subsequent to HME. The ^{13}C SSNMR data in the carboxyl region for samples made by HME are shown in Fig. 9.7. HME was initially attempted at 150 °C at two screw speeds 50 and 200 rpm to investigate whether the complex formation occurs below the melting point of sulindac (184.6 °C). Different screw speeds were used to change the residence time of the material in the extruder barrel. It is clear from the ^{13}C SSNMR data presented that no conversion was observed at this temperature, irrespective of the screw speed used. However, conversion of the crystalline material to the amorphous complex was observed when HME was conducted at a temperature of 200 °C (above the melting point of sulindac). Thus, temperatures above the melting point of sulindac are required to affect the conversion to the amorphous complex. Consistent with ball milling data, HME produced samples containing a mixture of the amorphous sulindac (Peak "b") and amorphous sulindac-Neusilin complex (Peak "c"). Notably, the amount of the complex formed was found to be greater than that achieved from ball milling for 60 min. The amount of complex formed, remained about the same irrespective of the starting composition of sulindac

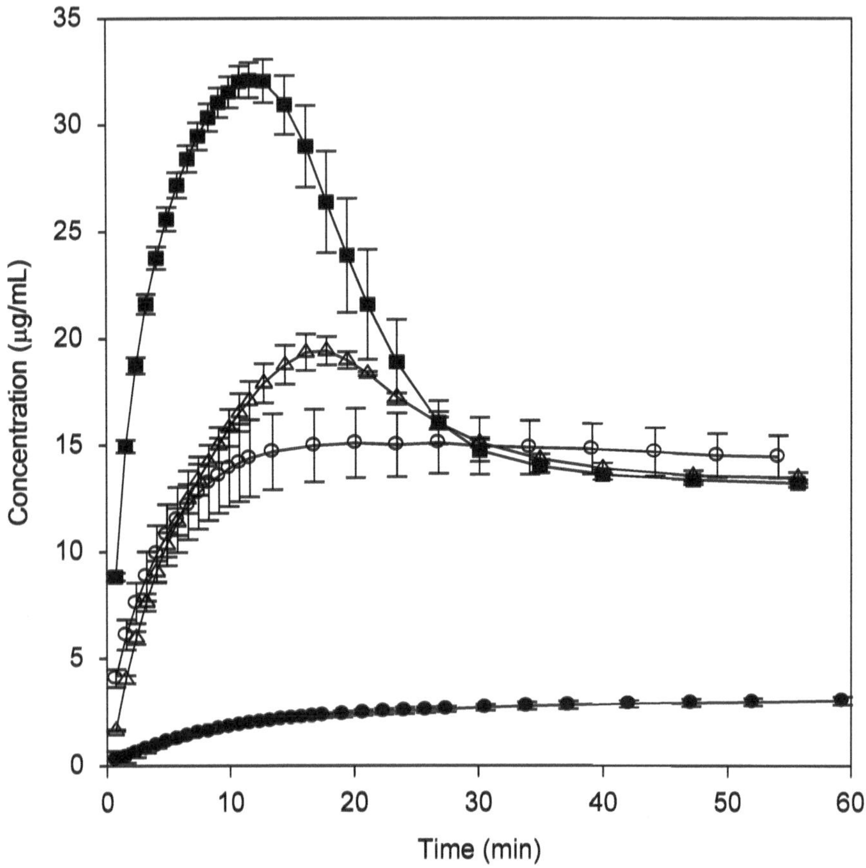

Fig. 9.8 Dissolution profiles in 0.1N HCl: (•) Crystalline sulindac; (○) Sulindac:Neusilin 1:1 ball-milled; (△) Sulindac:Neusilin 1:1 HME; and (■) Sulindac:Neusilin 1:2 HME

to Neusilin. This shows that Neusilin amounts greater than 1:1 did not serve to increase the amount of complex formed under these experimental conditions. About 100 g of 1:1 and 1:2 complexes were produced using HME at 200 °C with a screw speed at 50 rpm. The sample was recovered as a powder from the HME apparatus. The HME samples were analyzed by high-performance liquid chromatography (HPLC) and were found to be free of chemical degradation. HME thus provided an efficient and a convenient route to the continuous production of sulindac-Neusilin amorphous complexes. Moreover, all amorphous complexes produced by HME were found to be physically and chemically stable when stored at 40 °C/75 % RH for a period of 3 months.

Dissolution profiles of crystalline sulindac and sulindac-Neusilin amorphous complexes in 0.1N HCl are shown in Fig. 9.8. Crystalline sulindac is poorly soluble in the medium and reaches a solubility of 3 µg/ml in about 60 min. In contrast, the sulindac-Neusilin amorphous complexes showed faster dissolution rate and higher solubility. The HME samples show a peak concentration and a plateau concentration. For the 1:1 HME sample, the peak concentration was 19.4 μg/ml while the peak concentration for the 1:2 HME sample was 32.1 μg/ml. The plateau concentration for both samples was around 13.2–13.4 μg/ml. In contrast, the ball-milled 1:1 sample showed a peak concentration of 15.1 μg/ml and a plateau concentration of 14.4 μg/ml. The peak-to-plateau ratio followed the order 1:2 HME > 1:1 HME > 1:1 ball-milled. The peak concentration for all samples was attained within 10–17 min. In summary, amorphous sulindac-Neusilin complexes are able to provide enhanced dissolution rate and solubility when compared to the as-received crystalline material. Thus, HME was shown to be a scalable process to make physically and chemically stable amorphous complexes of sulindac and Neusilin that provide a dissolution advantage over the crystalline material.

9.5 Conclusions

The application of TSE/HME in the continuous production of cocrystals, salts, and amorphous complexes has been demonstrated. TSE provides highly efficient mixing and close material packing of components, which in turn lead to improved surface contact between components, thereby facilitating compound formation without the use of solvents. The presence of intimate mixing, which promotes surface contacts and accessibility to a wide range of temperatures, was found to be a critical factor that influenced the formation of pharmaceutically relevant systems during extrusion processing. The extruder configuration is customizable in terms of mixing and conveying elements and therefore can be optimized for yield and bulk properties of the final solid form. For the first time, liquid-assisted extrusion to form pharmaceutically relevant materials has also been demonstrated. The addition of small amounts of liquids adds another processing dimension to the extrusion process, thereby allowing for further flexibility in optimizing compound production using TSE. Moreover, liquid-assisted extrusion can promote compound formation at lower temperatures obviating the need of high-temperature processing. Unlike other mechanical mixing procedures, TSE is a continuous process and lends itself to scalability. As shown, extrusion can be considered an efficient, scalable, and environmentally friendly process for the manufacture of pharmaceutically relevant systems providing a viable alternative to solution crystallization processes.

Acknowledgments We wish to thank Cesar Medina, Jennifer Maclean, and Robert Saw for conducting some of the experiments described in this chapter. We wish to acknowledge Francisco Alvarez for providing the management support to conduct this research. We also wish to thank to following people for numerous contributions to this chapter: Julie Calahan, Kim Gochioco, Janan Jona, Chandra Ma, Eric Munson, Matt Peterson, Saroj Vangani, and Tian Wu.

References

Andrews GP, Jones DS, Diak OA, McCoy CP, Watts AB, McGinity JW (2008) The manufacture and characterisation of hot-melt extruded enteric tablets. Eur J Pharm Biopharm 69:264–273

Bahl D, Bogner RH (2006) Amorphization of indomethacin by co-grinding with neusilin US2: amorphization kinetics, physical stability and mechanism. Pharm Res 23:2317–2325

Bahl D, Hudak J, Bogner RH (2008) Comparison of the ability of various pharmaceutical silicates to amorphize and enhance dissolution of indomethacin upon co-grinding. Pharm Dev Technol 13:255–269

Beyer MK, Clausen-Schaumann H (2005) Mechanochemistry: the mechanical activation of covalent bonds. Chem Rev 105:2921–2948

Bond AD (2007) What is a co-crystal? Cryst Eng Comm 9:833–834

Chen AM, Ellison ME, Peresypkin A, Wenslow RM, Variankaval N, Savarin CG, Natishan TK, Mathre DJ, Dormer PG, Euler DH, Ball RG, Ye Z, Wang Y, Santos I (2007) Development of a pharmaceutical cocrystal of a monophosphate salt with phosphoric acid. Chem Commun (Camb) 28(4):419–421

Childs SL, Chyall LJ, Dunlap JT, Smolenskaya VN, Stahly BC, Stahly GP (2004) Crystal engineering approach to forming cocrystals of amine hydrochlorides with organic acids. molecular complexes of fluoxetine hydrochloride with benzoic, succinic, and fumaric acids. J Am Chem Soc 126:13335–13342

Chokshi RJ, Shah NH, Sandhu HK, Malick AW, Zia H (2008) Stabilization of low glass transition temperature indomethacin formulations: impact of polymer-type and its concentration. J Pharm Sci 97:2286–2298

Crowley MM, Zhang F, Repka MA, Thumma S, Upadhye SB, Battu SK, McGinity JW, Martin C (2007) Pharmaceutical applications of hot-melt extrusion: part I. Drug Dev Ind Pharm 33:909–926

Daurio D, Medina C, Saw R, Nagapudi K, Alvarez-Nunez F (2011) Application of twin screw extrusion in the manufacture of cocrystals, part I: four case studies. Pharmaceutics 3:582–600

Delori A, Friscic T, Jones W (2012) The role of mechanochemistry and supramolecular design in the development of pharmaceutical materials. Cryst Eng Comm 14:2350–2362

Desiraju GR (2003) Crystal and co-crystal. Cryst Eng Comm 5:466–467

Di Nunzio JC, Brough C, Hughey JR, Miller DA, Williams RO III, McGinity JW (2010) Fusion production of solid dispersions containing a heat-sensitive active ingredient by hot melt extrusion and Kinetisol dispersing. Eur J Pharm Biopharm 74:340–351

Dunitz JD (2003) Crystal and co-crystal. A second opinion. Cryst Eng Comm 5:506

Friscic T, Jones W (2012) Application of mechanochemistry in the synthesis and discovery of new pharmaceutical forms: co-crystals, salts and coordination compounds. RSC Drug Discovery Series 16:154–187

Friscic T, Trask AV, Jones W, Motherwell WDS (2006) Screening for inclusion compounds and systematic construction of three-component solids by liquid-assisted grinding. Angew Chem Int Ed 45:7546–7550

Ghosh I, Snyder J, Vippagunta R, Alvine M, Vakil R, Tong W-Q, Vippagunta S (2011) Comparison of HPMC based polymers performance as carriers for manufacture of solid dispersions using the melt extruder. Int J Pharm 419:12–19

Gupta MK, Vanwert A, Bogner RH (2003) Formation of physically stable amorphous drugs by milling with neusilin. J Pharm Sci 2(92):536–551

Hasa D, Perissutti B, Grassi M, Zacchigna M, Pagotto M, Lenaz D, Kleinebudde P, Voinovich D (2011a) Melt extruded helical waxy matrices as a new sustained drug delivery system. Eur J Pharm Biopharm 79:592–600

Hasa D, Voinovich D, Perissutti B, Grassi M, Bonifacio A, Sergo V, Cepek C, Chierotti MR, Gobetto R, Dall'Acqua S, Invernizzi S (2011b) Enhanced oral bioavailability of vinpocetine through mechanochemical salt formation: physico-chemical characterization and in vivo studies. Pharm Res 28:1870–1883

James SL, Adams CJ, Bolm C, Braga D, Collier P, Friscic T, Grepioni F, Harris KDM, Hyett G, Jones W, Krebs A, Mack J, Maini L, Orpen AG, Parkin IP, Shearouse WC, Steed JW, Waddell DC (2012) Mechanochemistry: opportunities for new and cleaner synthesis. Chem Soc Rev 41:413–447

Janssens S, Armas HN de, D'Autry W, Van Schepdael A, Van den Mooter G (2008) Characterization of ternary solid dispersions of Itraconazole in polyethylene glycol 6000/polyvidone-vinylacetate 64 blends. Eur J Pharm Biopharm 69:1114–1120

Janssens S, De Zeure A, Paudel A, Van Humbeeck J, Rombaut P, Van den Mooter G (2010) Influence of preparation methods on solid state supersaturation of amorphous solid dispersions: a case study with itraconazole and eudragit e100. Pharm Res 27:775–785

Jayasankar A, Somwangthanaroj A, Shao ZJ, Rodriguez-Hornedo N (2006) Cocrystal formation during cogrinding and storage is mediated by amorphous phase. Pharm Res 23:2381–2392

Kalivoda A, Fischbach M, Kleinebudde P (2012) Application of mixtures of polymeric carriers for dissolution enhancement of fenofibrate using hot-melt extrusion. Int J Pharm 429:58–68

Karki S, Friscic T, Jones W, Motherwell WDS (2007) Screening for Pharmaceutical cocrystal hydrates via neat and liquid-assisted grinding. Mol Pharm 4:347–354

Kennedy M, Hu J, Gao P, Li L, Ali-Reynolds A, Chal B, Gupta V, Ma C, Mahajan N, Akrami A, Surapaneni S (2008) enhanced bioavailability of a poorly soluble VR1 antagonist using an amorphous solid dispersion approach: a case study. Mol Pharm 5:981–993

Kinoshita M, Baba K, Nagayasu A, Yamabe K, Shimooka T, Takeichi YI, Azuma M, Houchi H, Minakuchi K (2002) Improvement of solubility and oral bioavailability of a poorly water-soluble drug, TAS-301, by its melt-adsorption on a porous calcium silicate. J Pharm Sci 91:362–370

Konno H, Handa T, Alonzo DE, Taylor LS (2008) Effect of polymer type on the dissolution profile of amorphous solid dispersions containing felodipine. Eur J Pharm Biopharm 70:493–499

Law D, Schmitt EA, Marsh KC, Everitt EA, Wang W, Fort JJ, Krill SL, Qiu Y (2004) Ritonavir-PEG 8000 amorphous solid dispersions: in vitro and in vivo evaluations. J Pharm Sci 93:563–570

Lyons JG, Hallinan M, Kennedy JE, Devine DM, Geever LM, Blackie P, Higginbotham CL (2007) Preparation of monolithic matrices for oral drug delivery using a supercritical fluid assisted hot melt extrusion process. Int J Pharm 329:62–71

MacLean J, Medina C, Daurio D, Alvarez-Nunez F, Jona J, Munson E, Nagapudi K (2011) Manufacture and performance evaluation of a stable amorphous complex of an acidic drug molecule and neusilin. J Pharm Sci 100:3332–3344

Mallick S, Pattnaik S, Swain K, De PK, Saha A, Ghoshal G, Mondal A (2008) Formation of physically stable amorphous phase of ibuprofen by solid state milling with kaolin. Eur J Pharm Biopharm 68:346–351

Maniruzzaman M, Boateng JS, Bonnefille M, Aranyos A, Mitchell JC, Douroumis D (2012) Taste masking of paracetamol by hot-melt extrusion: an in vitro and in vivo evaluation. Eur J Pharm Biopharm 80:433–442

Medina C, Daurio D, Nagapudi K, Alvarez-Nunez F (2010) Manufacture of pharmaceutical co-crystals using twin screw extrusion: a solvent-less and scalable process. J Pharm Sci 99:1693–1696

Miyazaki T, Yoshioka S, Aso Y, Kojima S (2004) Ability of polyvinylpyrrolidone and polyacrylic acid to inhibit the crystallization of amorphous acetaminophen. J Pharm Sci 93:2710–2717

Porter WW III, Elie SC, Matzger AJ (2008) Polymorphism in carbamazepine cocrystals. Cryst Growth Des 8:14–16

Remenar JF, Morissette SL, Peterson ML, Moulton B, MacPhee JM, Guzman HR, Almarsson O (2003) Crystal engineering of novel cocrystals of a triazole drug with 1,4-dicarboxylic acids. J Am Chem Soc 125:8456–8457

Repka MA, Battu SK, Upadhye SB, Thumma S, Crowley MM, Zhang F, Martin C, McGinity JW (2007) Pharmaceutical applications of hot-melt extrusion: part II. Drug Dev Ind Pharm 33:1043–1057

Roblegg E, Jaeger E, Hodzic A, Koscher G, Mohr S, Zimmer A, Khinast J (2011) Development of sustained-release lipophilic calcium stearate pellets via hot melt extrusion. Eur J Pharm Biopharm 79:635–645

Rumondor ACF, Marsac PJ, Stanford LA, Taylor LS (2009) Phase behavior of poly (vinylpyrrolidone) containing amorphous solid dispersions in the presence of moisture. Mol Pharm 6:1492–1505

Serajuddin ATM (1999) Solid dispersion of poorly water-soluble drugs: early promises, subsequent problems, and recent breakthroughs. J Pharm Sci 88:1058–1066

Serajuddin ATM (2007) Salt formation to improve drug solubility. Adv Drug Delivery Rev 59:603–616

Serajuddin ATM, Pudipeddi M (2011) Salt-selection strategies. In: Stahl PH, Wermuth G (Eds). Handbook of pharmaceutical salts 2nd edn. Wiley, New York.

Trask AV, Motherwell WDS, Jones W (2004) Solvent-drop grinding: green polymorph control of cocrystallisation. Chem Commun (Camb) (7):890–891

Trask AV, Haynes DA, Motherwell WDS, Jones W (2006a) Screening for crystalline salts via mechanochemistry. Chem Commun (Camb) (1):51–53

Trask AV, Motherwell WDS, Jones W (2006b) Physical stability enhancement of theophylline via cocrystallization. Int J Pharm 320:114–123

Variankaval N, Wenslow R, Murry J, Hartman R, Helmy R, Kwong E, Clas S-D, Dalton C, Santos I (2006) Preparation and solid-state characterization of nonstoichiometric cocrystals of a phosphodiesterase-IV inhibitor and L-tartaric acid. Cryst Growth Des 6:690–700

Watanabe T, Hasegawa S, Wakiyama N, Kusai A, Senna M (2002a) Prediction of apparent equilibrium solubility of indomethacin compounded with silica by 13C solid state NMR. Int J Pharm 248:123–129

Watanabe T, Hasegawa S, Wakiyama N, Usui F, Kusai A, Isobe T, Senna M (2002b) Solid state radical recombination and charge transfer across the boundary between indomethacin and silica under mechanical stress. J Solid State Chem 164:27–33

Wermuth CG, Stahl PH (2011) Selected procedures for the preparation of pharmaceutically acceptable salts. In: Stahl PH, Wermuth G. (Eds). Handbook of pharmaceutical salts 2nd edn. Wiley, New York.

Windbergs M, Strachan CJ, Kleinebudde P (2009) Tailor-made dissolution profiles by extruded matrices based on lipid polyethylene glycol mixtures. J Control Release 137:211–216

Yu L (2001) Amorphous pharmaceutical solids: preparation, characterization and stabilization. Adv Drug Delivery Rev 48:27–42

Chapter 10
Melt Extruded Controlled Release Dosage Forms

Justin M. Keen and James W. McGinity

Abstract Hot-melt extrusion is a proven pharmaceutical processing technology, enabling the formation of matrices and structures designed to control drug release. Controlled drug release formulations are important for reducing side effects, improving bioavailability, and patient compliance. Melt extrusion has been applied in the production of various controlled-release dosage forms: including pellets, tablets, films, and drug reservoirs. Controlled-release dosage forms prepared by extrusion have been demonstrated for oral delivery as well as in producing implants and transmucosal devices. This chapter reviews extruded controlled-release dosage forms, highlighting the importance of material properties and structure. The mechanisms by which drugs are released from different materials and structures are reviewed using examples from current literature.

10.1 Introduction

Over the last several decades, hot-melt extrusion has emerged as an effective formulation tool for enhancing and/or enabling drug therapy. Extrusion technology was developed many decades before its application in pharmaceutical technology and, certainly, the evolution of synthetic thermoplastic pharmaceutical-grade polymers has enabled the translation of the technology to drug delivery. The utility of polymers in controlled drug delivery was first realized in the 1960s (Hoffman 2008). However, widespread industry adoption has only recently occurred, which reflects advances over the last two decades in understanding the thermodynamic and kinetic factors responsible for the solid-state configuration of extruded compositions and, arguably more important, those that control their stability (Janssens and Van den Mooter 2009). In addition, over the last few decades the industry has been evolving toward continuous processing, which is inherent to extrusion processing and fully enables process analytical technology (PAT). The application of PAT is useful for fully controlling product quality and is considered a critical component of

J. W. McGinity (✉) · J. M. Keen
The University of Texas at Austin, Austin, TX 78712, USA
e-mail: mcginity.jw@austin.utexas.edu

J. M. Keen
DisperSol Technologies, LLC, Georgetown, TX 78626, USA

M. A. Repka et al. (eds.), *Melt Extrusion*,
AAPS Advances in the Pharmaceutical Sciences Series 9,
DOI 10.1007/978-1-4614-8432-5_10,

the quality-by-design (QbD) methodology encouraged by the Food and Drug Administration (FDA; Yu 2008). Furthermore, the drug molecules being presented to formulators are becoming increasingly difficult to solubilize (O'Driscoll and Griffin 2008), which is required prior to absorption, driving an interest in newer and more versatile technologies.

While a significant portion of recent research in hot-melt extrusion focuses on applications of the amorphous state for improving the bioavailability of poorly soluble drugs, early applications focused on controlled-release technology. Extruders were described a handful of times in the 1970s and 1980s for producing controlled-release dosage forms, with significant research activity in this area starting in the 1990s. Detailed reviews of these early experiments are available for the interested reader (McGinity and Zhang 2003; Breitenbach 2002; Crowley et al. 2007; Repka et al. 2007). Recent efforts in solubilizing drugs using hot-melt extrusion and other solvent-based techniques have made the evaluation of amorphous solid dispersions commonplace. As these therapies enter the market, there will be an increasing need to combine the amorphous form with a controlled release system to reduce dosing frequency, the strategies for accomplishing this are taking shape (Tran et al. 2011). As such, melt extrusion has come full circle and will become increasingly important as a tool to create formulations that combine the benefits of amorphous solid dispersions (Newman et al. 2012) with those of controlled release (Uhrich et al. 1999). On the whole, technologies such as melt extrusion are required to transform today's poorly water-soluble drug candidates, which should be designed to improve safety and efficacy by maximizing bioavailability, minimizing side effects, and minimizing dosing frequency. This chapter summarizes the various dosage forms prevalent in the literature and commercial products considering in detail the nature of the materials used to form them. In addition, the mechanisms by which a drug releases from controlled release melt extrudates are examined.

10.2 Melt-Extruded Dosage Forms for Controlled Release

In the simplest cases, melt extrusion processes intimately mix a drug and excipients. Most commonly, this involves grinding or milling the output of an extrusion process for preparation of a powder, which could be used directly or further processed into a dosage form, such as a capsule or tablet. Extrudates are readily formed directly into films or rods, which may be cut into pellets or tablets. Multiple extruders can be combined, known as coextrusion, to produce structured dosage forms having multiple layers. Following compounding in an extruder, the melt may be transferred to an injection-molding machine and formed into complex shapes. Together, the geometry, composition, and processing conditions can be manipulated to control the release of drug.

Melt extrusion has been utilized for a variety of controlled release applications in which the extrudate is ground or milled to produce granules or a process intermediate. Liu et al. (2001) prepared lipid matrix extrudates, which were then milled into granules and used in a conventional compression process to prepare sustained release tablets. When compared to a melt granulation process, the granules produced harder tablets and had improved drug content uniformity across the particle size distribution. The water-soluble drug, phenylpropanolamine hydrochloride, was released from Precirol® and Sterotex® K matrices over 12 h or more depending on the exact composition and pore former included in the tablets. Milled extrudates of enteric polymer formulations were prepared to facilitate a dry coating process by Sauer et al. (2007). In conventional film coating processes, solvents are beneficial to distribute plasticizers in the film-forming polymer, melt extrusion was demonstrated as an alternative approach to generating a plasticized coating material for a dry coating process. The resulting extrudate-coated tablets were found to exhibit pH-dependent release.

Early controlled release matrix tablets prepared by traditional processes, such as wet granulation, melt granulation, or direct compression, are often observed to have variable dissolution properties during aging due to curing of the matrix (Murthy and Ghebre-Sellassie 1993). Kidokoro et al. (2001) demonstrated the utility of melt extrusion in preparing acrylic matrix tablets by hot-melt extrusion, which showed, drastically reduced variation following curing than tablets prepared following melt granulation or direct compression processes. The tablets were prepared by the three processes using the same composition of ibuprofen in an Eudragit® RS PO matrix. Following treatment at 40 °C, the dissolution rate from the melt-granulated and direct compression formulations substantially decreased due to structural changes in the tablet as a result of fusion of the polymer network. The melt-extruded tablets, having an already fused structure, were observed at the initial time point to have a controlled release profile that changed minimally upon storage.

Tablets cut from polymeric extrudates have also been designed to control drug release within the stomach. In the plastics industry, foamed polymer matrices are routinely prepared. Foamed dosage forms have low densities and are hypothesized to float within the stomach, where a sustained release of a drug would continuously release the active into the upper small intestine. This approach would be advantageous for targeting drugs having an absorption window in the upper small intestine to increase bioavailability (Streubel et al. 2006). Extrusion processes have been demonstrated to prepare floating tablets through two mechanisms. Nakamichi et al. (2001) prepared foamed hydroxypropyl methylcellulose acetate succinate (HPMCAS) tablets by injecting water into the extrusion barrel, which mixed the formulation at high pressure and above the normal boiling temperature of water. At the die outlet, the water vapor expands and forms a porous structure; this foaming mechanism has also been explored in detail using supercritical carbon dioxide injection (Terife et al. 2012). Alternatively, thermal decomposition of sodium bicarbonate, resulting in the off-gassing, has been demonstrated in the preparation of acrylic controlled-release floating tablets (Fukuda et al. 2006a).

Young et al. (2005) investigated the addition of non-thermoplastic polymeric gelling agents, included at low levels in a thermoplastic matrix, on the sustained release properties of a pH-dependent acrylic matrix formulation. Carbopol® 974P was effective in preparing pH-independent sustained release profiles when incorporated at 2.5, 5, and 10 % by weight in rod-cut tablets (Fig. 10.1). Rod-cut tablets

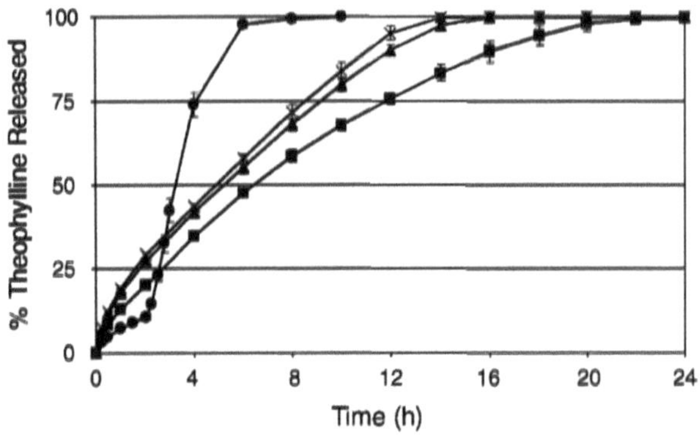

Fig. 10.1 Influence of 0 (●), 2.5 (■), 5 (▲), and 10 % (×) Carbopol 974P on theophylline release from acrylic melt-extruded tablets (2 h in 0.1 N HCl, followed by pH 6.8 PBS, 900 ml, 37 °C, 50 rpm). (Young et al. 2005)

have also been extensively studied using non-thermoplastic lipids. The temperature profile must be carefully controlled to prepare a solid extrudate as complete melting of the lipid matrix results in a pasty or liquid output (Reitz and Kleinebudde 2007). In addition, the temperature profile and processing parameters must be carefully considered to control the surface properties of lipid tablets (Reitz et al. 2008), control the polymorphic form of the lipid matrix (Windbergs et al. 2009), and to ensure repeatable sustained release dissolution profiles.

Controlled release films have been extensively studied for transmucosal applications. Extruded release hydroxypropylcellulose (HPC)- based films have been demonstrated in vivo as bioadhesive by Repka and McGinity (Repka and McGinity 2001). The preparation of pharmaceutically elegant films requires the careful selection of plasticizers to control the mechanical properties of the extrudate, such as tensile strength and percent elongation (Repka and McGinity 2000). HPC-based films, having a thickness of 600 μm, were observed to sustain the release of lidocaine over 12 h (Repka et al. 2005). The solubility of drugs in these compositions should be carefully considered. When molecularly dispersed, small-molecule drugs act as plasticizers for thermoplastic polymers, which in addition to altering the mechanical properties have been demonstrated to improve polymer stability during processing. However, if not fully miscible, the drug may recrystallize on the surface of the extrudate upon cooling of the melt (Crowley et al. 2004b).

Multiparticulate dosage forms can be prepared from melt extrudates through the careful application of downstream processing. Young et al. (2002) described a melt extrusion process in which 1.22-mm-diameter rods were cut to the same length using a strand pelletizer and then further processed in a spheronizer to produce round particles. Acrylic pellets prepared by this process were observed to sustain the release of theophylline (Young et al. 2003). Verhoeven et al. (2006) studied 3 mm diameter by 2 mm cylindrical particles comprising a water-swellable polymer, xanthan gum, and water-insoluble ethylcellulose in vivo. With increasing concentrations of xanthan gum, the drug release rate increased allowing more complete release of ibuprofen during transit along the gastrointestinal tract. The impact of xanthan gum concentrations of 10, 20, and 30 % in the ethylcellulose matrix significantly impacted plasma concentrations in dogs (Fig. 10.2).

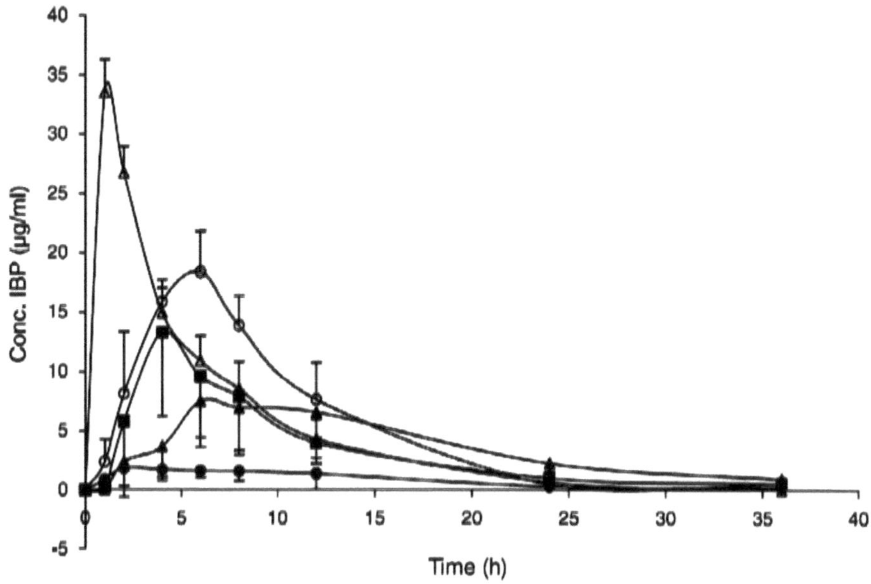

Fig. 10.2 Mean plasma concentration–time profiles after administration of 300 mg ibuprofen to dogs. Extrudates: (•) 10, (▲) 20, and 30 % (■) xanthan gum in ethylcellulose. Comparators: (○) Ibu-Slow 600 (half tablet) and (△) Junifen (15 ml). (Verhoeven et al. 2006)

Advanced, long-acting extruded dosage forms are capable of delivering drugs over several days to several weeks. Campbell et al. (2010) extruded ibuprofen containing poly(ε-caprolactone) (PCL) together with clay silicates (nanoplatelets). The clay silicates created a tortuous path for ibuprofen, which diffused over several days from the matrix. Cheng et al. (2010) evaluated PCL-based implants, which released praziquantel in vitro over 4 weeks, or much longer depending on the amount of plasticizer included in the formulation. The rod-form extrudates were implanted subcutaneously in rats and plasma concentrations were detected for more than 7 weeks (Fig. 10.3). In vitro release profiles lasting 40–60 days were observed for interferon-alpha (IFN-α) from an extruded lipid matrix (Schulze and Winter 2009). IFN-α was first complexed with hydroxypropyl-β-cyclodextrin (HP-β-CD) in solution and lyophilized. The complex was then extruded into rods containing a combination of low- and high-melting point acid triglycerides. The protein structure was not altered by the extrusion process and due to the large size of IFN-α, diffusion occurred over 1–2 months.

Complex structures can be prepared using coextrusion processes, typically resulting in a layered film or concentric cylinders, or through transferring the melt to an injection molder. Quenten et al. (2011) prepared ejection-molded tablets for sustaining the release of metoprolol tartrate from a xanthan gum and ethylcellulose matrix. In a dog study, one of the studied compositions was found to be statistically similar to a conventional marketed formulation. Coextruded compositions have been described and marketed to prepare reservoir devices. Vaginal inserts, such as NuvaRing®, are prepared using a coextrusion process (Groenewegen 1999). The

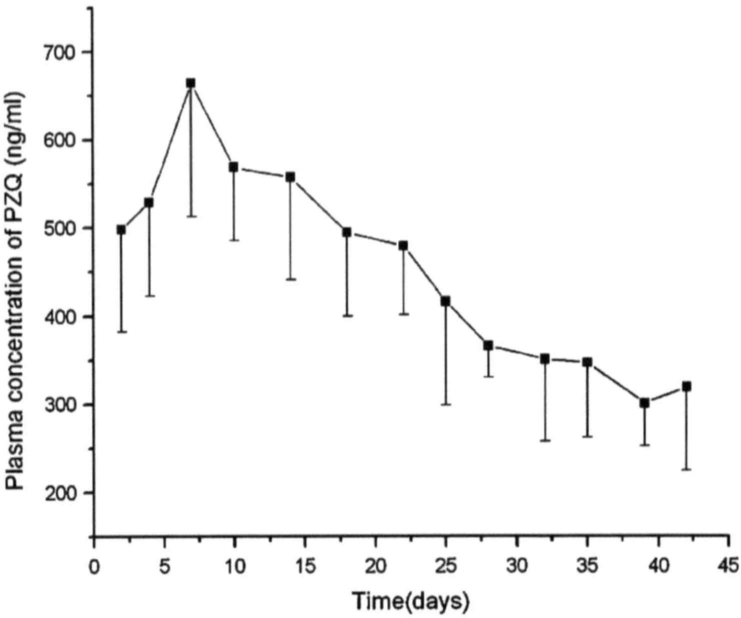

Fig. 10.3 Praziquantel plasma concentration in rats following subcutaneous implantation of PCL-extruded implants, $n = 4$. (Cheng et al. 2010)

hormone(s) is contained in the core of the extrudate, which is formed as a rod. The outer layer, coextruded with the core, forms a skin and acts as a diffusion barrier for drug release. The concentric rod is cut into cylinders, which are formed into a toroid by connecting the ends using an adhesive. Coextrusion processing for fixed dose sustained release compositions was studied by Dierickx et al. (2012). This system was a concentric coextruded rod, which was cut into minitablets. The core contained metoprolol tartrate in the insoluble polymer PCL. The outer layer dissolved during administration and released hydrochlorothiazide (Fig. 10.4). The formulation was determined to be statistically bioequivalent for both drugs to a commercial formulation following administration to dogs.

10.3 Drug Release Mechanisms

Drugs are released from extruded dosage forms by several different and well-known mechanisms. Drug release also occurs in three phases: a burst or lag phase, the early-time release phase, and the late-time or depletion phase. In reality, a combination of release mechanisms is typically responsible for the release profile and the mechanism of release may change depending on the release phase. Using recent melt extrusion literature examples, these mechanisms are reviewed.

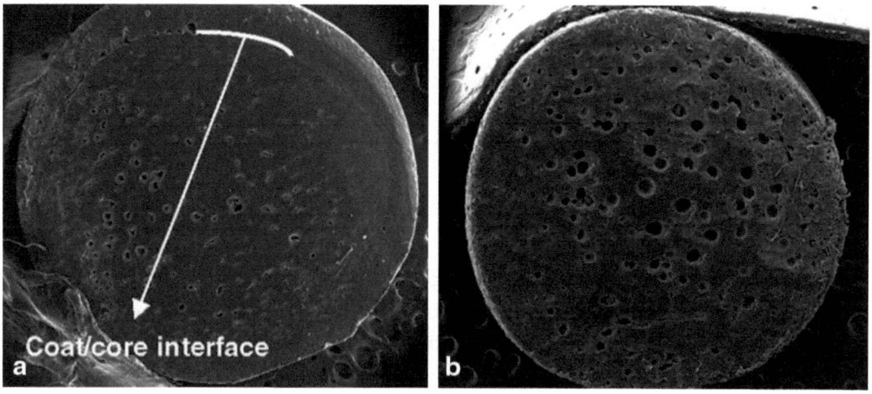

Fig. 10.4 SEM images of coextruded cut tablet with a core composed of 45/55 % MPT/PCL and a coat composed of 10/45/45 % HCT/PEO/PEG. (**a**) Immediately after production and (**b**) after fecal recovery from dogs. (Dierickx et al. 2012)

At the molecular level, four processes are responsible for drug release: dissolution, diffusion, erosion, and swelling. These processes are manipulated using excipients, the extrusion process, and the dosage form structure. Excipients may be selected to alter osmotic pressure, drug solubility, or create a porous network, to name a few examples. The extrusion process alters the molecular distribution of the formulation and may change the solid state of the drug. The dosage form structure may be chosen to create diffusion barriers or target a specific geometry.

Dissolution is readily modeled and described for simple systems using the well-known Noyes–Whitney equation, which at the most basic level describes the relationship of drug concentration, C, to time, t, and the saturation solubility, C_s, using a rate constant, k (Dokoumetzidis and Macheras 2006).

$$\frac{dC}{dt} = k(C_S - C)$$

Based on the work of Nernst and Brunner, we know that the rate constant is related to the diffusivity of the drug, D, in the diffusion layer of thickness, h, the exposed surface area, S, and the volume of the dissolution medium.

$$k = \frac{DS}{Vh}$$

Dissolution is rarely the only process involved in sustained release drug formulations; otherwise many drugs would simply exist as crystals of various sizes for administration. The dosage form may be designed to slowly release small particles of drug, which then undergo rapid dissolution. Alternatively, dissolution will occur within the dosage form and the drug released by diffusion. Diffusion is a stochastic process, which can be modeled using the following form of Fick's first law (in this case a one-dimensional system):

$$\frac{\partial C}{\partial t} = SD\frac{\partial C}{\partial x}$$

Table 10.1 Exponent, *n* of the power law and its relationship to the drug release mechanism for various dosage form geometries (Siepmann and Peppas 2001)

Exponent, n			Drug release mechanism
Slab	Cylinder	Sphere	
0.5	0.45	0.43	Diffusion
$0.5 < n < 1.0$	$0.45 < n < 0.89$	$0.43 < n < 0.85$	Anomalous
1.0	0.89	0.85	Case-II

This equation is applicable when the concentration gradient is modeled using a constant activity source, or in other words, when the concentration at high position is maintained at the initial level at all times (Baker 1987). When the initial concentration at the high position changes with time, Fick's second law also applies.

$$\frac{\partial C}{\partial t} = D\frac{\partial^2 C}{\partial x^2}$$

Degradation, often referred to as erosion, is applicable to systems containing biodegradable polymers and refers to the degradation of the polymer backbone. The rate of erosion, when compared to the rate of penetrant diffusion, will determine if erosion occurs at the surface or in the bulk. This, in turn, determines if the drug is sloughed off during release or diffuses from the matrix. Erosion is a complex process and requires modeling of the several mass transport processes along with reaction rates (Sackett and Narasimhan 2011).

Swelling is important for polymer-based sustained release systems, especially in those containing glassy water-soluble polymers. The Peppas research group has studied swelling in polymer-based systems extensively. Swelling controls the rate of water penetration into the dosage form for matrices formed from polymers exhibiting thermogelation. The relative rates of water penetration and polymer dissolution will control the thickness of the gel layer formed (Kaunisto et al. 2011). More importantly, the rates of these processes will determine if a crystalline drug is released intact or undergoes dissolution in the gel and is released by diffusion through the hydrated gel. Based on the dosage form geometry, the Peppas equation can be applied to identify if swelling and relaxation (Case-II Transport), diffusion, or both (anomalous) processes controlled drug release.

$$\frac{M_t}{M_\infty} = kt^n,$$

where M is the amount of drug released and n is the power law release exponent, which following fitting to a drug release data set can be interpreted according to Table 10.1 (Siepmann and Peppas 2001).

In the case of slab geometry, the Peppas equation reduces to the Higuchi equation, which resulted from modeling of a pharmaceutical ointment formulation (Siepmann and Peppas 2011).

Higuchi modeled the release of a low-concentration crystalline drug from a noneroding insoluble matrix, which slowly released drug by diffusion. In this case,

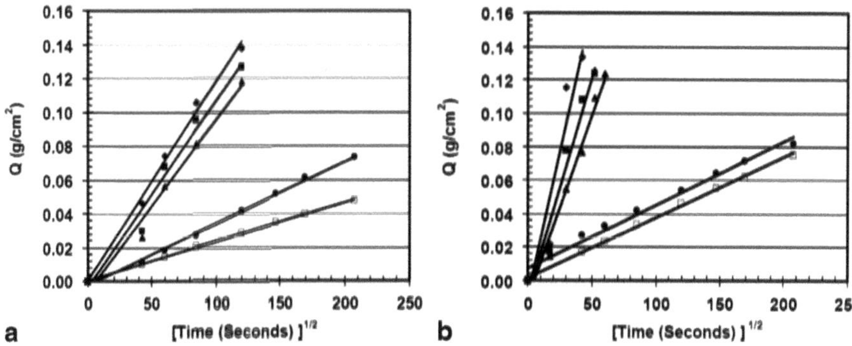

Fig. 10.5 Higuchi diffusion model fitting of the guaifenesin release data from matrix tablets prepared by direct compression and hot-melt extrusion. (**a**) Matrix tablets prepared using "fine" ethyl cellulose (325–80 mesh) and (**b**) matrix tablets prepared using "coarse" ethyl cellulose (80–30 mesh). (♦) Direct compression, 10 kN, (■) direct compression, 30 kN, (▲) direct compression, 50 kN, (•) hot-melt extrusion, 80, 85, 85, and 90 °C, (□) hot-melt extrusion, 90, 105, 105, and 110 °C (Crowley et al. 2004a)

drug diffusion from the matrix controls the rate at which the drug is dissolved in the ointment. As the drug is dissolved in the matrix, a moving boundary between crystalline and saturated drug continuously penetrates the slab. Practically speaking, a linear dependence of drug release rate on the square root of time is indicative of a diffusion boundary, although this should only be applied rigidly to slab geometry.

When the drug is dispersed in an insoluble matrix, drug diffusion occurs by two pathways, through the macromolecular network and through pores inherent to the dosage form or formed during the release process by the dissolving drug. Zhu et al. (2002) studied the release of chlorpheniramine maleate from an Eudragit® RS PO matrix. When tablets were prepared by direct compression, the release rate was rapid and decreased substantially following heat treatment. However, tablets prepared by melt extrusion exhibited sustained release properties with little effect from postprocess thermal treatment. Crowley et al. (2004a) demonstrated the influence of melt extrusion on the porosity and tortuosity of rod cut, ethylcellulose tablets, which were determined to have significantly lower pore sizes, lower percent porosity, and higher tortuosity than directly compressed tablets. Drug release from the ethylcellulose tablets occurred by diffusion as evidenced by the near linear dependence of release rate on the square root of time (Fig. 10.5).

Schilling et al. (2008) demonstrated the utility of water-soluble processing aids in altering drug release. Citric acid was found to act as a processing aid during extrusion, which facilitated the solubilization of diltiazem hydrochloride (DIL HCl) in an Eudragit® RS PO matrix. Drug diffusion was enhanced compared to formulations containing other water-soluble additives, such as sucrose and sodium chloride. The interactions, or lack thereof, of the additive with the polymer, impact the propensity of the drug to release through the macromolecular network or the porous network (Fig. 10.6). In this example, the formulation containing citric acid, released drug faster than the more porous composition containing sodium chloride.

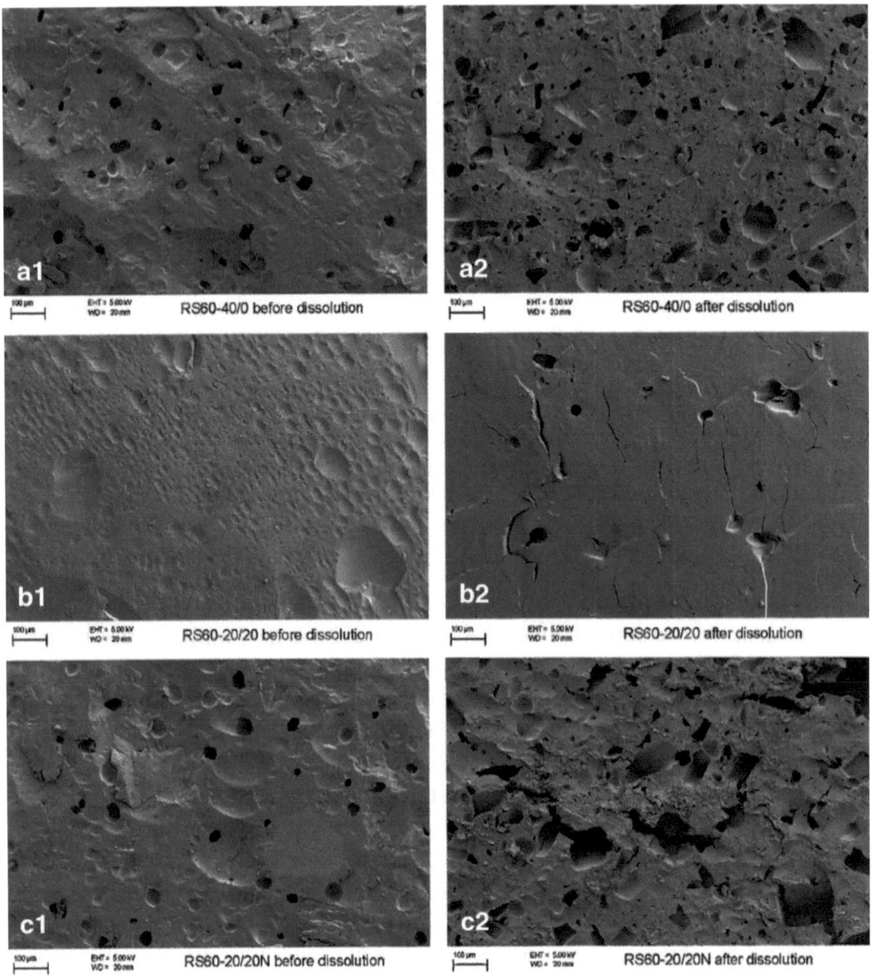

Fig. 10.6 Scanning electron microscope (SEM) of hot-melt extruded tablets composed of 60 % Eudragit® RS PO and (**a**) 40 % diltiazem hydrochloride (DIL HCl), (**b**) 20 % DIL HCl and 20 % citric acid monohydrate, and (**c**) 20 % DIL HCl and 20 % NaCl. Pictures were taken before (*a1, b1, c1*) and after (*a2, b2, c2*) dissolution in phosphate buffer pH 6.0 over 12 h. (Schilling et al. 2008)

Almeida et al. (2011) studied the release of metoprolol tartrate from the semicrystalline thermoplastic polymer polyethylene-co-vinyl acetate (EVA). When the vinyl acetate content in this polymer is increased, the percent crystallinity decreases. Crystalline domains act as diffusion barriers and, accordingly, the drug release rate was observed to increase as the vinyl acetate component of the matrix increased from 9 to 28 %. However, when the vinyl acetate concentration was 40 %, the molecular mobility of the polymer allowed for reforming of the matrix during dissolution, resulting in a decrease in the diffusion rate as pores were not as readily available for drug diffusion out of the matrix. Sax et al. (2012) studied the diffusion of large IFN-α protein molecules from a melt-extruded lipid matrix. The matrix was formed with a mixture of triglycerides having differing melting ranges. Using single-molecule wide-field fluorescence microscopy, single molecules were tracked during release. The authors observed two populations of diffusing protein molecules, one present in the solid fraction of the lipid matrix and the other relating to a molten lipid phase present due to partial melting of the implant as the temperature release was studied.

Polymer swelling is instrumental in altering release profiles and for some polymers, such as polyethylene oxide (PEO), the swelling properties control drug release, which results in zero-order release profiles. A common problem with diffusion-based drug formulations is that following transition from the early-time to the late-time release phase, or depletion phase, the drug release rate slows dramatically. This is a result of increasingly lower concentrations of drug within the matrix, which decreases the driving force according to Fick's law. This effect may be magnified in vitro when the drug accumulates in the release medium as opposed to being absorbed as would be expected in vivo. De Brabander et al. (2003) demonstrated the utility of adding a swelling polymer, such as xanthan gum or HPMC, to ethylcellulose matrices. The inclusion of soluble polymers not only affected the release rate, but altered the release mechanism from a diffusion-controlled half-time dependency to a zero-order swelling-controlled mechanism as determined by fitting to the Peppas equation.

The pH-dependent swelling properties of Eudragit® S 100 were utilized to develop a controlled release extruded composition of 5-aminosalicylic acid (5-ASA) for colonic delivery (Bruce et al. 2005). At pH 1.2, the formulation released little drug. At pH 7.4, cumulative drug release at 2 h was 20 % and at 12 h was 80 %. The release profile fit the Higuchi model, indicating that at colonic pH, the acrylic polymer undergoes swelling and dissolution. However, the dynamics of polymer hydration, polymer disentanglement, and drug diffusion created a transient hydrated network from which drug diffusion dominated release. Similar behavior was observed by Repka et al. (2005) for HPC-based films containing lidocaine. The polymer slowly swells and dissolves releasing the drug predominately by diffusion over approximately 12 h. Films produced from PEO demonstrate swelling- and relaxation-controlled release as demonstrated by zero-order release profiles of extruded films (Prodduturi et al. 2005) and tablets (Zhang and McGinity 1999). PEO-based extruded tablets produce very strong tablets, which have been utilized for the controlled release of abused drugs (Barkin et al. 2012; McGinity and Zhang 2002). The strength of the matrix prevents crushing, reducing the likelihood of dose dumping. In these cases, the rate of diffusion and dissolution are much faster than the rate of polymer hydration and relaxation, therefore the rate of polymer relaxation controls the drug release rate (Siepmann and Siepmann 2008). A graphic depicting the release mechanism from swelling-controlled systems is included for reference (Fig. 10.7).

Swelling-controlled formulations are susceptible to ionic strength-dependent drug release when the polymer gel strength is altered in the presence of ions (Joshi 2011). Fukuda et al. (2006b) studied this phenomenon in tablets containing the ionic polymers xanthan gum and chitosan. Extruded tablets prepared by direct compression

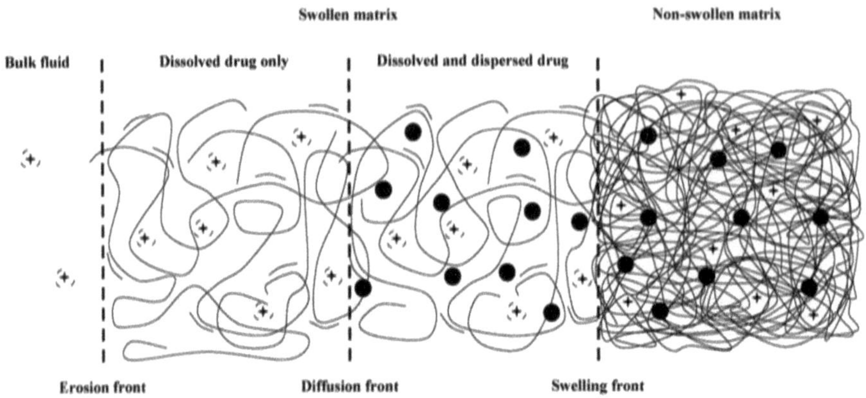

Fig. 10.7 Schematic presentation of a swelling-controlled drug delivery system containing dissolved and dispersed drug (*stars* and *black circles*, respectively), exhibiting the following moving boundaries: (1) an "erosion front," separating the bulk fluid from the delivery system; (2) a "diffusion front," separating the swollen matrix containing dissolved drug only and the swollen matrix containing dissolved and dispersed drug; and (3) a "swelling front," separating the swollen and nonswollen matrix. (Siepmann and Siepmann 2008)

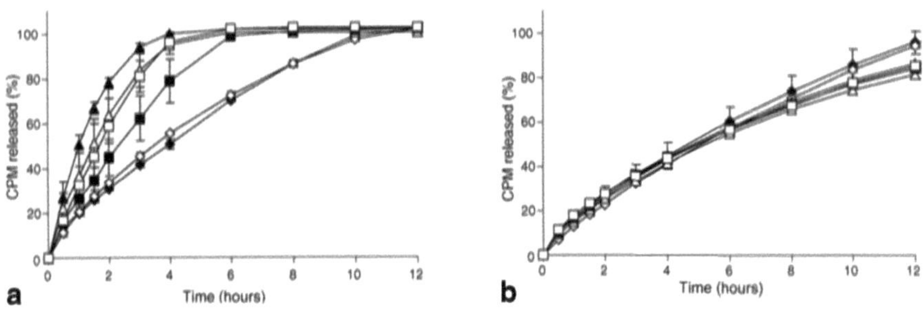

Fig. 10.8 CPM release profiles from HME tablets containing **a** PEO, Chitosan, and MCC or **b** PEO, Chitosan, and Xanthan Gum in 900 ml of either (♦) 0.1N HCl, (◇) pH 4.0 acetate buffer, (▲) pH 4.0 citrate buffer, (△) pH 4.0 phosphate buffer, (■) pH 6.8 phosphate buffer, or (□) pH 7.4 phosphate buffer at 37 ± 0.5 °C (USP 27 Apparatus 2, 100 rpm). (Fukuda et al. 2006b)

and containing PEO, xanthan gum, and chitosan were found to exhibit pH and ionic strength-independent dissolution. Interactions between the cationic chitosan and anionic xanthan gum were studied to explain this behavior. When xanthan gum was replaced with microcrystalline cellulose (MCC), the system exhibited pH- and buffer species-dependent release (Fig. 10.8).

Polymers that biodegrade, typically hydrolysable polyesters, have been extruded to prepare implantable dosage forms. PCL-extruded cylindrical implants, 3 mm diameter, containing praziquantel, were extruded and cut into 10 mm lengths and studied in vitro (Li et al. 2010). When included at 25 % loading, drug release was monitored for 150 days, over this time period approximately 60 % of the drug had been released. Drug was released from the matrix by diffusion. The morphology of the matrix before and after 150 days shows the formation of pores during release

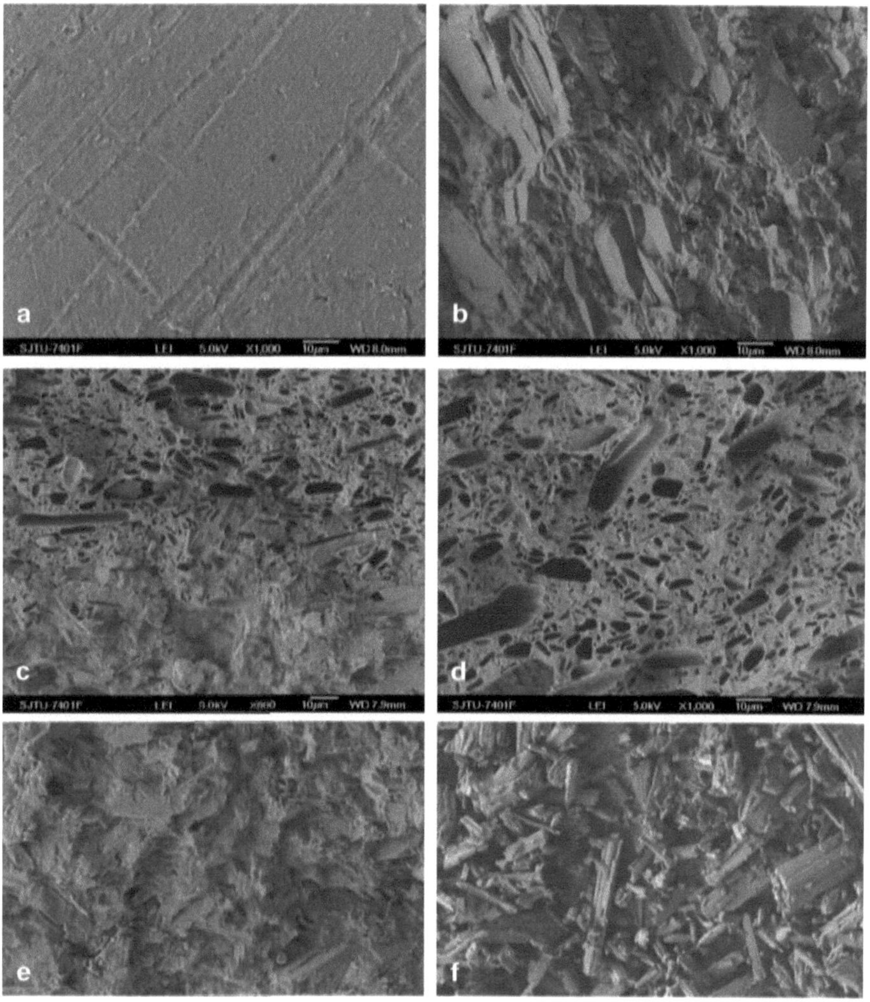

Fig. 10.9 SEM images of the CI-7 implant and PZQ: (**a** and **b**) surface and cross-section morphology of implant CI-7 before drug release, respectively; (**c**) cross-section morphology of implant CI-7 after 150-day drug release; (**d** and **e**) the exterior layer and interior layer in the cross-section of implant CI-7 after 150-day drug release, respectively; (**f**) PZQ crystals. (Li et al. 2010)

(Fig. 10.9). The molecular weight of the polymer decreased by approximately 20 % during the 150-day period was monitored and determined to occur throughout the matrix. Following the addition of PEG in the system, the release rate increased from 40 to > 80 days depending on the exact composition (Cheng et al. 2010). Following implantation in rats, elevated plasma concentrations were observed for over 40 days.

Campbell et al. (2010) studied the influence of nanostructures on release of ibuprofen from a PCL matrix. Ibuprofen-, PCL-, and silicate-based nanoclays were extruded, ground, and compression molded into disks. The nanoclay, present as platelets intercalated in the matrix, were found to alter the mechanical strength of the matrix, depending on the aspect ratio of the nanoclay. In addition to creating a

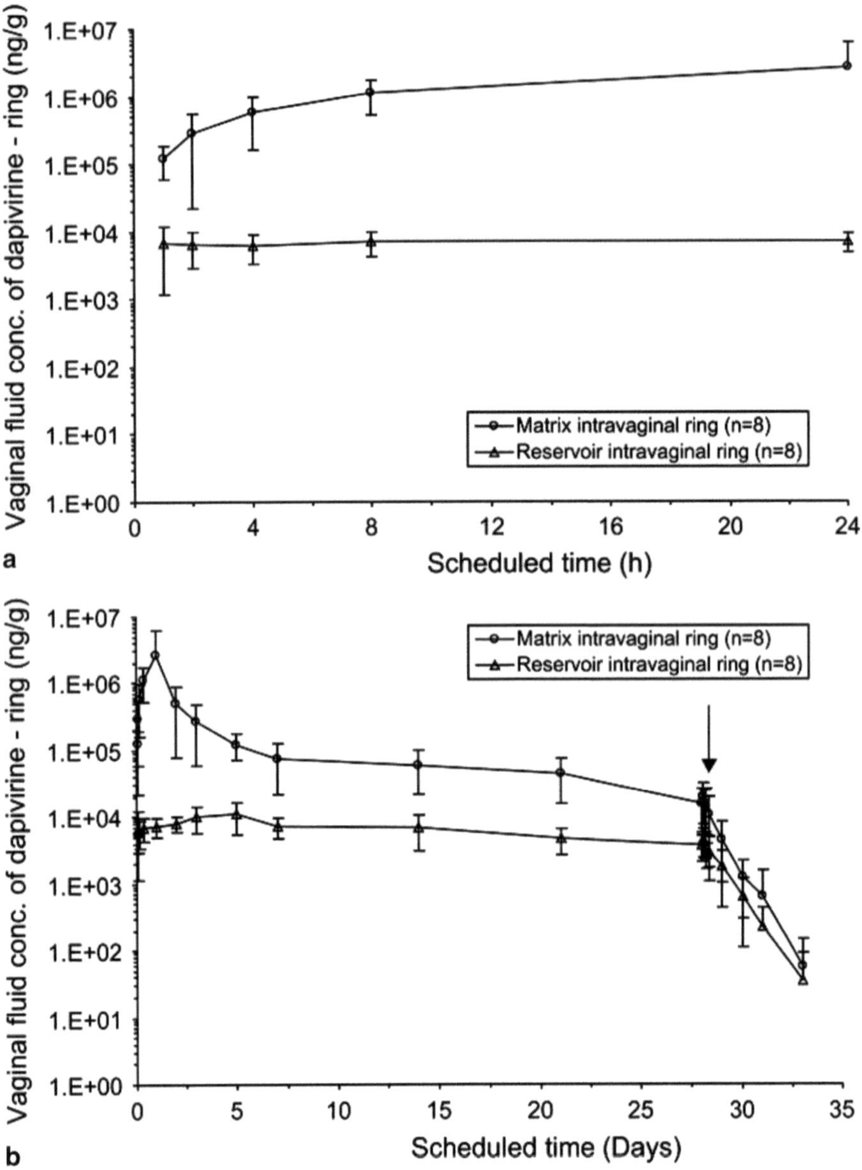

Fig. 10.10 Mean concentrations of dapivirine in vaginal fluids at the IVR site during the first 24 h postinsertion (**a**) and over 33 days (**b**). In (**b**), the *arrow* indicates time of removal of the IVR. (Nel et al. 2009)

tortuous path in the macromolecular network for ibuprofen diffusion, the nanoclays altered the degree of crystallization of the polymer. The compression-molded disks, having a large surface to area volume ratio (18 mm diameter by 1.2 in height), were found to release ibuprofen over 4 days.

Using conventional coating processes, constant activity reservoirs are readily prepared and known to exhibit zero-order release (Kaunisto et al. 2011). In these systems, the coating acts as a diffusion barrier membrane. Crystalline drug inside the membrane continually dissolves as drug diffuses out of the system, maintaining a steady concentration gradient/driving force across the membrane. EVA-based intravaginal devices operate similarly, with the exception that drug is contained in the core in a supersaturated state. The solubility of hormones in EVA and the resulting release mechanism of coextruded EVA rods were studied by van Laarhoven et al. (2002). The rods were formed to a diameter of 4 mm with a membrane thickness of 110 μm. The core contained etonogestrel and ethinyl estradiol in EVA 28 and was surrounded by an EVA 9 skin. The solubility of the hormones was observed to be 1–2 orders of magnitude greater in EVA 28 than in EVA 9. In addition, the drugs were dissolved in the matrix at elevated temperature, which resulted in supersaturation in the matrix following cooling. Partitioning of the drug occurs from the supersaturated core into the lower solubility skin, in this system the supersaturated core behaves as a constant activity reservoir. Due to equilibration following processing, the drugs were observed to partition into the skin on storage, which resulted in an initial burst release. Vaginal inserts are being studied for various antimicrobial drugs (Clark et al. 2011). Nel et al. (2009) studied reservoir devices and matrices containing dapivirine in women. The reservoir devices provided steady drug concentrations in vivo for 28 days and showed a reduced, but more constant, release rate than the matrix device (Fig. 10.10).

10.4 Summary

Melt extrusion processing is widely applicable in the preparation of controlled release oral medications, implants, and devices, which may take the form of simple matrix tablets or advanced structures. The pressure and heat applied to the materials results in molecular arrangements useful for manipulating diffusion or altering the solid state of the drug. At the molecular scale, various processes contribute to the release mechanism, including dissolution, diffusion, polymer degradation, and polymer swelling. Using various dosage form compositions and structure, controlled release compositions have been demonstrated to sustain the release of drug from several hours to several months.

References

Almeida A, Possemiers S, Boone MN, De Beer T, Quinten T, Van Hoorebeke L, Remon JP, Vervaet C (2011) Ethylene vinyl acetate as matrix for oral sustained release dosage forms produced via hot-melt extrusion. Eur J Pharm Biopharm 77(2):297–305. doi:10.1016/j.ejpb.2010.12.004

Baker R (1987) Controlled release of biologically active agents. Wiley, New York

Barkin RL, Bruckenthal P, Stanos SP (2012) Strategies to reduce the tampering and subsequent abuse of long-acting opioids: potential risks and benefits of formulations with physical or pharmacologic deterrents to tampering. Mayo Clin Proc 87(7):683–694

Breitenbach J (2002) Melt extrusion: from process to drug delivery technology. Eur J Pharm Biopharm 54(2):107–117. doi: 10.1016/s0939–6411(02)00061–9

Bruce LD, Shah NH, Waseem Malick A, Infeld MH, McGinity JW (2005) Properties of hot-melt extruded tablet formulations for the colonic delivery of 5-aminosalicylic acid. Eur J Pharm Biopharm 59(1):85–97. doi:10.1016/j.ejpb.2004.06.007

Campbell K, Craig D, McNally T (2010) Ibuprofen-loaded poly(E-Caprolactone) layered silicate nanocomposites prepared by hot melt extrusion. J Mater Sci: Mater Med 21(8):2307–2316. doi:10.1007/s10856-009-3963-2

Cheng L, Lei L, Guo S (2010) In vitro and in vivo evaluation of praziquantel loaded implants based on Peg/Pcl Blends. Int J Pharm 387(1–2):129–138. doi:10.1016/j.ijpharm.2009.12.010

Clark M, Kiser P, Loxley A, McConville C, Malcolm R, Friend D (2011) Pharmacokinetics of Uc781-loaded intravaginal ring segments in rabbits: a comparison of polymer matrices. Drug Deliv Transl Res 1(3):238–246. doi:10.1007/s13346–011-0032–4

Crowley MM, Schroeder B, Fredersdorf A, Obara S, Talarico M, Kucera S, McGinity JW (2004a) Physicochemical properties and mechanism of drug release from ethyl cellulose matrix tablets prepared by direct compression and hot-melt extrusion. Int J Pharm 269(2):509–522. doi:10.1016/j.ijpharm.2003.09.037

Crowley MM, Fredersdorf A, Schroeder B, Kucera S, Prodduturi S, Repka MA, McGinity JW (2004b) The influence of guaifenesin and ketoprofen on the properties of hot-melt extruded polyethylene oxide films. Eur J Pharm Sci 22(5):409–418. doi:10.1016/j.ejps.2004.04.005

Crowley MM, Zhang F, Repka MA, Thumma S, Upadhye SB, Kumar Battu S, McGinity JW, Martin C (2007) Pharmaceutical applications of hot-melt extrusion: Part I. Drug Dev Ind Pharm 33(9):909–926. doi:10.1080/03639040701498759

De Brabander C, Vervaet C, Remon JP (2003) Development and evaluation of sustained release mini-matrices prepared via hot melt extrusion. J Control Release 89(2):235–247. doi:10.1016/s0168–3659(03)00075–0

Dierickx L, Saerens L, Almeida A, De Beer T, Remon JP, Vervaet C (2012) Co-extrusion as manufacturing technique for fixed-dose combination mini-matrices. Eur J Pharm Biopharm 81(3):683–689. doi:10.1016/j.ejpb.2012.03.018

Dokoumetzidis A, Macheras P (2006) A century of dissolution research: from Noyes and Whitney to the biopharmaceutics classification system. Int J Pharm 321(1–2):1–11. doi:10.1016/j.ijpharm.2006.07.011

Fukuda M, Peppas NA, McGinity JW (2006a) Floating hot-melt extruded tablets for gastroretentive controlled drug release system. J Control Release 115(2):121–129

Fukuda M, Peppas NA, McGinity JW (2006b) Properties of sustained release hot-melt extruded tablets containing chitosan and xanthan gum. Int J Pharm 310(1–2):90–100. doi:10.1016/j.ijpharm.2005.11.042

Groenewegen RJJ (1999) Drug delivery system for two or more active substances. United States Patent 5989581

Hoffman AS (2008) The origins and evolution of "Controlled" drug delivery systems. J Control Release 132(3):153–163. doi:10.1016/j.jconrel.2008.08.012

Janssens S, Van den Mooter G (2009) Review: physical chemistry of solid dispersions. J Pharm Pharmacol 61(12):1571–1586. doi:10.1211/jpp.61.12.0001

Joshi SC (2011) Sol-gel behavior of hydroxypropyl methylcellulose (Hpmc) in ionic media including drug release. Materials 4(10):1861–1905

Kaunisto E, Marucci M, Borgquist P, Axelsson A (2011) Mechanistic modelling of drug release from polymer-coated and swelling and dissolving polymer matrix systems. Int J Pharm 418(1):54–77. doi:10.1016/j.ijpharm.2011.01.021

Kidokoro M, Shah N, Malick A, Infeld M, McGinity J (2001) Properties of tablets containing granulations of ibuprofen and an acrylic copolymer prepared by thermal processes. Pharm Dev Technol 6(2):263

Li C, Cheng L, Zhang Y, Guo S, Wu W (2010) Effects of implant diameter, drug loading and end-capping on praziquantel release from Pcl implants. Int J Pharm 386(1–2):23–29. doi:10.1016/j.ijpharm.2009.10.046

Liu J, Zhang F, McGinity JW (2001) Properties of lipophilic matrix tablets containing phenylpropanolamine hydrochloride prepared by hot-melt extrusion. Eur J Pharm Biopharm 52(2):181–190

McGinity JW, Zhang F (2002) Hot-melt extrudable pharmaceutical formulation. United States Patent US6,488,963B1

McGinity JW, Zhang F (2003) Melt-extruded controlled-release dosage forms. In: Ghebre-Sellassie I, Martin C (eds) Pharmaceutical extrusion technology, vol 133. Marcel Dekker, New York

Murthy KS, Ghebre-Sellassie I (1993) Current perspectives on the dissolution stability of solid oral dosage forms. J Pharm Sci 82(2):113–126. doi:10.1002/jps.2600820202

Nakamichi K, Yasuura H, Fukui H, Oka M, Izumi S (2001) Evaluation of a floating dosage form of nicardipine hydrochloride and hydroxypropylmethylcellulose acetate succinate prepared using a twin-screw extruder. Int J Pharm 218(1–2):103–112. doi:10.1016/s0378-5173(01)00617-2

Nel A, Smythe S, Young K, Malcolm K, McCoy C, Rosenberg Z, Romano J (2009) Safety and pharmacokinetics of dapivirine delivery from matrix and reservoir intravaginal rings to HIV-negative women. J Acquir Immune Defic Syndr Hum Retrovirol 51(4):416–423

Newman A, Knipp G, Zografi G (2012) Assessing the performance of amorphous solid dispersions. J Pharm Sci 101(4):1355–1377. doi:10.1002/jps.23031

O'Driscoll CM, Griffin BT (2008) Biopharmaceutical challenges associated with drugs with low aqueous solubility—the potential impact of lipid-based formulations. Adv Drug Deliv Rev 60(6):617–624. doi:10.1016/j.addr.2007.10.012

Prodduturi S, Manek RV, Kolling WM, Stodghill SP, Repka MA (2005) Solid-state stability and characterization of hot-melt extruded poly(ethylene oxide) films. J Pharm Sci 94(10):2232–2245. doi:10.1002/jps.20437

Quinten T, De Beer T, Onofre FO, Mendez-Montealvo G, Wang YJ, Remon JP, Vervaet C (2011) Sustained-release and swelling characteristics of xanthan gum/ethylcellulose-based injection moulded matrix tablets: in vitro and in vivo evaluation. J Pharm Sci 100(7):2858–2870. doi:10.1002/jps.22480

Reitz C, Kleinebudde P (2007) Solid lipid extrusion of sustained release dosage forms. Eur J Pharm Biopharm 67(2):440–448

Reitz C, Strachan C, Kleinebudde P (2008) Solid lipid extrudates as sustained-release matrices: the effect of surface structure on drug release properties. Eur J Pharm Sci 35(4):335–343. doi: 10.1016/j.ejps.2008.08.002

Repka MA, McGinity JW (2000) Influence of Vitamin E Tpgs on the properties of hydrophilic films produced by hot-melt extrusion. Int J Pharm 202(1–2):63–70. doi:10.1016/s0378-5173(00)00418-x

Repka MA, McGinity JW (2001) Bioadhesive properties of hydroxypropylcellulose topical films produced by hot-melt extrusion. J Control Release 70(3):341–351. doi:10.1016/s0168-3659(00)00365-5

Repka MA, Gutta K, Prodduturi S, Munjal M, Stodghill SP (2005) Characterization of cellulosic hot-melt extruded films containing lidocaine. Eur J Pharm Biopharm 59(1):189–196. doi:10.1016/j.ejpb.2004.06.008

Repka MA, Battu SK, Upadhye SB, Thumma S, Crowley MM, Zhang F, Martin C, McGinity JW (2007) Pharmaceutical applications of hot-melt extrusion: part II. Drug Dev Ind Pharm 33(10):1043–1057. doi:10.1080/03639040701525627

Sackett CK, Narasimhan B (2011) Mathematical modeling of polymer erosion: consequences for drug delivery. Int J Pharm 418(1):104–114. doi:10.1016/j.ijpharm.2010.11.048

Sauer D, Zheng W, Coots LB, McGinity JW (2007) Influence of processing parameters and formulation factors on the drug release from tablets powder-coated with Eudragit® L 100–55. Eur J Pharm Biopharm 67(2):464–475

Sax G, Feil F, Schulze S, Jung C, Bräuchle C, Winter G (2012) Release pathways of interferon A2a molecules from lipid twin screw extrudates revealed by single molecule fluorescence microscopy. J Control Release 162(2):295–302. doi:10.1016/j.jconrel.2012.07.014

Schilling SU, Bruce CD, Shah NH, Malick AW, McGinity JW (2008) Citric acid monohydrate as a release-modifying agent in melt extruded matrix tablets. Int J Pharm 361(1–2):158–168. doi:10.1016/j.ijpharm.2008.05.035

Schulze S, Winter G (2009) Lipid extrudates as novel sustained release systems for pharmaceutical proteins. J Control Release 134(3):177–185. doi:10.1016/j.jconrel.2008.11.026
Siepmann J, Peppas NA (2001) Modeling of drug release from delivery systems based on hydroxypropyl methylcellulose (Hpmc). Adv Drug Deliv Rev 48(2–3):139–157. doi:10.1016/s0169-409 × (01)00112-0
Siepmann J, Peppas NA (2011) Higuchi equation: derivation, applications, use and misuse. Int J Pharm 418(1):6–12. doi:10.1016/j.ijpharm.2011.03.051
Siepmann J, Siepmann F (2008) Mathematical modeling of drug delivery. Int J Pharm 364(2):328–343. doi:10.1016/j.ijpharm.2008.09.004
Streubel A, Siepmann J, Bodmeier R (2006) Drug delivery to the upper small intestine window using gastroretentive technologies. Curr Opin Pharmacol 6(5):501–508. doi:10.1016/j.coph.2006.04.007
Terife G, Wang P, Faridi N, Gogos CG (2012) Hot melt mixing and foaming of Soluplus® and Indomethacin. Polym EngSci 52(8):1629–1639. doi:10.1002/pen.23106
Tran P, Tran T, Park J, Lee B-J (2011) Controlled release systems containing solid dispersions: strategies and mechanisms. Pharm Res 28(10):2353–2378. doi:10.1007/s11095-011-0449-y
Uhrich KE, Cannizzaro SM, Langer RS, Shakesheff KM (1999) Polymeric systems for controlled drug release. Chem Rev 99(11):3181–3198. doi:10.1021/cr940351u
van Laarhoven JAH, Kruft MAB, Vromans H (2002) In vitro release properties of etonogestrel and ethinyl estradiol from a contraceptive vaginal ring. Int J Pharm 232(1–2):163–173. doi:10.1016/s0378-5173(01)00900-0
Verhoeven E, Vervaet C, Remon JP (2006) Xanthan Gum to Tailor Drug Release of Sustained-Release Ethylcellulose Mini-Matrices Prepared Via Hot-Melt Extrusion: In Vitro and in Vivo Evaluation. Eur J Pharm Biopharm 63 (3):320-330. doi:10.1016/j.ejpb.2005.12.004
Windbergs M, Strachan CJ, Kleinebudde P (2009) Understanding the solid-state behaviour of triglyceride solid lipid extrudates and its influence on dissolution. Eur J Pharm Biopharm 71(1):80–87
Young CR, Koleng JJ, McGinity JW (2002) Production of spherical pellets by a hot-melt extrusion and spheronization process. Int J Pharm 242(1–2):87–92. doi:10.1016/s0378–5173(02)00152–7
Young CR, Koleng JJ, McGinity JW (2003) Properties of drug-containing spherical pellets produced by a hot-melt extrusion and spheronization process. J Microencapsul 20(5):613–625. doi:10.3109/02652040309178350
Young CR, Dietzsch C, Cerea M, Farrell T, Fegely KA, Rajabi-Siahboomi A, McGinity JW (2005) Physicochemical characterization and mechanisms of release of theophylline from melt-extruded dosage forms based on a methacrylic acid copolymer. Int J Pharm 301(1–2):112–120. doi:10.1016/j.ijpharm.2005.05.025
Yu L (2008) Pharmaceutical quality by design: product and process development, understanding, and control. Pharm Res 25(4):781–791. doi:10.1007/s11095-007-9511-1
Zhang F, McGinity J (1999) Properties of sustained-release tablets prepared by hot-melt extrusion. Pharm Dev Technol 4(2):241
Zhu Y, Shah N, Malick A, Infeld M, McGinity J (2002) Influence of thermal processing on the properties of chlorpheniramine maleate tablets containing an acrylic polymer. Pharm Dev Technol 7(4):481

Chapter 11
Excipient or API Melt Processing via Injection Molding

Costas G. Gogos

Abstract This chapter first presents and discusses the physical phenomena taking place in both injection molding machines and the molds, which enable the injection molding process to produce shaped thermoplastic polymer products reproducibly, inexpensively and rapidly, with only minimum industrial waste stream. Industrial waste generated aside, it then addresses the question of whether the existing injection molding processes and equipment are appropriate for producing tablets of active pharmaceutical ingredient (API) solid solutions in polymer excipients, where the total dissolution of the API in the molten excipient, and the absence of API process-generated degradation must be guaranteed. The arguments presented indicate the need for specific modifications of the existing injection molding processes, which change and improve the melting and laminar mixing generated, to assure API dissolution, and decrease the chances of API degradation originating from the process.

11.1 Introduction

The casting, compression, and injection molding operations all entail forcing a molten polymer into a cold cavity, where it cools under pressure (or simply gravity) reproducing the cavity's shape. The *injection molding* process is carried out by an injection molding machine, where polymer solid particulates are fed into, conveyed, compressed, melted, mixed, and pressurized by a reciprocating *single screw* to very rapidly fill a single or multicavity *cold mold*. The entrance to each cavity, called the *gate*, has a small cross-sectional area. The positive displacement pressure generated by hydraulic pressure on the single screw in the injection molding machine is kept on after the mold cavity is filled until the melt in the gate freezes, thus sealing the molten polymer charge into the cavity. The process objective is that the thermal contraction of the melt in the cavity is balanced by its expansion during decompression so that the

C. G. Gogos (✉)
New Jersey Institute of Technology, The Polymer Processing Institute,
University Heights, Newark, NJ, USA
e-mail: costas.g.gogos@njit.edu

M. A. Repka et al. (eds.), *Melt Extrusion*,
AAPS Advances in the Pharmaceutical Sciences Series 9,
DOI 10.1007/978-1-4614-8432-5_11,

molded part has the exact shape of the mold cavity and is free of voids. In the *sandwich injection molding* process, two polymers generated by two injection molded machines are used *in sequence* to fill the mold forming molded articles with skin-core structure. Air can also be introduced in partially melt-filled molds, and pressurized to form a polymer skin-air core sandwich structure, through the *gas-assist injection molding* process. In all the above injection molding processes, the polymer is melted, mixed, and injected from single-screw injection molding units. Chabot and Malloy (1997) present the history of injection molding from the time of its invention by John Wesley Hyatt in 1872 to the invention of the single-screw reciprocating injection molding machine by Wilert in the 1950s. Comprehensive treatment of the injection molding process, mold filling flows and simulations, as well as mold design can be found in Rubin (1972) and Tadmor and Gogos (2006).

Compared with the widely studied area of pharmaceutical hot-melt extrusion (HME), there are relatively few references to be found on the use of injection molding machines for processing and direct tablet molding pharmaceutical API or excipient formulations, where the API forms a solid dispersion or solid solution. Rothen-Weinhold et al. (1989) were the first to report in 1989 an evaluation of injection molding as a pharmaceutical technology to produce pharmaceutical tablets, studying the drug release profiles via changing the macromolecular makeup and ratios of the mixtures of ethylcellose (EC) and hydroxylpropylmethylcellulose (HPMC) used as excipients. A 2005 patent application publication by Clarke (2005), assigned to GlaxoSmithKline discloses a process of first hot-melt extruding API or excipient formulations, where the API forms a solid solution. The extrudates are then cooled and pelletized. The resulting pellets are then fed into the hopper of a reciprocating single-screw injection molding machine, where they are remelted, with the resulting melt accumulating at the end of the screw. A gas soluble to the molten excipient, or a supercritical fluid, are injected under pressure in the latter part of the single screw, allowing for uniform mixing and dissolution. Upon the activation of the hydraulic piston, the molten charge is caused to flow rapidly into a number of cavities. The advancing flow region being characterized by a lower pressure, allows the gas to expand. As a result the molded tablet forms attain cellular or microcellular (foamed) structures. Quinten et al. (2009a, b, 2011) have reported in 2009 and 2011 on their work with injection molding using excipient mixtures of EC/L-HPC, and EC/PEO for sustained-release oral dosage forms. Finally, Cheng et al. (2009) report on their work investigating the properties of implants of ε-caprolactone with dispersed praziquantel, processed first in a batch Haake mixer to form the dispersion, cooled, and fed into a Minijet Haake to mold the implants.

In this chapter, we will first present and highlight the important physical phenomena taking place in both injection molding machines and the molds. We will then address the question of what are the advantages and disadvantages from producing an API solid solution in a polymer excipient and molding it into finished oral dosage pharmaceutical products, using injection molding processes. Finally, we will conclude the chapter by proposing modifications of the injection molding process, involving the use of a tandem HME corotating twin-screw extruder (TSE) delivering sequentially API solid solutions or dispersions to a number piston chambers which, in turn, fill a series of mold cavities on a rotating stage mold platform. Such a manufacturing scheme may improve the degree, rate, and uniformity of the API dissolution, and decrease the chances of API degradation originating from the process.

11.2 Injection Molding of Polymers

Injection molding involves two distinct processes. The first comprises the *elementary steps* of *solids transport, melt generation, mixing, and pressurization and flow*, which are *carried out in the injection unit of the molding machine*; the second is the product *shaping or molding*, which *takes place in the mold cavity*.

11.2.1 Injection Molding Machines

The majority of the currently used injection molding machines are of the *in-line, reciprocating single-screw* type, as illustrated in Fig. 11.1a. Less common, but of potential interest to the melt processing of pharmaceutical formulations, are the *two-stage injection molding machines*, shown in Fig. 11.1b, where the polymer melt stream is produced continuously by a single-screw extruder (SSE), delivering it into a reservoir connected to a hydraulic piston system, which is cyclically pressurized to deliver the melt into the cold mold cavity or cavities.

For in-line, reciprocating single-screw injection molding machine the injection "*molding cycle*" is shown schematically in Fig. 11.2, indicating the simultaneous positions and states/configurations of the screw, the mold and the process.

Referring to the above figure, *during the "screw rotation"* cycle period the injection molding machine *operates like a single-screw extruder* where the following "elementary steps" of polymer processing take place: (a) polymer and additives particulates are flood-fed with a hopper; (b) the particulates are conveyed and compressed by the frictional drag forces with the barrel wall; (c) the compressed particulate solid "bed" next to the hot barrel temperature begins the *gradual process of melting* aided by the viscous energy dissipation in that melted region; and (d) the molten polymer undergoes laminar distributive and dispersive mixing. The length of this injection molding- cycle period is dictated by the amount of polymer which is needed to fill the mold cavities, sprue, and runners. These elementary steps, which are described in detail by Tadmor and Gogos (2006) are reflected in the frozen carcass from a SSE shown on Fig. 11.3 below.

During the "*stationary screw*" cycle period only conductive heating takes place above by the hot injection molding machine barrel. After the part(s) are ejected and the mold is closed, the hydraulic system of the injection molding machine is activated and the single screw becomes a hydraulic piston causing positive high pressure and high flow rate flow filling the cavities. These high flows are necessary to avoid premature freezing of the melt by the cold mold, especially in the "gate"

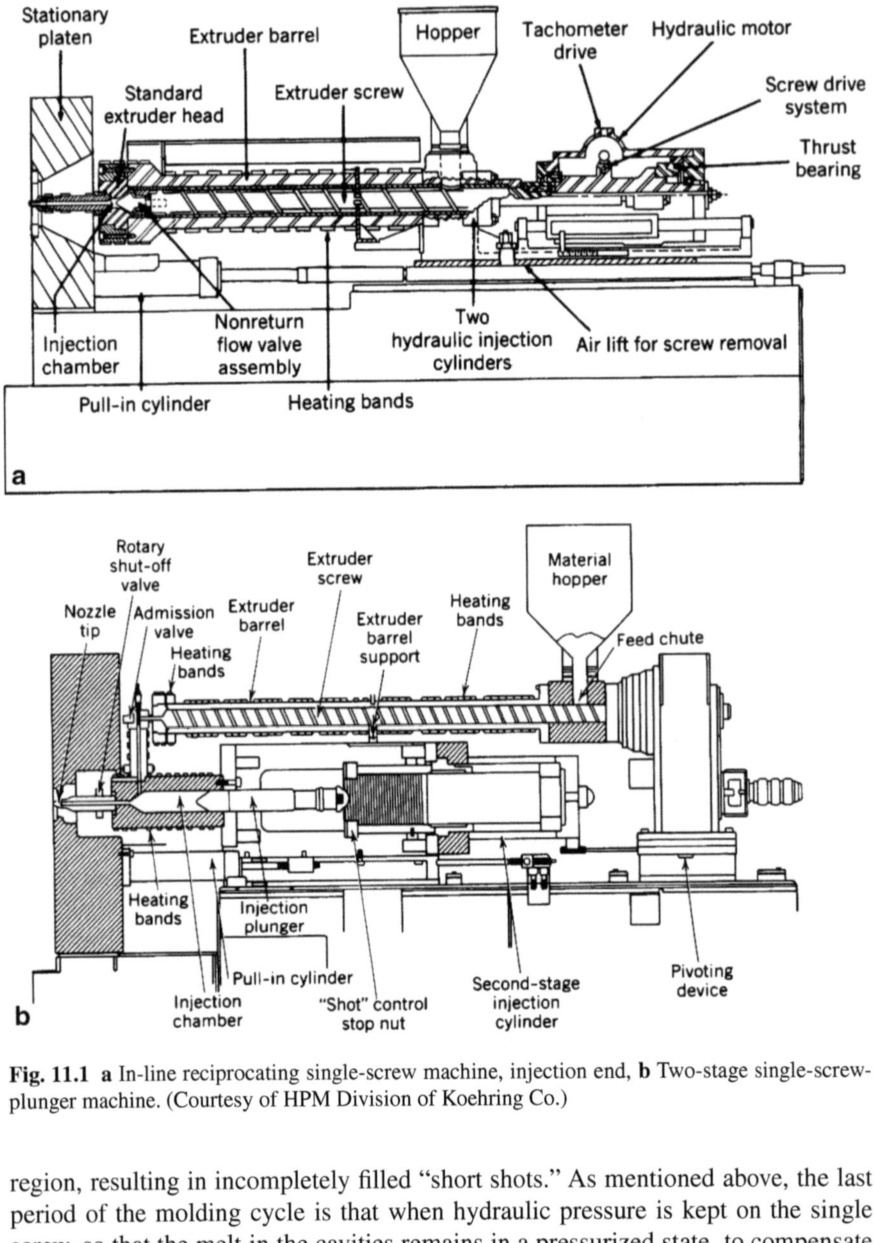

Fig. 11.1 **a** In-line reciprocating single-screw machine, injection end, **b** Two-stage single-screw-plunger machine. (Courtesy of HPM Division of Koehring Co.)

region, resulting in incompletely filled "short shots." As mentioned above, the last period of the molding cycle is that when hydraulic pressure is kept on the single screw, so that the melt in the cavities remains in a pressurized state, to compensate for the thermal contractions. The pressure is kept on till the narrow cross-section gate freezes, sealing the polymer charge into the mold.

There is no molding cycle for the SSE component of the two-stage injection molding machine, since it operates continuously. Thus all the elementary steps of polymer processing are identical to those discussed above. There is one important difference,

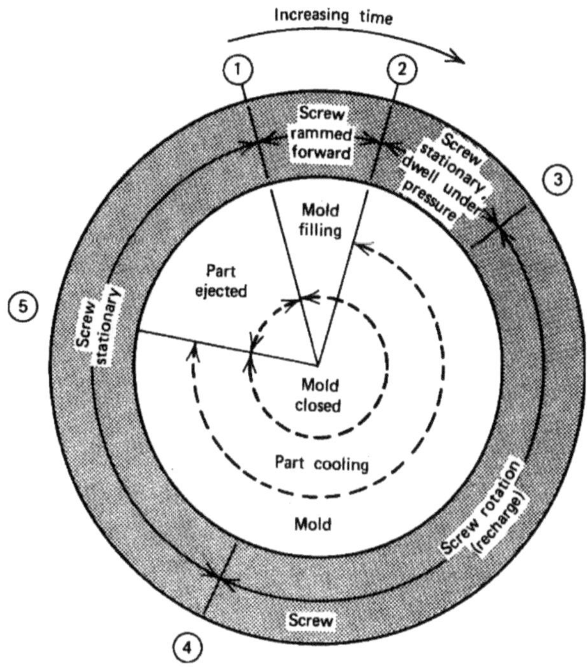

Fig. 11.2 The injection molding cycle of the in-line reciprocation single-screw injection molding machine. (Reprinted with permission from Donovan (1971))

which is also relevant to the melt processing of pharmaceutical formulations: since there is no "stationary screw" period, there is no time wasted for inefficient conductive heating of the polymer charge and, more importantly, the residence time in the single-screw unit is smaller to much smaller that than in the in-line, reciprocating single screw discussed above, which is beneficial to temperature sensitive APIs and polymer excipients. On the other hand, additional residence time is spent by the molten polymer in the injection chamber (Fig. 11.1b) above from the time the melt enters the chamber from the SSE to the time it gets injected into the mold cavities.

11.2.2 Injection Molds

A typical injection mold is made of at least two parts, one of which is movable so that it can open and close during different parts of the molding cycle as shown in Fig. 11.4a and 11.4b.

The entire mold is kept at a constant temperature below the glass transition, T_g or melting temperature, T_m of the polymer. The melt exits the nozzle of the injection unit and flows through the *sprue, runners,* and *gate* into the mold cavity. Each of these structural elements of the mold performs well-defined functions and affects the molding operation. Thus, the sprue forms the overall entrance into the mold. It should not generate large resistance to flow, yet at the same time the melt in it should

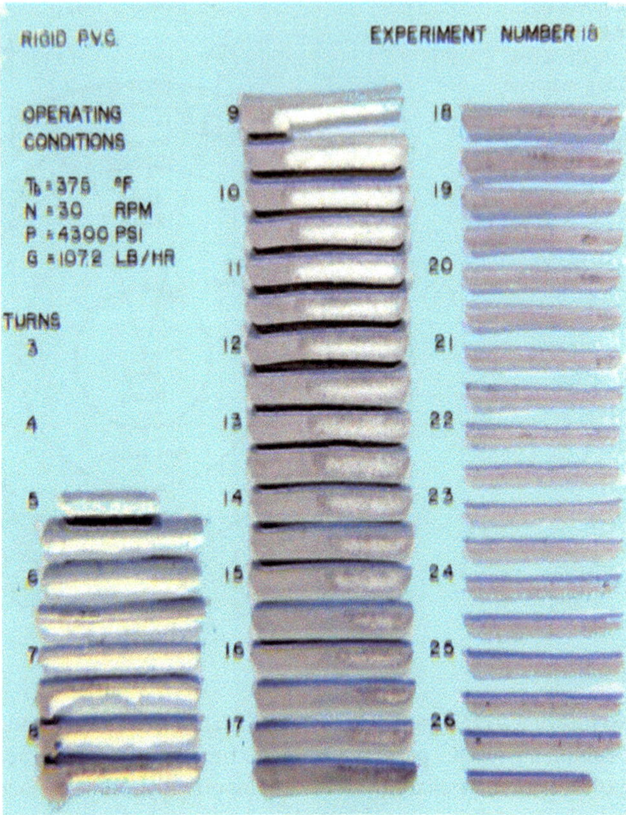

Fig. 11.3 Cross-sections obtained from frozen carcass experiments of a 2.5 in diameter 26.5 length-to-diameter ratio screw extruder. Material: rigid PVC powder. Operating conditions are listed in the figure: T_b *the* barrel temperature, N the screw speed, P the pressure at the die G the mass flow rate. Numbers denote turns from the beginning (*hopper* side) of the screw. The screw was of a metering type with a 12.5 turn feed section, 0.37″ deep, a 9.5 turn transition section, and a 4.5 turn metering section 0.127″ deep. (Reprinted by permission from Tadmor and Klein (1970))

quickly solidify upon completion of injection, so that it can be extracted from it without difficulty. The sprue should also form a streamlined transition between the nozzle and runner system. All these functions are attainable by a short flow channel with diverging conical shape.

The function of the runner system is to bring the hot melt to the cavities. This should be done with the minimum material and pressure drop "waste". Therefore, the runner conduit length must be kept to a minimum, and the cross-section should be *optimally* set for low pressure drop, low material waste, and relatively slow cooling, avoiding premature solidification and "short shots." Generally, the runner is about 1.5 times the characteristic thickness of the molded part, and it is of circular cross-section to minimize heat loss, as well as to facilitate easy machining. Polymer saving and faster cycles occasionally can be achieved by *hot runner systems*, where the polymer

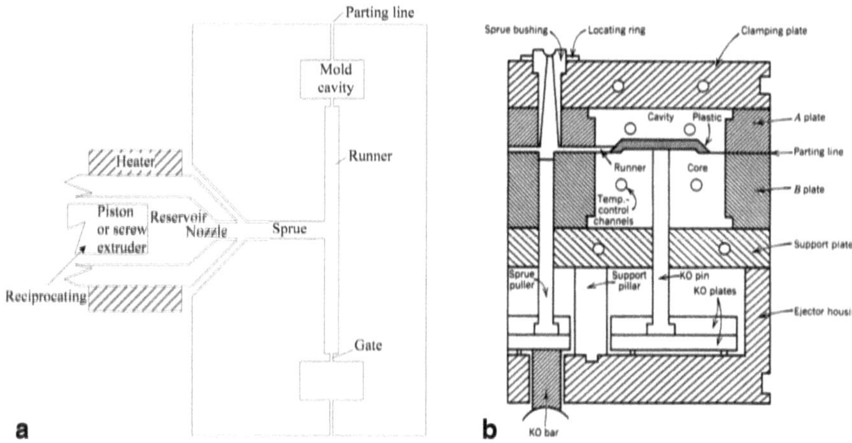

Fig. 11.4 **a** Schematic views of injection molding tooling showing the tip of the injection system and its connection to the mold. Melt passes from the reservoir through the nozzle, the sprue, and the runner system, entering the mold cavities through the gate, **b** two-plate mold. (Note: "parting line" is closed)

in the runners is prevented from solidifying through heating units built around them in the mold plate housing them in hot-runner molds. Alternatively, particularly with large parts, it is sufficient to insulate the runner system from the mold. In both cases the sprue can effectively be eliminated from the design.

The *gate* controls the flow of the polymer melt into the mold. Its size, shape, and position are affected by a number of considerations. First, a narrow gate is desirable from the standpoint of ease of separation of the molded part from the runner system, as well as solidification after the completion of melt injection and packing, to isolate and effectively seal the cavity from the rest of the system. Of course, early solidification must be prevented. Moreover, narrow gates may be detrimental to the finished product because they also bring about large shear rates and stresses (above the melt fracture region), and consequent excessive temperature rise by very high levels of viscous energy dissipation, albeit of very short duration. When the stress level must be reduced, divergent fan gates are used, spreading the flow. Generally, the gate length is about half the characteristic thickness of the section where the gate is attached (usually the heavy sections). The gate is positioned such that the emerging stream impinges on the opposite wall. Figure 11.5 presents typical gate designs and locations. In multiple cavity molds, gates (and runners) also serve the function of balancing flow such that all cavities fill simultaneously.

After the cavity has been filled, the injection pressure is maintained, to "*pack*" a small additional amount of melt into the cavity and to compensate for the thermal contraction of the polymer during the cooling and solidification stages. Packing increases the cavity pressure rapidly and appreciably. When the externally applied pressure is removed (by retracting the reciprocating screw or piston of the injection molding machine), backflow out of the cavity takes place, unless the polymer in the

Fig. 11.5 Typical gate designs and locations. (Reprinted with permission from Rubin (1972))

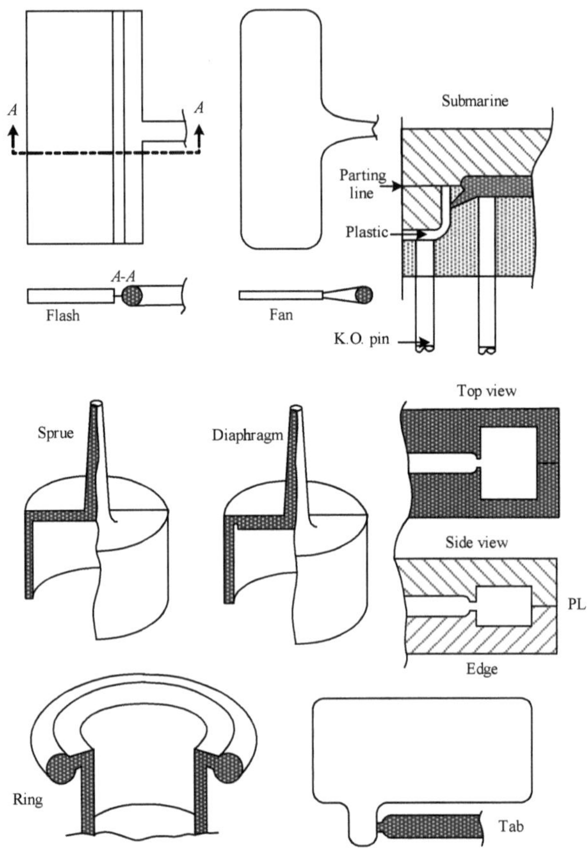

gate has solidified or unless such flow is prevented by a one-way valve. At the end of the backflow, if there is any, only cooling of the polymer takes place, together with minute contraction-induced local flows. When the polymer has solidified sufficiently to withstand the forces of ejection, the mold is opened and the molded article is removed from the cavity with the aid of the ejection (KO "knock out") pins.

From the short description of the molding cycle, it is clear that flow, viscous heat generation (filling flow rates are very high), heat transfer to cold molds, and melt stress relaxation occur to varying degrees of intensity simultaneously. The transport phenomena involved are coupled and, since the cooling or solidification times can be comparable to the polymer relaxation times from chain-oriented states caused by flow, molded articles solidify under strained conditions, i.e., they contain "frozen-in" strains. Such internal strains affect the properties and dimensional stability of molded articles. The detailed mathematical modeling of the flow of polymer melts through these conduits is not easy, and involves most of the complexities of the cavity filling problems, which we discuss below.

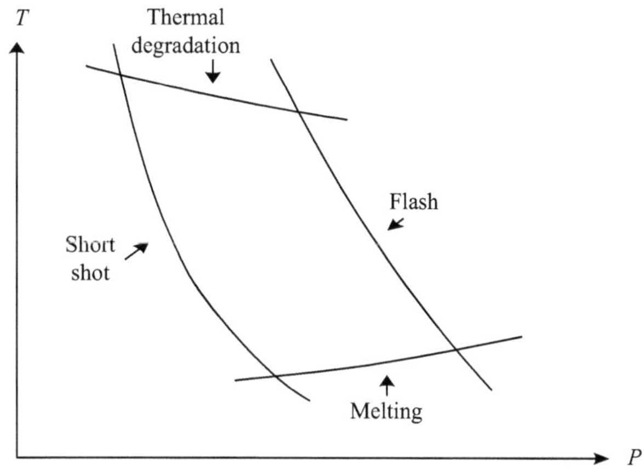

Fig. 11.6 Schematic representation of the "*molding area*" or "*molding window*" diagram that can be determined experimentally for a *given polymer and mold cavity* (3b). Another practical approach to the question of moldability, especially in comparing one polymer with another, is the use of a standard spiral mold cavity and the measurement, under prescribed molding conditions, of the spiral length filled (Caras 1963)

11.2.3 Mold Filling Flows

From the above short discussion it is reasonable to state that that there is no simple answer to the questions of what are the full details of the transient molding flows and heat transfer, and what are the optimal conditions for the proper molding of a specific polymer in a given mold cavity. Figure 11.6, however, illustrates an *empirical answer*, showing an experimentally determined "*molding area*" processing window on the melt temperature-injection pressure plane. Within this area the specific polymer is *moldable* in the specific cavity. The area is bounded by four curves. Below the bottom curve, the polymer is either a solid or will not flow. Above the top curve, the polymer degrades thermally, an issue which is of cardinal importance to the polymer melt processing of pharmaceutical formulations. To the left of the "short shot" curve, the mold cannot be completely filled, and to the right of the "flash" curve, the melt flows in the gaps formed between the various metal pieces that make up the mold, creating thin webs attached to the molded article at the parting lines. Below the "melting" line incomplete melting may present charge homogeneity problems

As mold filling is a complex process, flow visualization studies have been useful and necessary, both for the actual mold design and for the mathematical simulation of the process. An example of such experimental studies can be found in the work of Schmidt (1974). Such studies revealed that the mode of filling at moderately high flow rates is an orderly forward flow, as shown schematically in Fig. 11.7a for a constant depth thin rectangular cavity.

In the "fully developed region" above the melt flow is affected by the cold mold walls, slowing it down near the walls as shown schematically. During the filling process most of the melt flows in an almost fully developed flow in a narrow gap configuration between cold walls. The nature of this flow determines filling time and part core orientation, as well as the occurrence of short shots. A great deal of insight can be obtained by analyzing one-dimensional flow (either radial, spreading disk, or rectilinear) of hot melt between cold walls.

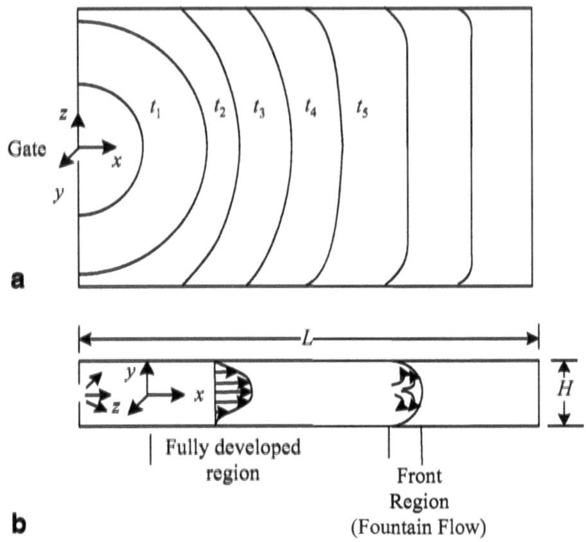

Fig. 11.7 Schematic representation of the flow patterns during the filling of an end gated rectangular mold whose width and length is much greater than its thickness. **a** *Top* view of the flow fronts at various mold filling times. **b** Velocity profiles in the fully developed region, and schematic representation of the "fountain flow" in the front region (3b)

It has been widely established that the *flow front advances with the average cross-sectional speed* of the melt filling the mold. On the other hand, in the front region just behind the front, the melt at the center of the thickness direction moves with the maximum velocity of the parabola-like velocity profile. Thus, it is forced to spill out or *fountain out* toward the mold wall *to form the surface of the molded article at that location*, as in Fig. 11.7b and more clearly as in Fig. 11.8 (Quinten et al. 2011).

This is the only way of filling the region near the wall of the mold, if there is no slip. Thus, in the front region, the central core decelerates from the maximum velocity at the centerline upstream the front, to the mean velocity at which the front advances. As it decelerates in the direction of flow x, it acquires a velocity component in the thickness direction y.

The term "fountain effect" or "*fountain flow*" was coined and first discussed by Rose (1961), and it is essentially the reverse of the flow observed near a plunger emptying a fluid out of a channel of the same cross section. The two-dimensional flow in the front region is important in determining the quality and morphology of the surface of molded articles. It is also the flow phenomenon which makes the *sandwich*, *coinjection*, and *gas-assisted* injection molding possible.

In 1967, Imperial Chemical Industries (ICI) developed the "*sandwich*" molding process for producing "structural" foam products, where the skin is made of unfoamed and the core of the same foamed polymer. In this process, the first melt is injected into the mold until it fills it partway. At this point the second melt, which is the same polymer as the first but contains a foaming agent, is injected behind the first, pushing its core forward and, *because of the fountain flow*, creating a skin made of the first unfoamed melt. When the mold is filled and the polymers undergo cooling, the internal pressure is reduced and foaming takes place, resulting in the formation of a structural foam product—foamed core and solid skin. Before the gate

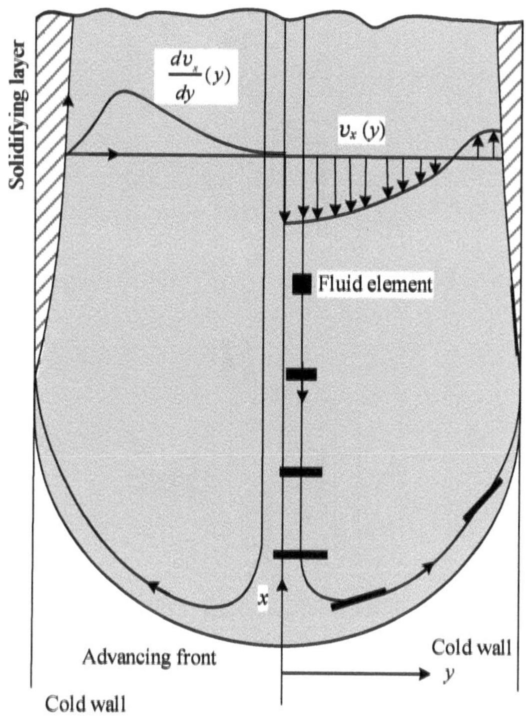

Fig. 11.8 Schematic representation of the flow pattern in the advancing front between two parallel cold walls. *Black rectangles* denote the stretching and orientation of a fluid particle approaching the central region of the front. The curved shape of the front causes fluid particles initially oriented in the y-direction to end up on the wall, oriented in the x-direction. The velocity profile upstream the front is in the x-direction and is viewed from a coordinate system located on the front (Tadmor 1974)

freezes, the first polymer is injected again to purge the sprue of the second melt, in preparation for the next molding cycle. These sequential steps of the process are shown schematically in Fig. 11.9, where the changeover from one melt to the other is carried out by shifting a valve. Two melt generating devices are needed. This can be achieved either with two in-line injection molding machines or with two extruders feeding into the reservoir of a ram positive displacement hydraulic injection unit. *Coinjection molding* has found application in "green" products where the second, foaming agent-containing melt is a recycled grade of the "virgin" melt. Two different polymers can be used in this process to take advantage of the following benefits: lower cost parts, higher strength, reinforcing agent-containing core, sound absorbing core and reduced cooling times, especially in thick parts, when the injection temperature of the core is lower. However, at least one of the two different polymers must contain a compatibilizing agent with the first in order to achieve needed interfacial bonding.

The gas-assisted injection molding process involves the high pressure injection of a gas into a partially filled mold. Under the gas pressure, the polymer *core* melt is driven downstream the mold and, again because of the fountain flow, a continuous melt skin is formed until the mold is filled. The result of this process is a product with a polymer skin and a gas core. In this process, the gas pressure is maintained while the polymer skin melt is cooling, transmitting the required packing pressure

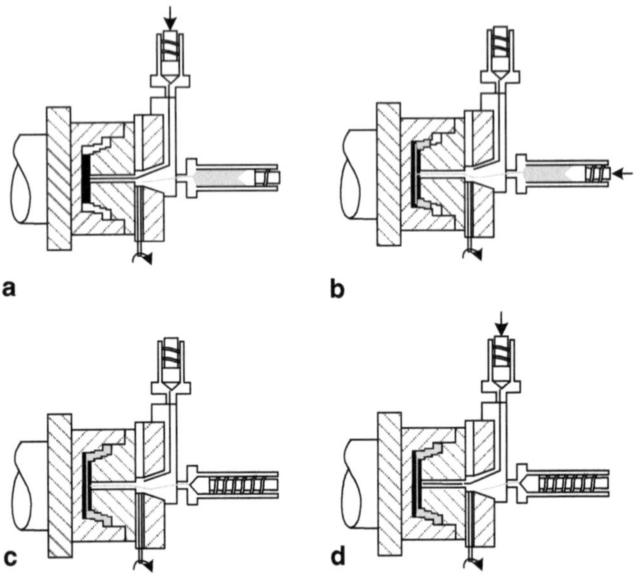

Fig. 11.9 Four stages of coinjection molding. **a** Short shot of skin polymer melt (shown in *dark* shade) is injected into the mold. **b** Injection of core polymer melt until cavity is nearly filled, as shown in **c**. **d** Skin polymer is injected again, to purge the core polymer away from the sprue. (Reprinted with permission from Design Center, School of Engineering, Santa Clara University)

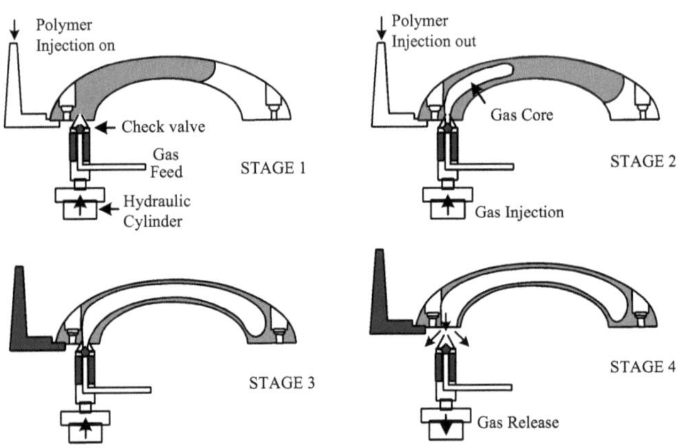

Fig. 11.10 Schematic representation of the stages of the gas-assisted injection molding process for a container handle

to the skin. Having gates which are not polymer filled, this process is more effective than the application of the packing melt pressure in conventional injection molding, especially for thick product parts, resulting in the absence of surface sink marks. Once the polymer skin has completely solidified, the gas pressure is released, the mold opened and the product ejected. The process stages are shown in Fig. 11.10.

In visualizing and understanding the gas-assisted injection molding process, it is important to note that the *viscosity of the gas is immensely lower* than that of the polymer melt it is displacing, and that the gas is *compressible,* while polymer melts are practically incompressible.

11.3 Melt Processing of Polymer Excipient-API Pharmaceutical Formulations

Melt processing of polymer excipient-API pharmaceutical formulations is carried out most commonly by HME, which has been explored and studied in the last several decades by both industrial and academic investigators, because of its potential of rendering poorly water-soluble APIs readily bioavailable to patients through oral dosages. The HME field is currently being investigated even more intensively, because of recent discoveries of families of large number of potent and promising, but essentially water-insoluble APIs. HME is a term that the pharmaceutical sector adopted to differentiate it from traditional oral dosage producing techniques, such as direct compression and tableting. It involves the use of mostly corotating twin rotor extruders for the processing of usually water-soluble polymeric excipients, mixing them with APIs, while molten, to *affect partial or total API dissolution*, and pumping the homogeneous mixture through a die to form an extrudate, where the API exists in a totally or partially dissolved, but in both cases, stable form. Compared with the traditional drug production processes, HME is a solvent-free continuous process and it may lead to fewer required processing steps. However, degradation of the drug (API) and excipient may occur during HME due to the relatively high processing temperatures needed to melt the excipient and to laminar flow heating, due to viscous energy dissipation. This may limit "universal" application of HME for all excipient or API pairs. The same comments apply to the *virtually unexplored Hot Melt Injection Molding (HMIM) of excipient-API formulations to produce directly oral dosage products, such as pills.*

As mentioned above, both HME and HMIM involve the five elementary steps of polymer processing or compounding: handling of particulate solids, melting, mixing, devolatilization, and pressurization and pumping, the two most important being *melting and dispersive and distributive mixing of the additives* in the polymer matrix. On the other hand, as shown conceptually in Fig. 11.11, for the HME and HMIM pharmaceutical melt processing, *dissolution of the API in the molten excipient is an additional and most significant elementary step*, along with *melting* which precedes it and *mixing,* which assists and speeds up dissolution.

The following discussion of *melting*, mixing, and dissolution as they may occur in corotating twin-screw extruders (Co-TSEs), contrasted to SSEs, which include in-line single-screw injection molding machines, is based on the text of Tadmor and Gogos (2006) and the recent work of Gogos et al. (2012)

The physical mechanisms available for melting polymer systems in polymer processing equipment are: (a) In *SSEs (and in-line reciprocating single-screw injection*

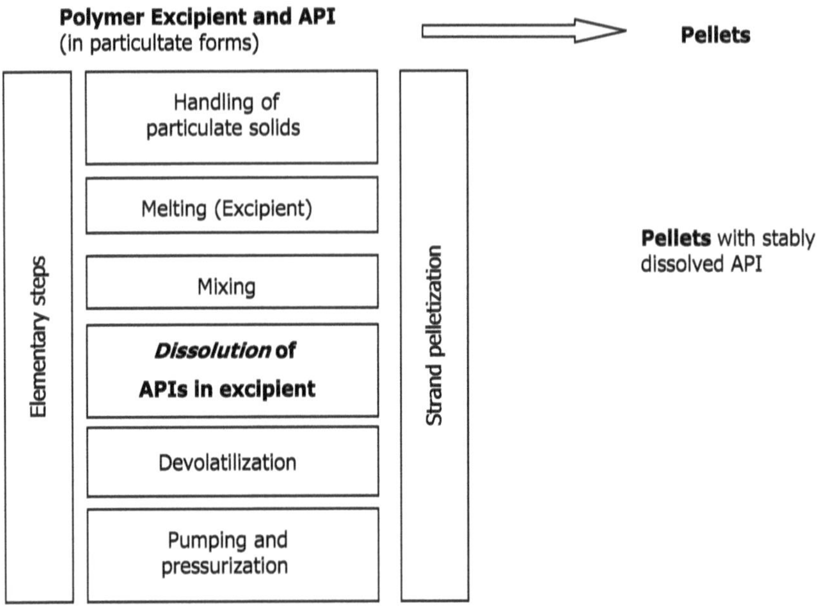

Fig. 11.11 Conceptual structural breakdown of pharmaceutical melt processing by HME and HMIM

molding)—conductive melting of the packed particulate bed surface next to the hot barrel surface, and, after the thickness of the melted polymer layer exceeds that of the barrel-screw tip clearance, also *viscous energy dissipation* during the drag flow causing the removal of the melt generated to the trailing end of the bed. Thus, melting is localized at the barrel surface and is gradual, requiring much of the extruder length to complete. Consequently, the "age distribution" of the melt generated in a typical SSE is of the order of its average residence time. This fact may result in adverse consequences for HME processing: First, since dissolution and mixing takes place primarily when the excipient is molten, there will be only limited dissolution of the API in that fraction of the excipient, which melted late, resulting potentially in a *wide distribution of the percentage of API dissolved*. Second, the portion of the excipient which melted early will be more susceptible to *thermal degradation* (3c). Thus *single-screw* HME and HMIM suffer from these two disadvantages; (b) Conversely, in *Co-TSEs* the dominant available melting mechanism is not *conductive melting* of the starve-fed loose particulates by the hot barrel. Rather, in such processing equipment reverse screw or reverse kneading elements are used to create "material holdback" and a *filled* section, where the packed particulates undergo repeated *volume-wide* deformations before exiting the fully filled region, during which the very *powerful melting mechanism of plastic energy dissipation (PED)* takes place. This is shown schematically in Fig. 11.12 below.

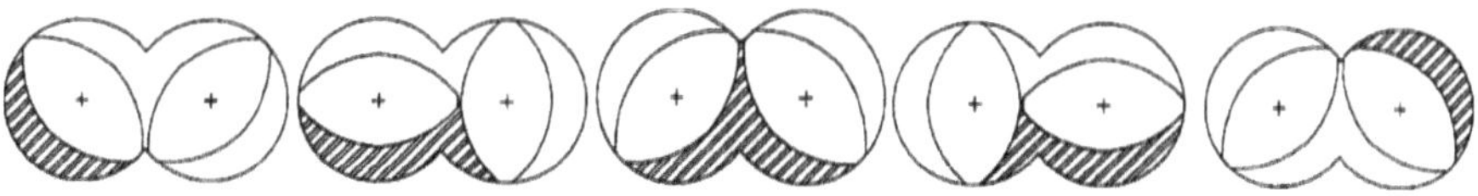

Fig. 11.12 Snapshots of the repetitive expansion/contraction cycle of each of the cross-sectional area pockets between a pair of fully filled kneading disks and the barrel of fully intermeshing, corotating extruders (3d)

It is also worth pointing out that the repeated large compressive deformations imposed by the corotating kneading elements in full kneading blocks induce particle-to-particle surface frictional heating *frictional energy dissipation (FED)* resulting in additional localized melting (3d, Gogos et al. 1998). With a appropriately selected reverse kneading element or screw element sequences, PED and FED are capable of completely melting the entire charge within an axial length of one-to-two diameters, giving rise to a melt stream which has almost the same "age," which is very helpful for uniform API dissolution and for operating at short mean residence times in the Co-TSEs thus reducing the chances of thermal degradation.

The mixing processes in single- and twin-screw extruders are generally categorized into two types: *dispersive* mixing and *distributive* mixing (3e). *Dispersive mixing* refers to the process involving the particle size reduction of cohesive components such as solid fillers (by deagglomeration) or immiscible liquid droplets (by droplet deformation and break-up). *Distributive mixing* refers to distributing deagglomerated particulates uniformly throughout space, or stretching the interfacial area between the components lacking a cohesive resistance and distributing them uniformly throughout the product volume. *Dispersive mixing requires high flow stresses* (either through high viscosity of high shear or elongational rates) in order to provide the dispersive forces to overcome the cohesive forces of the agglomerates or immiscible droplets; *distributive mixing is dictated only by the flow-generated strain and does not require high stresses*. According to these definitions, the *mixing of miscible and dissolving liquids is regarded as distributive mixing*; mixing of hard solid agglomerates, immiscible liquids and soft *agglomerates is regarded as dispersive mixing*. SSEs and reciprocating single-screw injection molding machines apply only shear (not extensional) flows, which result in slow (linear with time) and nonuniform laminar distributive mixing. Fully filled Co-TSE kneading elements, on the other hand, produce rapid (exponential with time) and efficient chaotic laminar distributive mixing, composed mostly extensional and "folding" flows. Thus, they contribute to efficient and volume-wide dissolution, as we see next paragraph. Consequently, Corotating twin rotor extruders result in very rapid, efficient and space- and time-uniform melting, distributive mixing, as well as dissolution, compared with single-screw processing devices.

As described earlier, let us look at a "cartoon" of the sequence of the physical phenomena taking place during the Co-TSE HME process, on Fig. 11.13 (Gogos et al. 1998), of melt processing of an API or polymer excipient system, in which the processing temperature is above the melting temperature (semi-crystalline polymer)

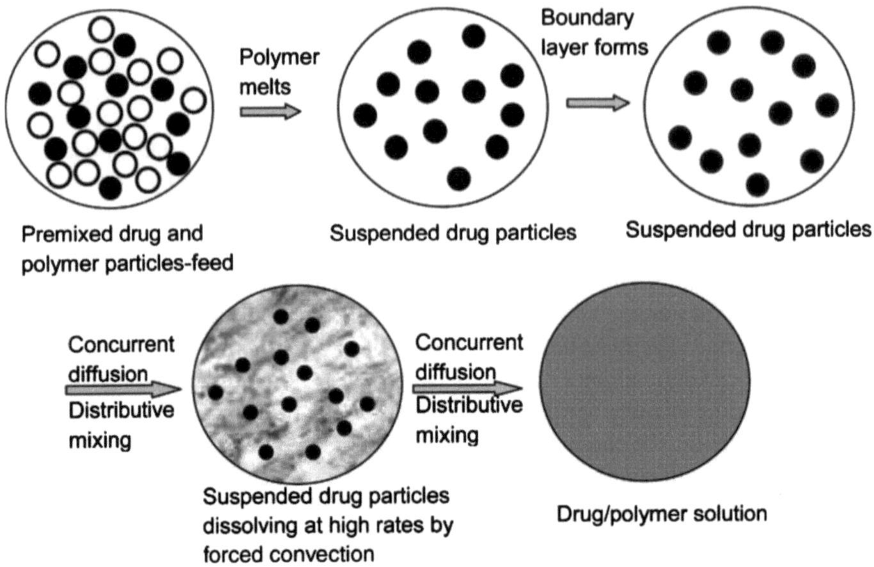

Fig. 11.13 Schematic representation of the morphological changes of the drug and polymer system during the dissolution process. The HME/HMIM processing temperature is below the API melting point (Gogos 2012)

or the softening temperature of an amorphous polymer ($Tg > 50$–$100\,°C$), but below the melting point of a crystalline API.

Processing below the melting point of the API provides a viable dissolution path, which minimizes or circumvents the thermal degradation issue. The API is processed below its melting point and mixed with a polymer melt and the solid drug particles gradually dissolve into the polymer excipient melt, resulting in a desirable polymer–drug solid dispersion or solid solution. In this case, the solid API and the polymeric melt act as a solute and a highly viscous solvent, respectively, during HME. Firstly, the premixed drug (black) and polymer particles (white) are fed into the melt processing equipment. The polymer particles then start melting due to the conductive heat from the SSE barrel, from the frictional and plastic energy dissipation for co-TSEs, leading to the solid drug particles being suspended in the molten polymer melt matrix. While suspended at the processing temperature, which favors dissolution assuming intermolecular forces compatibility between the API and the excipient (i.e., miscibility), the drug molecules start dissolving and create a mass-transfer boundary layer around each drug particle. This boundary layer is continuously wiped away and replaced by fresh polymer melt around each API particulate by the laminar distributive flow of the mixer. The same laminar mixing flow helps the drug molecules to diffuse and mix distributively into the molten excipient. The more efficient the laminar distributive mixing (as is the case with Co-TSEs), the more efficient and rapid the boundary layer wiping and, thus the rate of dissolution. Distributive mixing can homogenize the drug concentration in the polymeric melt through extensional flows or reorientation and bring more polymer melt into contact with the suspended drug particles, thus leading to dissolution rate enhancement. The size of suspended drug particles diminishes as the diffusion continues until the particles disappear and a homogeneous solution is formed or until the limit of API solubility at the processing temperature is reached. In the latter case, they reach a minimum average size and remain suspended.

11.4 Melt Injection Molding of Polymer Excipient-API Pharmaceutical Formulations

11.4.1 Pharmaceutical Injection Molding with Existing Injection Molding Equipment and Processes

The *principal advantage* of the injection molding process for thermoplastic polymers and thermoplastic polymer or additives systems is that it is capable of producing *shaped polymer products reproducibly, inexpensively and rapidly, helping polymers* replace metals and glass in broad categories of products. There are some limitations regarding the product thickness, which is practically limited to a maximum of 0.5 cm to keep the molded part cooling time economically short. Because, as discussed above, the mixing in both the in-line reciprocating single-screw machine and the two-stage single-screw-plunger machines is slow, additives are incorporated by feeding with the "virgin" polymer a concentrated masterbatch (usually at the 10–20 % level), which alleviated the distributive and dispersive mixing limitations of single-screw devices. The sprue and runner polymers can be and are recycled after being ground up to levels of 10–15 % of the total feed, since they represent industrial reprocessed material. With proper gate design the separation of the molded parts from the runner is accomplished automatically, thus, there are no postmolding steps, and the process can be considered to be a net-shape manufacturing one.

The ability of the injection molding process to shape API or polymer excipient solid solution oral dosage products (such as tablets), directly and in a *net-shape manufacturing* fashion is *a great advantage*, compared with producing such tablets by a multistep process which includes HME, followed by grinding of the extrudate, mixing the ground particulates with necessary tableting ingredients, and finally producing tablets. Furthermore, the existence of the additional processes of sandwich, coinjection and gas-assist injection molding net shaping, present opportunities of having different skin and core formulation compositions, foamed core and solid skin tablets, or tablet cores which contain only inert gas.

Nevertheless, injection molding of API or polymer excipient formulations (HMIM) with *existing* injection molding equipment, which is based on *single screws,* presents the following difficulties:

- The *total time* a representative formulation element spends in the *molten state is higher than in HME,* which is in the vast majority of cases carried out in Co-TSEs, increasing the chance of thermal degradation

- The *age distribution of the formulation in the molten state is very wide*, making the fraction which was melted first susceptible to degradation
- The laminar distributive mixing and, consequently, the *extent and rate of dissolution* of the API in the molten polymer excipient matrix will be lower than in HME processes which are carried out in Co-TSEs
- It is difficult to imagine that the sprue and runner API or excipient can be ground and used as industrial reprocessed material at any concentration level—the chances of degradation will be very high.

11.4.2 Concluding Proposal for Modifications to the Existing Injection Molding Equipment to Safely and Properly Melt-process Pharmaceutical Formulations

Thus, one may conclude that the probability of success of melt processing pharmaceutical oral dosage tablet-like products with the existing polymer processing machinery is very small. This is primarily because single-screw processors are used in conventional injection molding machinery.

Thus, we conclude this chapter by proposing that it is possible to envision the *following needed modifications to the two-stage single-screw-plunger machine* which would address several of the shortcomings mentioned above:

1. The *First modification* is to *replace the SSE with a corotating, self-wiping twin-screw exrtruder*, which has an appropriately designed forward/reverse screw element sequence which will melt the entire polymer charge within 1–5 s and which will enhance the rate of API dissolution; the residence time in such Co-TSE need not be more than 30–60 s, with reasonable flow rates, which are possible by increasing to Co-TSE screw/barrel diameter
2. With reasonable flow rate generated by the Co-TSE in the first stage of the machine, the injection chamber will fill in short times; when full and before the plunger cylinder is activated a valve will be activated to seal this chamber from the melt generating Co-TSE; the same valve will connect the Co-TSE with a *second injection chamber* (*the second equipment modification*) to receive molten API/Excipient formulation, while the first chamber is emptied filling a large number of single dosage product mold cavities
3. When the second chamber is filled at the end of the injection molding cycle, the mold is opened, the product, sprue and runners are removed; the mold is then closed and filled with the molten formulation of the second chamber; the process is repeated.
4. It may also be necessary to have two molds, the second receiving the molten formulation of the second injection chamber, without waiting for the full cycle time of the first; this will result in shorter times of the pharmaceutical formulation in the molten state.

References

Caras FC (1963) "Spiral mold for thermosets." Mod Plast 41:140

Chabot J, Malloy RA (1997) "A history of thermoplastics injection molding. Part I—the birth of an industry". J Inj Molding 1:1

Cheng L, Guo S, Wu W (2009) "Characterization and *in-vitro* release of praziquantel from ε−caprolactone implants." Int J Pharm 377:112–119

Clarke AJ (2005) US Patent Application Publication No US2005/0202090 A1, assigned to GlaxoSmithKline

Donovan RC (1971) Polym Eng Sci 11:353

Gogos CG, Liu H, Wang P (2012) Chapter 13, "Laminar dispersive and distributive mixing and applications to HME". In Douroumis D (ed) Hot-melt extrusion. Wiley, Weinheim

Gogos CG, Tadmor Z, Kim MH (1998) Melting phenomena and mechanisms in polymer processing equipment". Adv Polym Technol 17:285–305

Quinten T, Vervaet C et al (2009a) "Evaluation of injection molding as a pharmaceutical technology to produce matrix tablets." Eur J Pharm Biopharm 71:145–154

Quinten T, Vervaet C et al (2009b) "Development of injection molded matrix tablets based on mixtures of EC and L-HPC." Eur J Pharm Sci 37:207–216

Quinten T, Vervaet C et al (2011) "Development and evaluation of injection-molded sustained-release tablets containing EC and PEO." Drug Devel Ind Pharm 37(2):149–159

Rose W (1961) Fluid—fluid interfaces in steady motion". Nature 191:242–243

Rothen-Weihold A et al (1989) "Injection molding versus extrusion as manufacturing technique for the preparation of biodegradable implants." Eur J Pharm Biopharm 48(2):113–121

Rubin I. I (1972) Injection molding—theory and practice. Wiley—Interscience, New York

Schmidt LR (1974) "A special mold and tracer technique for studying shear and extensional flows in a mold cavity during injection molding." Polym Eng Sci 14:797

Tadmor Z, Gogos GG (2006) Principles of polymer processing, 2nd edn. Wiley—Interscience, New York ((a) Chapter 1; (b) Chapter 13; (c) Chapter 5 and Section 9.3; (d) Sections 5.9 and 10.3; (e) Chapter 7)

Tadmor Z, Klein L (1970) Engineering principles of plasticating extrusion. Van Nostrand Reinhold Book Co., New York

Tadmor Z (1974) J Appl Pol Sci 18:1753

Chapter 12
Devices and Implants Prepared Using Hot Melt Extrusion

Andrew Loxley

Abstract The development of medical devices (MD) and implants presents unique challenges and requirements not seen in the development of oral solid dosage forms. Through the use of hot melt extrusion (HME), it is possible to engineer novel systems capable of achieving specific geometries and release characteristics that enable the therapy. With extrusion, it is also possible to create systems containing multiple active ingredients leading to treatments for some of the most pressing diseases today, including HIV/AIDS. This chapter discusses the formulation of and processing technologies used for the production MD and implants prepared using HME. Recent examples, including inserts and vaginal rings, are also discussed to illustrate these principles.

12.1 Introduction

Hot melt extrusion (HME) has recently seen increased use in the development and manufacturing of solid oral dosage forms, where it is largely used to mix active pharmaceutical ingredients (API) with water-soluble or biodegradable thermoplastic polymers to form homogenous blends. The blends are usually then further processed to form tablets; however, recent work has been carried out to prepare solid oral dosage forms from nondegradable polymers such as ethylene vinyl acetate copolymer.[1]

While the use of HME in the oral dosage arena is relatively new, it has long been the process choice for preparing thermoplastic medical devices (MD) such as orthopedic implants, medical tubing, and catheters. The role of HME in the manufacture of these products varies from providing efficient mixing of additives or APIs into polymers used to make the devices to shaping the polymers into the form suitable for the intended application of the device.

This chapter will describe some of the medical devices and implant systems that involve an HME step in their manufacture, and will focus on intravaginal rings (IVRs) made using an extrusion to illustrate key concepts related to drug-eluting devices (DED).

[1] Belgian work.

A. Loxley (✉)
Particle Sciences Inc., Bethlehem, PA, USA
e-mail: andrew@particlesciences.com

M. A. Repka et al. (eds.), *Melt Extrusion*,
AAPS Advances in the Pharmaceutical Sciences Series 9,
DOI 10.1007/978-1-4614-8432-5_12,

12.2 Devices

FDA defines a medical device as:

> an instrument, apparatus, implement, machine, contrivance, implant, in vitro reagent, or other similar or related article, including a component part, or accessory which is:
> - recognized in the official National Formulary, or the United States Pharmacopoeia, or any supplement to them,
> - intended for use in the diagnosis of disease or other conditions, or in the cure, mitigation, treatment, or prevention of disease, in man or other animals, or
> - intended to affect the structure or any function of the body of man or other animals, and which does not achieve any of its primary intended purposes through chemical action within or on the body of man or other animals and which is not dependent upon being metabolized for the achievement of any of its primary intended purposes.[2]

All such MD and DED products are regulated by FDA Office of Medical Products and Tobacco—The Center for Devices and Radiological Health (CDRH) regulates MD, and the Center for Drug Evaluation and Research (CDER) is also involved for DED since they contain API.

Devices, necessarily made from thermoplastic polymers, fall into three general groups:

- Mechanical devices designed for ex vivo use, such as valves or closures
- Objects designed for in vivo implantation that provide mechanical functions, such as orthopedic implants and catheters
- DED designed for in vivo application that release APIs for therapeutic purposes, such as IVRs, subcutaneous implants, and transdermal patches.

The types of MDs and DEDs that involve an HME step are numerous. In fact, almost any thermoplastic component of any device in the market is likely to have passed through an extruder for either compounding or shaping purposes. MD such as valves and closures will likely have been colored by compounding the thermoplastic with pigments using a twin-screw extruder in order to achieve consistent color. Similarly, APIs would have been incorporated into thermoplastics for DED by twin-screw extrusion since homogenous API distribution throughout the device is critical to drug elution kinetics. Catheters are made by extrusion of polyurethanes through an appropriate tube die using a single-screw extruder. The simpler, lower cost single-screw equipment suffices here, since neat polymer is used and no mixing with excipients or API is required.

12.3 Hot Melt Extrusion

HME is a plastics processing technology in which a thermoplastic polymer is heated in a metal barrel containing one or two screws and the molten material is conveyed

[2] http://www.fda.gov/MedicalDevices/DeviceRegulationandGuidance/Overview/ClassifyYourDevice/ucm051512.htm

Fig. 12.1 Schematic of a single screw and twin-screw extruder

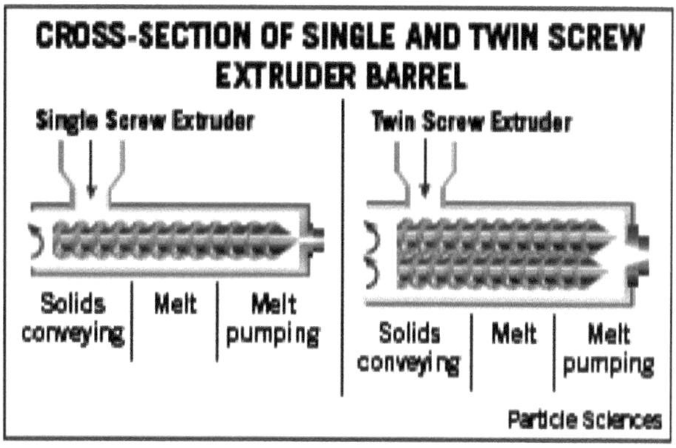

by the rotating screw(s) to the end of the barrel where it is shaped as it exits through a die. In some cases HME is a process step in a more complex manufacturing process that involves further downstream processing steps to prepare the device, such as pelletizing followed by injection molding. In other cases, the HME process produces the extruded product essentially in its final form, which is the case for several tubing and filament products.

12.3.1 *Extrusion Equipment*

Hot melt extruders are generally of two types: single-screw and twin-screw (Fig. 12.1), and come in a variety of sizes suitable for the stage of development of manufacturing of the device. Extruders with screws having cross-sectional diameters as small as 3 mm have been made for early feasibility work, especially useful when valuable raw materials are in scarce supply, and may be as large as 40 mm or more in diameter for commercial manufacturing.

12.3.1.1 Single-Screw Extruders

A commercially available single-screw extruder useful for the development of devices is shown in Fig. 12.2. Single-screw extruders are essentially used as simply conveyors of molten polymer, and are used almost exclusively to extrude molten polymer through appropriately shaped dies fitted to the output end of the barrel to form parts with a desired cross-sectional shape. One recent study demonstrated mixing arguably at least as good as that of twin-screw extruders, using a single-screw extruder with a specially designed screw that provides extensional mixing (Costeux et al. 2011).[3]

[3] Costeux et al. (2011)

Fig. 12.2 Single-screw extrusion—*left*: Thermo Fisher Scientific PolyLab OS Torque Rheometer System coupled with a single-screw extruder and film die for sheet profile manufacture; *right*: A close-up of film profile extrusion with a Post-Extrusion Film Haul-Off System (reproduced with permission from Thermo Fisher Scientific)

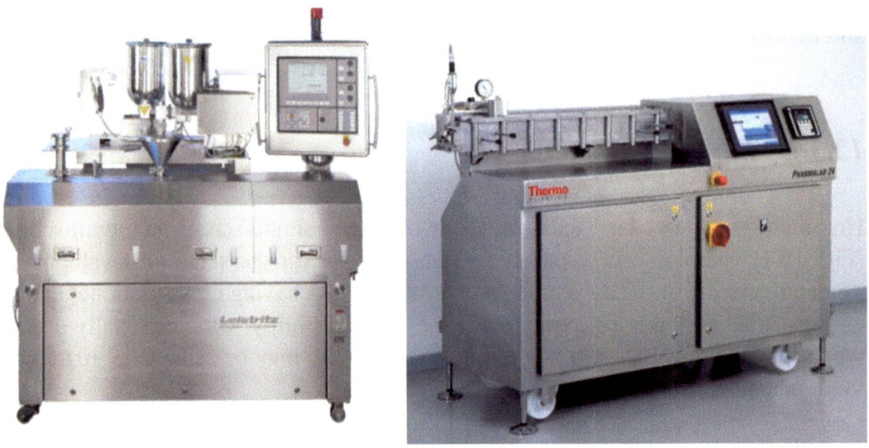

Fig. 12.3 Twin-screw extrusion—*left*: A Leistritz ZSE-18PH 18 mm Pharma twin-screw extruder; *right*: A Thermo Fisher Scientific PharmaLab 24 mm GMP, pharmaceutical-grade twin-screw extrusion system (reproduced with permission from Leistritz Corp and Thermo Fisher Scientific, respectfully)

12.3.1.2 Twin-Screw Extruders

Multiple commercially available twin-screw extruders are useful for the development of devices, with representative systems shown in Fig. 12.3. Twin-screw extruders have barrels that contain two intermeshing corotating screws often with complex custom designs made from the arrangement of various elements (such as those shown in Fig. 12.4) fitted onto a pair of shafts. Mixing of polymers with colorants, additives,

Fig. 12.4 Elements of a modular screw

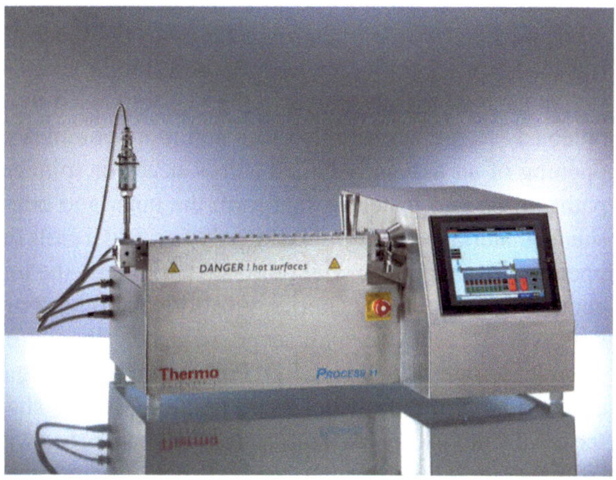

Fig. 12.5 A Thermo Fisher Scientific Pharma 11 mm twin-screw extrusion system—a parallel, fully intermeshing twin-screw extruder with modular barrel and screw as well as clamshell design (reproduced with permission from Thermo Fisher Scientific)

or API is best achieved with twin-screw extruders, as a result of the combination of distributive (extensional) and dispersive material flow fields provided by the action of the screws as the molten material is conveyed down the barrel (Martin 2008).[4] An 11 mm twin-screw extruder that has a clamshell design for easy cleaning, is shown in Fig. 12.5 and shows the intermeshing screws fitting snuggly within the barrel. This intermeshing nature of the equipment results in self-wiping aids in the processing heat-sensitive materials by improving residence time distribution during processing.

[4] Martin (2008)

Fig. 12.6 Schematic representation of various extrusion die shapes

12.3.1.3 Dies

The shape of the extruded material is determined by the output die and the cross-sections of the three commonest shapes are shown in Fig. 12.6. The gray areas illustrate the openings through which the extrudate exits the extruder.

The circular die shown in Fig. 12.6a is used to form rods, which may be cut to length and used directly, but are most often pelletized for further processing, such as injection molding, to manufacture the device. This approach is also currently the most common method for preparing oral drug products, where the extrudate is milled for further downstream processing. This type of die is most commonly used at the output of a twin-screw extruder, but may also be used on a single screw when the rod product did not require mixing.

The slot die shown in Fig. 12.6b is used to extrude sheets or films of polymer, and is the die of choice for fabricating the hot-melt adhesive layer of dermal patches. The adhesive, which may be API-loaded, is directly extruded from the slot die onto a release liner which is then laminated to the patch backing film in a continuous roll-to-roll process. Figure 12.2 shows a slot die fitted to a single-screw extruder.

The coaxial die presented in Fig. 12.6c is used to extrude two materials simultaneously to prepare simple tubes and more complex coaxial fibers. The die is the opening of an assembly called a crosshead (or spinneret), which controls the flow paths of the two materials that form the inner and outer parts of the extrudate. The sheath material is a thermoplastic polymer, and it exits the die via the outer concentric opening. To prepare coaxial (core-sheath) fibers, the inner (core) material is usually another thermoplastic material, most often another polymer and can be formulated with API since, as we shall see, core sheath fibers are an excellent way to control the release rate of API in DED. For simple tubes, a gas is fed to the spinneret and exits the core of the coaxial die to help support the shape of cooling tube as it exits the die during extrusion. Thus a crosshead die is fed with two material streams each requiring their own separate feeds. If both streams are thermoplastics, each stream requires its own separate extruder, and they are usually fitted to the crosshead at right angles to each other as a setup illustrated in Fig. 12.7.

12.3.1.4 Downstream Equipment

The tension on the extruded material is often controlled by a puller, which feeds a cutter, pelletizer, or uptake roll. The cross-sectional diameter of the extrudate is controlled by the die size, the extent of "die swell" that the material experiences on exiting the die due to the pressure drop, and the force that is exerted on the extrudate by the puller. The latter ultimately dictates the cross-sectional diameter, and when it

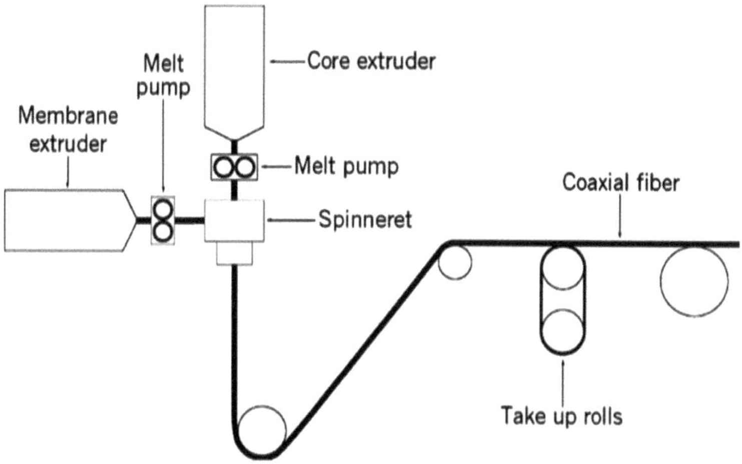

Fig. 12.7 Schematic of coextrusion line for producing coaxial fibers

is important to control this, the puller force can be regulated by a laser-micrometer that measures the outer diameter of the extrudate and feeds this information to the puller that then self-adjusts the pulling force.

The extrudate must be cooled and conveyed away from the die. Cooling may be by air—either passively or by blowers or air-knifes or air-rings that surround the extrudate, or by passing the extrudate under water contained in a temperature-controlled trough.

The cooled extrudate may be cut into small pieces by a rotary pelletizer for feeding an injection molding process, or into longer tubes or rods by an automatic puller-cutter set to the desired length. In the case of dermal patches, the coated film is usually cut using punches fitted with appropriately shaped cutting dies, rather like cookie cutters.

Torroidal devices (o-rings), especially those made from coaxial fibers, are sometimes prepared by bonding the ends of a cut extrudate, in which case an induction welder is generally the tool of choice. The ends are held in place and reheated to melt, fuse, and form the bond (Amanat et al. 2010).[5]

12.3.1.5 Injection Molding

Injection molding is a process that incorporates an HME step, in which a thermoplastic material is melted and conveyed by a type of single-screw extruder, but instead of the extrudate being collected, the output orifice of the barrel, called the nozzle in this case, is directly attached to a mold whose cavity is in the shape of the part to be manufactured. The molten material flows into the mold where it takes the shape

[5] Amanat et al. (2010)

Table 12.1 Polymers used to make devices by HME

Polymer	Water-solubility	Example device use
Polyolefins (PE, PP)	Insoluble	Orthopedic implants Device components Transdermal patch adhesives
Poly(ethylene-co-vinylacetate)	Insoluble	Intravaginal rings Tubing Transdermal patch adhesives
Poly(styrenics)	Insoluble	Device components
Polyacrylates	Insoluble Swellable (depends on chemistry)	Device components Transdermal patch adhesives
Polyether block amide	Insoluble Swellable (depends on chemistry)	
Polyesters	Biodegradable (hydrolysis, enzyme accelerated)	Biodegradable implants sutures
Polyamides	Insoluble Swellable (depends on chemistry)	
Polyanhydrides	Biodegradable (hydrolysis)	Biodegradable implants
Polyurethanes	Swellable (depends on chemistry) Biodegradable (enzymatic)	Catheters Intravaginal rings
Silicones	Insoluble	Tubing Catheters Intravaginal rings (injection molded)

of the mold cavity, and the mold opens to eject the formed part after the polymer has cooled sufficiently. Most thermoplastic devices with shapes more complex than those that can be prepared by direct extrusion are prepared by injection molding. Both MD and DED may utilize injection molding.

12.4 Polymers for Devices

All polymers for MD and DED made by HME are thermoplastics, meaning the solid state melts to form a viscous liquid when heated, and the melt solidifies when cooled, in a reversible process that can be cycled indefinitely when material degradation is neglected.

Unlike many solid dosage forms made by HME, the polymers used for MD and DED are generally not readily water soluble, but may be biodegradable by either hydrolysis or enzyme-mediated processes, water-swellable, or water-insoluble. Some examples of thermoplastic polymer classes that have physical properties suitable for device manufacture using HME are given in Table 12.1, with some example devices indicated where known.

All polymers for *in vivo* use must be biocompatible, and must pass ISO10993 testing to demonstrate this. The ISO10993 tests are shown in Table 12.2.

Table 12.2 ISO 10993 tests

ISO 10993 BIOCOMPATIBILITY TESTS

Device Type	Device category: Body Contact	Device category: Contact Duration (Limited □ Less than 24 hours; Prolonged □ 24 hours to 30 days; Permanent □ Over 30 days)	Cytotoxicity	Sensitization	Irritation or Intracutaneous Reactivity	Systemic Toxicity (acute)	Pyrogenicity	Sub-acute and Sub-chronic Toxicity	Genotoxicity	Implantation	Hemocompatibility	Chronic Toxicity	Carcinogenicity
Surface	Skin	Limited	•	•	•								
		Prolonged	•	•	•								
		Permanent	•	•	•								
	Mucosal Membrane	Limited	•	•	•								
		Prolonged	•	•	•								
		Permanent	•	•	•			•	•				
	Breached or compromised surfaces	Limited	•	•	•								
		Prolonged	•	•	•								
		Permanent	•	•	•			•	•				
External Communicating	Blood Path, Indirect	Limited	•	•	•	•	•				•		
		Prolonged	•	•	•	•	•				•		
		Permanent	•	•		•	•	•	•		•	•	•
	Tissue/Bone/ Dentin	Limited	•	•	•								
		Prolonged	•	•	•	•	•	•	•	•			
		Permanent	•	•	•	•	•	•	•	•		•	•
	Circulating Blood	Limited	•	•	•	•	•				•		
		Prolonged	•	•	•	•	•	•	•	•	•		
		Permanent	•	•	•	•	•	•	•	•	•	•	•
Implant	Tissue/Bone	Limited	•	•	•								
		Prolonged	•	•	•	•	•	•	•	•			
		Permanent	•	•	•	•	•	•	•	•		•	•
	Blood	Limited	•	•	•	•	•	•		•	•		
		Prolonged	•	•	•	•	•	•	•	•	•		
		Permanent	•	•	•	•	•	•	•	•	•	•	•

*Additional tests may be required to satisfy FDA requirements

12.5 Devices Made by HME

Although HME is a process step in the production of many MD, this section will focus on its use in DED. There is a wide variety of APIs used in DED, from hormones to antiretrovirals to antibiotics. Some examples are shown in Table 12.3.

Table 12.3 Drugs used in various drug-eluting devices (DED)

API name	Structure	Device
Dapivirine	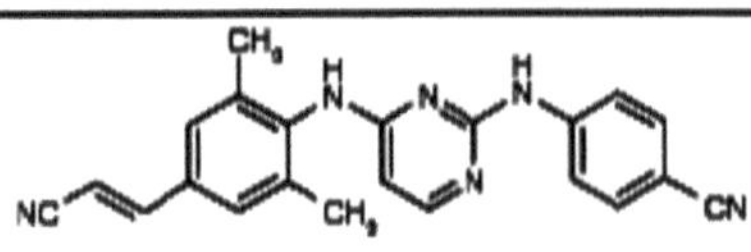	IVR (Loxley 2008)[a]
Maraviroc	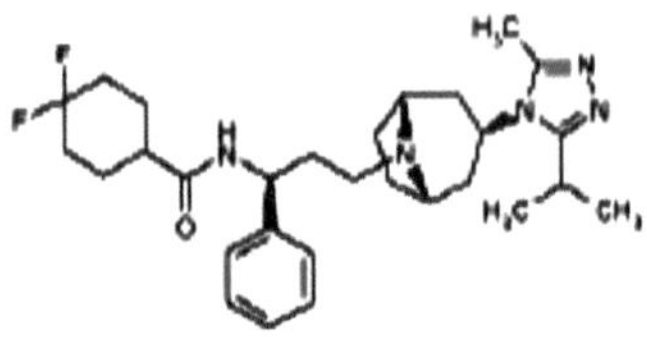	IVR
UC781	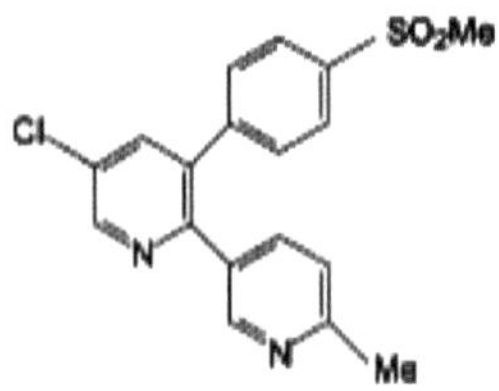	IVR (Loxley 2011; Clark 2012)[b,c]
Tenofovir	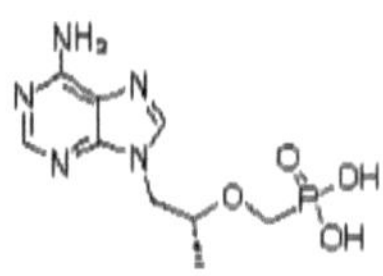	IVR (Johnson 2010)[d]
Etonogestrel	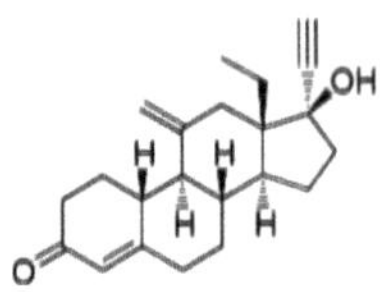	IVR[e]
Ethylene estradiol	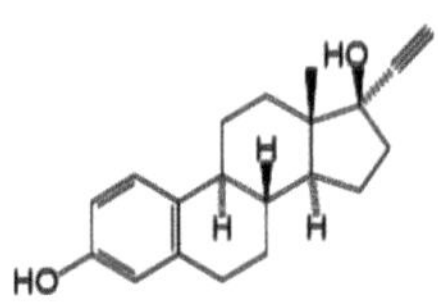	IVR[e]
Levonorgestrel	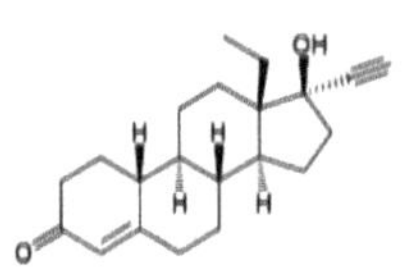	IVR (Loxley 2011)[b] Subdermal implant[f]
Nicotine	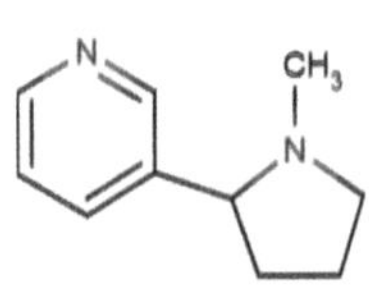	Transdermal patch

Table 12.3 (continued)

API name	Structure	Device
Testosterone	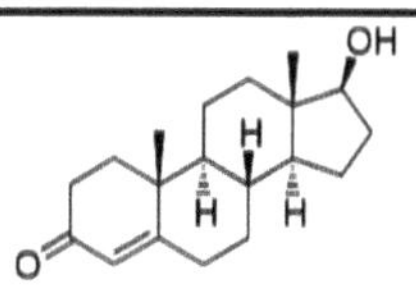	Transdermal patch
Fluocinolone acetonide	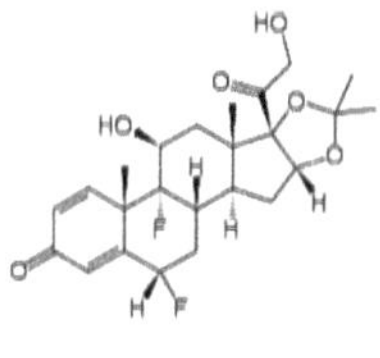	Intraocular implant[g]
Tetracycline hydrochloride	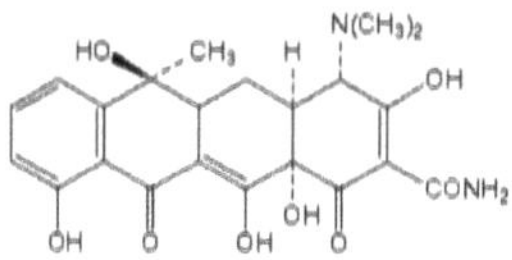	Dental implant[h]
Oxybutynin	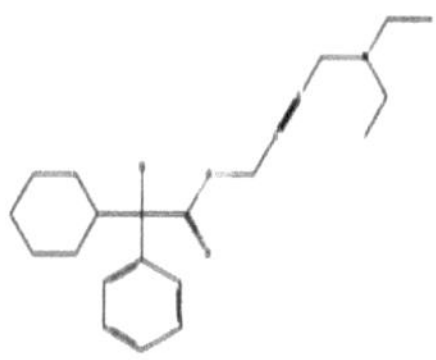	IVR (Woolfson 2003)[i]

[a] Loxley et al. (2008)
[b] Loxley et al. (2011)
[c] Clark et al. (2012)
[d] Johnson et al. (2010)
[e] http://www.nuvaring.com
[f] http://www.rxlist.com/jadelle-drug.htm
[g] http://www.alimerasciences.com/Products/iluvien.aspx
[h] http://www.rxlist.com/actisite-drug.htm
[i] Woolfson et al. (2003)

12.5.1 *Transdermal Patches*

Transdermal patches are sandwich devices applied to the skin, and usually designed for systemic or local delivery to an API via the skin (for example nicotine, testosterone, and estrogen). A common method of preparation involves roll-to-roll coating of a solvent-borne adhesive formulation containing API onto a siliconized release liner which is then passed through a heated drying tunnel before being laminated to a backing film (fabric, polyester, etc.) then cut to size. The solvents in the roll-to-roll process are used to reduce the viscosity of the adhesive polymer to allow it

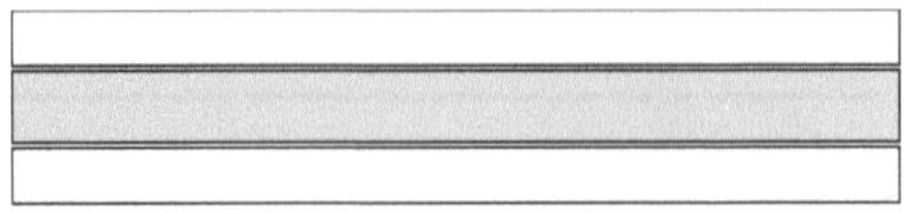

Fig. 12.8 Schematic of a transdermal film/patch (*top*: backing film, *middle*: (*shaded*) medicated adhesive, *bottom*: release liner)

to be coated to the release liner. The final transdermal patch device is illustrated in Fig. 12.8 showing the shaded API-loaded adhesive layer sandwiched between the release liner (*lower*) and the backing film (*upper*).

Since HME uses heat instead of solvents to reduce the viscosity of polymers for extrusion, it is used in place of roll-to-roll coating to eliminate the organic solvents from the coating process of transdermal patches. The solvent-free HME adhesive formulation is based on a thermoplastic elastomer with a very low glass transition temperature, which is tacky at room temperature either due to its inherent chemical structure, or after formulation with additional low molecular weight compounds that impart tackiness to a polymer film and are commonly referred to as tackifiers. The API and the solid adhesive polymer are fed to a twin-screw extruder for intimate mixing, then extruded through a slot die with an appropriate gap size onto the release liner. The coating is cooled sufficiently, then laminated with the backing film, cut to size, and packaged.

12.5.2 Subcutaneous, Ocular, and Dental Implants

Implants are usually fibers or small rods cut from fibers that require a surgical procedure to implant via a small incision in the skin, cornea, or gum. These devices are made simply by feeding API and polymer to a twin-screw extruder fitted with a circular die, then cutting the extrudate to the required product length with a pelletizer or puller/cutter. Actisite® is a long fiber designed to be packed by a dentist into the periodontal pocket after surgery where it releases antibiotic to prevent infection. Jadelle® is a silicone coaxial fiber contraceptive device with a core loaded with the progestogen levonorgestrel. After insertion of two Jadelle® rods under the skin, the slow release of the hormone enables controlled systemic circulation and provides contraception for up to 5 years. Another example of an extruded product for implantation is Ozurdex, a poly(lactic-co-glycolic acid) solid dispersion containing dexamethasone for intravitreal implantation. Design and development of these type of systems requires special considerations beyond release performance related to mechanical integrity of the device and sterility of the system.

12.5.3 Intravaginal Rings

IVRs are torroidal-shaped devices designed to elute API into the vaginal cavity to prevent pregnancy or HIV transmission, or provide hormone replacement therapy

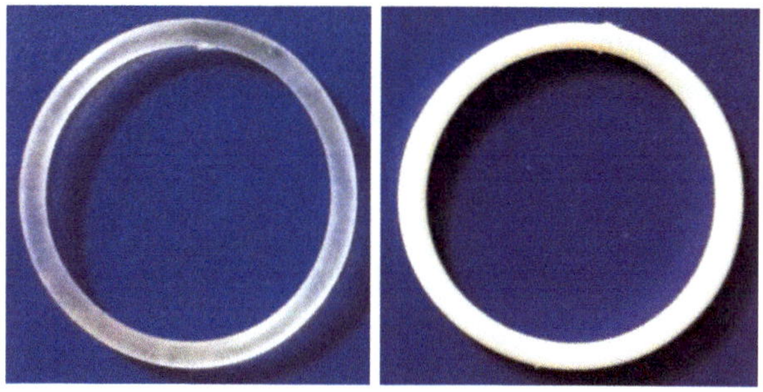

Fig. 12.9 Photographs of IVRs—*left*: placebo EVA IVR, *right*: drug-loaded IVR (dapivirine/maraviroc APIs)

for menopausal women. This type of dosage form has been reviewed extensively in the literature and relies extensively on thermal processing methods for production of the drug products (Malcolm 2003).[6]

12.5.3.1 Matrix IVR

HME may be used to prepare "matrix" type IVRs where the API is uniformly distributed throughout the polymer by a twin-screw extruder and the extrudate passes through a circular die to provide a circular cross-section. The extruded rod of the proper cross-sectional diameter can then be cut to appropriate length (given by $\pi \times$ desired IVR diameter) and the ends bonded to form the IVR using adhesive or induction welding. Matrix IVRs can also be prepared by pelletizing the API/polymer extrudate from a twin-screw extruder, and using injection molding to form the pellets into the IVR using an appropriately shaped o-ring mold. A placebo EVA IVR and a drug-loaded EVA matrix IVR made by mixing API and EVA in a twin-screw extruder, followed by injection molding, are shown in Fig. 12.9.

Matrix devices are of two types depending on the concentration of the API in the device compared to the solubility of the API in the polymer. If the device contains less API than the saturation concentration in the polymer (all API in solution), the device is a monolithic solution device. If the device contains API above the solubility limit in the polymer (API in saturated solution plus particles of undissolved API), then the device is referred to as monolithic dispersion device.

Cumulative release of API from matrix IVRs is generally first order (i.e., proportional to time$^{1/2}$), so matrix IVRs are used when daily release from the device does not need to be constant as long as a minimal therapeutic dose is achieved, and the high initial release of API that is typical of matrix devices is safe. For example, many antiretroviral IVRs in development for HIV prevention are matrix IVRs. If constant (zero order) release kinetics are required, for example, for hormone-releasing devices, more complex IVR designs are required.

---

[6] Malcolm (2003)

Fig. 12.10 Core-sheath reservoir IVR

12.5.3.2 Reservoir IVR

"Reservoir" type IVRs (as shown in Fig. 12.10) can be prepared by coextrusion with a drug-loaded core extruded from the inner orifice of a coextrusion die, and a drug-free polymer extruded over it from the outer orifice of the die. The extruded coaxial fiber is cut and the ends bonded to form the IVR as before. Alternatively, reservoir IVRs can be prepared by a two-step injection molding process where the sheath is molded over the drug-free core in a second molding step (Malcolm et al. 2005).[7] Reservoir IVRs have the advantage that if properly designed, release of the API from the core of the device is zero order. Kinetics of API release from devices will be considered in a later section.

12.6 Drug Release Kinetics from Devices

API is released from a device by either erosion/dissolution, or by diffusion through the device polymer(s). Biodegradable devices tend to release API by erosion of the polymer, devices highly loaded with water-soluble API tend to release API by dissolution of the API, and nonerodable polymer devices with poorly water-soluble APIs release API by solid-state diffusion, which is controlled by Fick's Law:

$$J = D(dC/dt)$$

where J is the mass flux, D is the diffusion coefficient for API in the device polymer, and dC/dt is the concentration gradient.Depending on the mechanism of release, the associated behaviors are described by different kinetics, as illustrated in the subsequent sections.

[7] Malcolm et al. (2005)

12.6.1 Modeling API Release from Slabs

Considering release controlled only by API diffusion, a simple model for the release of API from a monolithic solution slab device of thickness l containing M_0 of API at time $t = 0$ and M_t at time t is described by:

$$\frac{M_t}{M_0} = 4\sqrt{\frac{Dt}{\pi l^2}}$$

for < 60 % drug release and

$$\frac{M_t}{M_0} = 1 - \frac{8}{\pi^2} \exp\left(\frac{-\pi^2 Dt}{l^2}\right)$$

for > 40 % drug release. D is The drug diffusion coefficient of the API in the polymer.

API release from monolithic dispersion slab devices is initially from the API in solution in the polymer near the device surface. As this is depleted, drug particles near the surface dissolve to maintain the dissolved drug concentration. For devices with up to approximately 5 % API, release kinetics are described by Higuchi's equation (Higuchi 1961, 1963) [8,9]

$$\frac{dM_t}{dM_0} = \frac{A}{2}\left[\frac{DC_{s(m)}\left(2C_0 - C_{s(m)}\right)}{t}\right]^{1/2}$$

where C_0 and $C_{s(m)}$ are the total API concentration in the device dissolved/dispersed at time $t = 0$ and the API solubility in the polymer, respectively. A is the total surface area of the device slab. Plots of release rate versus $t^{-0.5}$ or cumulative release versus $t^{0.5}$ exhibit linearity.

12.6.2 Modeling API Release From More Complex Device Architectures

More complex models have been developed to consider a wide variety of device architectures, for both solution and dispersion devices (Siepmann and Siepmann 2012; Helbling et al. 2011).[10,11] Using these models as a starting point, the author commissioned software for describing and predicting API release from IVRs developed in our laboratories. For example, we predict the differences in release kinetics of a model API having a nominal solubility of 5 % in the device polymer from IVRs

[8] Higuchi (1963)

[9] Higuchi (1961)

[10] Siepmann and Siepmann (2012)

[11] Helbling et al. (2011)

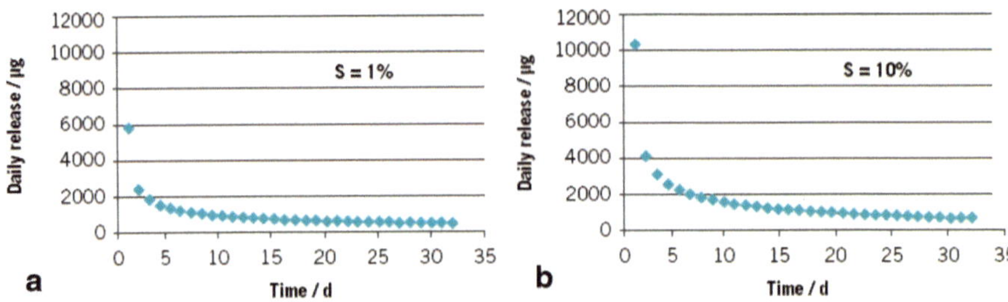

Fig. 12.11 API release from devices containing 5 % API, with **a** API solubility in polymer = 1 %, **b** API solubility in polymer = 10 %

made with drug loadings 1 % (below polymer saturation solubility) and 10 % (above polymer saturation solubility) as shown in Fig. 12.11.

For reservoir devices, the release kinetics depends on the API loading of the core, the relative solubility of API in the core and sheath polymers, and the thickness of the sheath layer. In this case, the drug release rate can be essentially zero-order over extended periods if there is sufficient API in the core. From an extension of Fick's law for core-sheath cylindrical devices, (ignoring end effects, and therefore applicable to torroids such as vaginal rings also) the literature reports (Baker and Lonsdale 1974) [12]:

$$\frac{dM}{dt} = \frac{2\pi l K C}{\left(\ln \frac{r_o}{r_i}\right)}$$

where K is the partition coefficient of the API between sheath and core polymers (the ratio of the solubility of the API in the respective polymer), C is the concentration of drug in the core of the reservoir, r_o and r_i are the outer and inner radii of the device, and l is The length of the device.

12.7 Example

Particle Sciences prepared Dapivirine-loaded EVA IVRs by injection molding of a compound prepared by twin-screw extrusion of Dapivirine and EVA (28 % VA content) for a project funded by the International Partnership for Microbicides.[13] These IVRs were under investigation as possible HIV-preventative devices for use in sub-Saharan Africa where the incidence of HIV/AIDS is very high. During development, monolithic (matrix) IVRs were prepared with 54 mm diameter and 4 mm cross-sectional diameter, containing 25 mg or 100 mg Dapivirine. The 25 mg API devices appeared transparent, and for API release modeling were considered as monolithic solution devices, whereas the 100 mg API-loaded IVRs were white since the API

[12] Baker and Lonsdale (1974)

[13] www.ipmglobal.org

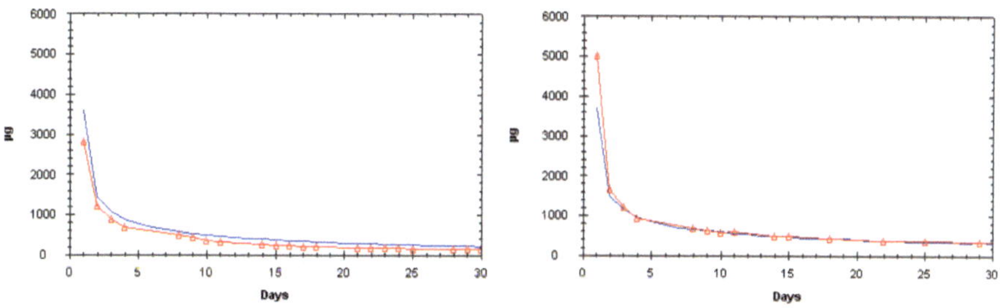

Fig. 12.12 In vitro drug elution from Dapivirine-loaded EVA IVRs (*left*: 25 mg Dapivirine; *right*: 100 mg Dapivirine; solid line model prediction, open triangles data)

was above its solubility limit in the EVA at this loading, and were therefore treated as monolithic solution devices.

The diffusion coefficient (D) and solubility (S) of Dapivirine in 28 % VA EVA was determined by thin film permeation measurements using Franz cells, giving D $\sim$ 2E-9 cm^2s^{-1} and S $\sim$ 0.2 %. The *in vitro* release of Dapivirine from both devices was measured by incubating IVRs at 37 °C under sink conditions in aqueous surfactant solution and sampling daily for drug concentration in the medium. Figure 12.12 shows the predicted in vitro drug elution curve (*solid line*) generated using the modeling software with the values of S and D from the film diffusion experiment, along with the measured drug release data (*open triangles*), for 25 mg and 100 mg IVRs, showing good agreement for both loadings (i. e., solution devices and dispersion devices).

12.8 Conclusions

HME is a key step in the preparation of various MD and DED and implants. The process may be used to homogenously mix API with polymer for further processing, or maybe used to form the final device shape, or both. Devices in the form of rods, films, and torroids fabricated using HME processes have been successfully commercialized.

Models for the *in vitro* elution of drugs from devices are well established, and can provide starting points for formulation of new devices after straightforward measurements of drug solubility and diffusion coefficients in the polymer(s) of the device.

Polymer suppliers are gaining a better understanding of the field of DED, and working closely with device manufacturers to provide appropriate materials that meet the strict specifications and regulatory requirements required for such products.

It is apparent that these recent developments in the manufacturing, modeling, and materials fields enable the development of more effective delivery platforms. As these areas continue to grow, so too will the use of HME for the production novel drug delivery devices.

References

Amanat N et al. (2010) Welding methods for joining thermoplastic polymers for the hermetic enclosure of medical devices. Med Eng Phys 32:690–699

Baker RW, Lonsdale HK (1974) Controlled release: mechanisms and rates. In: Tanquarry AC, Lacey RE (eds) Controlled release of biologically active agents. Plenum, New York, pp 15–71

Clark MR, Johnson TJ, Mccabe RT, Clark JT, Tuitupou A, Elgendy H, Friend DR, Kiser PF (2012) A hot-melt extruded intravaginal ring for the sustained delivery of the antiretroviral microbicide UC781. J Pharm Sci 101(2):576–587

Costeux S et al (2011) Facile TPO dispersion using extensional mixing, Podium presentation, ANTEC (May 2nd 2011, Boston). http://www.randcastle.com/ANTEC11.pdf

Helbling IM, Cabrera MI, Luna JA (2011) Mathematical modeling of drug delivery from one-layer and two-layer torus-shaped devices with external mass transfer resistance. Eur J Pharm Sci 44:288–298

Higuchi T (1961) Rate of release of medicaments from ointment bases containing drugs in suspension. J Pharm Sci 50:874–875

Higuchi T (1963) Mechanism of sustained action medication: theoretical analysis of rate of release of solid drugs dispersed in solid matrices. J Pharm Sci 52:1145–1149

Johnson TJ, Gupta KM, Fabian J, Albright TH, Kiser PF (2010) Segmented polyurethane intravaginal rings for the sustained combined delivery of antiretroviral agents dapivirine and tenofovir. Eur J Pharm Sci 39(4):203–212

Loxley A, Ghokale A, Kim Y, McConnell J, Mitchnick M (2008) Ethylene-vinylacetate intravaginal rings for zero-order release of an antiretroviral drug. 2008 Controlled Release Society Annual Meeting, NYC, Poster Presentation

Loxley A, Mitchnick M, Okoh O, McConnell J, Goldman L, Morgan C, Clark M, Friend DR (2011) Ethylene vinyl acetate intravaginal rings for the simultaneous delivery of the antiretroviral UC781 and contraceptive levonorgestrel. Drug Deliv and Transl Res 1:247–255

Malcolm RK (2003) Modified-release drug delivery technology. In: Rathbone M, Hadgraft J, Roberts M (eds) The intravaginal ring. Marcel Dekker, NY, pp 775–790

Malcolm RK, Woolfson AD, Toner CF, Morrow RJ, McCullagh SD (2005) Long-term, controlled release of the HIV microbicide TMC120 from silicone elastomer vaginal rings. J Antimicrob Chemother 56(5):954–956

Martin C (2008) Continuous mixing of solid dosage forms via hot-melt extrusion. Pharma Tech 32(10):76–86

Siepmann J, Siepmann F (2012) Modeling of diffusion controlled drug delivery. J Control Release 161(2):351–362

Woolfson AD, Malcolm RK, Gallagher RJ (2003) Design of a silicone reservoir intravaginal ring for the delivery of oxybutynin. J Control Release 91:465–476

Chapter 13
Solid Lipid Extrusion

Peter Kleinebudde

Abstract Solid lipid extrusion takes place some Kelvin below the melting point or melting range of the lipid. Depending on the melting point of the lipid, the extrusion can be performed at room temperature or up to approximately 70 °C. Plasticity is achieved by a thermomechanical treatment of the lipids without melting the bulk part of the lipid. The extrusion process is simple and robust and can be integrated with several downstream processes into a continuous production. Solid lipid extrusion is a solvent-free process, which is suitable for thermosensitive drugs. Solid lipid extrudates and solid dosage forms made by these extrudates can be used to modify the dissolution profile of the active pharmaceutical ingredient (API). The dissolution profiles of the cut, milled, or spheronised extrudates can be tailored in a wide range and are mechanistically understood. In particular, a prolonged release can be achieved without any further coating of the particles. The topics formulation, processing, physical stability, dissolution profiles including their modelling and pharmaceutical applications of solid lipid extrudates are addressed.

13.1 Ways of Extrusion

Extrusion is the application of pressure to a mass, until the mass flows through dies of defined diameters. The mass requires certain rheological properties for extrusion. Two dimensions of the extrudate are defined, only the length can vary. Depending on the die cylindrical, flat or other shaped extrudates like films can be produced.

Different ways of extrusion are used in pharmaceutical technology for the production of solid dosage forms. In *wet extrusion*, the required plasticity of the mass is achieved by adding a suitable liquid to the powders in order to produce a paste. In most cases, water is used as liquid. Wet extrusion is performed at room temperature. The solidification of the mass is achieved by drying of the liquid. Some authors include lipids in wet extrusion (Roblegg et al. 2010).

P. Kleinebudde (✉)
Institute of Pharmaceutics and Biopharmaceutics,
Heinrich-Heine-University, Düsseldorf, Germany
e-mail: kleinebudde@hhu.de

M. A. Repka et al. (eds.), *Melt Extrusion*,
AAPS Advances in the Pharmaceutical Sciences Series 9,
DOI 10.1007/978-1-4614-8432-5_13,

Melt extrusion uses meltable binders. Small molecules like sugars, sugar alcohols, and lipids can be used as meltable binders. In most cases, polymers are used for this purpose. Extrusion is performed above the melting point of the binder, if the binder is crystalline, or above the glass transition temperature, if the binder is amorphous. At the elevated temperature, the binder is a more or less viscous liquid and acts in the same way as the liquid in wet extrusion: it converts the powder mass into a plastically deformable mass, which can be extruded. Depending on the chemical nature of the meltable binder and the active pharmaceutical ingredient (API) it can be possible to dissolve the API in the meltable binder during the extrusion process. APIs with a low melting point can melt themselves and sometimes act as a plasticizer for the used polymer. Solidification of the extrudate is achieved by cooling. The binder recrystallises or changes into the glassy state. Unlike the liquid in wet extrusion, which is removed during drying, the meltable binder stays in the extrudate and is part of the final extrudate. Often the meltable binder constitutes a kind of matrix. Depending on the hydrophilic or lipophilic nature of the binder, many different dissolution profiles are achievable from melt extrudates.

Solid lipid extrusion, sometimes also termed cold extrusion or gentle extrusion, uses lipids in order to achieve the plasticity for extrusion. Sometimes hydrophilic molecules like macrogols can also be used or added for this purpose. In these cases, the term solid lipid extrusion may be misleading, but will nevertheless be used throughout the following chapter. In contrast to melt extrusion, the lipid is not molten (at least not the bulk part of the lipid) and extrusion takes place some Kelvin below the melting point or melting range of the lipid. Depending on the melting point of the lipid, the extrusion can be performed at room temperature or up to approximately 70 °C. The plasticity is achieved by a thermomechanical treatment of the lipids. A solidification of the extrudate is not necessary; however, in many cases cooling of the extrudate increases its brittleness. Since neither the lipid nor the API is molten during the extrusion process the formation of a solid solution or eutectic mixture is unlikely during solid lipid extrusion. The lipid is not removed and usually forms a matrix, which can influence the drug release from the extrudate. In the simplest case, binary mixtures of lipid and API are extruded. The API load in the extrudate can be as high as 80 % or even higher in some cases.

Solid lipid extrusion was described already in 1976 (Schmidt and Prochazka 1976). A gear wheel press was used for extrusion. A few products based on solid lipid extrusion are on the German market since a long time, but newer developments are missing. By using twin-screw extruders it was possible to achieve a much better temperature control during extrusion. The first papers were published in 2002 using macrogols as hydrophilic binders (Breitkreutz et al. 2003a) and hard fat among other lipids as lipophilic binders (Breitkreutz et al. 2003b). Since then a number of papers were published on the topic, which are discussed in detail later.

Sometimes lipids are also used as excipients in wet extrusion (Roblegg et al. 2010; Siepmann et al. 2006). These papers are not discussed, because the primary type of extrusion is different and the resulting product properties are not necessarily comparable to those of solid lipid extrudates.

Table 13.1 Advantages of solid lipid extrusion

Solvent-free process, no drying, no residual solvents
Low temperature
Physically stable extrudates possible (if done properly)
Wide range of dissolution profiles
Simple, robust, continuous process

13.2 Advantages and Disadvantages

13.2.1 Advantages

The advantages are listed in Table 13.1. Solid lipid extrusion is a solvent-free process. A drying step is not required as in wet extrusion. This allows the extrusion of hydrolytically sensitive APIs. By avoiding organic solvents there is no issue with residual solvents and their analytical determination.

Solid lipid extrusion can be performed at comparably low temperatures compared with melt extrusion. It is thus suitable for thermally sensitive APIs. Furthermore, it is unlikely that the API itself melts and issues regarding recrystallisation of metastable modifications of the API can be avoided.

The lipid itself is completely or at least to a high degree in the solid state during the extrusion process. Thus recrystallisation of the lipid can be avoided or takes place in the presence of the bulk lipid in its current modification. Thus it is possible to keep the lipid in its stable modification during processing and storage and avoid issues concerning the complex recrystallisation of molten lipids. This is a critical prerequisite for the physical stability of extrudates during storage. Changes in the solid state of lipids can affect critical product properties like the dissolution profile. The 'ageing' of lipids during storage is a major issue for many formulations based on solid lipids (Choy et al. 2005). Therefore, solid lipid extrusion has a major advantage over techniques working with molten lipids. However, the advantages are only valid, if the starting material is the stable modification of the lipid.

Solid lipid extrudates and solid dosage forms made by these extrudates can be used to modify the dissolution profile of the API. In particular, a prolonged release can be achieved without any further coating of the particles. By selecting the appropriate lipid, the loading of the matrix, the size of the extrudates and the addition of release modifiers it is possible to achieve a wide range of dissolution profiles. Even a two-step extrusion can be used to modify the release profile (Windbergs et al. 2010a).

The extrusion process is simple and robust and can be integrated with several downstream processes into a continuous production.

13.2.2 Disadvantages

In some cases, lipid-based dosage forms are used in order to improve the bioavailability of APIs (Porter and Charman 2001; Trevaskis et al. 2008; Chakraborty et al. 2009).

Such effects are not described for solid lipid extrusion. The used lipids remain solid during processing, storage and application. It is not possible to dissolve an API in the lipid during processing. Thereby solid lipid extrudates are not suitable to increase the bioavailability by this mechanism.

To achieve the beneficial product properties described in the previous section it is mandatory to control the extrusion temperature carefully. In case of uncontrolled melting during the process some problems regarding physical stability might occur during storage.

Some combinations of lipids and APIs can result in electrostatic charging during the extrusion process (Witzleb et al. 2011a).

13.3 Equipment

The main piece of equipment in solid lipid extrusion is the extruder. Principally, many types of extruders can be used like a ram extruder (Hasa et al. 2011), a gear wheel press (Schmidt and Prochazka 1976), a ring die press (Breitkreutz et al. 2003b), or a flat die press.

However, the preferred type of extruder is a twin-screw extruder. The main advantage is the possibility to control the temperature during extrusion. A tight temperature control is achieved by heating or cooling the different segments of the extruder barrel. Heat can be supplied or removed in order to keep the temperature constant. In the die itself the temperature is not controlled. High friction in the die can generate a temperature increase, which can lead to a partial melting of the lipid at the surface.

Other advantages of twin-screw extruders are the good mixing properties in the barrel, the self-cleaning properties due to intermeshing screws and the possibility of designing the modular screw using sequences of elements suitable for the particular product. These advantages make the twin-screw extruder the most flexible, versatile and robust type of extruder for solid lipid extrusion. The remaining part of the chapter focuses on the use of twin-screw extruders. Since twin-screw extruders are described in detail in Chap. 2 the focus here is on solid lipid extrusion. So far all reports refer to co-rotating twin-screw extruders.

Figure 13.1 shows the set-up for solid lipid extrusion with a twin-screw extruder. The two screws are located in the extruder barrel. The screws consist of a sequence of different elements like transport elements, kneading elements, or mixing elements. A powder mixture consisting of the API, the lipid and sometimes other additives is supplied continuously by a volumetric or gravimetric powder feeder. At the other end of the barrel the plastified mass is extruded through a die plate. In most cases, multi-hole dies are used for extrusion. Usually cylindrical dies are used but different and sometimes complicated shapes are possible (Hasa et al. 2011). The die is of major importance regarding the properties and the quality of the extrudate. The diameter of the die determines the size of the extrudate. The length of the die and the ratio of length to radius are relevant. The quality of the surface in the die influences the possible friction and, consequently, the temperature during extrusion.

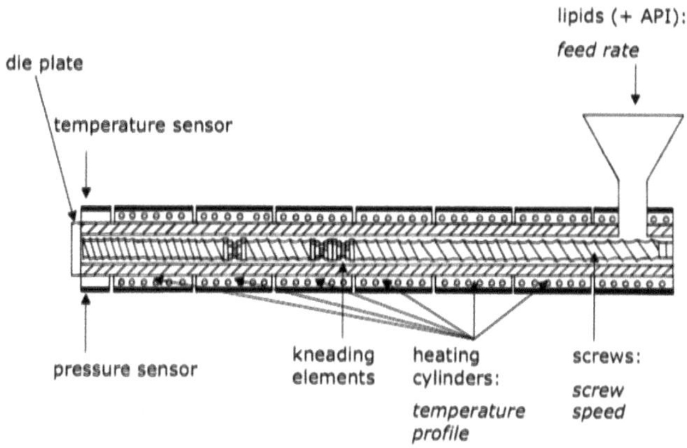

Fig. 13.1 Scheme of a twin-screw extruder with important parts and process variables

13.4 Extrusion Process

The powdered lipid or lipid mixture is pre-blended with the API and possibly other ingredients. The pre-blended mixture is fed into the barrel at a fixed feed rate by using a gravimetric powder feeder (loss in weight powder feeder). The extruder operates with a fixed screw speed. The screw speed needs to be high enough to draw the powder mixture into the barrel. The barrel is not completely filled, i.e. the twin-screw extruder runs in a starved mode. Depending on the screw configuration, further mixing, kneading, densification and transportation takes place inside the barrel. During the transport the temperature of the mass can be adjusted due to friction and the contact with the preheated barrel wall. The residence time of the mass depends on the length of the extruder, the filling degree and the screw speed. Typically the mean residence time is between 30 and 120 s. In front of the extruder, die plate pressure is build up for extrusion. The pressure depends on the filling degree and the number, diameter and length of the holes in the die plate. The pressure is important for the densification of the mass and the porosity of the final extrudates. The die plate itself is not actively cooled or heated. Thus, friction during the short die passage can result in a temperature increase and partial melting of components at the die surface supporting the flow through the die and the extrusion.

The suitable temperature range depends on the properties of the selected lipid. With increasing barrel temperature the pressure at the die plate decreases (Fig. 13.2). For each lipid the large symbols represent optically homogeneous extrudates. With increasing temperature the solid fat fraction decreases gradually. Above a certain temperature the extrusion pressure is very low because the extrudate liquefies and the shape of the extrudate is no longer retained. Below a certain temperature the pressure is high and the extrudate is not compact and optically homogeneous. The temperature range for successful solid lipid extrusion is for all lipids below the melting range of the respective lipid, generally 5–15 K below. The pressure range is comparable for

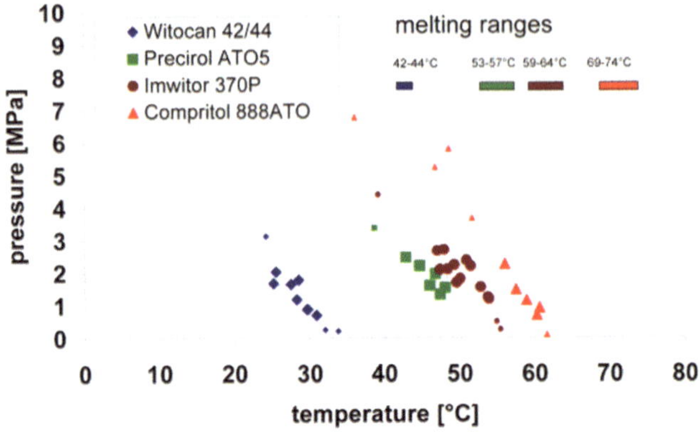

Fig. 13.2 Relation between temperature and extrusion pressure

all tested lipids; however, the pressure depends also on the selected die plate, which was constant for the lipids in Fig. 13.2. In all cases of successful extrusion, the solid fat content exceeded 90 % (Reitz and Kleinebudde 2007a), indicating that the bulk of the lipid is in the solid state.

13.5 Materials

13.5.1 Lipids

The lipids should be solid at room temperature. The melting temperature is in the range from about 40–90 °C. A second prerequisite is the availability of the lipid in powdered form. This is required since the powdered lipid is pre-blended with the API and other excipients. In order to assure a constant feed rate of a homogeneous mixture the lipid must be in powdered form. Only a limited number of lipids are commercially available in powdered form, which is one limitation for solid lipid extrusion. However, it is possible to mill flakes or pastilles of a solid lipid to the desired particle size. Milling of lipids with low melting points is possible by decreasing the temperature during milling.

Among the lipids glycerides, waxes or paraffins have been used. Glycerides are the most important group. They differ in their melting range and the degree of esterification. Solid glycerides mainly contain saturated fatty acids. The chain length mainly determines the melting range. The chain length distribution in a glyceride can be narrow or broad.

One group of lipids are monobasic triglycerides. These lipids consist to a high extent of triglycerides of one defined fatty acid. Examples are glyceryl trimyristate, glyceryl tripalmitate and glyceryl tristearate. These lipids have a narrow melting range, which requires a good temperature control during extrusion. It is more difficult

Fig. 13.3 Dissolution profiles from matrices of pure triglyceride (TG), pure macrogol (PEG) and mixtures of TG+PEG. (Reproduced from Windbergs et al. 2009b, with permission from Elsevier)

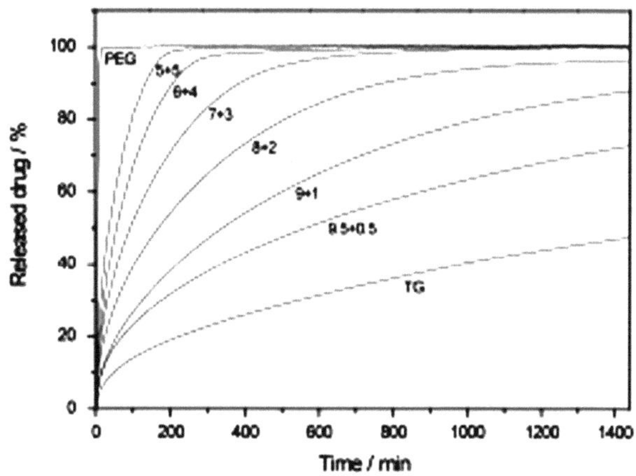

to obtain a suitable extrudate. Another type of lipids contains mixtures of mono-, di- and triglycerides composed of different fatty acids. The melting range of such more heterogeneous lipids is broader and the extrusion process is more robust with respect to small variations in the temperature. An example for such a lipid is glyceryl palmito stearate.

13.5.2 Mixtures of Lipids

For some applications, mixtures of lipids are used (Reitz and Kleinebudde 2009; Krause et al. 2009). Usually lipids with distinctly different melting ranges are mixed. A particular application is the pelletisation of lipid extrudates.

13.5.3 Release Modifier

One approach to design the dissolution profile is the addition of release modifiers. The most prominent examples are macrogols (Breitkreutz et al. 2003a; Windbergs et al. 2009a, b; Schulze and Winter 2009; Gures and Kleinebudde 2011). Macrogols can be used instead of a lipid since they also show thermomechanical plasticisation (Breitkreutz et al. 2003b). The dissolution profile can be varied in a broad range. An increasing fraction of the highly water-soluble macrogol results in a faster dissolution. The macrogol dissolves during dissolution leaving pores in the lipid matrix. One example is shown in Fig. 13.3.

Other *pore formers* like sugars, sugar alcohols, or water-soluble salts can be used as well (Gures and Kleinebudde 2011).

Other groups of release modifiers are hydrophilic polymers or disintegrants (Gures and Kleinebudde 2011). The 5 % of such release modifiers in the formulation can

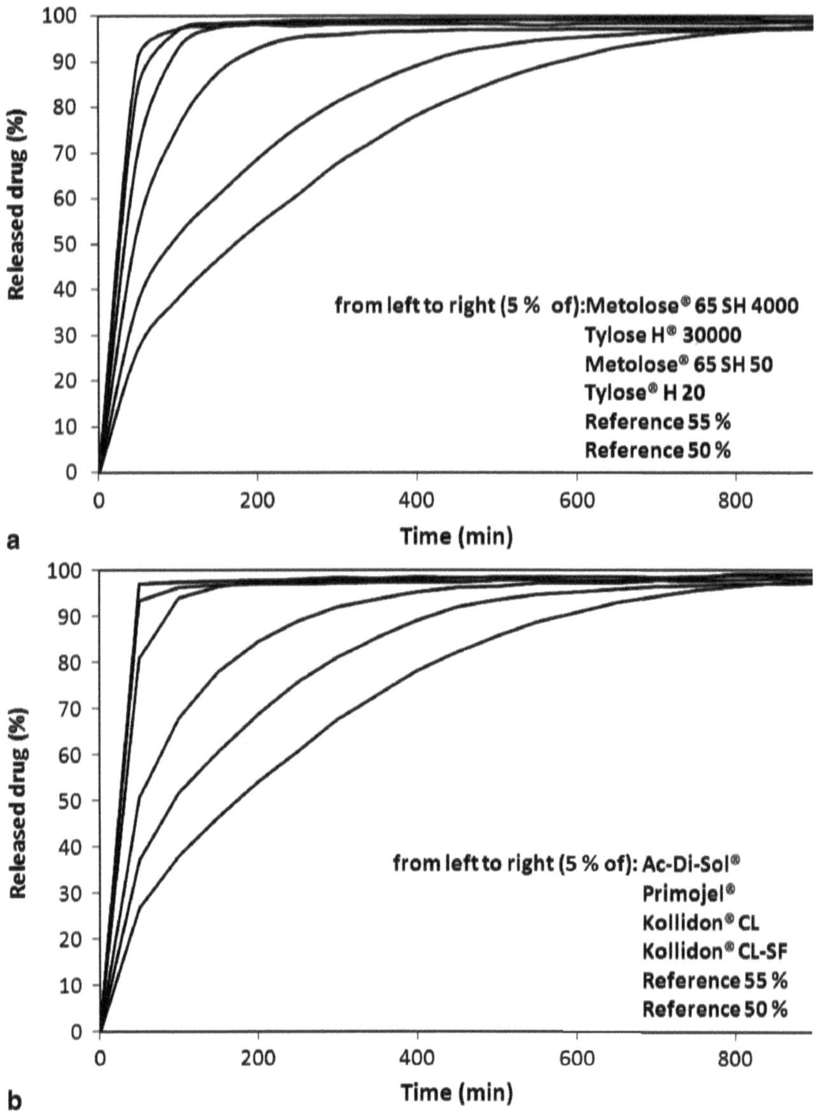

Fig. 13.4 Effect of 5 % release modifier on drug release. **a** Hydrophilic soluble polymers. **b** Disintegrants. (Reproduced from Gures and Kleinebudde 2011, with permission from Elsevier)

lead to drastic changes in the release profile (Fig. 13.4). *Hydrophilic polymers* like cellulose ethers with higher molecular weights resulted in a complete disintegration of the matrix due to extended swelling. This disintegration led to fast drug release. Lower molecular weight hydrophilic polymers tended to swell, but the lipid matrix stayed intact. The dissolution rate was higher compared to the pure lipid matrix,

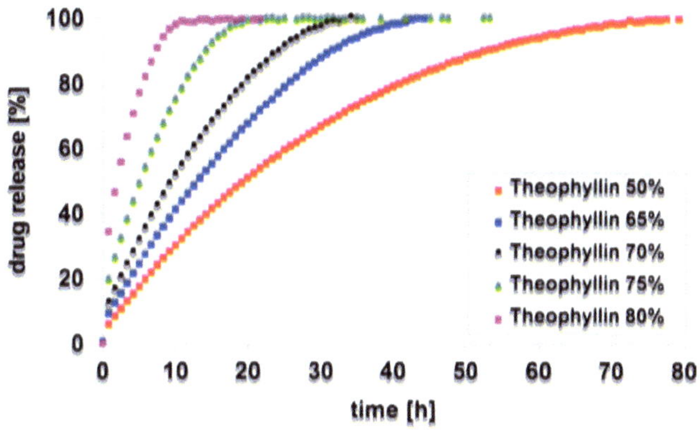

Fig. 13.5 Active pharmaceutical ingredient (API) release profile in dependence of fraction of API in the formulation

but slower compared to the disintegrating extrudates. Similar results were found for *disintegrants* as release modifiers. Some of the tested disintegrants resulted in the disintegration of the matrix, which in turn led to fast drug release. Other disintegrants like crospovidone of a small particle size resulted only in a local swelling and an increase in pore size and roughness of the matrix, which resulted in a moderate enhancement of the dissolution rate.

By adjusting type and fraction of release modifiers it is possible to tailor release profiles.

13.5.4 Active Pharmaceutical Ingredient

Many APIs can be incorporated into solid lipid extrudates. The loading of API can vary in a broad range. In many cases, loadings of 80 % API can be achieved (Breitkreutz et al. 2003a; Witzleb et al. 2011a; Krause et al. 2009). The dissolution profile from the lipid matrix depends on the loading with API (Fig. 13.5). With increasing API fraction in the extrudate the release rate will become higher. Thus, a simple method to vary the dissolution rate is to adjust the loading with API.

In some cases, the solid lipid extrusion is apparently difficult. The main reason can be associated with the shape of the drug crystals: needle-shaped API crystals resulted in an increase of extrusion pressure over time and at the end of a blockage of most die holes (Fig. 13.6). This was proven with the needle-shaped APIs praziquantel, caffeine and mesalazine (Witzleb et al. 2011a). Similar difficulties were reported in wet extrusion with a ram extruder (Di Pretoro et al. 2010). Recent investigations confirmed these difficulties during extrusion for nimesulide and praziquantel in wet and solid lipid extrusion using the same type of equipment (Di Pretoro et al. 2012). The extrudability could not be achieved by changing the screw configuration or

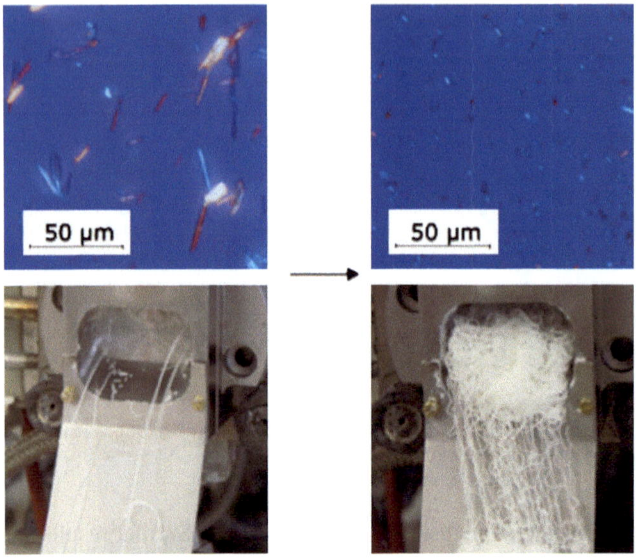

Fig. 13.6 Extrusion with unmilled needle-shaped praziquantel particles (*left*) and milled particles (*right*). (Reproduced from Witzleb et al. 2011b with permission of Elsevier)

the extrusion temperature, by decreasing the loading with API down to 10 %, by changing the lipid, by adding other excipients or by changing the die diameter.

A simple way of overcoming the problems is milling of the needle-shaped crystals to more isometric particles. This is done by air jet milling. The isometric particles are suitable for solid lipid extrusion and loadings up to 80 % could be achieved. The die holes do not block any more and the extrusion pressure is constant (Fig. 13.6). The size of the particles is less important compared with the shape of the particles.

13.5.5 Other Excipients

Sometimes it is useful to include further excipients in the formulation. One case is given, if electrostatic charging occurs during solid lipid extrusion (Witzleb et al. 2011a). For certain combinations of lipid and API, e.g. praziquantel with glyceryl trimyristate or cetyl palmitate, electrostatic charging is observed (Figs. 13.7 and 13.8). This problem is mainly observed at low relative humidity in the production environment. An air condition resulting in a relative humidity of 58 % drastically reduced the electrostatic charging. Alternatively, the addition of an anti-static agent can reduce the charging and the related difficulties (Fig. 13.9). Macrogols, which are molten during the extrusion process were found to be effective anti-static agents. If the macrogol was solid during extrusion it has no anti-static properties. Macrogol 1500 was effective at the chosen extrusion temperature since it was molten during

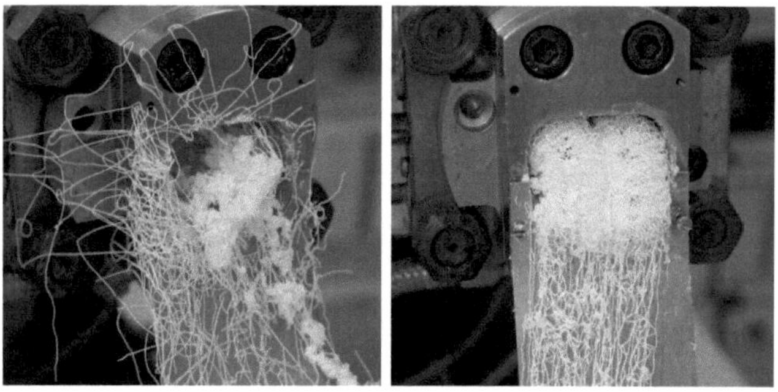

Fig. 13.7 Extrusion without (*left*) and with (*right*) electrostatic agent. (Reproduced from Witzleb et al. 2011a, with permission from Elsevier)

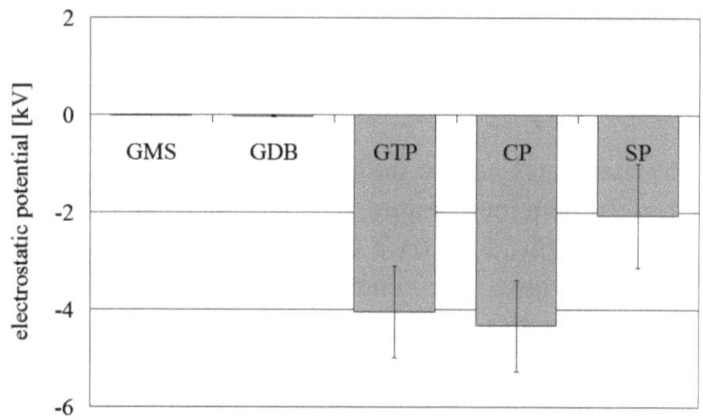

Fig. 13.8 Electrostatic charging with 50 % milled praziquantel and different lipids: glyceryl tripalmitate (*GTP*), glyceryl dibehenate (*GDB*), glyceryl monostearate (*GMS*), cetyl palmitate (*CP*) and solid paraffin (*SP*), mean ± SD. (Reproduced from Witzleb et al. 2011a, with permission from Elsevier)

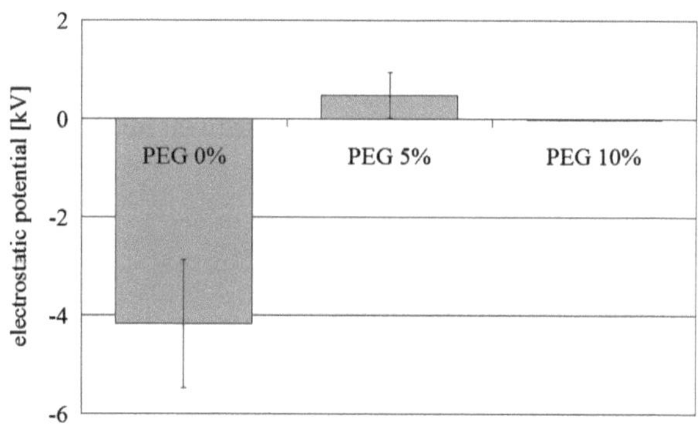

Fig. 13.9 Electrostatic charging during extrusion with 50 % praziquantel/glyceryl tripalmitate (49, 44 and 39 %)/PEG 1500 (0, 5 and 10 %)/1 % silicium dioxide, mean ± SD. (Reproduced from Witzleb et al. 2011a, with permission from Elsevier)

extrusion. The extrusion temperature was varied when using macrogol 6000 as antistatic agent. At an extrusion temperature above the melting point of macrogol 6000 no electrostatic charging was observed and the extrusion process runs smoothly. However, a decrease of the extrusion temperature below the melting point of macrogol 6000 resulted in increasing electrostatic charge and the extrusion process became more difficult indicated by blocking of some die holes. It can be concluded that only molten macrogol can effectively act as anti-static agent in solid lipid extrusion. The selection of the appropriate macrogol depends on the extrusion temperature and thus on the chosen lipid for the extrusion.

13.6 Downstream Processing

13.6.1 Cutting

In the simplest case, extrudates are cut into short cylinders of defined length. This can be done manually or automatically.

13.6.2 Milling

The extrudates can be milled directly after extrusion. If the extrusion temperature exceeds room temperature, a cooling step can be incorporated prior to milling. This can be simply done on a conveyor belt. One type of mill, which has been used for this purpose, is a centrifugal mill. The residence time of the extrudate in the milling chamber is short and the milled particles are directly withdrawn from the milling part. The particle size distribution can be widely determined by the diameter of the extrudate, the screen size in the mill and the processing parameters. By the use of an air stream the milled granules can be separated in a cyclone from the air and collected in a container. This set-up allows a continuous production by using a gavimetric powder feeder, an extruder, a conveyor belt, a vibrating feeder to the mill, a centrifugal mill and a cyclone combined with a container (Witzleb et al. 2011a) (Fig. 13.10). Instead of using a centrifugal mill the granules can also be granulated by other means.

13.6.3 Spheronisation

For some applications, spherical pellets are required instead of irregular granules. This requires a spheronisation process. During spheronisation the extrudates break down into small pieces, which are later spheronised. The spheronisation requires a certain plasticity of the mass. This is obtained by raising the temperature during spheronisation in order to plastify the mass by melting small amounts of suitable components. The temperature can be raised by friction between the particles during spheronisation, by heating the spheroniser wall using a double jacket wall, or by increasing the temperature from outside, e.g. by using infrared light. At a suitable temperature a minor component in the formulation melts partly or completely. For this

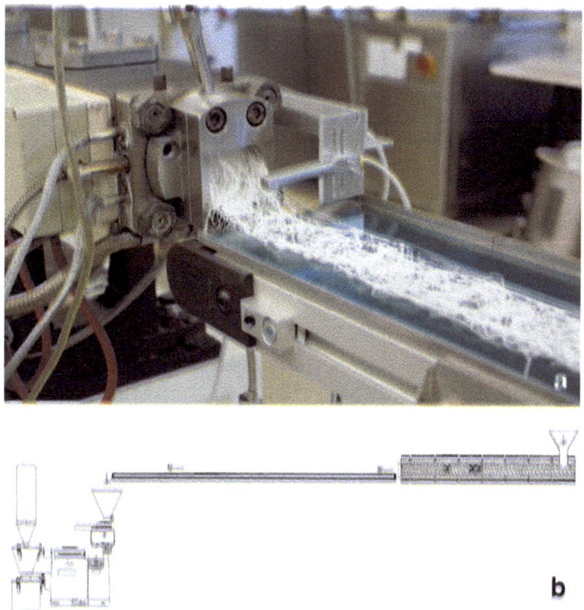

Fig. 13.10 Photo (**a**) and scheme (**b**) for continuous extrusion. (Reproduced from Witzleb et al. 2011a, with permission of Elsevier)

purpose mixtures of high and low melting lipids can be used (Reitz and Kleinebudde 2009; Krause et al. 2009).

Proper pellets can be achieved by selecting an appropriate formulation and by assuring a careful process control. The evolution of a spheronisation process is shown in Fig. 13.11 (Reitz and Kleinebudde 2009). After 17 min round pellets were obtained

Process time [s]	15	300	540	780	1020
Process temperature [°C]	25.4	34.8	36.6	37.6	38.1
Median Aspect ratio	2.2	2.0	1.6	1.2	1.1
Median equivalent diameter [mm]	1.85	1.73	1.68	1.56	1.51
10%-Intervall	46.8	58.6	70.2	87.6	87.4
Mean mass per pellet [mg]	2.50	2.24	2.18	2.15	1.95

Pictures of spot samples

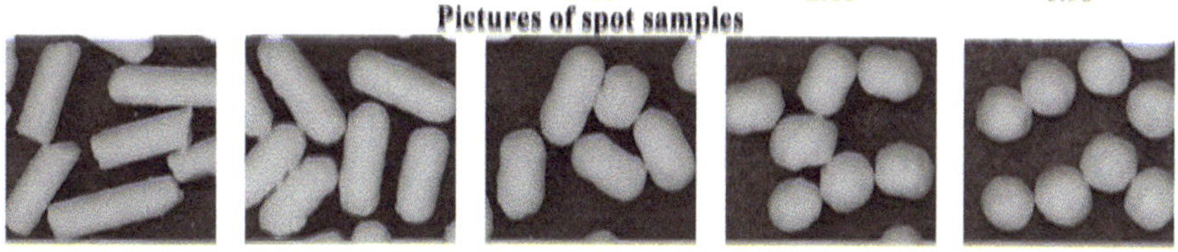

Fig. 13.11 Spheronisation of an extrudate containing 15 % hard fat and 31 % glyceryl trimyristate at 31 °C spheroniser temperature. (Reproduced from Reitz and Kleinebudde 2009, with permission of Elsevier)

at a spheroniser temperature of 31 °C. The process temperature exceeded 38 °C at the end due to friction. A hard fat was the low melting lipid in this case, which was combined with glyceryl trimyristate.

13.7 Characterisation of Solid-State Product Properties

The lipids used are mainly polymorphic substances. The recrystallisation behaviour of molten lipids is complicated (Sato 2001). If the lipids are not in their thermodynamically stable form from the beginning or after processing, the extrudates are prone to changes in the solid state during storage or administration. This can affect a number of product properties, especially the dissolution profile (Reitz and Kleinebudde 2007b; Witzleb et al. 2012). Thus, it is of essential importance to determine the solid-state properties of the starting materials as well as the extruded products. Beside the information about the modification of the lipid, solid-state analysis allows to detect possible interactions between the different components in the extrudate. Solid-state analysis is an important tool during stability studies. It helps to understand the physical state and the properties of the extrudate.

X-ray diffraction is a major tool to detect the state and number of phases in a starting material or a product. It allows statements about the nature of the solid, i.e. whether the solid is amorphous or crystalline and the degree of crystallinity. If the solid is crystalline, the type of modification can be detected. If several modifications are present it allows estimating the ratio of modifications.

Differential scanning calorimetry (DSC) is another important method for the characterisation of solid lipid extrudates. Information about the crystalline or amorphous state of the components is available. The number and offset of melting points and the heat of fusion can be used to detect modifications and to characterise the extent of interactions between components in the solid lipid extrudate. However, in some cases artefacts are observed, e.g. an API having a high melting point can be partly or completely dissolved in a low melting lipid or release modifier like Macrogol during the heating cycle. This may be interpreted as an interaction in the extrudate, while during extrusion and storage all components are solid and dissolution of the API is not taking place. Thus, it is required to use more than one method to generate valid information about the solid-state properties of a solid lipid extrudate.

Other methods for the solid-state characterisation include spectroscopic methods, e.g. Raman spectroscopy.

For solid lipid extrusion it is preferred to use starting materials in their thermodynamically stable form, which is the β-modification in case of glycerides. If metastable modifications are present the transition to the stable modification during storage is possible. This can be enhanced by storage at elevated temperature, denoted as tempering or curing. When comparing a lipid with a broad distribution of mono-, di- and triglycerides with a pure triglyceride (Table 13.2), distinct differences can be observed.

DSC measurements of extrudates containing the two lipids show a huge difference in the change of heat of fusion during storage time (Reitz and Kleinebudde 2007a).

Table 13.2 Lipid with narrow and wide distribution of components

	Glycerol trimyristate (GTM)	Glycerol palmitostearate (GPS)
Fatty acids	Myristic acid	40–60 % stearic acid 40–60 % palmitic acid
Degree of esterification	> 90 % triester	25–35 % triester 40–60 % diester 8–22 % monoester
Melting range	55–58 °C	53–57 °C

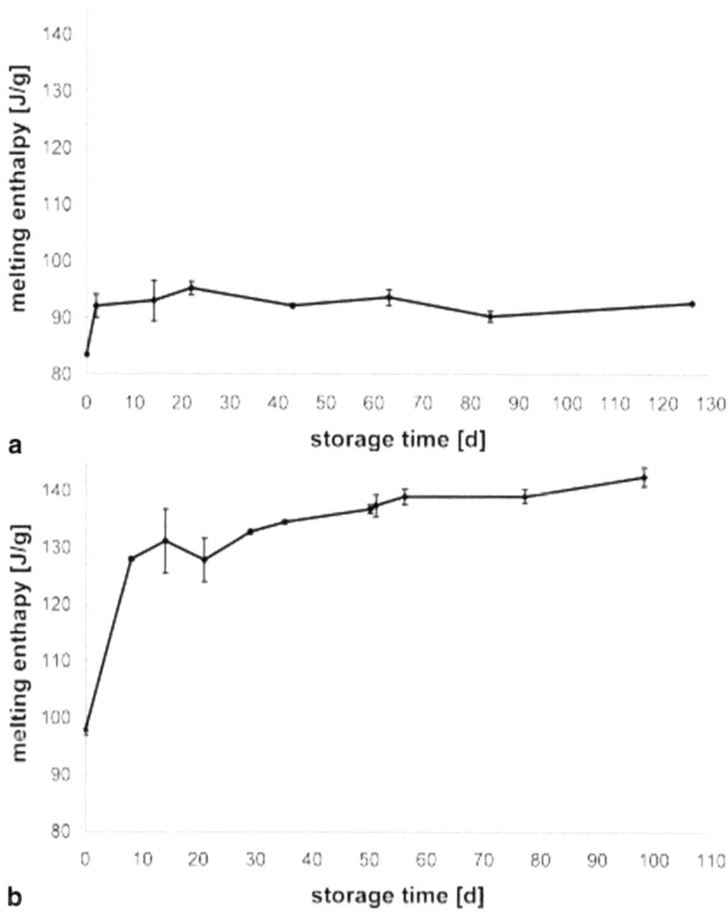

Fig. 13.12 Differential scanning calorimetry (DSC) measurements after storage at 40 °C. **a** Glycerol trimyristate/theophylline (50:50) extrudates. **b** Glycrol palmito stearate/theophylline (50:50) extrudates. (Reproduced from Reitz and Kleinebudde (2007a, with permission from Elsevier)

A minor increase is observed for the well-defined glycerol trimyristate while a significant increase is observed for the glycerol palmito stearate, which has a broad composition (Fig. 13.12). The changes can be seen in the DSC thermograms (Fig. 13.13) (Reitz and Kleinebudde 2007b). While the melting point for glycerol

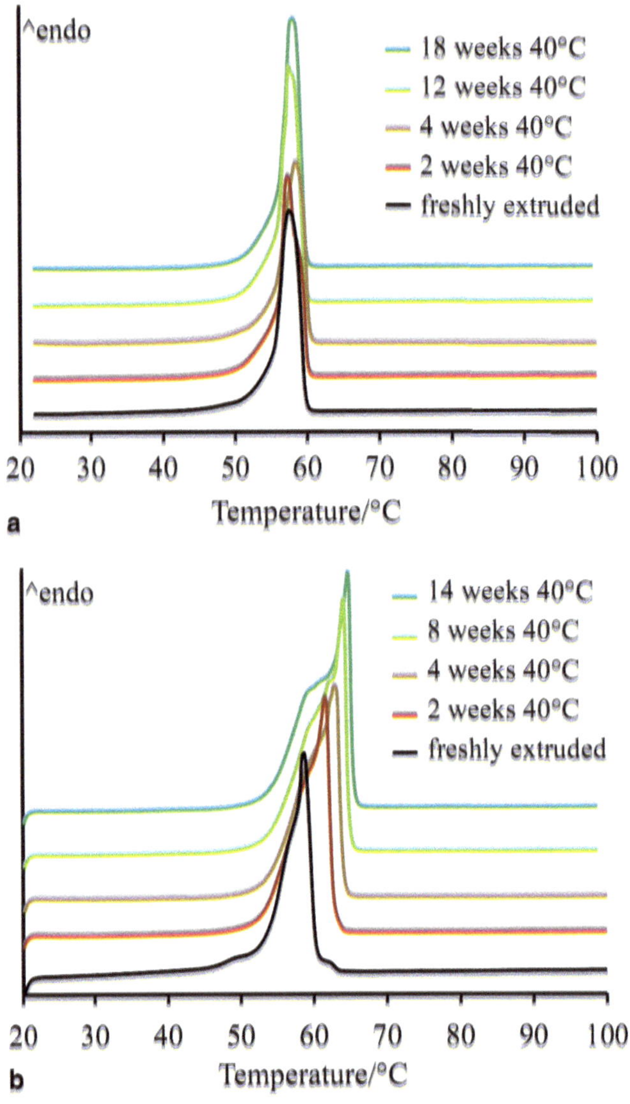

Fig. 13.13 Differential scanning calorimetry (DSC) measurements of extrudates of **a** glycerol trimyristate and **b** glycerol palmito stearate after different storage periods at 40 °C. (Reproduced from Reitz and Kleinebudde 2007b, with permission from Springer)

trimyristate is stable a clear shift towards higher melting points can be seen for extrudates with glycerol palmito stearate, which is associated with changes in the solid state of the lipid.

The changes in solid-state properties of the lipid are of relevance concerning the dissolution profile. While the dissolution from the glycerol trimyristate extrudate is stable over storage time, significant fluctuations are observed for the extrudate

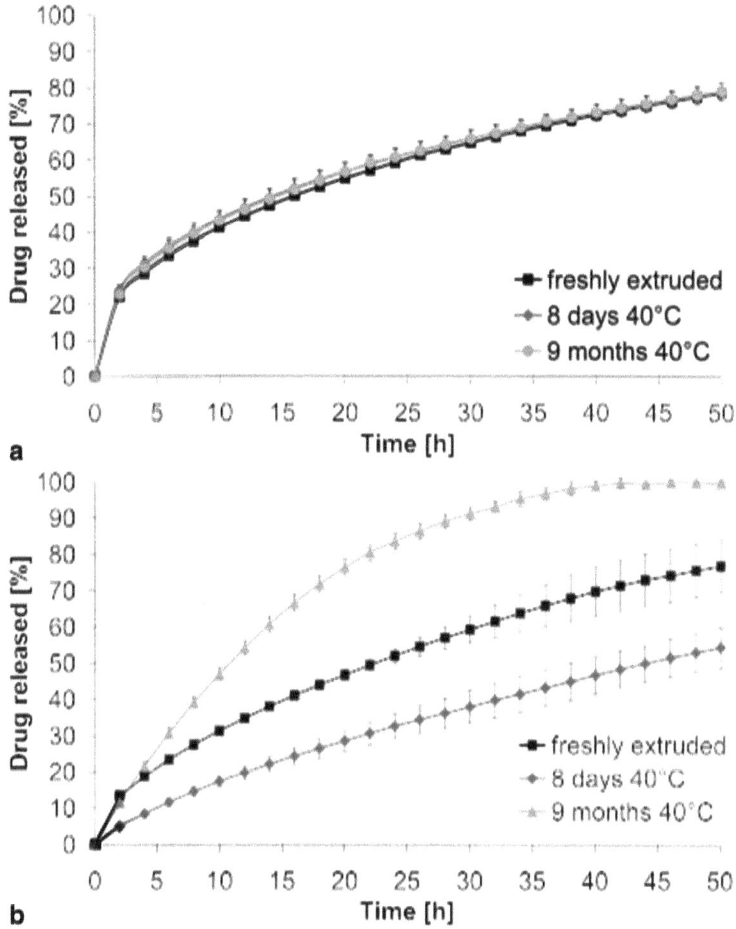

Fig. 13.14 Drug release studies after storage. **a** Glycerol trimyristate/theophylline (50:50) extrudates. **b** Glycerol palmito stearate/theophylline (50:50) extrudates, $n = 3$. (Reproduced from Reitz and Kleinebudde 2007a, with permission from Elsevier)

containing the glycerol palmito stearate (Fig. 13.14) (Reitz and Kleinebudde 2007a). Lipids with well-defined compositions seem to be more suitable for the production of physically stable extrudates. The recrystallisation behaviour of partial glycerides is much more complex (Windbergs et al. 2009c).

Solid lipid extrusion with lipids of well-defined and narrow composition is more demanding since the melting range is narrow and this requires a good temperature control during extrusion. Extrusion with glycerol trilaurate and glycerol tripalmitate has been possible without complications (Windbergs et al. 2009d). However, the production of solid lipid extrudates with glycerol tristearate resulted in different solid-state properties of the lipid depending on the extrusion temperature (Fig. 13.15).

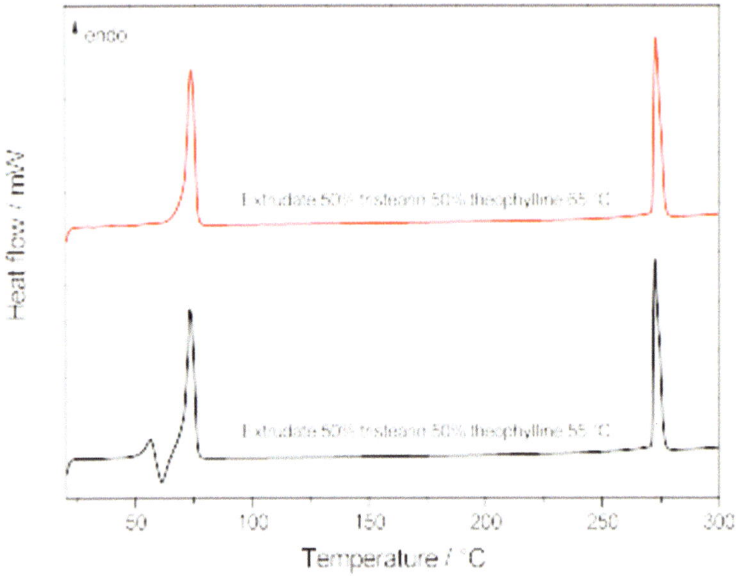

Fig. 13.15 Differential scanning calorimetry (DSC) thermograms of extrudates with glycrol tristearate produced at different extrusion temperatures

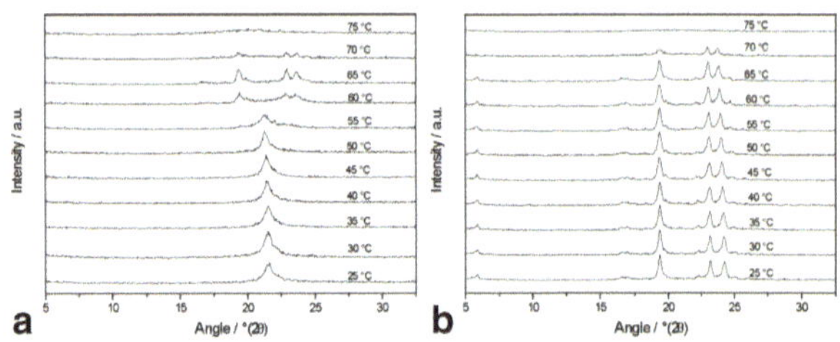

Fig. 13.16 Temperature-dependent X-ray powder diffraction (XRPD) patterns of glycerol tristarate powder. **a** α-modification, **b** β-modification. (Reproduced from Windbergs et al. 2009d, with permission from Elsevier)

When extruded at 65 °C only the β-modification of the lipid can be observed. However, at an extrusion temperature of 55 °C the α-modification can be observed beside the β-modification. At first this is surprising since the higher extrusion temperature may lead to a higher molten fraction, which might be able to recrystallise in the α-modification.

The phenomena can be explained by looking at the temperature-dependent solid-state behaviour of the α-modification and β-modification of glycerol tristearate. When the β-modification is heated up from room temperature the modification is stable until the melting point of about 73 °C, where it melts into an amorphous liquid (Fig. 13.16). When heating the α-modification it melts at its melting point of about 55 °C and recrystallises in the β-modification, which then melts at its own melting temperature of 73 °C.

At an extrusion temperature of 55 °C, the glycerol tristearate can recrystallise in the α-modification during cooling. If the extrusion temperature is above the melting

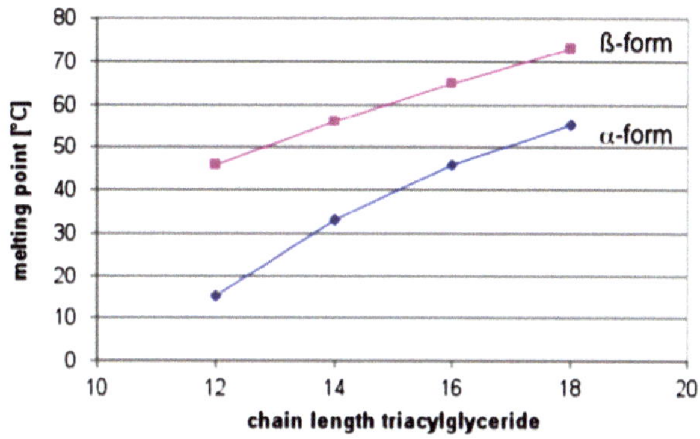

Fig. 13.17 Melting points of triacylgycrides

point of the α-modification, the lipid recrystallises directly in the β-modification. Thus, it is recommended to extrude at temperatures above the melting point of the α-modification but below the melting point of the β-modification (Fig. 13.17).

The combination of temperature and friction is a key point for the lipid polymorphic composition after extrusion. As some lipid melting always occurs while extruding triglycerides, the temperature of the extruder die plate should always be above the melting point of the unstable α-form of the extruded triglyceride to avoid subsequent alteration of the product (Windbergs et al. 2009d).

If some α-modification is produced during extrusion the lipid can convert into the β-modification during storage. The β-modification is crystallising as small needles on the surface of the extrudates. This phenomenon is called 'blooming'. It can affect the dissolution profile from solid lipid extrudates in an undesired way.

13.8 Dissolution

13.8.1 Standard Dissolution Media

In many cases, the dissolution from solid lipid extrudates is tested in standard dissolution media including water, hydrochloric acid solution, or buffer solutions. Many lipids do not dissolve or swell in these dissolution media. In this case, the lipid acts as an inert matrix resulting in a dissolution process, which can be described by the Higuchi equation. Often the Korsmeyer–Peppas equation is used for further analysis of the dissolution curve. The API dissolves from the matrix leaving a porous system.

The addition of release modifiers can have various effects. If the release modifier is a low molecular weight water soluble substance like a salt, sugar or low molecular Macrogol, it will dissolve in the same manner as the API leaving pores (Windbergs et al. 2010b; Haaser et al. 2011). However, high molecular swellable release modifiers can change, disrupt or even disintegrate the matrix structure resulting in strong deviations from the release profile described before (Gures and Kleinebudde 2011).

13.8.2 Biorelevant Media

Many lipids like triglycerides can be digested in the gastrointestinal (GI) tract. This may influence the dissolution profile. These lipids act as an inert matrix formed in standard dissolution media, but this can be different in the natural environment. Therefore, in vitro dissolution tests can be performed using biorelevant dissolution media (Reitz and Kleinebudde 2007a; Witzleb et al. 2012). Such media contain different physiologic components like lipase, phospholipids or bile salts in certain concentrations. Biorelevant media for the fasted and the fed state are described in the literature (Jantratid et al. 2008).

For some lipids, the dissolution profile in the biorelevant medium is similar to the standard medium (Fig. 13.18). This is the case for solid paraffin as expected. In other cases, especially for glycerol monostearate, significant differences are observed indicated by low f_2-values. Changes in the dissolution profiles during storage are associated with the solid state of the lipids. The faster dissolution in the biorelevant medium might be associated with lipolysis of the lipid. In a lipolysis experiment, a certain lipolysis of glyceryl monostearate could be detected (Fig. 13.19). However, the lipolysis of the solid lipid is much less than that of liquid glycerol tricaprylate. The other tested lipids show no lipolysis. There might be reasons beside the lipolysis responsible for the enhanced dissolution rate of other solid lipids in biorelevant media. One effect might be a solubilisation of the lipid by surface-active agents in the biorelevant medium like phospholipids or bile salts. It can be shown that cetyl palmitate is solubilised in solutions with higher polysorbate 20 concentrations while solid paraffin is not solubilised. Thus, different mechanisms might explain a faster dissolution from some solid lipid extrudates in biorelevant media. The lipolysis of the solid lipids seems to be less relevant in several cases (Witzleb et al. 2012).

13.8.3 Spectroscopic Studies

Raman mapping of extrudates after a period of dissolution gives no evidence of a uniformly receding drug boundary in the extrudates during drug release (Haaser et al. 2011). Raman microscopy can be used to examine soluble and insoluble excipient distribution during dissolution testing, which, in turn, affects drug release behaviour. The Raman mapping has to be done offline.

In situ spectroscopy is possible by using coherent anti-Stokes Raman scattering (CARS) microscopy to visualise the solid-state properties of lipid-based oral dosage forms in real time (Windbergs et al. 2009e). The drug release from the dosage form matrix is monitored with a spatial resolution of about 1.5 micron (Fig. 13.20). In addition, CARS microscopy allows the solid-state transformation of a drug to be spatially visualised. The results obtained by CARS microscopy revealed that the method used to combine lipid and active ingredient into a sustained release dosage form can influence the physicochemical behaviour of the drug during dissolution. In the investigated case, formation of theophylline monohydrate on the surface was

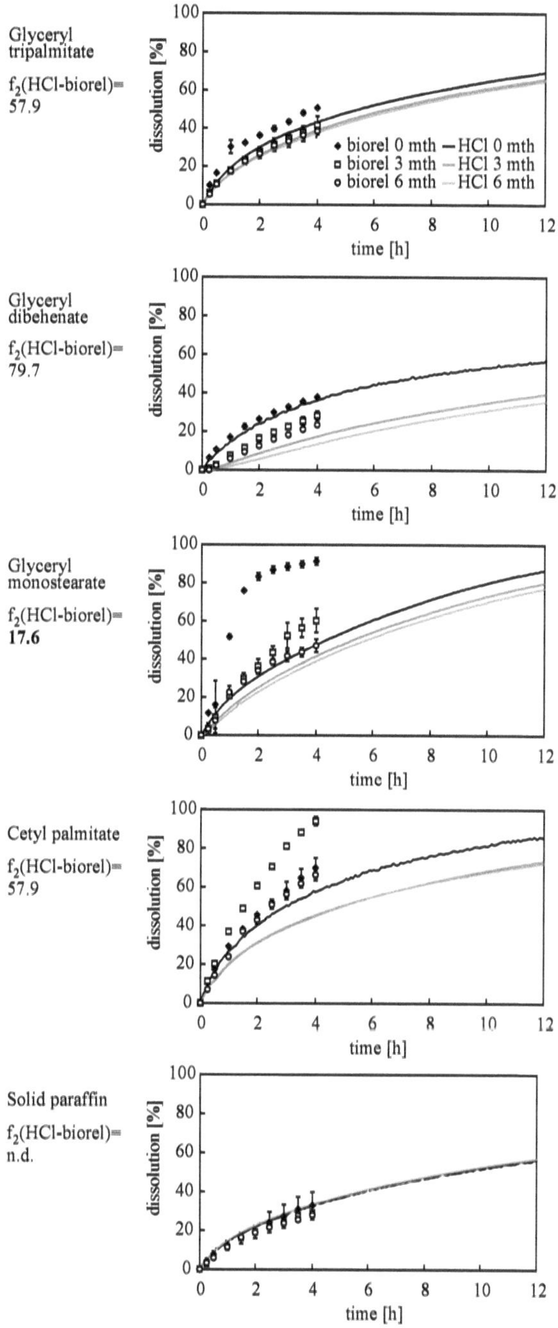

Fig. 13.18 Drug dissolution in biorelevant medium (_) and in standard medium (—) and f2 values for similarity analysis. n.d.: not determined, because CV > 15 %, biorelevant medium: mean ± SD, standard medium: SD not shown (< 3 % at all time points), $n = 3$, formulation: 50 % praziquantel/49 % lipid/1 % silicium dioxide. (Reproduced from Witzleb et al. 2012, with permission from Elsevier)

visualised during dissolution with tablets compressed from powdered mixtures but not with solid lipid extrudates (Fig. 13.21). CARS microscopy is a powerful tool to follow the dissolution from solid lipid extrudates.

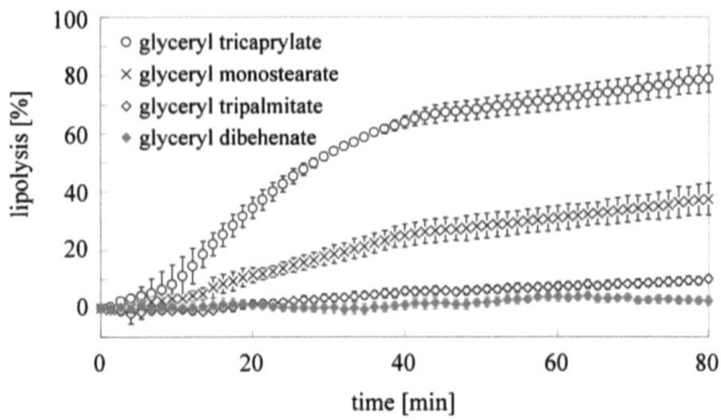

Fig. 13.19 In vitro lipolysis of lipids in extrudates with 50 % praziquantel each, expressed as percentage of titrated fatty acids, mean ± SD, $n = 3$. (Reproduced from Witzleb et al. 2012, with permission from Elsevier)

13.8.4 Modeling of Dissolution

In some cases, the dissolution from solid lipid extrudates can be described by Fick's diffusion (Gures et al. 2012). In order to better understand the mass transport mechanisms controlling drug release from the solid lipid extrudates, an analytical solution of Fick's second law of diffusion was used to quantify the release kinetics.

The model is based on the following assumptions:

- Perfect sink conditions are maintained during dissolution.
- The matrices stay intact during the drug release measurements and their dimensions do not significantly change.
- Diffusional mass transport (in radial and axial direction) is the release rate controlling mass transport step: either of water into the solid lipid extrudates or of diprophylline out of the systems (Siepmann and Siepmann 2012).
- The drug, lipid and the pore former are homogenously distributed in extrudates before exposure to the release medium.
- Erosion of the lipid matrix is negligible during the observation period.
- Limited drug solubility effects within the cylinders are negligible.

In order to quantify diffusional mass transport in axial and radial direction within the solid lipid extrudates, Fick's second law of diffusion for cylindrical geometry was used:

$$\frac{\partial_c}{\partial_t} = \frac{1}{r}\left\{\frac{\partial}{\partial r}\left(rD\frac{\partial c}{\partial r}\right) + \frac{\partial}{\partial \theta}\left(\frac{D}{r}\frac{\partial c}{\partial \theta}\right) + \frac{\partial}{\partial z}\left(rD\frac{\partial c}{\partial z}\right)\right\}$$

where, c is the concentration of the diffusing species (drug or water); t represents time; r denotes the radial and z the axial coordinate of the cylinder and θ the angle perpendicular to the r-z-plane; D is the apparent diffusion coefficient of water or drug.

This partial differential equation can be solved under the given initial and boundary conditions (homogeneous initial compound distribution within the extrudates and perfect sink conditions/constant water concentration at the extrudates' surface) using the method of Laplace transformation, leading to:

$$\frac{M_t}{M_\infty} = 1 - \frac{32}{\pi^2}\cdot\sum_{n=1}^{\infty}\frac{1}{q_n^2}\cdot\exp\left(-\frac{q_n^2}{R^2}\cdot D\cdot t\right)\cdot\sum_{p=0}^{\infty}\frac{1}{(2\cdot p+1)^2}\cdot\exp\left(\frac{(2\cdot p+1)^2\cdot\pi^2}{H^2}\cdot D\cdot t\right)$$

Here, M_t and M_∞ represent the absolute cumulative amounts of drug released or water taken up at time t and infinite time, respectively; q_n are the roots of the Bessel function of the first kind of zero order; R and H denote the radius and the height of the cylinder. (Gures et al. 2012).

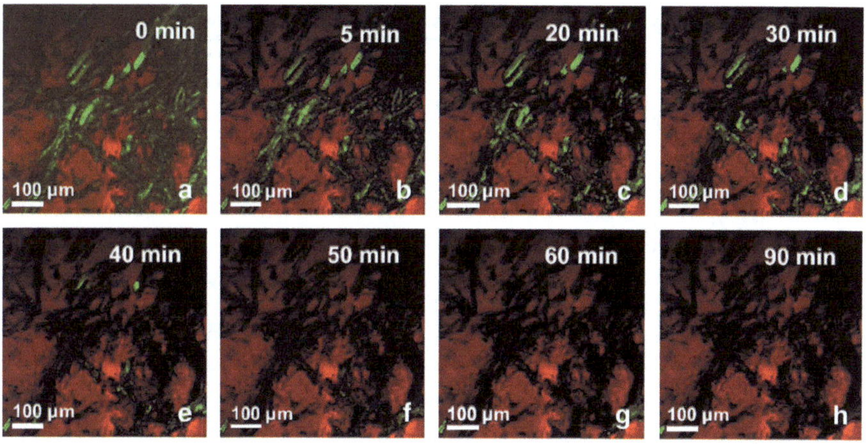

Fig. 13.20 In-line visualisation of drug release from a tablet of tripalmitin (*red*) and theophylline monohydrate (*green*) during dissolution testing. (Reproduced from Windbergs et al. 2009e, with permission from ACS)

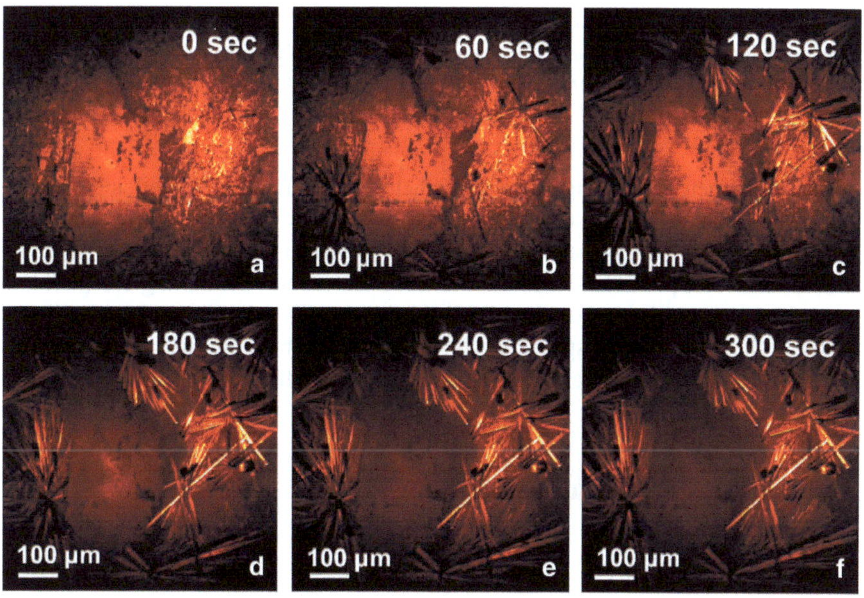

Fig. 13.21 In-line visualisation of theophylline monohydrate crystal growth on the surface of a tablet in water consisting originally of tripalmitin/theophylline anhydrate. (Reproduced from Windbergs et al. 2009e, with permission from ACS)

Extrudates of different lengths and diameters were produced and tested. The apparent diffusion coefficient for the drug was estimated by fitting the dissolution curve of 1 mm extrudates to the equation. Then, the dissolution curves for extrudates of

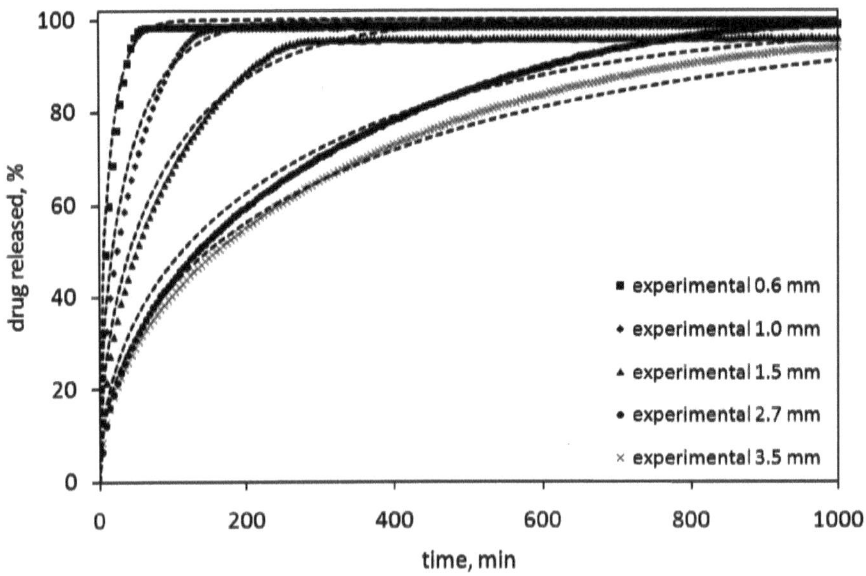

Fig. 13.22 Theoretically predicted (*dotted curves*) vs. experimental (*symbols*) dissolution curves of extrudates with different diameters. The extrudates consisted of 50 % diprophylline, 45 % glycerol tristearate and 5 % macrogol 20000. (Reproduced from Gures et al. 2012, with permission from Elsevier)

different diameters and lengths were predicted by using the obtained apparent diffusion coefficient. The experimental curves could be described well by the model (Fig. 13.22). The length was not of influence. When the assumptions are violated, the prediction is not possible. This was shown for extrudates containing crospovidone instead of Macrogol as release modifier. While Macrogol dissolves in water leaving pores in the extrudate crospovidone is swelling and changing the matrix structure itself. This results in a faster release than predicted by the model.

13.9 Pharmaceutical Applications

13.9.1 Granules

In most cases, the extrudates have a cylindrical shape. The diameter is usually in the range from 0.5 to 2 mm, but it can be as low as 0.3 mm (Michalk et al. 2008) or even 0.2 mm (Witzleb et al. 2011a). Extrudates of a more complicated shape (Fig. 13.23) can be cut into pieces of the desired length (Hasa et al. 2011). Helical extrudates with two-, three- and four-blades structure were produced and compared with respect to the modified dissolution behaviour.

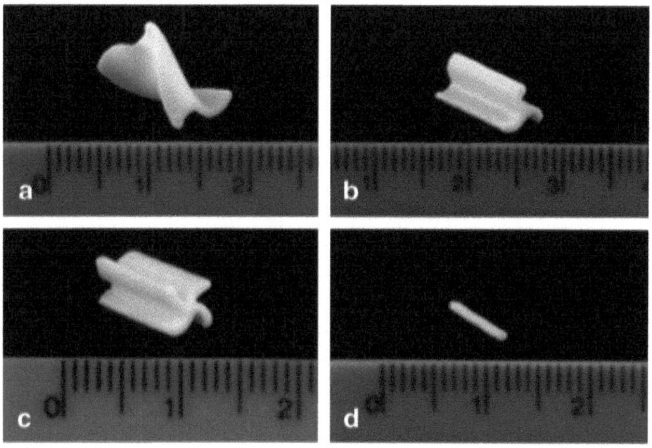

Fig. 13.23 Images of helical extrudates with two blades (**a**), three blades (**b**), four blades (**c**), and of cylindrical extrudates (**d**). (Reproduced from Hasa et al. 2011, with permission from Elsevier)

The most common application of solid lipid extrusion is the production of granules. The granules are obtained by cutting or milling the cylindrical extrudates. In some cases, the granules can be used as such, but it is possible to coat the granules (Breitkreutz et al. 2003a, b), e.g. in order to obtain gastric resistance. The granules can be filled into capsules. So far there are no studies concerning the compression of lipid granules into tablets.

A major application of the granules is the modified release of the incorporated API, especially the extended release. A second application is the taste masking of an API.

For extrudates cut into short cylinders the release rate per surface area is higher from the cylindrical surface than from the cut surface (Reitz et al. 2008). Dissolution curves from extrudates with 2.7 mm diameter are shown in Fig. 13.24. One type of surface was coated with a lipid when testing the dissolution from the other type of surface. Both types of surface show a burst effect in the release profile. The burst effect is higher for the cut surfaces of the cross sections.

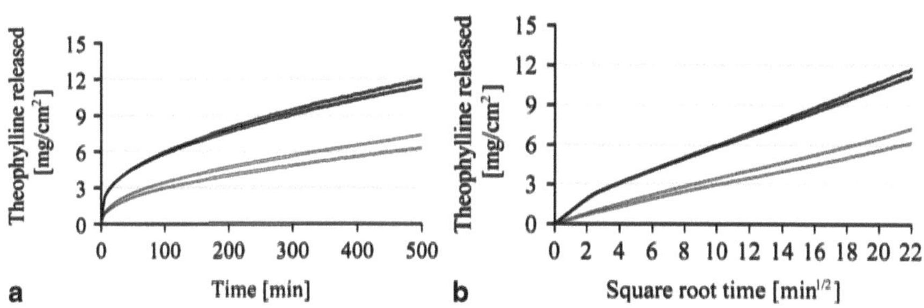

Fig. 13.24 Drug release out of defined surfaces of Dynasan 114/theophylline (50:50) extrudates (Ø 2.7 mm): *black*: cut cross-sections; *grey*: extruded surfaces; on *x*-axis: coated blank ($n = 2$). (Reproduced from Reitz et al. 2008, with permission from Elsevier)

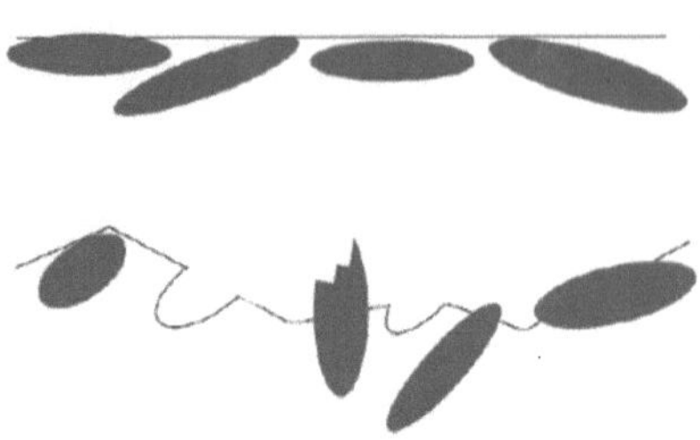

Fig. 13.25 Schematic representation of extruded surfaces and fractured/cut areas. (Reproduced from Reitz et al. 2008, with permission from Elsevier)

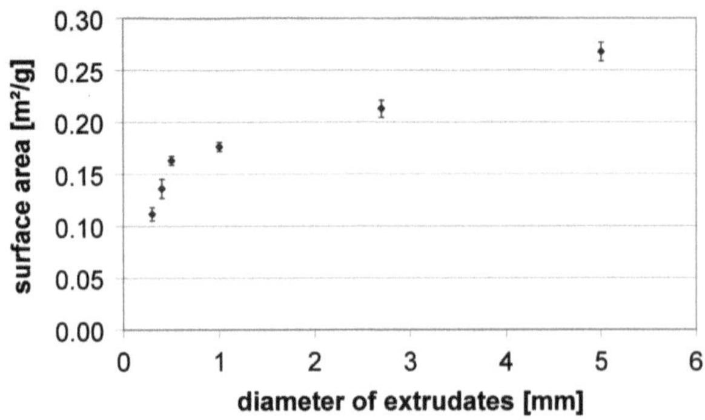

Fig. 13.26 Specific surface area of milled granules in the size fraction of 315–400 μm in dependence of the extrudate diameter. (Reproduced from Michalk et al. 2008, with permission from Elsevier)

The different dissolution behaviour can be explained with the morphology of the surfaces. The cylinder surface of the extrudate is comparably smooth. During extrusion a thin layer of the softened lipid is at the surface of the extrudate reducing the friction in the die hole. During milling or cutting, surfaces with a higher roughness are produced and more API particles are directly exposed at the surface. This explains the higher burst effect and the different dissolution rates (Fig. 13.25).

The distribution of the drug in the extrudate can be analysed by Raman mapping (Windbergs et al. 2010b; Haaser et al. 2011). Raman mapping of the extrudate cores revealed that drug particles of diverse size were dispersed in a continuous lipid phase with or without polyethylene glycol. At the surface, there was evidence of more mixing between the components. Previous characterisation by other methods suggested that the extrudate surface is covered predominantly by lipid, and the Raman mapping suggested that such a layer is in general less than a few micrometres thick. Nevertheless, the lipid layer dramatically reduced the drug dissolution rate.

These findings were used in the context of a taste masking application (Michalk et al. 2008). A fast dissolution of the API in the mouth should be avoided. It is thus intended to increase the ratio of cylinder surface area to surfaces produced during milling. In order to influence this ratio, extrudates with diameters from 0.3 to 5 mm were produced and milled to a target particle size fraction of 315–400 μm. With increasing extrudate diameter the ratio of cylinder to milled surfaces decreases. This was confirmed by Brunauer-Emmett-Teller (BET) measurements. Smaller diameters result in lower BET surface area in the given granule size fraction (Fig. 13.26).

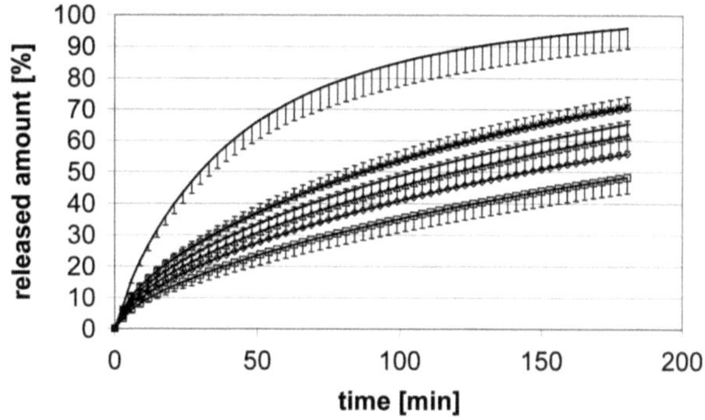

Fig. 13.27 Release profiles of the extruded batches versus time. Legend: '□' 0.3 mm, '◇' 0.4 mm, '△' 0.5 mm, '+' 1.0 mm, '○' 2.7 mm, '°—' 5.0 mm, '–' precipitate. (Reproduced from Michalk et al. 2008, with permission from Elsevier)

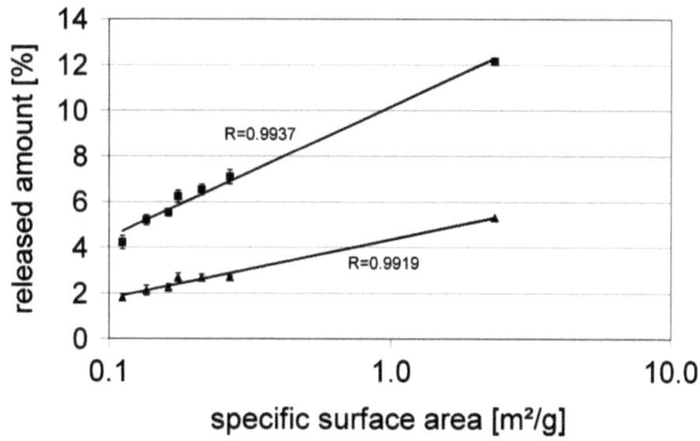

Fig. 13.28 Short-time dissolution test after 15 s (▲) and 1 min (■). (Reproduced from Michalk et al. 2008, with permission from Elsevier)

The release of the milled granules at pH 7.4 depends on the specific surface area (Fig. 13.27). Granules obtained from small diameter extrudates showed a slower dissolution. This is favourable for a taste masking application. A short time dissolution test confirmed these findings (Fig. 13.28). In the context of a taste masking application the use of extrudates with a diameter as small as 0.3 mm was preferred.

13.9.2 Pellets

Extrudates can be spheronised into pellets (Reitz and Kleinebudde 2009; Krause et al. 2009). The spheronization of lipid extrudates is more demanding than that of wet-extrudates. Depending on the type of API and the API loading it is possible to achieve extended release or immediate release pellets. Some pellets with a loading of 80 % of sodium benzoate showed > 90 % release within 40 min (Krause et al. 2009) (Fig. 13.29). Regarding the shape of the pellets, a mixture of 15 % hard fat, 2.5 % glycerol trimyristate and 2.5 % glycerol palmito stearate turned out to be preferable. Such a mixture of lipids with different melting points is a good starting point for pelletisation. The physical stability of extrudates with glycerol palmito stearate was problematic, which can be seen in a change in the dissolution profile with storage time.

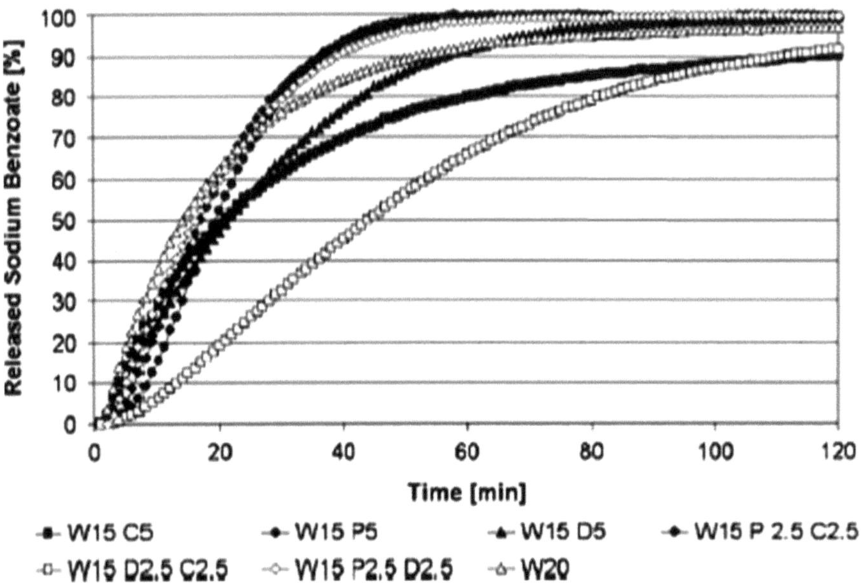

Fig. 13.29 Sodium benzoate release from lipid pellet formulations over time. Dissolution media: purified water (arithmetic mean, $n = 6$), temperature: $37 \pm 0.5\,°C$, SD < 7 %. (Reproduced from Krause et al. 2009, with permission from Elsevier)

13.9.3 Implants

Lipid extrusion was evaluated for the production of implants for the delivery of proteins (Schulze and Winter 2009). Extrudates of 0.5, 1 and 2 mm diameter with interferon alpha were produced on a mini-scale twin-screw extruder. The aim of the study was to achieve dissolution over 30 days. The 10 % of a 1:1 lyophilisate of interferon alpha with cyclodextrin was mixed with 10 or 20 % Macrogol 6000 and a lipid mixture of 1 part of a hard fat and 4 parts of glycerol tristearate. The hard fat was molten during extrusion at 40 °C. The melting of the low melting point lipid during extrusion was essential for the mass flow through the barrel. The model protein was stable at the extrusion conditions. The dissolution profile could be varied by the diameter of the extrudate and the fraction of Macrogol added. A prolonged dissolution in vitro up to 60 days could be realised.

13.10 Conclusion

Solid lipid extrusion is a comparable new application of extrusion. It offers the advantages of simple, robust and continuous processes. The dissolution profiles of the cut, milled or spheronised extrudates can be tailored in a wide range and are mechanistically understood. Solid lipid extrudates are especially useful for sustained release purposes.

One hurdle is the requirement for powdered lipids. Only few lipids are available as powder. Furthermore, the polymorphic behaviour of lipids can affect the dissolution profile during shelf life. It is required to obtain the stable modification of the lipid after extrusion. Lipids with defined and narrow composition seem to be less prone to problems. Since the behaviour during extrusion is well understood, robust products can be made.

References

Breitkreutz J, Bornhöft M, Wöll F, Kleinebudde P (2003a) Pediatric drug formulations of sodium benzoate: I. Coated granules with a hydrophilic binder. Eur J Pharm Biopharm 56:247–253

Breitkreutz J, El-Saleh F, Kiera C, Kleinebudde P, Wiedey W (2003b) Pediatric drug formulations of sodium benzoate. Eur J Pharm Biopharm 56:255–260

Chakraborty S, Shukla D, Mishra B, Singh S (2009) Lipid—An emerging platform for oral delivery of drugs with poor bioavailability. Eur J Pharm Biopharm 73:1–15

Choy YW, Khan N, Yuen KH (2005) Significance of lipid matrix aging on in vitro release and in vivo bioavailability. Int J Pharm 299:55–64

Di Pretoro G, Zema L, Gazzaniga A, Rough SL, Wilson DI (2010) Extrusion-spheronisation of highly loaded 5-ASA multiparticulate dosage forms. Int J Pharm 402:153–164

Di PG, Zema L, Gazzaniga A, Kleinebudde P (eds) (2012) Impact of needle-like crystals on wet and solid-lipid extrusion proceses. 8th World Meeting on Pharmaceutics, Biopharmaceutics and Pharmaceutical Technology, Istanbul

Gures S, Kleinebudde P (2011) Dissolution from solid lipid extrudates containing release modifiers. Int J Pharm 412:77–84

Gures S, Siepmann F, Siepmann J, Kleinebudde P (2012) Drug release from extruded solid lipid matrices: theoretical predictions and independent experiments. Eur J Pharm Biopharm 80:122–129

Haaser M, Windbergs M, McGoverin CM, Kleinebudde P, Rades T, Gordon KC et al (2011) Analysis of matrix dosage forms during dissolution testing using raman microscopy. J Pharm Sci 100:4452–4459

Hasa D, Perissutti B, Grassi M, Zacchigna M, Pagotto M, Lenaz D et al (2011) Melt extruded helical waxy matrices as a new sustained drug delivery system. Eur J Pharm Biopharm 79:592–600

Jantratid E, Janssen N, Reppas C, Dressman JB (2008) Dissolution media simulating conditions in the proximal human gastrointestinal tract: an update. Pharm Res 25:1663–1676

Krause J, Thommes M, Breitkreutz J (2009) Immediate release pellets with lipid binders obtained by solvent-free cold extrusion. Eur J Pharm Biopharm 71:138–144

Michalk A, Kanikanti VR, Hamann HJ, Kleinebudde P (2008) Controlled release of active as a consequence of the die diameter in solid lipid extrusion. J Control Release 132:35–41

Porter CJH, Charman WN (2001) In vitro assessment of oral lipid based formulations. Adv Drug Deliv Rev 50:127–147

Reitz C, Kleinebudde P (2007a) Solid lipid extrusion of sustained release dosage forms. Eur J Pharm Biopharm 67:440–448

Reitz C, Kleinebudde P (2007b) Influence of thermal and thermo-mechanical treatment Comparison of two lipids with respect to their suitability for solid lipid extrusion. J Therm Anal Calorim 89:669–673

Reitz C, Kleinebudde P (2009) Spheronization of solid lipid extrudates. Powder Technol 189: 238–244

Reitz C, Strachan C, Kleinebudde P (2008) Solid lipid extrudates as sustained-release matrices: the effect of surface structure on drug release properties. Eur J Pharm Sci 35:335–343

Roblegg E, Ulbing S, Zeissmann S, Zimmer A (2010) Development of lipophilic calcium stearate pellets using ibuprofen as model drug. Eur J Pharm Biopharm 75:56–62

Sato K (2001) Crystallization behaviour of fats and lipids – a review. Chem Eng Sci 56:2255–2265

Schmidt PC, Prochazka J (1976) Über die herstellung von retardgranulaten durch granulatformung. Pharmazeutische Industrie 38:921–926

Schulze S, Winter G (2009) Lipid extrudates as novel sustained release systems for pharmaceutical proteins. J Control Release 134:177–185

Siepmann J, Siepmann F (2012) Modeling of diffusion controlled drug delivery. J of Control Release 161:351–362

Siepmann F, Muschert S, Flament MP, Leterme P, Gayot A, Siepmann J (2006) Controlled drug release from gelucire-based matrix pellets: experiment and theory. Int J Pharm 317:136–143

Trevaskis NL, Charman WN, Porter CJH (2008) Lipid-based delivery systems and intestinal lymphatic drug transport: a mechanistic update. Adv Drug Deliv Rev 60:702–716

Windbergs M, Strachan CJ, Kleinebudde P (2009a) Influence of structural variations on drug release from lipid/polyethylene glycol matrices. European J Pharm Sci 37:555–562

Windbergs M, Strachan CJ, Kleinebudde P (2009b) Tailor-made dissolution profiles by extruded matrices based on lipid polyethylene glycol mixtures. J Control Release 137:211–216

Windbergs M, Strachan CJ, Kleinebudde P (2009c) Investigating the principles of recrystallization from glyceride melts. AAPS PharmSciTech 10:1224–1233

Windbergs M, Strachan CJ, Kleinebudde P (2009d) Understanding the solid-state behaviour of triglyceride solid lipid extrudates and its influence on dissolution. Eur J Pharm Biopharm 71: 80–87

Windbergs M, Jurna M, Offerhaus HL, Herek JL, Kleinebudde P, Strachan CJ (2009e) Chemical imaging of oral solid dosage forms and changes upon dissolution using coherent anti-Stokes Raman scattering microscopy. Anal Chem 81:2085–2091

Windbergs M, Gueres S, Strachan CJ, Kleinebudde P (2010a) Two-step solid lipid extrusion as a process to modify dissolution behavior. AAPS PharmSciTech 11(1):2–8

Windbergs M, Haaser M, McGoverin CM, Gordon KC, Kleinebudde P, Strachan CJ (2010b) Investigating the relationship between drug distribution in solid lipid matrices and dissolution behaviour using Raman spectroscopy and mapping. J Pharm Sci 99:1464–1475

Witzleb R, Kanikanti VR, Hamann HJ, Kleinebudde P (2011a) Solid lipid extrusion with small die diameters–electrostatic charging, taste masking and continuous production. Eur J Pharm Biopharm 77:170–177

Witzleb R, Kanikanti V-R, Hamann H-J, Kleinebudde P (2011b) Influence of needle-shaped drug particles on the solid lipid extrusion process. Powder Technol 207:407–413

Witzleb R, Mullertz A, Kanikanti VR, Hamann HJ, Kleinebudde P (2012) Dissolution of solid lipid extrudates in biorelevant media. Int J Pharm 422:116–124

Chapter 14
Formulation, Bioavailability, and Manufacturing Process Enhancement: Novel Applications of Melt Extrusion in Enabling Product Development

Jay P. Lakshman

Abstract Melt extrusion is increasingly popular in pharmaceutical applications due to a host of benefits it confers. However, due to the technology's relatively more recent introduction into pharmaceutical processing, challenges in the development of extrusion based formulations/processes as well as the strategies to resolve them may not be as widely understood as the benefits. This chapter illustrates strategies to resolve challenges in the development of extruder-based formulations/processes using case studies on poorly soluble drugs (a) that are thermosensitive, (b) that exhibit high, or (c) low melting points and glass transition temperatures. Additional case studies also illustrate strategies to enable (d) very high drug loads of up to 95 % for highly water-soluble drugs and (e) moisture-sensitive drugs. As these case studies represent a small part of the expanding universe of melt extrusion applications and a ready-to-use solutions that address the challenges become increasingly available, it is fair to expect that twin-screw extrusion will continue to emerge as a powerful technology solution to a number of pharmaceutical formulation/process challenges.

14.1 Introduction

Melt extrusion applications may have begun with metals such as aluminum (Bramah 1802), and developed for plastics and foods in the 1930 and 1940s but were extended to therapeutic applications such as sutures (Schneider 1972) and pharmaceutical dosage forms such as buccal films later (Tsuk 1976). Pharmaceutical applications have expanded ever since. Versatility of melt extrusion offers much to be gained from and to be applied in pharmaceutical product development and manufacturing. Building on the previous chapters of this book that have covered multiple materials for extrusion and the process for formation of dispersions/extrudates, this chapter will focus on specific novel applications of melt extrusion that directly address the weaknesses of the melt extrusion technology. The intent behind the

J. P. Lakshman (✉)
Novartis Pharmaceuticals Corporation, East Hannover, NJ, USA
e-mail: jay.lakshman@novartis.com

M. A. Repka et al. (eds.), *Melt Extrusion*,
AAPS Advances in the Pharmaceutical Sciences Series 9,
DOI 10.1007/978-1-4614-8432-5_14,

choice of application examples in this chapter is to serve as a learning experience for the formulator/scientist who is looking to apply melt extrusion for the first time as well as to serve as a point of trigger that helps expand the applications space for the more sophisticated specialist.

One of the key areas of pharmaceutical development, formulation of poorly water-soluble drugs for enhanced solubility and enhanced therapeutic efficacy, is highly challenging to work on (Bosselmann and Williams 2012). Up to 40 % of new chemical entities evaluated for pharmaceutical use are poorly water-soluble and exhibit poor oral bioavailability, resulting in increased risk for product development if not complete lack of developability or abandonment of the compound. Enhanced solubility and/or dissolution are highly desirable for such poorly soluble compounds, as any resulting enhancement of in vivo exposure can be used to elicit an appropriate pharmacological/therapeutic effect. Some of the commonly used techniques for the improvement of oral bioavailability of drugs are:

- Chemical modification
  - Salts, prodrugs
- Physical modification
  - Particle size reduction, pH adjustment, and solid dispersions
- Carrier systems/interface modification
  - Wetting agents, micelles/liposomes, emulsions, and cyclodextrins.

However, despite the simplicity of some of these techniques, practical limitations tend to preclude wider use of some of these methods. Salt formation, for example, is limited to non-neutral compounds and screening for the appropriate salt form with desirable physical attributes (e.g., crystal shape/size that aids powder flow/compaction, hygroscopicity) or chemical properties (e.g., stability) is not always successful. Even when a salt with appropriate physicochemical properties is identified, its in vivo reconversion to free form or Hydrochloride form in the gastric fluid may not be avoidable. Particle size reduction suffers from its own limitations such as high cohesiveness, agglomeration, poor wettability, poor flow, and handling difficulties. These problems tend to become exponentially difficult below 10 μm of particle size. Drugs solubilized using organic solvents, cosolvents, or amphiphilic surfactants end up being liquid formulations that are less desirable both from a commercial view point and stability perspective, while cyclodextrins tend to be highly expensive. On the other hand, despite a few disadvantages, solid dispersions provide some new opportunities to handle poorly water-soluble drug candidates.

14.1.1 Solid Dispersions for Poorly Water-Soluble Compounds

Apart from the increase in the number of poorly soluble compounds being developed, heightened interest in pharmaceutical solid dispersions in the past two to three decades has been due to the wider realization of the potential for amorphous

Table 14.1 Solubility advantage of amorphous drug over its crystalline state. (Modified from Hancock and Parks 2000)

Compound	Forms	Solubility ratios (calc)	Temperature (°C)
Indomethacin	Amorphous/γ-crystal	25–104	25
Glibenclamide	Amorphous/crystal	112–1,652	23
Glucose	Amorphous/crystal	16–53	20
Griseofulvin	Amorphous/crystal	38–441	21
Hydrocholorothiazide	Amorphous/crystal	21–113	37
Iopanoic acid	Amorphous/crystal	12–19	37
Polythiazide	Amorphous/crystal	48–455	37

drugs to provide significantly high multiples of saturation solubility during dissolution compared with other techniques (Hancock and Zografi 1997; Serajuddin 1999). Compared with their crystalline form, solubility enhancements of 10–1,600-fold have been reported for amorphous drugs as summarized in Table 14.1. Both the aspects, supersaturation achieved by amorphous solid dispersions and the ability to sustain the supersaturation that is achieved through careful selection of crystallization inhibiting polymeric materials, effectively remove the solubility/dissolution bottlenecks for the typical BCS class 2 drugs. Since BCS class 2 drugs exhibit high permeability, once the rate-limiting step of solubility/dissolution is removed, substantial enhancement of bioavailability can be realized. One of the hurdles in realizing the potential of solid dispersions, the need for further understanding the physical stability of amorphous dispersions and the potential risk for recrystallization, continues to be an area of critical need (Hancock et al. 1995; Serajuddin 1999; Bhattacharya and Suryanarayanan 2009). However, other past hurdles such as lack of suitable manufacturing technologies for manufacturing solid dispersions, have been well addressed using spray congealing (Killeen 1993), spray drying (Paudel et al. 2012), and melt extrusion (Breitenbach 2002) technologies. These technologies will be discussed in the next section in more detail.

14.1.2 Manufacture of Solid Dispersions

Comparing spray congealing, spray drying, and melt extrusion, solid dispersion manufacturing techniques that are most developed commercially, it is apparent that spray congealing is a variation of spray drying that involves cooling instead of drying and that congealing would be more suitable for thermolabile compounds. This also means the excipients that could be used in spray congealing are predominantly limited to waxy or lipidic materials with low melting/solidification points. Given that stabilization of amorphous drugs generally depends on materials with high glass transition temperatures (Tg) for forming high-Tg solid dispersions, waxy/lipidic materials with low melting points, and low Tg have significantly less applicability in formulating amorphous drugs into solid dispersions. Spray drying on the other hand does not suffer from the same disadvantage, as it can use solvents that solubilize hydrophilic materials/polymers with high Tg to form a solution that can be spray dried. In general, spray dried or not, solvent-based methods for generating solid dispersions suffer from similar advantages and disadvantages. Advantages of solvent-based methods in general and spray drying in particular include availability of high-throughput systems for screening solvents or solvent mixtures for solubility and the high-throughput technique's amenability to use small amounts of drug or sample preparation techniques for assessing feasibility. Although this means laboratory work is simple and efficient, disadvantages are much more significant for routine manufacturing in large scale. Disadvantages of solvent-based methods include:

- Lack of suitable solvents that are able to dissolve both the active pharmaceutical ingredient (API) and the polymer.
- Complex solvent residue removal process, both from in-process material or from the end product.
- Need for handling large volumes of solvents and the associated occupational hazards and exposure.
- Safety issues relevant to explosion hazard.
- Solvent disposal/environmental issues.
- Overall high cost of acquiring solvents, handling solvents, and disposing solvents.
- Highly variable incidence of phase separation during evaporation of the solution containing the drug and polymer(s).
- Scale-up risk from difficulties in reproducing quality of solid dispersion across multiple scales.

Despite the ability of high-throughput methods for screening for suitable solvents, finding solvent(s) that can solubilize both the hydrophobic drug and the hydrophilic, high-Tg polymers is a challenge. Even after the identification of suitable solvent(s), industrial use of organic solvent comes with its own set of safety and environmental hazards apart from the need to minimize residues in the final dosage form to below acceptable limits for human consumption. Consequently, spray drying tends to be the most expensive of the three processes being compared. Hot-melt extrusion on the other hand avoids the need for use of solvents and is well capable of handling high-Tg materials that are typically required for stabilizing amorphous drugs in a solid dispersion formulation. Though use of high temperatures continues to be a limiting factor in expanding applications of melt extrusion, melt extrusion's ability to substantially minimize the residence time (Gao et al. 2000) to a few minutes or a fraction of a minute, over which materials are exposed to high process temperatures helps diminish the severity of the limitation. Versatility of melt extrusion that stems from its ability to combine multiple unit operations in a single step (Chap. 2) also makes it economically more desirable. Considering the overall benefits and disadvantages, it is straightforward to see that melt extrusion is commercially preferred over other methods for the manufacturing of solid dispersions. Throughout this chapter, mention of extrusion or extruder in general specifically relates to the use of corotating twin-screw extruders (TSE). Compared with the other two methods for manufacture of solid dispersion, spray congealing and spray drying that have been

Table 14.2 Advantages of melt extrusion over spray drying/spray congealing and the general challenges for melt extrusion process

Benefits	Challenges
Short residence time	Thermolabile APIs
High-mixing efficiency	Very high-melting APIs
Small equipment, highly flexible	Very low-melting APIs
multiunit operations, well-controlled	Number of patents
Good scalability	
High volume/low production cost	

API Active Pharmaceutical Ingredient

discussed, melt extrusion's advantages can be summarized as shown in Table 14.2. Challenges for melt extrusion in general are also listed in the same table.

14.2 Overcoming Challenges in Melt Extrusion

Although the advantages of melt extrusion are substantial and have been discussed elsewhere in substantial detail (Chap. 1 and Brietenbach 2002), it is worthwhile to discuss the disadvantages outlined in Table 14.2 and how best these can be addressed. Taking the first disadvantage outlined, thermolabile drugs, it is a fairly common notion that drug substances can be destroyed due to thermal exposure during hot-melt extrusion process and that this is especially true for thermally sensitive drug substances. Much of this perception tends to arise from the conventional thinking and the current way thermal sensitivity of API is assessed. Drug substances are typically characterized for thermal sensitivity in the solution state where the kinetics and dynamics are very different from that of the solid state. Even when the characterization is performed in the solid state, for example, during studies evaluating the differences in degradation of an API between 25 °C and 40 °C/75 % RH or 50 °C over 4–8 weeks, thermal sensitivity is over estimated. It is hard to extrapolate these accelerated stability data meant for assessing room temperature shelf life over days, weeks, or months to shorter timescales of minutes or seconds relevant to understanding the sensitivity of an API to thermal exposure during melt extrusion at high temperatures. Exposures, for example, of 130 °C over 3 min can provide a very different picture and even if there is substantial degradation at these conditions, there are good strategies to minimize the degradation by optimizing the hot-melt extrusion process parameters.

Considering the second disadvantage, very high-melting drugs, it can be conceived that at the high temperature where the crystalline drug can melt and become miscible with the polymer, the polymer may be degraded or that the drug itself may get degraded in the process of being molecularly dispersed in the hydrophilic polymer matrix. A subset of this challenge is the case of a drug degrading before, during, or immediately after melting. The issue for very high-melting drugs is how does one disrupt crystalline structure and get to amorphous material so as to form an amorphous solid dispersion if one cannot melt the drug? Is the only option then to use a solvent-based solid dispersion manufacturing technology? A few strategies to address this challenge using a model drug will be discussed shortly in the next section.

The third disadvantage, very low-melting API, is really much more of a formulation challenge. The bigger challenge in this case is to generate a solid dispersion with a single high Tg (i.e., full miscibility of API and polymer) in order to minimize the risk of physical instability and the lesser challenge would be in optimizing the process or manufacturing of the solid dispersion. That is, the issue is to choose polymer(s) with high enough of a Tg to form a more stable solid dispersion rather than to optimize the processing conditions for hot-melt extrusion. This aspect will be considered in more detail as the last case study in the next section.

The fourth challenge listed, number of patents, is perhaps the most difficult of challenges for the scientist with minimal training in the legal aspects to overcome but is a subject matter for more detailed discussion elsewhere and is covered only superficially here. A large number of patents pepper the area of hot-melt extrusion, especially covering manufacture or processing of solid dispersions. It is difficult for a new practitioner to familiarize him/herself with the patent literature and its legal implications on his/her work without appropriate guidance. In general, the difficulties in working in the area of hot-melt extrusion stem from the fact that hot-melt extrusion is a newer area in pharmaceutical development and pharmaceutical manufacturing, compared with perhaps other technologies such as high-shear granulation or extrusion-marumerization. Consequently, a number of strategies or techniques in hot-melt extrusion are still protected as intellectual property by the early adopters of the technology. Unlike the situation with the more conventional pharmaceutical manufacturing technologies where a serious development issue or processing challenge can be dealt with using a wealth of techniques and strategies that are well known from expired patents and/or is readily available in the public domain, much of the know-how in hot-melt extrusion is only now slowly becoming available in the public domain. As some of the early key patents on strategies to deal with challenging development or manufacturing difficulties start to expire and as some of the proprietary knowledge slowly starts to materialize in presentations and publications, it would be reasonable to expect that it would become easier for new practitioners to work in the area of hot-melt extrusion and that application of the technology in a wider range of pharmaceutical development projects and product manufacturing will expand significantly.

14.3 Solid Dispersions for Bioavailability Enhancement

A number of excellent review articles (Serajuddin 1999; Leuner and Dressman 2000; Janssens 2009; Newman et al. 2012) cover this area well and it is not in the scope of this chapter to recap this information. Specific challenges in the application of melt extrusion to the manufacture of solid dispersions and strategies to overcome them are illustrated using case studies covered in this section and the reader is referred to the articles cited for more general information on bioavailability enhancement.

Table 14.3 Physicochemical properties of drug A

Physicochemical parameter	Approximate value
Molecular weight	810 g/mole
cLog P	7.5
Melting point	150 °C
Aqueous solubility	0.1 mg/ml
Solubility in DMSO	25 mg/ml
Glass transition temperature	75 °C

DMSO dimethyl sulfoxide

14.3.1 Application to Thermolabile Drugs

A drug presumed to be thermolabile based on data typically available during early development such as degradation rate of drug in solution at room temperature versus higher temperatures of 50 or 80 °C often ends up being not thermolabile in the solid state. Before such a drug is considered to be thermolabile and unsuitable for melt extrusion or unsuitable for other melt-based solid dispersion forming techniques such as spray congealing, it is worthwhile to study degradation kinetics in the solid state at multiple temperatures. The more rational route to assess the thermal sensitivity of the drug as it relates to melt extrusion would be to use differential scanning calorimetry (DSC) or Thermogravimetric analysis (TGA). DSC/TGA methods that could mimic the potential melt extrusion process' temperature-time profile could provide excellent guidance on thermal liability of the process for a given drug. When the potential temperature-time profile process is unknown, a reasonable DSC/TGA condition to use could involve approximately 10 °C above the phase transition temperature (melting point or Tg) of the drug with a holding duration of about 3 min or twice the approximate expected residence time. Data complimentary to DSC or TGA could also be obtained by subjecting the thermally scanned sample to HPLC or LC-MS analysis. Repeating the DSC or TGA under a nitrogen blanket instead of air or vice versa can indicate the extent of improvement that would be possible to gain through the use of an inert gas blanket.

A case study involving a heat-sensitive poorly soluble immunomodulator drug, A, illustrates the potential strategies that can be put to practice in situations where the drug is thermolabile. Apart from heat, degradation of this drug A is also known to be accelerated by light, oxygen, and water. Its physicochemical properties are summarized in Table 14.3.

Due to solubility-limited absorption exhibited by the drug, formulation into a solid dispersion was considered and due to the drug's tendency to crystallize easily, after extensive screening studies using DSC, hydroxypropylmethyl cellulose (HPMC) was chosen to be used as the polymeric matrix based on its compatibility and miscibility with the drug. A drug load of approximately 20 % was used during the first feasibility trial using hot-melt extrusion. Processing temperature of about 165 °C, clearly above the melting point of the drug 150 °C was selected to ensure complete melting of the drug as well as to ensure adequate mixing that would allow molecular dispersion of the drug into the viscous molten HPMC. The torque values of about 235 Ncm were

Table 14.4 Drug degradation observed for a thermolabile drug A upon melt extrusion (3.5 min at 165 °C and 235 Ncm)

Sample	Assay (%)	Total impurity (%)
Extrudate 1 (start)	79.8	20.3
Extrudate 4 (middle)	75.8	24.8
Extrudate 7 (end)	68.0	32.1

Table 14.5 Drug degradation seen for a thermolabile drug A upon melt extrusion (1 min at 130 °C and 250 Ncm) under nitrogen while using a processing aid

Sample	Assay (%)	Total impurity (%)
Extrudate 1 (start)	98.4	0.6
Extrudate 4 (middle)	102.4	0.7
Extrudate 7 (end)	99.7	0.6

observed and an approximate residence time of 3.5 min was recorded. The extrudates obtained were clear and uniform without any visual evidence (opacity tends to reflect lack of miscibility or two phases) of crystalline drug in the extrudates or any apparent color change. HPLC analysis of the extrudate samples obtained, however, showed a different picture that is more consistent with the nature of the drug that is prone to degradation from the influence of light, heat, oxygen, and water. Degradation levels as measured using HPLC reached levels of about 20–32 % with a corresponding loss of assay as summarized in Table 14.4. A clear pattern of increasing levels of degradation was also seen over time despite the short total process run time of about 15 min indicating escalating thermal exposure.

Upon observing high levels of degradation of about 20–30 %, such a feasibility study could be considered a failure and a reasonable justification to consider alternate methods that do not involve thermal stress. However, a few modifications were capable of bringing down the degradation levels to within typically acceptable levels. In this particular case, three modifications were made that worked together to make an initial feasibility failure ultimately into a successful acceptable process. First, it was decided to use a plasticizer as a processing aid to reduce the temperature at which extrusion was performed. Use of around 2 % of a polyol was adequate to reduce the processing temperature from about 165 to 130 °C and still allow formation of a solid dispersion without residual crystalline drug. Lack of crystallinity in the solid dispersion could be confirmed by powder X-ray diffraction as well as from the formation of clear extrudates without any visual evidence of white distributed crystalline material in them. To further reduce thermal degradation, extrusion was performed under a blanket of inert gas. Specifically, argon blanket was used to take advantage of argon's high density that supports downward flow and effective displacement of air. Finally, the process was optimized to reduce the residence time from about 3.5 min to just under a minute by increasing the feed rate and rpm of the extruder screws. The torque remained nearly unchanged at around 250 Ncm, but the thermal degradation had dropped down to below 1 %. Granted, since the availability of drug substance was very limited, as is often the case during early development, these changes could not be systematically evaluated one at a time and only the combined impact of all the three changes could be evaluated and is summarized as such in Table 14.5.

Notable observations from the study were, total degradation levels fell to 0.7 % or lower from about 20–30 % before process adaptations were made. The assay values rose from 68–80 % values to 98–102 %. Also, the trend of increasing degradation over time from the beginning to the end of the process run that was seen earlier was no longer seen indicating better thermal control. This case study illustrates the point that with an appropriate choice of a small amount of plasticizer, it is possible to convert crystalline drug into amorphous state below its melting point through the interplay between solubilization, viscosity reduction, and high shear. In selecting the appropriate plasticizer to use, multiple materials would need to be screened for their ability to reduce the viscosity of the polymeric material used in the solid dispersion. Viscosity reduction studies using binary mixtures of the polymer and multiple plasticizing materials, as well as multiple concentrations of plasticizers are much easier to perform using a rheometer with a thermally jacketed sample chamber and the capability to perform thermal scans. Viscosity-temperature profiles beyond the temperatures immediately relevant to melt extrusion conditions can open up further options to lower thermal exposure especially where shear-thinning/thixotropic materials are involved. Performing the rheometric study under a blanket of inert gas would also provide additional insight as well as mimic the conditions of melt extrusion process better. The ability of a plasticizer to dissolve the drug may be studied using a binary mixture separately at elevated temperatures as well, to allow better understanding of the underlying mechanisms of the interplay between solubility and viscosity reduction. Based on the rheometric studies of the polymer-plasticizer and the solubility studies of the drug-plasticizer, a final confirmatory rheometric study using the drug-plasticizen-polymer ternary system could open up further avenues to reduce the thermal exposure. A small factorial or half-factorial experimental study design can allow optimization of the formulation composition both qualitatively and quantitatively in a short duration. Overall, the case study elaborated here bears the point that seemingly insurmountable issues of thermal sensitivity of a drug could be addressed through systematic characterization and ultimately by a judicious combination of plasticizer use, inert gas blanketing, and reduction of residence time.

A less dramatic reduction in drug degradation levels but a necessary fine tuning was reported recently (Ghosh et al. 2012) and was achieved using full-scale process equipment and a simple statistical experimental design instead of rheometric characterization. A separate plasticizer was not used in the study, but it illustrates that degradation profile of a thermosensitive drug substance can be controlled simply by proper design of extrusion screw assembly and by optimization of screw rotations per minute, process temperature, and feed rate during development and scale-up of a pharmaceutical product. A different study (Crowley et al. 2002) focused on polymer degradation, instead of drug degradation, during melt extrusion at different temperatures as well as during storage of dosage form at multiple temperatures. The chemical stability of polyethyleneoxide (PEO) was found to depend on both the storage and processing temperature, and the molecular weight of the polymer. Lower molecular weight PEO MW = 100,000 da (PEO 100K) was demonstrated to be a suitable processing aid for the high-viscosity PEO 1M (Molecular weight = 1000,000 Da). Incorporation of PEO 100K as a plasticizing agent reduced the viscosity of and the degradation of PEO 1M. Vitamin E, Vitamin E Succinate, and Vitamin E TPGS were also found to be suitable stabilizers for PEO during extrusion processing. Binary mixtures of PEO 1M and Vitamin E Succinate or Vitamin E TPGS exhibited only the melting point of PEO 1M post-extrusion indicating that they were solubilized and molecularly dispersed in PEO 1M during extrusion. As plasticizers, they also suppressed the melting point of PEO 1M by about 10 °C when used at 15 % w/w indicating the ability to allow processing at a lower temperature.

14.3.2 Application in High-Melting Drug

Contrasting the difficulties of defining "thermolabile," a property that is context dependent and so has to be accompanied with descriptions of the temperature range where it is labile or the time frame over which it is labile at a given temperature or under what conditions such as presence of oxygen, moisture, or other materials and process conditions it is labile, defining a high-melting drug is relatively simple. Dealing with a high-melting drug and to process it into a solid dispersion using melt extrusion, however, is not. Poorly water-soluble drug substances that exhibit high melting points are difficult to successfully process by fusion-based techniques. In addition, materials with high melting points also tend to have high Tg (Kanno 1981) and consequently, tend to offer comparatively better opportunities to formulate them in their amorphous form without the need for combining them with polymeric matrices into solid dispersions. If stability of the drug and that of the polymer is not an issue above the melting point of the drug, and they are compatible with each other, melt extrusion can be conducted at temperatures above the melting point of the drug and a solid dispersion can be formed to provide additional assurance regarding physical stability. Extrusion processing temperatures below the melting point of the drug may be possible using adequate shear and/or miscibility and if necessary, long residence time (Liu et al. 2010, Hughey 2011). Limitations do exist, however, in the level of shear that can be generated or in identifying a material that can help improve miscibility or issues that arise from high thermal exposure and degradation from the long residence time. If a solid dispersion does need to be formed for a high-melting drug due to its high tendency for recrystallization, the choice of polymers available are typically smaller and the need for thermal stability of the polymer becomes more important.

A case study involving high-melting antineuropathic agent, B, can illustrate the strategy adapted to form a solid dispersion of a high-melting drug. The drug is poorly soluble and had a melting point of greater than 180 °C and, as expected, a correspondingly high Tg of about 93 °C. Wettability and solubility of the drug were both extremely poor. Recrystallization tendency of the drug was also high. A solid dispersion was considered necessary to enhance solubility/bioavailability of the drug and to ensure physical stability of the amorphous drug. Physical chemical properties of drug B are summarized in Table 14.6.

Table 14.6 Physicochemical properties of B

Physicochemical parameter	Approximate value
Molecular weight	570 g/mole
cLog P	4.9
pKa	3 and 10
Melting point	180 °C
Aqueous solubility	0.001 mg/ml
Solubility at pH 1	0.03 mg/ml
Glass transition temperature	93 °C

A number of polymers were screened for miscibility with the drug using DSC and Hot-Stage Microscopy. Polyvinylpyrrolidone K30 (PVP K30) was chosen as a suitable polymeric matrix material for forming a solid dispersion with the drug, as zero or only slight miscibility was seen between the drug and other polymers evaluated. A single-phase system was formed between the drug and PVP, as confirmed by the observation of a single Tg in DSC thermogram for a 1:20 drug polymer blend obtained at a scan rate of about 10 °C/min. The drug and polymer were then identified to be miscible between all the concentrations of 15, 20, 30, and 40 % by preparing the solid dispersions using a solvent evaporation method and characterizing them again for single Tg. It was possible to achieve good Tg values of about 127 to 140 °C for the PVP-based solid dispersion depending on the drug load. Over the drug loads of 15–40 %, Tg of the solid dispersion increased slightly with the increasing concentration of drug as shown in Fig. 14.1.

Based on these positive feasibility studies, solid dispersions needed to be manufactured using a melt extruder for further studies. However, given the process parameter

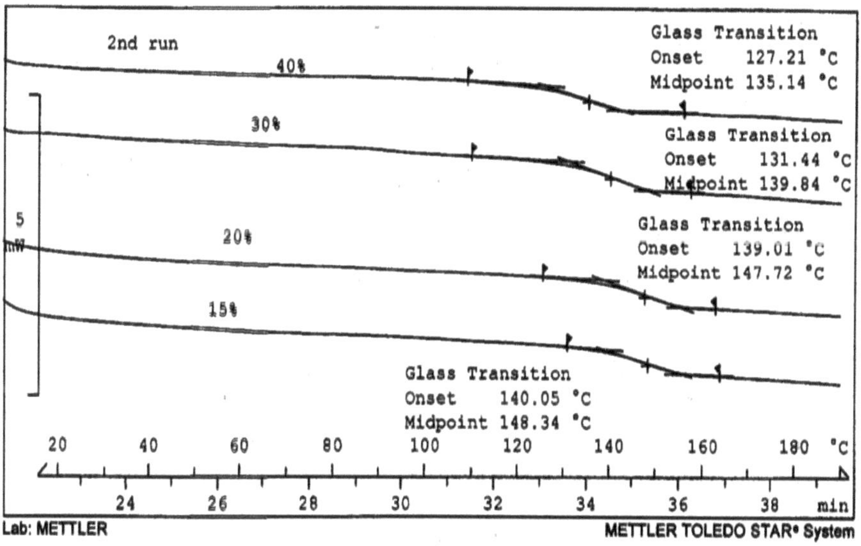

Fig. 14.1 Differential scanning calorimetry of solid dispersions at multiple drug loads, showing raising glass transition temperature

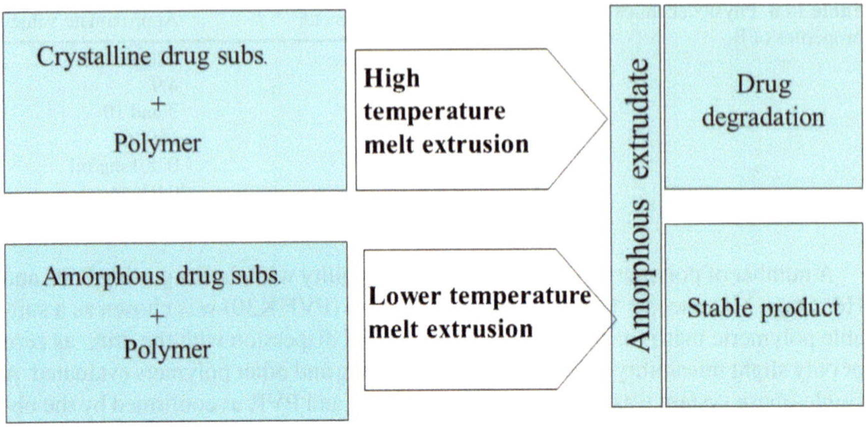

Fig. 14.2 Comparison of extrusion process using crystalline and amorphous drug as a starting material

conditions and the materials involved, drug degradation was an issue. To avoid thermal degradation issues a novel strategy was adapted (Lakshman et al. 2008). Instead of starting with crystalline drug, as it is conventionally done in the manufacture of solid dispersions, this strategy calls for using amorphous drug substance as a starting material for hot-melt extrusion. This unconventional beginning affords significant benefits as extrusion could be performed just above the Tg of the drug, a temperature much lower than the temperature conventional method requires—a processing temperature that is above the melting point of the drug. The strategy is illustrated in Fig. 14.2. Figure 14.2 illustrates that the crystalline drug substance requires a high processing temperature that could lead to drug degradation issues but if the process employed amorphous drug substance as a starting material, lower processing temperatures become a possibility. Drug degradation could thus be reduced or avoided.

Among the benefits that can be reaped from this novel strategy, the primary one is the ability of the amorphous drug to soften significantly as it transitions from glassy state to rubbery state at or above its Tg. Allowing the drug to be dispersed and dissolved into the polymeric matrix material above the Tg where the drug is in a "rubbery" state is an effective way to achieve molecular dispersion, especially to take advantage of the comparatively lower temperature needed for melt extrusion. Starting with crystalline drug to generate an amorphous solid dispersion, on the other hand, requires the drug to be heated through its melting point that is substantially higher than its Tg. Granted, drug in the molten liquid (above its melting point) state has a lower viscosity compared with "rubbery" state of the drug, and is easier to disperse compared with softened amorphous material. However, given the high-shear capabilities of the hot-melt extruder and the substantial reduction in viscosity/softening that is already achieved above Tg, shear in the extrusion is easily adequate to achieve a molecularly dispersed amorphous solid dispersion as long as miscibility is not a specific constraint.

One of the benefits of this strategy is also that the drug substance itself starts to act as a plasticizer. Since the glass transition temperature of the amorphous drug is in most cases lower than that of the high-Tg polymer used in solid dispersion formulations, the drug substance gains the role of plasticizer during melt extrusion by melting/softening ahead of the polymer. With the drug acting as a plasticizer of the polymeric matrix, it is often easier to preclude or minimize the need for additional processing aids as well as avoid or reduce the physical stability disadvantages inherently associated with low molecular weight, low-melting plasticizers. Further details regarding the strategy and its benefits are also described in more detail in a patent filing (WO2004069138).

Returning to the preparation of solid dispersion of B using melt extrusion, increasing the drug load from 20 to 30 % helped reduce the overall torque seen in the process as well as the total degradation levels, illustrating the effect of drug acting as a plasticizer. Supplementing this plasticizing effect by using an additional plasticizer helped much more, even in small amounts.

As can be seen in Table 14.7, addition of small amounts (5 %, 10 %) of plasticizer caused torque levels to drop by an order of magnitude (from 200–500 to 27–43 Ncm) at 20 % drug load and nearly as substantial a drop was seen at 30 % drug load. Total amount of drug degradation also decreased by an order of magnitude, from 15 to 1.5 %, at both the drug loads studied upon introduction of small amounts of plasticizer. A drop in the solid dispersion's Tg is a consequence of using plasticizer in the formulation as well, resulting in potentially a slight increase in the risk for recrystallization. However, this was acceptable in this case as the Tg of 20 and 30 % solid dispersions were quiet high in the range of 130–140 °C before use of the plasticizer.

Subsequent to bench-scale studies, solid dispersions of B were manufactured at multiple drug loads using a 18 mm Thermoprism hot-melt extruder at a process temperature of 175 °C utilizing the plasticizer concentrations identified in Table 14.7. The extruded solid dispersions were subsequently milled using an oscillator and screened through a 0.5 mm screen. The milled solid dispersion was then filled in size 0 capsules at a dose of 50 mg as a powder blend containing 100 mg of Avicel and 56 mg of Poloxamer per capsule. The solid dispersions were evaluated for dissolution under nonsink conditions using USP apparatus paddles at 50 rpm and 1,000 mL of pH 2 buffer at 37 °C. Solid dispersions showed a clear increase in dissolution compared with the crystalline drug control. Among the formulations, the ones with 20 or 30 % drug load tended to achieve higher final concentrations at a faster rate. Whereas at 40 % drug load, the dissolution profile was significantly slower and it reached a plateau concentration that is significantly lower than that reached by 20 and 30 % drug-loaded solid dispersions as illustrated in Fig. 14.3. However, all the solid dispersions were clearly superior in dissolution compared with the control sample containing the crystalline drug substance.

Based on this promising in vitro dissolution data, additional solid dispersion formulations were prepared and subject to in vivo evaluation in fasted beagle dogs

Table 14.7 Effect of drug load and plasticizer level on the extrusion processing conditions of the solid dispersion of drug B

Drug load, % w/w	20 %	20 %	30 %	30 %
Plasticizer level, % w/w	0 %	10 %	0 %	5 %
process temperature (°C)	178–184	180	180–184	178
Torque (Ncm)	200–500	27–43	130–450	51–70
Total degradation product level, % w/w	15 %	< 1.5 %	10 %	1.5 %

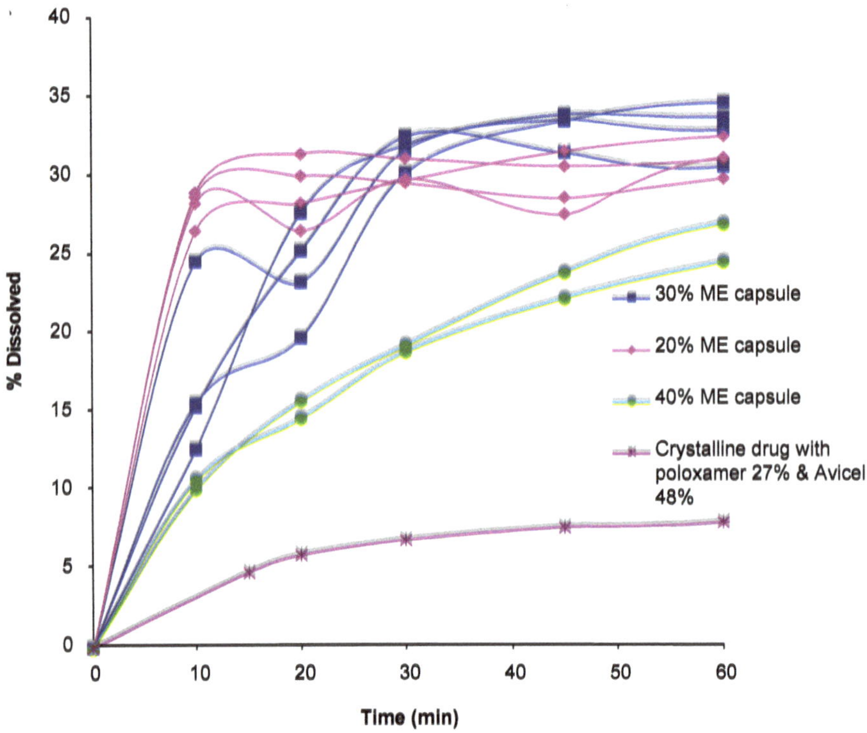

Fig. 14.3 In vitro dissolution profiles for 50 mg capsules containing solid dispersions at 20, 30, and 40 % drug loads compared against the control capsule containing crystalline drug B in a similar formulation

($n = 4$) using a randomized cross-over pharmacokinetic study. Two of the formulations containing melt extruded PVP K30-based solid dispersions with drug loads of 20 % (Formulation A) and 30 % (Formulation B) were tested for relative bioavailability using fasted beagle dog as the animal model. A separate formulation containing triturates of the crystalline drug substance along with poloxamer 188 (Formulation C) served as the control. No attempt was made to determine absolute bioavailability since an acceptable intravenous formulation at the relevant dose could not be formulated due to poor solubility of the drug. As can be seen from Fig. 14.4, the plasma concentration profiles achieved by the 20 % drug load formulation A, as characterized by earlier C_{max} and larger AUC, was better than that for the 30 % drug load formulation B and both of these formulations were clearly better than the crystalline control formulation C. The variability for the solid dispersions also appeared to be less than that for the crystalline control formulation.

Subsequently, a separate attempt was made to determine whether methods of preparing amorphous solid dispersions other than melt extrusion could provide comparable bioavailability or if they would have different impact on the bioavailability of the drug. Commonly used techniques such as spray-drying, spray-granulation, and solvent evaporation by rotary evaporator were evaluated. The same external phase composition was used in preparing the solid dispersion-filled capsule formulation in each case. A melt extruded formulation using PVP-K30 was also developed to understand the effect of the higher polymer viscosity/molecular weight on dissolution and bioavailability of the drug. Summary of pharmacokinetic results from this study in dogs is shown in Fig. 14.5. The results show that the 20 % drug-loaded melt extrudate using PVP-K30 provided a seven-fold greater bioavailability than the control formulation made of crystalline drug triturated with poloxamer 188. The solvent-evaporated solid dispersion made using PVP-K30 and processed using a rotary evaporator provided the second best improvement in bioavailability. Among the other formulations, PVP-K30-based spray-dried, spray-granulated, and the PVP-K90-based solid dispersion produced relatively lower bioavailability enhancement of about 2–3-fold. The T_{max} values for these three formulations and the crystalline control sample were also slightly delayed compared with the PVP K30-based Rotovap and extrusion formulations, indicating slower dissolution rates from these formulations. Solid dispersion based on PVP-K30 clearly outperformed the solid dispersion based on PVP-K90. Use of the surfactant sodium lauryl sulfate (SLS) with PVP-K90 improved the bioavailability significantly but not farther than that of PVP-K30 without the surfactant.

To elucidate the mechanisms that underlie the differences in dog bioavailability among the different formulations, further microscopic evaluation of the solid dispersions was performed. Based on observations under a polarizing filter, it was inferred that spray-dried and spray-granulated formulations were not intimately or molecularly dispersed as the melt extrudates that were exposed to high-shear mixing. The rate of dispersion and the degree of supersaturation of drug substance in the gastrointestinal fluids could also be different for samples prepared by the different methods.

Having mentioned hot-melt extrusion's ability to avoid solvents as a key advantage in the beginning of the chapter, it behooves the author to discuss why this novel strategy for high-melting drugs call for use of solvent-based method to generate amorphous drug substance prior to melt extrusion. This may be less of a problem than it seems at first glance. In implementing the strategy during the development of the compound B under discussion, it became obvious that generating amorphous drug substance using a Rotovap apparatus for solvent evaporation was a bottleneck and a tedious process. Considering that PVP and the water associated with PVP were at least in part to blame for the drug degradation during melt extrusion, an

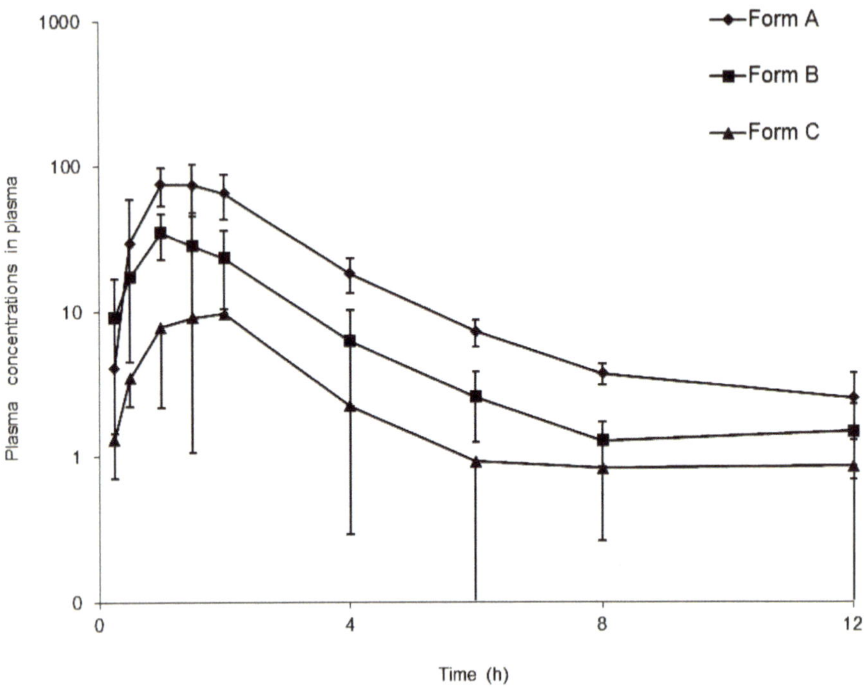

Fig. 14.4 Mean pharmacokinetic profiles obtained for solid dispersion formulations A, B, and C in fasted beagle dogs ($n = 4$) using a cross-over study design

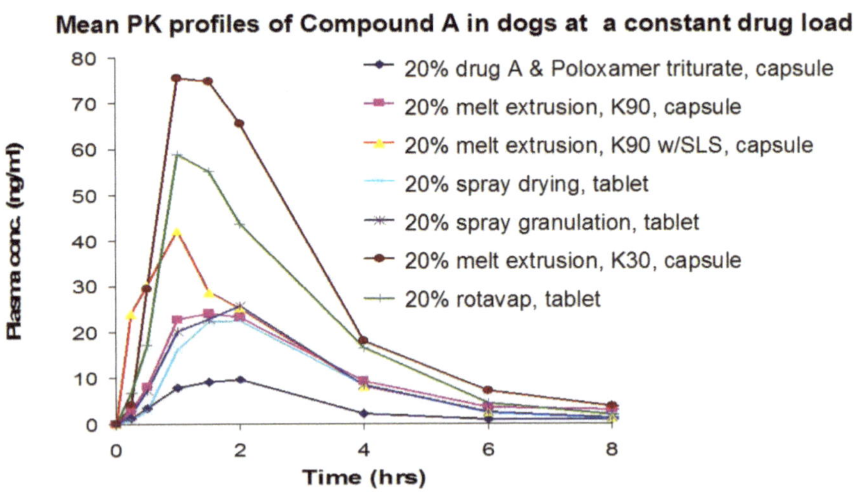

Fig. 14.5 Mean pharmacokinetic profiles of solid dispersion formulations of drug B in dogs ($n = 4$) at a constant drug load prepared using different manufacturing technologies. Data pooled from two dog studies

attempt was made to generate amorphous drug substance using the melt extruder by quench cooling neat molten API. Operating at a barrel temperature of about 3 °C above the melting point of the drug, using only transfer elements in the screw setup, it was possible to melt the drug and rapidly cool the material to − 78 °C using dry ice in small scale. Large amounts of amorphous API could thus be generated rapidly to perform further studies and the amorphous API thus generated had barely measureable levels of drug degradation. Amorphous API then could be used as a starting material and processed in presence of PVP in a melt extruder at a lower process temperature into amorphous solid dispersion. The resultant amorphous solid dispersion exhibited lower degradation level ($< 0.5\,\%$) compared to the original process that started out with the crystalline drug substance. It is unusual that melt processing the drug substance twice, once to melt and cool the crystalline drug to form the amorphous drug substance and a second time with PVP to form a solid dispersion, results in a lower degradation than melt processing the crystalline drug once with PVP. This may not be the case for drugs where compatibility is a non-issue. In this case, despite the need for a second lower temperature processing, with APIs where the interaction between polymeric material and drug substance is substantial at high temperatures, amorphous conversion of the drug without the presence of polymeric material provided an opportunity to reduce drug degradation.

Another practical option for generating amorphous drug is to obtain the drug substance at the end of the chemical synthesis process, prior to the usual final crystallization step, in its amorphous form. During drug substance manufacturing process development, typically significant effort is spent on identifying the conditions and optimizing the process to get appropriate crystal form and habit as a last step. This last step of drug substance manufacturing may be avoided or diverted to generate amorphous drug substance using a different set of solvents and conditions for rapid precipitation instead of slow crystallization. Understandably, this option could work only if the amorphous drug substance is adequately stable for a practical duration of time such as a few weeks, unless drug substance manufacturing and drug product manufacturing can be synchronized as in the case of a continuous manufacturing process.

14.3.3 Application in Low-Melting/Low-Tg Drug

Drugs with low melting points tend to have low-Tg values (Kanno 1981) and consequently exhibit higher risk of recrystallization due to their higher molecular mobility (Yoshida 1995). An often quoted rule of thumb is that solid dispersions stored at 50 °C below their Tg are unlikely to pose risk for recrystallization (Hancock et al. 1995). Considering room temperature storage conditions, the baseline Tg for solid dispersions should be 75 °C or higher. This target Tg is significantly harder to achieve for low-Tg drugs, as the formulation composition choices become limited to high-Tg polymers as well as large amounts of the polymer (i.e., lower drug load). A good case study in the development of such a low-Tg compound is drug C, an antihyperalgestic

Table 14.8 Physicochemical properties of drug C

Physicochemical parameter	Approximate value
Molecular weight	370 g/mole
cLog P	7
pKa	No ionizable groups
Melting point	< 80 °C
Aqueous solubility	0.001 mg/ml
Glass transition temperature	1 °C

agent that has a melting point of about 80 °C and a Tg of about 1 °C. The drug is neutral and also highly lipophilic as indicated by the high-log *P* value of about 7. Amorphous form of this drug was a sticky fluid at room temperature with a high risk for recrystallization at room temperature storage or under refrigeration. The physical chemical properties of drug C are summarized in Table 14.8.

Based on extensive screening studies using DSC and hot stage microscopy to find a suitable polymeric matrix that would be miscible with the API, PVP K30 was chosen for the final formulation. However, it was not possible to get full miscibility even with this polymer despite using low drug loads of about 10 % w/w. Immiscibility was likely due to the high lipophilicity of the drug. Polymer blends with more lipophilic polymers were evaluated and a copolymer evaluated in the study that incorporated a small amount of a lipophilic polymeric material appeared to help achieve full miscibility as characterized by a single Tg at 10 % drug load (Fig. 14.6). Apart from helping achieve a single Tg, moisture sorption of the solid dispersion was also reduced significantly by the addition of a small quantity of hydrophobic polymer. The net result was good stabilization of a challenging drug without issues. The DSC thermograms obtained for the PVP-based formulation (left) and PVP copolymer-based (right) formulations are shown in Fig. 14.6.

It can be seen from the DSC thermogram depicted in the left panel of the figure that the solid dispersion exhibited three phases of materials. The first one with a Tg of 1 °C is the amorphous drug or the mostly drug-rich phase. The middle part

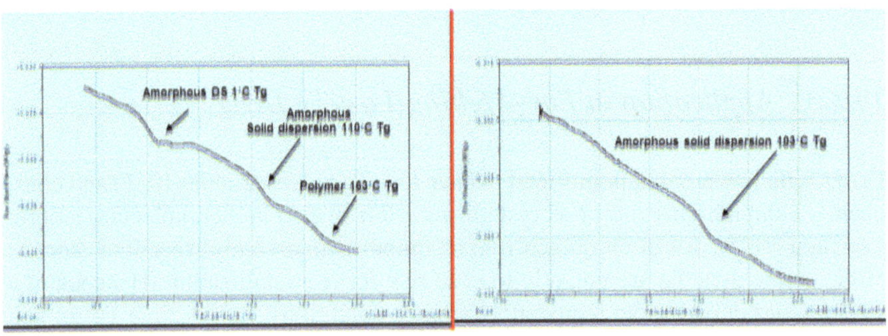

Fig. 14.6 Solid dispersion exhibiting three Tg values (*left*): one for the amorphous drug, one for the polymeric matrix, and one for the solid dispersion. A fully miscible solid dispersion exhibiting single Tg (*right*)

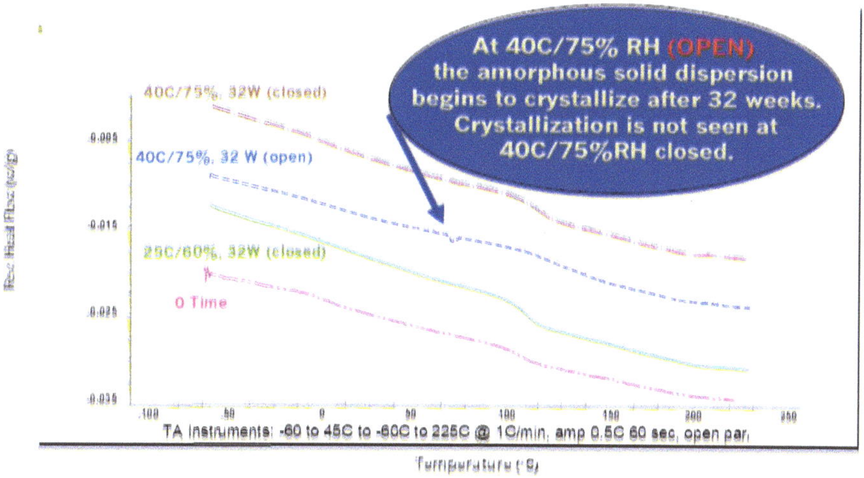

Fig. 14.7 Initial indication of physical instability in a single-phase/single-Tg solid dispersion of drug C under harsh open storage conditions and lack of instability issues under more realistic storage conditions over 32 weeks

of the thermogram with a Tg of 110 °C represents the solid dispersion where the API is molecularly dispersed. The highest Tg is exhibited by the polymeric matrix or the mostly polymer-rich phase. The disadvantage in having such a multiphasic system is that the drug-rich phase is likely to exhibit a high risk for recrystallization and once seed crystals start to form, with time, the system could cascade down to become a fully crystalline dispersion. Such a transformation would reverse the solubility/dissolution advantage and bioavailability enhancement the solid dispersion was designed to provide. The panel to the right on the other hand depicts a single-phase solid dispersion where the entire drug is fully miscible and molecularly dispersed in the polymeric matrix. This solid dispersion's composition included a copolymer of PVP with a small amount of a more hydrophobic polymer and this system is unlikely to have a similar high risk of recrystallization even at room temperature when it is appropriately protected from moisture. It is to be noted here that it is possible to artificially induce instability in the solid dispersion exhibiting single-Tg system that is shown in the left panel here as well. Without the benefit of a moisture-protective packaging or moisture barrier coatings, under open conditions of 40 °C/75 % RH due to the hygroscopicity of PVP, moisture level in the solid dispersion can increase significantly and drastically bring down the Tg (Oksanen and Zografi 1990). Under such harsh conditions, phase separation into drug-rich and polymer-rich phases can occur and DSC thermograms for such a phase-separated system would look similar to the left panel of the figure under discussion. The single-Tg system for drug C was intentionally subject to harsh open storage at 40 °C/75 % RH conditions and in this study the beginning of phase separation or crystallization could be evidenced using DSC as shown in Fig. 14.7. However, at all other conditions, including 40 °C/75 %

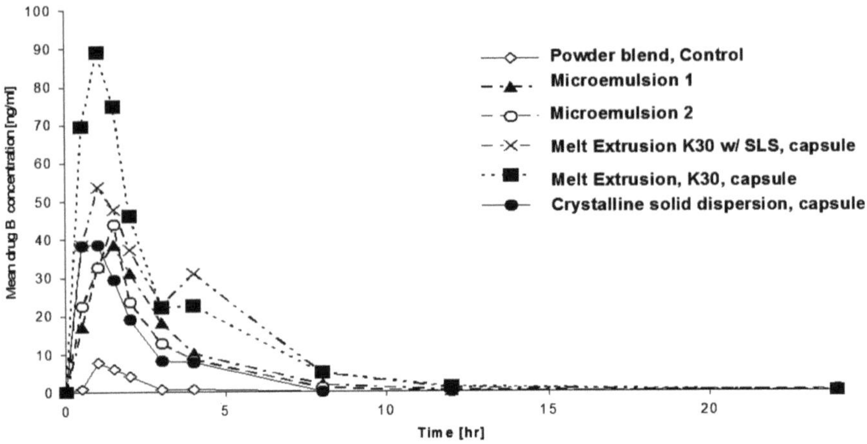

Fig. 14.8 Mean pharmacokinetic profiles of drug C as amorphous solid dispersion, crystalline solid dispersion, and microemulsion formulations in fasted beagle dogs using a cross-over study design

under closed conditions (HDPE bottle), a single Tg could be observed with no indication of new Tg values or melting endotherms that could mean phase separation or potential recrystallization. Based on this initial indication of slight risk under open conditions, a range of backup concepts and a strategy for their application had been developed, mostly based on moisture barriers and hydrophobic coatings, but the need for their application did not arise. Stability of the "single Tg" solid dispersion of drug C was well acceptable in a conventional closed HDPE bottle with induction sealed bottle closure.

Apart from obtaining a stable solid dispersion formulation for this low-Tg drug, the main goal of development was also to enhance dissolution and bioavailability of the drug. Microemulsion formulations had initially been developed, but high risk for recrystallization was already seen with them. Consequently, the microemulsion formulations were used as experimental controls or as a bench mark during in vivo evaluation of the solid dispersion formulations in fasted dog model. Based on the low Tg of the drug and the potential high risk for recrystallization, a separate effort to formulate a crystalline solid dispersion, i.e., a solidified suspension instead of solidified solution, was undertaken. In this approach instead of quench cooling the solid dispersion, the rate of polymer solidification was allowed to be significantly slower than the rate of drug crystallization, thus allowing for the drug to crystalize under slow controlled cooling. The resultant crystalline dispersion formulation was also evaluated in vivo in dogs. It was understood that crystalline dispersion's dissolution or solubility enhancement advantage would likely be comparatively less as its advantage stems primarily from the increased surface area of the API and not from the loss of crystallinity—the more effective approach to increase kinetic solubility. The outcomes of the evaluation did confirm this aspect of the crystalline dispersion as summarized in Fig. 14.8.

An eight-fold increase in C_{max} and a correspondingly similar eight-fold increase in bioavailability were seen for the melt extruded PVP K30-based amorphous solid dispersion of drug C compared with the powder blend of the same components. Addition of SLS to the solid dispersion reduced the bioavailability instead of improving it but this was not surprising, given the high lipophilicity of the drug and the likely lack of required partitioning of the drug from the SLS micelles to the gastrointestinal fluids prior to absorption. Both the solid dispersions produced a small shoulder in the plasma concentration profiles that was considered inconsequential for the comparative evaluation but was attributed to likely enterohepatic recirculation of the drug. Crystalline solid dispersion and the microemulsions produced similar middling bioavailability enhancement compared to the melt extruded amorphous solid dispersions. As would be expected, solubility and dissolution enhancement from the crystalline solid dispersion was lower than that for the amorphous solid dispersion. Regardless, the crystalline dispersion was considered a backup formulation option that could be used in case the amorphous solid dispersion developed physical stability issues over time in ongoing stability studies. Given that the drug was already in solution in the microemulsion formulation, it was surprising that the microemulsion produced lower bioavailability enhancement than the amorphous solid dispersions. However, it should be noted that PVP in the solid dispersion formulation played a significant role as a crystallization inhibitor in the case of drug C and the microemulsion formulation components on the other hand did not appear to have a similar effect. This is further corroborated by the fact that T_{max} values for all the formulations were between 1 and 2 h indicating the likelihood that the formulations did not differ substantially in their rates of dissolution but rather in their ability to reach higher supersaturation levels. As discussed earlier, lower bioavailability seen when SLS was used with the solid dispersion (likely due to lack of complete API partitioning out of the SLS micelles), is indicative of a similar phenomenon occurring with microemulsion micelles and could explain the lower bioavailability seen from the microemulsion formulation.

In summary, despite the very low Tg of 1 °C and the high lipophilicity presented by the drug C it was possible to formulate C into a stable amorphous dispersion by carefully selecting a copolymer composition that supported bioavailability enhancement and the formation of a single-phase/single-Tg solid dispersion. It should also be noted here that although addition of surfactants to solid dispersions helps dissolution/solubility of drug in vitro, surfactants may trigger partitioning issues that decrease overall bioavailability in vivo. A micellar system that does not yield drug to gastro intestinal fluids due to low partitioning, in effect, creates a new restriction to absorption or bioavailability instead of the original solubility/dissolution limitation. This aspect needs to be considered with care especially when it comes to highly lipophilic compounds such as drug C.

Overall, based on the three case studies discussed thus far, it could be concluded that the challenges encountered in developing solid dispersion formulations based on melt extrusion, namely (a) thermolabile drugs, (b) high-melting drugs, (c) low-melting/low-Tg drugs, can be managed for a number of APIs with carefully thought out strategies. It would be advisable not to take initial negative results as lack of viability of the technique. With a growing body of literature on melt extrusion and ready to adopt solutions becoming more available, the path towards a successful formulation or product development using melt extrusion could only get easier going forward.

14.4 Formulation and Process Enhancement

14.4.1 Application to High-Dose and Moisture-Sensitive Drug

Although melt extrusion has been extensively adapted for manufacturing solid dispersions in the recent past, and the question "Don't existing technologies already handle water soluble drugs very well?" is often asked, there are a few unique applications that melt extrusion makes possible. These applications may be very challenging if not impossible to handle for conventional processes such as direct compaction, roller compaction, and high-shear granulation. A case study that could illustrate this point involves metformin Hydrochloride, a highly water-soluble drug with a high dose of up to 1,000 mg.

The typical drug load for conventional dry or wet-granulated tablets is usually around 50 % or less; drug loads greater than 50 % are typically referred to as high drug load (Sun et al. 2009). At high drug loads, especially exceeding 75 % (w/w), inadequate tablet compaction properties and process robustness tend to become major impediments. For high-dose drug products, it is important to minimize the amounts of excipients in the formulation to keep the tablet size small enough for patients to swallow. Given that the desirable upper limits of the largest reasonable tablet mass and size for oral administration/swallowing is around 1,000 mg, formulating tablets of metformin Hydrochloride at a dose of 1,000 mg starts out quiet unfavorably. A few commercial products do stretch the tablet size/mass limit up to 1,350 mg because of similar clinical need to deliver high doses. Examples include niacin extended-release (ER), lovastatin immediate-release 1,000/40 mg tablet (Advicor®, Abbott, Abbott Park, Illinois); ranolazine ER tablet 1000 (Renexa®, Gilead, Foster City, California), and nutritional supplements (Vitamin C, Vitamin C with bioflavonoids, multivitamin and mineral combinations, amino acid supplements, etc.).

In addition to the need for a reasonably swallowable tablet size at 1,000 mg dose, the development of metformin Hydrochloride is challenging from other perspectives such as its powder properties in presence of moisture (Desai and Li 2003). Due to its high aqueous solubility even a small amount of atmospheric moisture adsorbed into metformin Hydrochloride powder can dissolve significant amounts of the drug. Consequent moisture migration or desorption would leave behind metformin Hydrochloride particles with solid bridges, similar to the phenomenon that has been studied more extensively in the case of sucrose, lactose, and mannitol (Bika 2005; Billings et al. 2006). Formation of solid bridges results in poor powder flow as well as sticking and picking problems during tabletting of metformin Hydrochloride. In worst cases, extensive formation of solid bridges could result in, a drum of free-flowing metformin Hydrochloride transforming overnight into a single solid block. Due to these moisture effects, the drying end point for the wet granulates of metformin Hydrochloride needs to be in a fairly tight range of around 2 % moisture level in the current conventional wet/moist granulation (WG) processes to avoid subsequent flow and tablet compaction issues. Other than controlling the granulation moisture level tightly during drying, the storage time and conditions for granules between completion of granulation and completion of tablet compaction also needs to be closely controlled due to the potential for atmospheric moisture-induced changes in metformin Hydrochloride granules.

In order to develop a robust manufacturing process for high-dose metformin Hydrochloride in combination with a second moisture-sensitive drug with hydrolytic liability, various processing techniques such as WG, solvent granulation (SG), and melt granulation (MG) were evaluated. MG technique that used the twin-screw melt extruder was found to overcome several of the issues with metformin Hydrochloride and the stability issue for the second drug in a fixed dose combination product development effort (Lakshman et al. 2011). To provide context to the MG process discussed here, it is to be noted that there are reports of MG performed using a high-shear granulator for different drugs by typically employing low-melting waxy materials (Royce et al. 1996; Evrard et al. 1999) as binders for formulation of tablets; however, the technique has not gained wide acceptance as the processing of materials is at a relatively high temperature and the maintenance of uniform temperature in a high-shear granulator is difficult due to inefficient heat transfer and mass transport. More recently, a melt extrusion process that employed low-melting waxes or polyethylene glycols (Liu et al. 2001; Van Melkebeke et al. 2006; Windbergs et al. 2009) as binders was used to overcome some of the deficiencies of the high-shear granulator. The extruder-based MG process used for metformin Hydrochloride under discussion is relatively different. It differs from the literature citations of MG in the current paragraph in that instead of low-melting lipids, waxes or polyethylene glycol, a high-melting conventional binder, hydroxypropyl cellulose (HPC) is used to avoid forming comparatively soft tablets. It also clearly differs from extruder-based solid dispersion manufacture in that MG is performed at temperatures below the melting point of the drug substance but higher than the melting or softening point for the binder. Manufacture of amorphous solid dispersions on the other hand needs to be performed at a temperature above the melting/softening point of the drug and the polymer. In addition, in the currently discussed MG process using an extruder, as a die plate is not employed in the extruder, the process produces granulates, as agglomerates or powder instead of the typical semi-molten shaped strands. The composition of the formulation that was evaluated using MG and compared against WG and SG is summarized in Table 14.9. The formulation is extremely simple and consisted of only drug and binder in the internal phase. Any added water or solvent added to the internal phase being mostly removed by drying. The external phase of the formulation was comparatively small, with only drug D and lubricant.

It is to be noted that MG tended to produce highly reproducible very low Loss on drying levels at around 0.2 % and that, unlike with the moist aqueous granulation and SG processes, such low moisture levels were adequate to produce better than

Table 14.9 Formulation composition used for evaluation of wet, solvent, and melt granulation processes

Ingredients	Amount per tablet	w/w %
Granulation components		
Metformin Hydrochloride	1000.0	88.2
Hydroxypropylcellulose	98.9	8.7
Purified water[a] or ethanol/isopropanol mixture (95:5 v/v)[b]	q.s.	0[c]
Extragranular components		
Second drug substance, D	25.0	2.2
Magnesium stearate	10.1	0.9
Total	1,134	100

[a] Used in wet/moist granulation process only
[b] Used in solvent granulation process only
[c] Water removed as a part of drying process

Table 14.10 Process conditions for melt granulation

Process parameter	Set point (range)
Hydroxypropylcellulose (HPC) level	9 % (5–15 %)
Feed rate	30–35 g/min (20–45 g/min)
Extruder screw speed	210 rpm (120–300 rpm)
Extruder temperature[a]	160 °C (140–180 °C)
Magnesium stearate level	1 % w/w (0.75–1.25 %)

[a] The melt extruder used five different heating zones. Temperatures for each zone at low condition: 40, 90, 110, 130, and 140 °C. For high condition: 40, 110, 130, 170, and 180 °C

acceptable tablet hardness and friability. This was a definitive advantage over both wet/moist aqueous granulation and SG processes where moisture or Loss on drying levels were difficult to control and the moisture level had a major influence on tablet characteristics. These difficulties were seen despite the fact that, to avoid Loss on drying variability, WG had been specifically adapted to use the exact amount of water needed for granulation (referred to as wet/moist granulation), instead of using excess water and drying the excess away as reported elsewhere (Railkar and Schwartz 2000). More importantly, the low moisture level of the melt granules allowed incorporation of the moisture-sensitive drug D simply by mixing it in as a part of the external phase prior to tablet compaction. Accelerated stability studies indicated that due to faster degradation of drug D, wet/moist granulation of metformin Hydrochloride could only support a substantially less than 2-year shelf life for the combination product. The MG processing parameters investigated in this study are detailed in Table 14.10.

Heat transfer through the extruder barrel and frictional heat developed by the rotation of the extruder screws both contributed to the softening or melting of the HPC polymer above its Tg of about 130 °C. During the extrusion process, the preblend of metformin Hydrochloride and HPC was initially conveyed through the screws while being preheated and then mixed with the molten polymer before finally being subjected to compressive forces within the extruder barrel to form dense granules. Overall, granule formation in the extruder occurred as a result of shear energy imparted to the material. The energy input to the molten mass of HPC and metformin

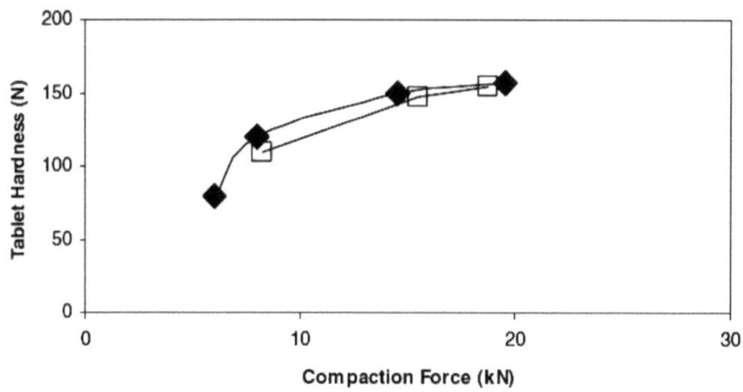

Fig. 14.9 Tablet compaction profiles for melt-granulated metformin Hydrochloride with and without a second drug D. *Solid diamond* metformin Hydrochloride, 250 mg; *open square* metformin Hydrochloride + second drug D at 250 mg/50 mg

Hydrochloride mixture was modulated by increasing the screw speed to increase energy input or decreasing the screw speed to reduce the energy input. Conversely, feed rate could be increased to reduce specific energy input or feed rate could be decreased to increase the energy input. Hot granulate obtained was cooled to near room temperature before milling and blending it with drug D as a part of the external phase. The granules obtained were evaluated for compaction properties using a Korsch tablet press with 21 × 8.3 mm ovaloid tooling at 60 rpm at different compaction forces.

To investigate whether and to what extent the second drug substance D would have an impact on the compactibility of metformin granules; tablets were prepared with and without the second drug substance D at a ratio of 250 mg metformin Hydrochloride to 50 mg of the second drug substance D (5:1). The 5:1 ratio was a much higher level of the second drug substance D than what was expected in the final formulation (40:1). The compression profile data in Fig. 14.9 indicate that a maximum tablet hardness of approximately 160 N was attainable at a compression force of 20 kN for tablets prepared with and without the second drug substance component.

Figure 14.9 illustrates that addition of drug D did not bring down the compaction profile of metformin Hydrochloride melt granulates. The compaction profiles practically continue to overlap despite addition of drug D. Based on these data, it could be seen that metformin Hydrochloride granulation has the capacity to accommodate various loads of the second drug substance D between 5:1 and 40:1, without any negative impact on the compactibility of the metformin Hydrochloride granules.

Overall, it was possible to obtain robust tablet hardness profiles by using melt granulate manufactured with a combination of high temperature, high feed rate, low screw speed, and low magnesium stearate level. Comparing the MG against WG and SG, the study (Lakshman et al. 2011) concluded that wet/moist granulation was unacceptable due to effects of moisture-induced physical transformation of metformin Hydrochloride as evidenced by (a) inadequate hardness profile, (b) changing tablet hardness with time, (c) higher than acceptable friability, (d) sensitivity of granulation flow to room humidity, (e) a narrow Loss on drying target range for manufacture of acceptable tablets, and (f) highly variable Loss on drying.

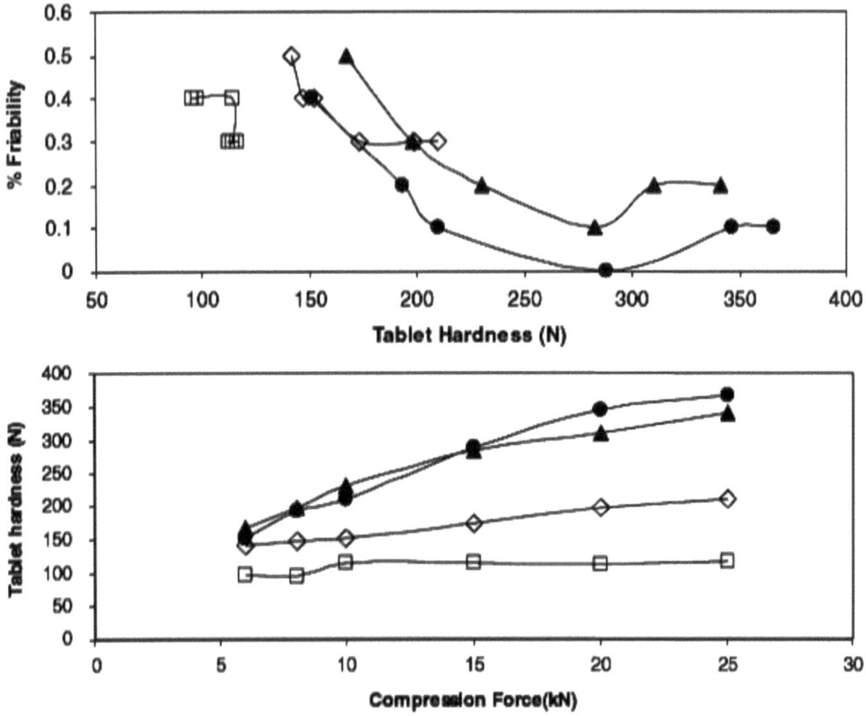

Fig. 14.10 Comparison of tablet compaction properties between SG and MG. *Open symbols* representative solvent granulations prepared at 1.7 % Loss on drying; *closed symbols* representative melt granulations prepared at 40 g/min, 120 rpm

This is not to say that manufacturing of metformin Hydrochloride tablets is not possible using WG, but that such a process is logistically demanding, highly variable, and prone to failure without stringent process controls. In comparison to the moist aqueous/WG, the SG was seen to offer advantages such as no change in tablet hardness over time, no granulation flow problems, no sticking/picking issues, and acceptable levels (<0.5 %) of friability. However, the SG still required a narrow Loss on drying target range for manufacture of acceptable tablets and the Loss on drying obtained continued to be highly variable between batches. It was postulated that the lower solubility of metformin Hydrochloride in alcohol (compared with water) diminished the extent of physical transformation of metformin Hydrochloride by allowing less dissolution and minimizing the consequent recrystallization and the formation of solid bridges between the drug particles.

Figure 14.10 compares the best compaction profiles from a Design of experiments for SG achieved at 1.7 % w/w Loss on drying against that for MG with an Loss on drying of 0.2 % w/w achieved at 40 g/min feed rate and 120 rpm screw speed. It is clear that MG allowed for a substantially better compaction profile and low friability over a much wider range of tablet hardnesses despite the low inherent moisture levels of about 0.2 % w/w. In providing context for this low level of moisture content achieved, it should be noted that even direct compaction formulations and dry granulated (roller compaction) formulations can easily contain an order of magnitude higher moisture levels. Microcrystalline cellulose, a key component in many direct compaction formulations, has an equilibrium moisture content of about 5 % w/w at room conditions of 25 °C and 50 % RH. In this case study, MG enabled the formulation to start out at a low moisture level and thus made it possible to achieve a reasonable shelf life despite the high moisture sensitivity for drug D by employing a moisture barrier packaging for the prevention of moisture incursion over time. It is expected that the low moisture content of melt extrudates can be further exploited in formulating tablets of other moisture-sensitive drugs in a similar manner.

Apart from achieving adequate shelf life, the moisture level of MG was reported to be stable, and no aging of tablets or change of tablet properties was observed with respect to tablet hardness or dissolution over time—a particular problem for tablets of metformin Hydrochloride prepared using WG. The melt granulate continued to be free flowing even after several weeks of storage, having avoided the moisture-induced physical transformation of metformin Hydrochloride. More importantly, metformin granules produced using MG, due to inherent low moisture content (~0.2 %) and high compactibility was amenable to combination with the moisture-sensitive second drug D. Based on these outcomes, a high (90 %) drug-load tablet formulation, containing 1,025 mg of active pharmaceutical ingredients and 109 mg of excipients, was successfully scaled up and manufactured.

In summary, this case study shows that MG technology can be used to formulate solid oral dosage forms containing as high as 90 % (w/w) of active ingredients. MG was compared against wet/moist granulation that did not require drying as well as an SG process. Under identical compression force, the hardness of tablets produced was best with MG and other techniques followed the order MG > SG > WG and the friability was also best with MG and followed the order MG < SG < WG. MG was successfully applied in averting the moisture-induced physical transformations in metformin Hydrochloride tablets, thus avoiding complicated issues such as change of tablet hardness with time, variable powder flow behavior, and issues with dissolution stability. The moisture content of the granules was also low enough (only 0.2 % w/w) to be conducive to the stability of the moisture-sensitive drug substance D. The granules contained only the drug substance and the polymer, thus obviating the need for other excipients to enhance compactibility of drug substance. The simplicity of the formulation that uses only two excipients—a binder and a lubricant—perhaps also exemplifies the point that the technique is well suited for highly soluble drugs.

14.4.2 Application in Modified Release

Modified release (MR) dosage forms, in general, aim to avoid peaks in drug plasma concentration, and increase patient convenience by prolonging the release of drug. MR formulations are commonly developed using matrix or membrane-based systems but the membrane-based systems tend to be less preferred due to the potential risk for dose dumping. On the other hand, polymers often in large amounts are needed to

Table 14.11 Composition of modified release formulations studied

Formulation name	Drug load, w/w (%)	Drug release-controlling polymer	Polymer level, w/w (%)	Tablet weight (mg)
MR1	94	HPC (Klucel HF)	5	1,017
MR2	89	HPC (Klucel HF)	10	1,074
MR3	89	Ethyl cellulose 100cP	10	1,074
MR4	89	Hydroxypropylmethyl cellulose K100M + Ethyl cellulose 100cP	5 + 5	1,074

enable the matrix-based systems. For highly water-soluble drugs, it is not uncommon (Sung et al. 1996; Huang et al. 2004) for formulations to require > 50 % w/w amounts of the release-controlling polymers. For the same reasons outlined in the previous case study on metformin Hydrochloride, such as swallowability for a patient, the challenge of an MR formulation becomes much more difficult for a high-dose drug as both compaction properties and drug release retardation need to be achieved using only a small amount of excipients.

For example, 500 and 750 mg Glucophage XR (sustained-release metformin Hydrochloride; Bristol- Myers Squibb, New York) were developed for doses up to 2,000 mg for the treatment of diabetes by using multiple tablets per dose, and the extended release formulation of zileuton (Zyflo CR; Cornerstone Therapeutics, Cary, North Carolina) has been developed as the 600 mg tablet, which is administered as two tablets twice daily (b.i.d.) for a total dose of 2.4 g in the treatment of asthma. Because the amount of release-controlling materials and other excipients have to be kept to a minimum in high-dose tablets, it is important to enhance efficiency of the release-controlling polymer using a suitable processing technique. A case study involving imatinib mesylate describes how hot-melt extrusion-based technology may be applied to the development of sustained-release dosage forms for a high-dose drug that is freely water soluble. By using this technology, up to 95 % of imatinib mesylate in a formulation was granulated with 5 or 10 % of a release-controlling polymer. A number of release-controlling polymers were evaluated and several offered widely differing release profiles for imatinib mesylate despite being present at low levels in the formulation. Multiple formulation variants were prepared using different high-viscosity polymers and the extrusion-based MG method described earlier under Sect. 4.1. The tablet variants thus prepared were subject to in vitro dissolution evaluation. Based on dissolution evaluation, selected formulations were ultimately scaled up and tested in humans to evaluate the in vivo performance of the formulations. Reports of a melt extrusion-based process being used for the formulation of sustained release products is known at lower drug loads, for example, at 60 % for ibuprofen (De Brabander et al. 2003) and at 20 % for theophylline (Young et al. 2005) but not at high drug loads close to or above 90 % for modified release formulation of freely water-soluble drugs.

The formulations evaluated in this study all included nearly 90 % drug load, about 9–10 % polymer, and 1 % magnesium stearate as shown in Table 14.11.

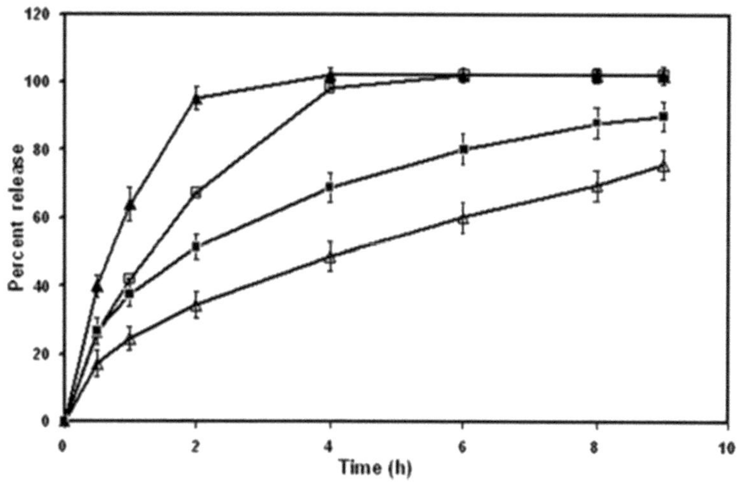

Fig. 14.11 Release profiles for tablet formulations prepared using melt granulation. *Solid triangle* MR1, 94 % API + 5 % HPC; *open square* MR2, 89 % API + 10 % HPC; *solid square* MR3, 89 % API + 10 % EC 100cP; and *open triangle* MR4, 89 % API + 5 % HPMC K100M + 5 % EC 100cP

The formulations were processed using a 16 mm corotating twin-screw melt extruder (Thermo Prism) at about 20 g/min feed rate and 50 rpm. Zone 1–5 temperatures were set up at about 50, 110, 130, 170, 185 °C and torque levels were at 40 %. The granulate obtained was screened using a 850 μm sieve before lubricating with magnesium stearate and then compressed into tablets using 21 × 8.3 mm ovaloid tooling.

Tablets obtained were subject to dissolution evaluation using 900 mL of 0.1M Hydrochloride at 37 °C in a USP I dissolution apparatus with baskets at 100 rpm. Results in Fig. 14.11 indicate release over 2 to > 8 h. Water-insoluble Ethylcellulose (MR3) retarded drug release to a greater extent than water-soluble HPC (MR2). By combining the insoluble and soluble polymers (MR4), release retardation could be extended significantly. Ethyl Cellulose-based tablets exhibited diffusion-controlled drug release, as evident from the intact drug-free matrix at the end of dissolution, while HPC and other water-swellable polymers, such as HPMC, eroded eventually after initially forming a gel. It was proposed that when used in combination, HPMC swells within the porous shell of the ethyl cellulose, thereby creating a more effective barrier for drug release.

Based on in vitro dissolution, MR2 and MR3 were selected for further studies. The selected formulations were evaluated in healthy human volunteers ($n = 12$) to assess in vivo release profiles and to compare the oral bioavailability with immediate release administered b.i.d as the control. The plasma concentrations of imatinib were determined using LC–MS/ MS. Figure 14.12 shows the mean pharmacokinetic profiles for the modified release and immediate release (IR) formulations. B.i.d. administration of the control IR formulation resulted in two peaks, at about 4 and 18 h,

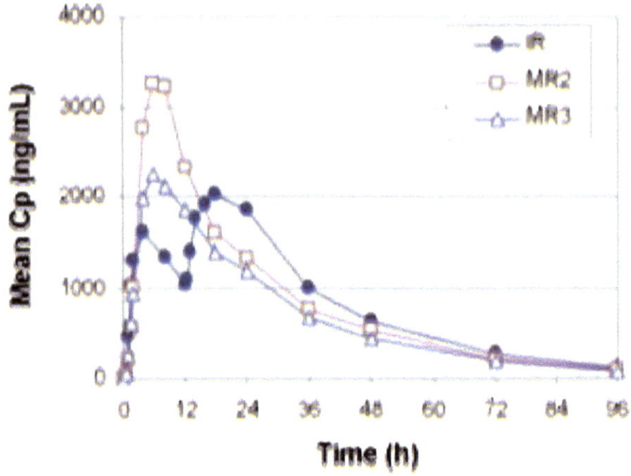

Fig. 14.12 Mean plasma concentration profiles for 800 mg imatinib mesylate tablet formulations MR2 and MR3, each dosed once, and the 400 mg IR tablet dosed twice at 12-h interval

with a trough at 12 h when the second 400 mg dose was administered. Administration of a single dose of imatinib 800 mg MR2 resulted in a peak concentration nearly two-fold higher than that of the first peak and more than 1.5-fold higher than that of the second peak seen with the control IR formulation. The peak concentration for 800 mg MR2 occurred at 6 h, later than the peak concentration for the first IR tablet. MR3 demonstrated a similar PK profile to MR2, but with a lower peak concentration compared with the first IR peak but very similar to the second IR peak demonstrating that both MR2 and MR3 formulations had characteristics of a controlled delivery of drug, with the peak concentration being delayed for approximately 2 h. As expected from in vitro dissolution, in vivo release rate of MR2 appeared to be faster than MR3. This is evident from the different C_{max} values for the two formulations. Drug released after 6 h was possibly not well absorbed as illustrated by the lowered (79 %) relative bioavailability for MR3—the slowest of the formulations—compared with the 100 % bioavailability observed for the faster MR3 or the IR control formulations.

In summary, MG using the extruder emerged as a robust technology for the development of MR oral formulation for a high-dose, highly soluble drug—a very challenging goal that was well accomplished by the extrusion-based MG technique. A modified-release tablet containing about 90 % drug load was successfully developed with excellent physical attributes as well as dissolution and stability characteristics. Human PK studies also demonstrated achievement of the desired modified release goal for both the formulations tested. MG was considered to have produced tablet physical properties including dissolution properties superior to a brief WG study that was performed subsequently to understand the differences between the two techniques.

In considering the last two case studies on metformin Hydrochloride and imatinib mesylate, although there may be other factors that explain the differences in properties of tablets prepared by melt and wet/moist granulation technologies, one of them could be the manner in which the drug and the polymer interact with each other during MG. MG process operates at a temperature above the Tg of the polymer, and the granulation occurs under the intense shear generated in the small clearance between the screws and barrel wall of the extruder. The polymer, which is in its rubbery state above its Tg, is sheared to flow, mix with, coat, and aggregate the drug substance particles. The granules so formed are compacted and they exit the extruder before being subject to cooling at room temperature. As the extruded material cools, the polymer is transitioned back to its amorphous state, and it is possible for such amorphous polymer that is coating the surface of drug particles (preliminary Confocal Raman Microscopy evidence discussed in (Vasanthavada et al. 2011)) to facilitate much stronger interparticulate bonding during tabletting, thereby resulting in harder tablets. Though water added during the wet/moist granulation process could also lower the Tg of polymers, the change in fluidity of hydrated/swollen polymers may not be comparable to that produced in the molten and sheared polymer during MG. Comparatively lower shear during wet/moist granulation may not provide as intimate a mixing as that offered by MG.

On a different note, considering productivity benefits, the MG technology described here may be applicable not only to high-dose drugs such as metformin Hydrochloride or imatinib mesylate, but also for reducing tablet sizes of drugs at other lower doses. This can substantially reduce the manufacturing cost of tablets as lower amounts of excipients would be used and more tablets per batch, at a fixed batch size, could be manufactured. For example, if the current weight of a 200-mg dose tablet is 400 mg (50 % drug load), then 33 % more tablets can be produced from the same batch size by weight, if the weight of each tablet can be reduced from 400 to 300 mg (67 % drug load) by using the MG process. Productivity by number of tablets per batch would increase by 60 % if the tablet weight can be reduced to 250 mg, which is only 80 % drug load in contrast to greater than 90 % drug load discussed here. In addition, MG using twin screw extruder is a continuous process, as opposed to other granulation techniques, such as aqueous high-shear WG, fluid bed granulation, SG, etc., which are batch processes. Because time is the only variable in determining batch sizes in the case of continuous manufacturing, the scale-up of the process is easier—which helps minimize risk in development. Due to its ability to daisy chain multiple unit operations into single equipment, the extruder-based MG process is also more amenable to quality by design with fewer variables to monitor and control. It is thus expected that MG using twin screw extruder has the potential to emerge as a powerful technology for developing highly efficient and robust manufacturing processes, especially for the high-dose drugs.

14.5 Conclusion

Applications of melt extrusion have expanded a great deal in other industries and it is likely that this will continue to happen in the pharmaceutical industry as well. The unique advantages the technology offers such as high-efficiency high-shear mixing, ability for water avoidance, continuous manufacturing capability, low scale-up risk, and ability to daisy chain multiple unit operations into one, all distinctively position melt extrusion to expand into further applications in the future. Good mixing being one of the fundamental requirements for pharmaceutical processes in ensuring dose accuracy and dose uniformity, strength of melt extrusion in this area is a highly desirable attribute in pharmaceutical applications. Other than achieving dose accuracy/uniformity, application of this attribute in enabling manufacture of solid dispersions (e.g., drugs A, B, and C), achieving high-dose immediate release formulations (metformin Hydrochloride) and high-dose sustained release formulations (imatinib mesylate) were successfully illustrated in this chapter. Ability for water avoidance, in specific instances even better than direct compaction/roller compaction is a very important attribute that enables formulation of highly water-sensitive drugs (drug D) and, in the case of highly soluble drugs that undergo capillary condensation, solid bridges, and caking (metformin Hydrochloride). The main disadvantage, use of heat, is appropriately countered by the technology's ability to minimize thermal exposure through reduction of residence time and a number of applications have shown that residence time control and plasticizers can enable heat-sensitive compounds (drug A and B) to be dealt with by melt extrusion. Short residence time also happens to be of significant importance in continuous manufacturing applications. Process control, of both upstream and downstream unit operations in a continuous process is significantly easier with short residence time and smaller variability around the mean residence time, as the uncertainty around the ability to track material throughout a process becomes lesser. A recent development, KinetiSol®, is also a step in the same direction of shorter residence time and reduced thermal exposure for dealing with heat-sensitive drugs (DiNunzio et al. 2010). Minimization of thermal exposure is also achievable by using transient plasticizers, materials that are used during processing to reduce operating temperature but that do not remain in the final dosage form (Lakshman 2005), examples of transient plasticizers include materials such as water or water-generating hydrates such as sorbitol hydrate or supercritical fluids such as CO_2.

Acknowledgments The author wishes to acknowledge several colleagues from Novartis for their experimental support, many discussions, and insights: Al Hollywood, Arun Patel, Jim Kowalski, Madhav Vasanthavada, Vivian Georgousis, Yu Cao, Jennifer Snyder, Marilyn Alvine, Bal Reddy, and Raju Vegesna; and Mike Motto, Colleen Ruegger, Steve Krill, Tony Tong, Ann Comfort, Alan Royce, Abu Serajuddin, and Yatindra Joshi for their support and encouragement.

References

Bhattacharya S, Suryanarayanan R (2009) Local mobility in amorphous pharmaceuticals—characterization and implications on stability. J Pharm Sci 98(9):2935–2953

Bika D, Tardos GI, Panmai S, Farber L, Michaels J (2005) Strength and morphology of solid bridges in dry granules of pharmaceutical powders. Adv Powder Technol 150:104–116

Billings SW, Bronlund JE, Paterson AHJ (2006) Effects of capillary condensation on the caking of bulk sucrose. J Food Process Eng 77:887–895

Bosselmann S, Williams RO (2012) Route-specific challenges in the delivery of poorly water-soluble drugs. In: Williams RO, Watts AB, Miller DA (eds) Formulating poorly water soluble drugs, AAPS advances in the pharmaceutical sciences series. Springer, New York

Bramah J. (1802) UK Patent No. 2652. 1802. Per Wikipedia article on Joseph Bramah. http://en.wikipedia.org/wiki/Joseph_Bramah. Accessed 18 July 2012

Breitenbach J (2002) Melt extrusion: from process to drug delivery technology. Eur J Pharm Biopharm 54(2):107–117

Crowley MM, Zhang F, Koleng JJ, McGinity JW (2002) Stability of polyethylene oxide in matrix tablets prepared by hot-melt extrusion. BioMaterials 23(21):4241–4248

De Brabander C, Vervaet C, Remon JP (2003) Development and evaluation of sustained release mini matrices prepared via hot melt extrusion. J Control Release 89:235–247

Desai D, Li D (2003) Evaluation of anti-caking agents for metformin hydrochloride and metformin-glipizide granulations. http://www.aapsj.org/abstracts/Am_2003/AAPS2003-002575.PDF. Accessed 26 Mar 2013

Dinunzio JC, Brough C, Hughey JR, Miller DA, Williams RO 3rd, McGinity JW (2010) Fusion production of solid dispersions containing a heat-sensitive active ingredient by hot melt extrusion and Kinetisol dispersing. Eur J Pharm Biopharm 74(2):340–351

Evrard B, Amighi K, Beten D, Delattre L, Moes AJ (1999) Influence of melting and rheological properties of fatty binders on the melt granulation process in a high shear mixer. Drug Dev Ind Pharm 25:1177–1184

Gao J, Walsh GC, Bigio D, Briber RM, Wetzel MD (2000) Mean residence time analysis for twin screw extruders. Polym Eng Sci 40:227–237

Ghosh I, Vippagunta R, Li S, Vippagunta S (2012) Key considerations for optimization of formulation and melt-extrusion process parameters for developing thermosensitive compound. Pharm Dev Technol 17(4):502–510

Hancock BC, Zografi G (1997) Characteristics and significance of the amorphous state in pharmaceutical systems. J Pharm Sci 86(1):1–12

Hancock BC, Parks M (2000) What is the true solubility advantage for amorphous pharmaceuticals? Pharm Res 17(4):397–404

Hancock BC, Shamblin SL, Zografi G (1995) Molecular mobility of amorphous pharmaceutical solids below their glass-transition temperatures. Pharm Res 12:799–806

Huang YB, Tsai YH, Yang WC, Cang JS, Wu PC, Takayama K (2004) Once-daily propranolol extended-release tablet dosage form: formulation design and in vitro/in vivo investigation. Eur J Pharm Biopharm 58:607–614

Hughey JR, Keen JM, Brough C, Saeger S, McGinity JW (2011) Thermal processing of a poorly water-soluble drug substance exhibiting a high melting point: the utility of KinetiSol® dispersing. Int Pharmaceutics 419(1–2):222–230

Killeen MJ (1993) The process of spray drying and spray congealing. Pharm Eng 13:56–64

Kanno H (1981) A simple derivation of the empirical rule TG/TM = 2/3. J Non-Cryst Solids 44:409–413

Lakshman JP (2005) Process for making pharmaceutical compositions with a transient plasticizer. Patent number WO2007106182A3

Lakshman JP, Cao Y, Kowalski j, Serajuddin ATM (2008) Application of melt extrusion in the development of a physically and chemically stable high-energy amorphous solid dispersion of a poorly water-soluble drug. Mol Pharm 5(6):994–1002

Lakshman JP, Kowalski J, Vasanthavada M, Tong W, Joshi Y, Serajuddin ATM (2011) Application of melt granulation technology to enhance tabletting properties of poorly compactible high-dose drugs. J Pharm Sci 100(4):1553–1565

Leuner C, Dressman J (2000) Improving drug solubility for oral delivery using solid dispersions. Eur J Pharm Biopharm 50:47–60

Liu J, Zhang F, McGinity JW (2001) Properties of lipophilic matrix tablets containing phenylpropanolamine hydrochloride prepared by hot-melt extrusion. Eur J Pharm Biopharm 52: 181–190

Liu H, Wang P, Zhang X, Shen F, Gogos CG (2010) Effects of extrusion process parameters on the dissolution behavior of indomethacin in Eudragit® E PO solid dispersions. Int J Pharm 383:161–169

Newman A, Knipp G, Zografi G (2012) Assessing the performance of amorphous solid dispersions. J Pharm Sci 101(4):1355–1377

Oksanen CA, Zografi G (1990) The relationship between the glass transition temperature and water vapor absorption by poly (vinylpyrrolidone). Pharm Res 7(6):654–657

Paudel A, Worku ZA, Meeus J, Guns S, Van den Mooter G (2012) Manufacturing of solid dispersions of poorly water soluble drugs by spray drying: formulation and process considerations. Int J Pharm 2012 Jul 20. [Epub ahead of print]

Railkar AM, Schwartz JB (2000) Evaluation and comparison of a moist granulation technique to conventional methods. Drug Dev Ind Pharm 26(8):885–889

Royce A, Suryawanshi J, Shah U, Vishnupad K (1996) Alternative granulation technique: melt granulation. Drug Dev Ind Pharm 22:917–924

Schneider AK (1972) Polylactide sutures. US Patent 3, 636, 956. 1972

Serajuddin A (1999) Solid dispersion of poorly water-soluble drugs: early promises, subsequent problems, and recent breakthroughs. J Pharm Sci 88(10):1058–1066

Sung KC, Nixon PR, Skoug JW, Ju TR, Gao P, Topp EM, Patel MV (1996) Effect of formulation variables on drug and polymer release from HPMC-based matrix tablets. Int J Pharm 142:53–60

Sun CC, Hou H, Gao P, Ma C, Medina C, Alvarez FJ (2009) Development of a high drug load tablet formulation based on assessment of powder manufacturability: moving towards quality by design. J Pharm Sci 98(1):239–247

Tsuk AG, Martin FH (1976) Dosage form. US Patent 3(972):995

Van Melkebekea B, Vermeulena B, Vervaet CRemonJP (2006) Melt granulation using a twin-screw extruder: a case study. Int J Pharm 326:1–2

Vasanthavada M, Wang Y, Haefele T, Lakshman JP, Mone M, Tong W, Joshi YM, Serajuddin ATM (2011) Application of melt granulation technology using twin-screw extruder in development of high-dose modified-release tablet formulation. J Pharm Sci 100(5):1923–1934

Windbergs M, Strachan CJ, Kleinebudde P (2009) Understanding the solid-state behavior of triglyceride solid lipid extrudates and its influence on dissolution. Eur J Pharm Biopharm 71:80–87

Young CR, Dietzsch C, Cerea M, Farrell T, Fegely KA, Rajabi-Siahboomi A, McGinity J (2005) Physicochemical characterization and mechanism of release of theophylline from melt-extruded dosage forms based on methacrylic acid copolymer. Int J Pharm 301:112–120

Chapter 15
Hot Melt Extrusion as a Continuous Pharmaceutical Manufacturing Process

Daniel Treffer, Patrick Wahl, Daniel Markl, Gerold Koscher, Eva Roblegg and Johannes G. Khinast

Abstract The implementation of continuous manufacturing in the pharmaceutical industry has been of increasing interest over the last years. This chapter focuses on continuous hot melt extrusion (HME) processing as well as on the continuous downstream options that are available. Furthermore, process analytical technology (PAT) tools and the integration of such tools in process control environment are presented. In general, real-time pharmaceutical process verification is accomplished by monitoring univariate (temperature, pressure, etc.) and multivariate (spectra, images, etc.) process parameters and quality attributes, to provide an accurate state estimation of the process, required for advanced control strategies. This chapter describes the development and use of such tools for a continuous HME process, monitored with generic sensors and a near-infrared (NIR) spectrometer in real time, using SIPAT (Siemens platform to collect, display and extract process information) and additional components developed as needed.

15.1 Continuous Processing in the Pharmaceutical Industry

In the last years, the interest in continuous manufacturing has increased significantly, albeit for a wide variety of reasons. These include the smaller scale of operations, the lack of scale-up-related problems, or the fact that by using a small container-based plant, drugs can be easily manufactured at different locations (which may be a significant issue for the permission to sell drugs in certain countries, such as China). Moreover, the advent of individualized or personalized drugs requires the development of robust and flexible continuous manufacturing methods. From a chemical-engineering point of view, one major advantage of continuous manufacturing is the real-time quality control of the products manufactured, eliminating the need for end-of-pipe testing. Hot melt extrusion (HME) is an ideal platform for such a manufacturing concept.

J. G. Khinast (✉) · D. Treffer · P. Wahl · D. Markl · G. Koscher · E. Roblegg
Institute for Process Engineering, Pharmaceutical and Process Engineering,
Graz University of Technology, Graz, Austria
e-mail: khinast@tugraz.at

M. A. Repka et al. (eds.), *Melt Extrusion*,
AAPS Advances in the Pharmaceutical Sciences Series 9,
DOI 10.1007/978-1-4614-8432-5_15,

The pharmaceutical industry, however, is still manufacturing in batch mode, mostly for historical reasons. Batch quality can be controlled, and thus, accepted or rejected, which is seen as an advantage over continuous manufacturing, especially since regulations are exclusively based on batch concepts. Nevertheless, batch production presents many disadvantages, some of them being:

- defined batch sizes, thus output is driven by batch size
- batch-to-batch variability
- many and long interruptions between process steps
- numerous transport steps between process steps (e.g., leading to segregation)
- long throughput times from start to finish
- large raw material and intermediate inventories required
- extensive validation and scale-up activities needed
- quality measured by in-process sampling and end-product testing.

In the past, regulatory uncertainty and the perception that the regulatory environment is rigid and opposed to innovation have been cited as the main reasons why pharmaceutical companies have been slow in introducing innovative production technology. This regulatory environment, however, has significantly changed over the last decade. Indeed, the regulatory bodies now actively encourage the development and implementation of innovative pharmaceutical development, manufacturing and quality assurance, among them continuous manufacturing. The advantages are obvious:

- integration of compliance or quality within the process
- reduction of systems' footprint, capital costs, and operational costs
- reduction of raw material and intermediate inventories
- less complex scale-up
- reduced time to market.

Nevertheless, continuous manufacturing poses significant challenges that should not be underestimated, including:

- Continuous operations are not suitable for every process.
- Dedicated equipment, facilities, and periphery are required.
- Process technology is often underdeveloped, especially with regard to secondary manufacturing and at the interface between primary (drug substance, DS) and secondary manufacturing (drug product, DP).
- Advanced and robust process analytical technology (PAT) and control approaches are required.
- Many critical sensors simply do not exist (e.g., powder flow rate, impurities, etc.).
- A single failure that can bring down the entire plant.
- Definition of a batch is complex, i.e., making recalls more difficult to handle.

Continuous systems, depending on the amount of the product made, may run for a few days to a few weeks, before the system is shut down, cleaned, and set up for a new campaign with a different product. Thus, continuous manufacturing in the pharmaceutical industry is different from other industries, such as the petrochemical industry, requiring a stronger focus on start-up and shut-down sequences.

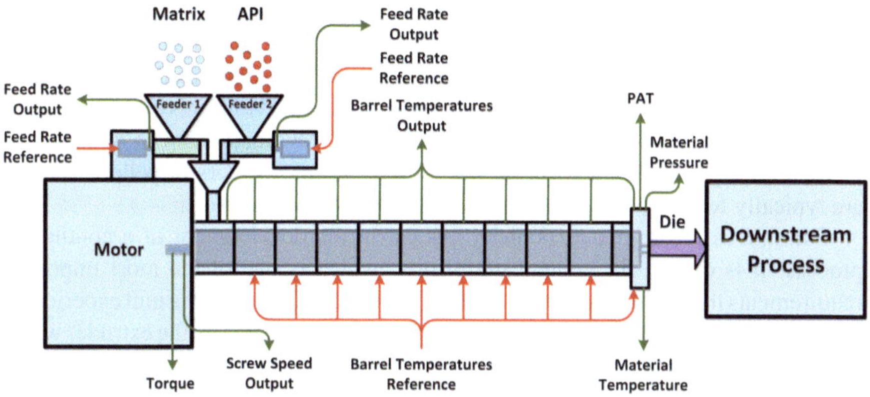

Fig. 15.1 Schematic view of an extruder and periphery

15.2 Continuous HME

HME is generically a continuous process for making a homogeneous material and offers several advantages over traditional processing techniques for pharmaceutical applications. A formulation for hot melt extruded pellets typically comprises a matrix and an active pharmaceutical ingredient (API), as well as other components, such as plasticizers, antitacking agents, pore formers, colorants, stabilizers, and others. Carriers that are generally used are (thermoplastic) polymers and waxes or wax-based materials, which control the release (mechanism) of the drug of the final dosage form. The process must be controlled in such a way that the softening temperature of the carrier is exceeded, but the degradation temperature of the API is not reached. The development of HME formulations has been described in much detail earlier in the book. Apart from the drug-release aspects, functional additives significantly impact the process (Crowley et al. 2007). Thus, formulation aspects also impact the processability and robustness of the process, which is a prerequisite for continuous manufacturing. Specifically, the formulation impacts

- the process temperature
- the required screw configuration and screw speed
- the torque
- the transition time
- and the solidification of the strand at the outlet (and possible die swelling)

Thus, a close understanding of the rheological behavior of the material-mixture prior to the experiment and a tight monitoring and control of the earlier-mentioned factors during extrusion are essential for establishing a robust continuous process. Figure 15.1 shows a schematic of a typical continuous extrusion process, which generally involves (1) the feeding unit, (2) the process unit, (3) a continuous downstream process, and (4) the monitoring system. In many new applications it is possible to set (red) and monitor (green) most of the process parameters.

1. Feeding The feeding unit consists of one or more feeders for powders or liquids. If there is one feeder the formulation must be premixed. If multiple feeders are used, the components (or premixes thereof) are fed by a dedicated feeder, resulting in higher design flexibility and eliminating the need for a previous mixing process (and problems related to poor mixing and segregation). Raw materials can be fed volumetrically or gravimetrically (preferred) in the case of powders or pellets. Liquids are typically fed volumetrically or by using a mass flow controller.

Continuous feeding is particularly critical in the development of a continuous process. It is crucial for content uniformity, which is one of the most important requirements that a pharmaceutical product must meet. Each feeding unit experiences fluctuations in the fed material stream, which can be dampened by the extruder which has a certain mixing ability. Extruders can generally dampen high-frequency or short-time fluctuations. Low-frequency fluctuations pass through the extruder almost undamped (Schenck et al. 2011). Thus, development of optimal feeding strategies (especially for low powder flow rates) is an important prerequisite (and a common problem). The key to the highest product uniformity is selecting an appropriate combination of the feeding equipment and screw. The feeder design is usually chosen based on empirical knowledge, as there is no perfect feeding setup for all powders due to the wide variety of factors, e.g., particle size distribution (PSD), cohesion, electrostatic forces, screw design, etc. As such, the right combination of the raw material and the feeder design is required. A comprehensive list of factors that impact feeding was provided by Schenk et al. (2011). Engisch and Muzzio (2012) suggested a method for characterizing feeding equipment, based on a feeder and a scale with data logging. In their contribution, the feeder feeds the powder to the scale, and the mass increase is measured. Subsequent data analysis delivers statistical information that helps to select the most suitable setup for specific powders.

In our experience, most content uniformity issues are caused by problems in the feeding section. An investigation of the feeding accuracy should be conducted in such cases. Target values are the composition, in case of premixes, and the mass flow rate in case of pure substances. Pure substances with poor flow properties typically cause avalanching at the feeding screw outlet. Premixes can (and often do) show significant segregation in the feeder hopper, usually leading to a concentration drift over time. Another critical issue during continuous feeding is the refilling of hoppers as during this time, the weight signal of the loss-in-weight (LIW) feeders is not available. Special protocols are necessary to ensure constant powder feeding during hopper refilling.

2. Extrusion From a process technological point of view the HME process is a continuous manufacturing process that combines multiple batch unit operations in one single process. Different process steps, such as mixing, melting, homogenization, and shaping, can be performed sequentially, offering the opportunities for automation of the manufacturing plant to limit material loss, increase the throughput, decrease energy input, and yield a product with high quality. Here, areas of research include the design of screw assemblies and extruder dies, mixing in the extruder, long-term operational stability, controlled powder and liquid feeding, and the simulation of the flow in unfilled screws (more detail about extruder design and operation is provided in other chapters of this book).

3. Downstream Continuous extrusion produces a homogeneous material. Thus, a continuous downstream process is needed to form a final product (pellets, sheets, powders, etc.) of the extrusion line. An appropriate downstream process is selected based on the targeted dosage form, the material's rheology, the product's purity, and the production costs (more detail is given in Sect. 15.3).

4. Monitoring API-concentration uniformity of the extruded material (in contrast to content uniformity of a final product) can be ensured by several tools. The simplest approach is to monitor closely the actual feed rates of the feeding units (Fig. 15.1). This, however, is only suitable for processes in steady state. Nonsteady-state processes and phases, such as start-up, concentration change, or simple process disturbances and instabilities, require a PAT analyzer at the end of the extruder and a control method to ensure product quality (i.e., the desired concentration uniformity; an overview of the state of the art PAT analyzers is given in Sect. 15.4).

15.3 Downstream Processing

An extruder produces a homogeneous material as an intermediate, in pharmaceutical industry typically in the range of a few kilograms per hour (e.g., 18 mm extruder 1–6 kg/h and 27 mm extruder $\sim$ 15 kg/h). The downstream process is usually coupled to the extrusion die to form a final product of the extrusion line. The final continuous downstream process is selected based on the target dosage form, the material's rheology, the product's purity, and the production costs. It can either be the final dosage form, as in case of injection molding or direct shaping, or an intermediate product, such as sheets, or pellets/granules for capsule filling, tablet compaction or other processes. An overview of common downstream processes is given in Fig. 15.2.

The rheology of the emerging melt is an important factor with regard to the selection. It mainly depends on the product formulation but is also influenced by almost all of the extruder's process parameters and their interactions during a coupled downstream process. Rheological parameters are typically a function of temperature, pressure, shear, and in some cases time history (Mezger 2011).

There are no general rules for selecting a downstream process. Rather, it is based on empirical knowledge and trial and error principles. Often, a formulation can be adjusted to be suitable for a desired downstream process. For example, in the case of die-face pelletization, anti-tacking components have to be added to reduce the stickiness of the material. Various downstream processes are described in more detail.

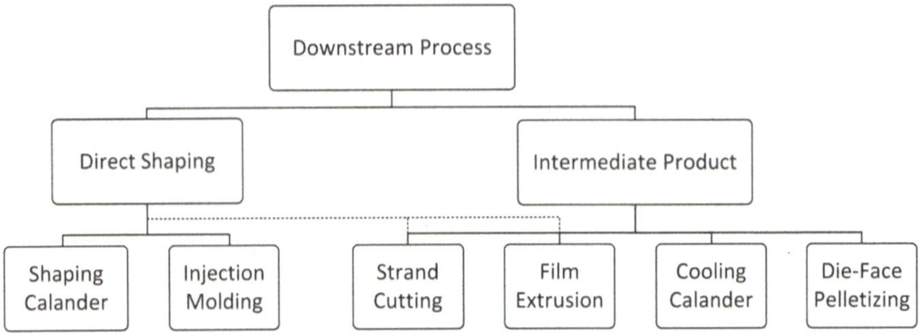

Fig. 15.2 Overview of common downstream processes

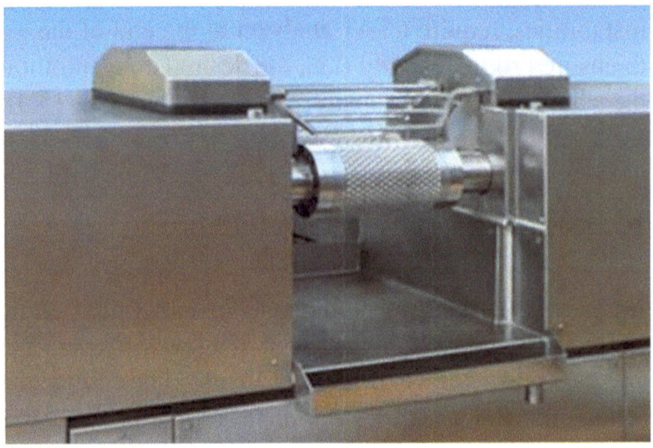

Fig. 15.3 Shaping calander (picture courtesy of Dr. Collin GmbH)

15.3.1 Direct Shaping of Final Product

Two continuous processes are established in the pharmaceutical industry for direct shaping, where the extruded molten material is formed into a directly applicable dosage form.

The first process involves a shaping calander (see Fig. 15.3), which consists of two narrow counterrotating rolls with forming cavities. A homogeneous melt band is fed through a slit die of an extruder. The band passes through the forming calander, where the melt is forced into the cavities to form tablets. A thin film that remains between the tablets can be removed via a subsequent separation process (e.g., using a rotating drum). In some cases, the film may be ground and again fed to the extruder. The challenges of this process are the breakage during the operation due to stickiness of the material as well as product weight uniformity caused by pulsating exit velocity of the melt strand.

The second process, which is semicontinuous, involves injection molding. It is one of the primary shaping processes in the plastics industry, during which plastic parts weighing from a gram up to several kilograms are produced in high volume.

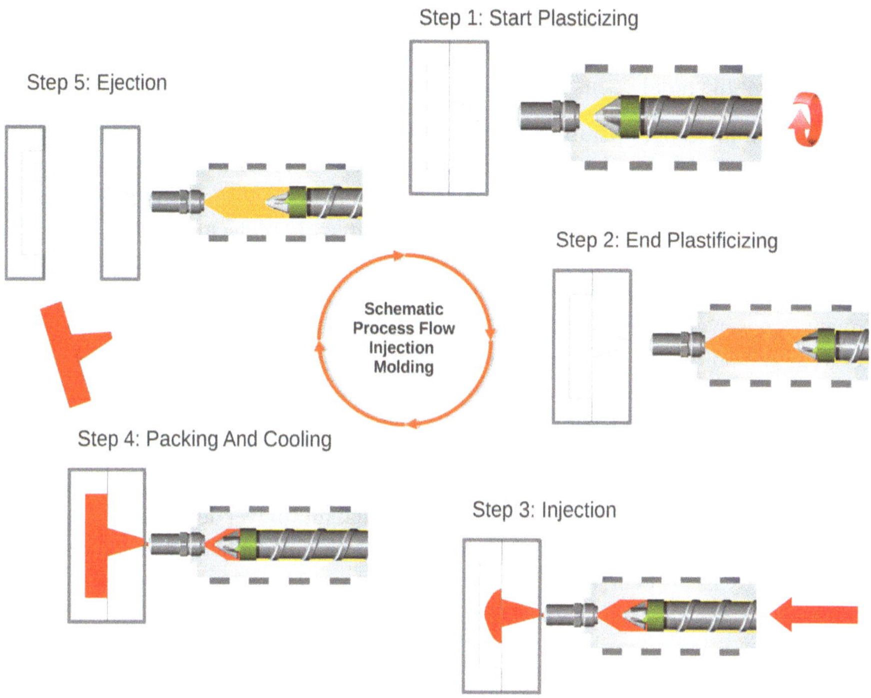

Fig. 15.4 Illustration of an injection molding process (picture courtesy of Engel Austria GmbH and IPIM, Johannes Kepler University Linz)

In comparison, the product weight is low in the pharmaceutical industry: a tablet, for example, typically weighs less than 1 g. Implants, however, may weigh several grams.

A conventional injection molding machine, consisting of a two-section mold with cavities, may be coupled to the end of a single-screw extruder. The feature of the single screw is the axial movability of the screw, which ensures high pressures during the injection. Generally, a single-screw extruder has a limited mixing capability compared with a twin-screw extruder. Thus, homogeneous materials, such as pellets produced by a preceding twin-screw extrusion, are typically used as starting material. Injection molding with a twin-screw extruder requires more complicated equipment and is therefore an exception to the rule.

Figure 15.4 illustrates an injection molding process. It is a cyclic process that is divided into five parts:

- *Start plasticizing*: the screw is in the front position and material is molten due to internal friction and heat transfer.
- *End plasticizing*: the screw moves backward and the melted material is accumulated in front of the screw.
- *Injection*: the molten material is pushed through the so-called "gate" into the cavity by the forward moving screw.

- *Packing and cooling*: the pressure is retained during the cooling and associated volume shrinkage to supply the cavity with fresh material.
- *Ejection*: the mold opens and the product is ejected.

In injection molding, operating pressures of up to several thousand bars during the filling stage are standard (thus not suitable for all APIs). This pressure is created during the injection either using an axial displacement of the screw or an additional piston. The required pressure depends on the complexity of the mold cavity, which is defined by the dosage form and the quantity per cycle. The production capacity of an injection molding machine depends on the quantity per cycle and the cycle time. The quantity per cycle is limited by the mold area and the shape of the product. In case of a simple tablet, over 100 pieces per cycle are possible. The cycle time depends on the formulation and its thermal characteristics and lies in range between 3 and 10 s. In the case of 100 pieces per cycle and a cycle time of a few seconds up to 100,000 tablets can be made per hour.

Injection molding can be applied to the production of various dosage forms, e.g., tablets, capsules, implants, intravaginal inserts, and multilayer devices for controlled release applications. A comprehensive review of pharmaceutical injection molding was given by Zema et al. (2012).

15.3.2 Intermediate Products

In many cases an extrusion line produces an intermediate product. Thus, the continuous extrusion process is used as a means for making new materials (solid solutions, solid dispersions, or for embedding nanoparticles in a melt, see Khinast et al. (2012, 2013)). Alternatively, the extrusion process functions similar to a granulation process, i.e., concentration uniformity and flowability of the ground material are achieved and segregation is suppressed. Finally, pellets can be made, which may directly be filled into capsules. Four downstream options for the production of intermediates are commonly used in the pharmaceutical industry and are chosen based on requirement for the next production step and the cuttability above the softening point.

The first method (Fig. 15.5) is strand cutting, which is used for sticky materials. Here, the extruder produces cylindrical strands. A conveyer belt with air knives cools the strand, leading to solidification, and transports the material to the strand cutter, which consists of a cutting rotor, a bed knife, and two feeding rolls. The strands pass between the feeding rolls and are subsequently cut by the cutting rotor. The obtained pellets are small cylinders with a relatively narrow size distribution. Typically, the throughput is not rate-limiting, but must match closely the extrusion flow rate. Thus, control of the strand cutter speed (both feeding and cutting) is critical for obtaining a stable continuous process for making uniform pellets. The pellet size uniformity depends strongly on constant operating conditions of the extrusion system and on the brittleness of the material; many amorphous dispersions are brittle glasses and can cause fracturing of the pellets during the cutting process. In rare cases, strand

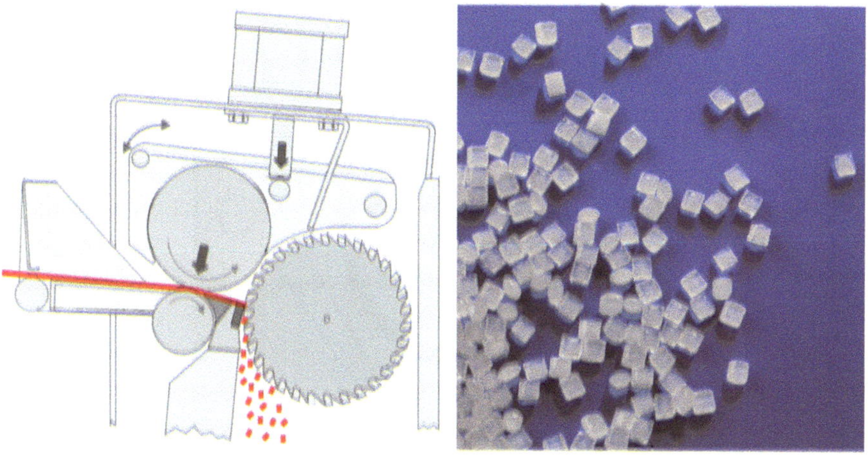

Fig. 15.5 Strand cutting (picture courtesy of Automatik Plastics Machinery GmbH)

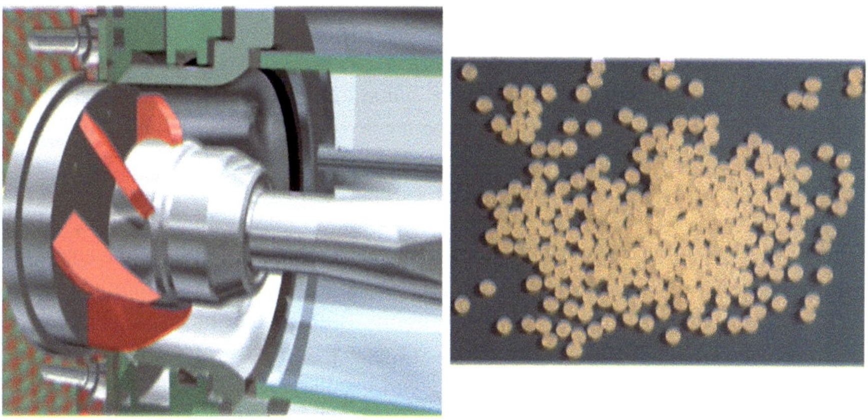

Fig. 15.6 Hot die-face cutting (picture courtesy of Automatik Plastics Machinery GmbH)

cutting is used to form final products like flat-faced plain tablets (Kipping and Rein 2013). The diameter of the tablet is equal with the stand diameter. However, the optical appearance still is inferior to conventional tablet production.

The second method is film extrusion, which is used, for example, for the production of transdermal, transmucosal, or transungual films (Repka et al. 2008). In this process, the extruder is equipped with a sheet die. An extrudate is formed into a thin film of molten material. The film is cooled between counterrotating chill rolls and subsequently coiled. These coils are the base material for the final production step, such as cutting or attaching to a patch.

The third method is hot die-face cutting (Fig. 15.6). A hot die-face cutter is directly connected to the extrusion die. An extruder generates cylindrical strands, and a cutting rotor cuts the emerging strands directly at the die plate into pellets.

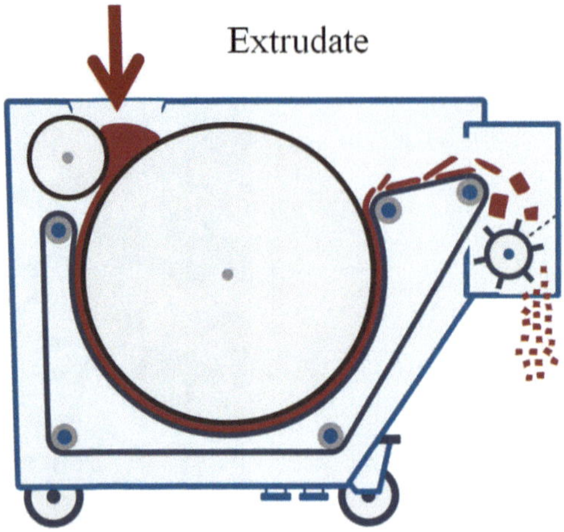

Fig. 15.7 Cooling calander principle (picture courtesy of BBA Innova AG)

The pellets are transported pneumatically into a product container. Several authors obtained almost spherical pellets via hot die-face cutting (Bialleck and Rein 2011; Roblegg et al. 2011). The spherical shape is due to surface tension and viscoelastic behavior of the material and results in a better flowability of the pellets, which may eliminate a subsequent spheronization process. The challenge is the cutting above the softening point of the formulation. The formulation must have a suitable rheology for the hot cutting, otherwise the product smears up knives and die plate, which leads to lumps and sometimes to burnt material as well. The cooling of the emerging material has strong impact on the rheology and therewith on the cuttability (Mürb 2012). The material is usually cooled with compressed air. The plastics industry uses water to achieve high cooling rates. However, water is in most cases not suitable for pharmaceuticals due to solubility of APIs and products purity requirements. Thus, the design and control of the cooling air flow is critical for achieving a robust process.

The third method involves a cooling calander, also termed a drum cooler or chill roller (Fig. 15.7). In this process, the extrudate is formed into a thin sheet by passing between a cooling drum and a counterrotating in-feed roll. A belt presses and redirects the sheet to the cooling drum. The sheet solidifies on the cooling drum and detaches from the belt. During the detachment it breaks, and a subsequent crusher reduces the fragments into small particles. Although the particle size is not uniform, the material may be milled again and may be used for direct compaction.

15.4 PAT Analyzers and Integration

Powder or liquid feeding, HME, and most of the downstream processes can be utilized in a continuous production line. However, these processes must be constantly monitored to ensure the design quality of the final product. Disturbances in the input parameters (e.g., during hopper recharging or due to segregation of a premix) or the process parameters (e.g., pressure build up), even only for a few seconds, can lead to significant output fluctuations (Schenck et al. 2011). Therefore, an adequate control strategy of the entire process is necessary to ensure the appropriate product quality. The toolbox provided by PAT can provide the necessary link between monitoring and control. This link is of importance, especially since the US Food and Drug Administration (FDA) increasingly requires inline monitoring and process control of manufacturing processes, as prescribed by the PAT initiative (US Department of Health 2004) and the subsequent ICH guidelines (ICH Harmonised Tripartite Guideline 2009). This includes:

- *Monitoring*: Collecting and aggregating real time, inline data of all kinds of analyzers.
- *Supervision*: Extracting information from the data regarding the state of the system. The analysis is based on the process knowledge obtained from previous runs. This knowledge is applied to statistical models, ideally combined with mechanistic models.
- *Diagnosis*: Check for possible deviations from the desired state of the system. This must include automatic root cause identification.
- *Control*: Decide which process settings must be manipulated to eliminate the root cause of the deviation and move back to the desired state of the system.

The next sections focus on commercially available PAT tools, their integration into the extrusion process and into modern IT infrastructure.

15.4.1 Overview of Available PAT Analyzers

Most of the available PAT analyzers for extrusion originate from the plastics or chemical industry. Some of the tools are well known in the pharmaceutical industry, e.g., near-infrared (NIR) spectroscopy for monitoring moisture, drug content, and uniformity, and Raman for polymorph detection, as reviewed in literature (Roggo et al. 2007; De Beer 2011; Vankeirsbilck et al. 2002). Some applications are used for validated systems (Peinado et al. 2011). However, the requirements for an extrusion PAT sensor are higher than for other purposes. High pressure, high temperature, and possibly abrasive materials require robust sensor design and materials. For optical interfaces quartz glass is the standard choice. Moreover, the location of the sensing device is critical for obtaining accurate information of the process status (see Chap. 15.3.2).

Next is an overview of commercially available PAT tools for inline monitoring of the extrusion process. The overview focuses on NIR and Raman spectroscopy and particle size measurements since these PAT analyzers have significant potential for standard production monitoring. A recent review of available sensors and their applications can be found in reference (Alig et al. 2010; for additional reading, see Chaps. 16 and 17 extrusion).

15.4.1.1 Near-Infrared Spectroscopy

NIR spectroscopy is based on the attenuation of light by absorption. To avoid complicated sample preparation, diffused reflected light is often used. The sample is irradiated with light in the wavelength range of about 1,000–2,500 nm, depending on the spectrometer system. The light photons can be absorbed by the sample's molecules, and the energy of the photons is used to excite higher vibrational states. As the wavelengths of the incident light that excite such vibrational states are absorbed, less intensity at this wavelength is reflected back to the detector. Only molecular vibrations, which result in a change of dipole moment, can be excited (Reichenbächer and Popp 2007). Therefore, polar bonds show stronger absorption of NIR radiation. Additionally, bonds with a pronounced difference in atomic mass have stronger NIR absorption and the most prominent bands belong to -CH, -OH, and -NH functional groups (Siesler 2007). Within the NIR region mainly overtones and combination vibrations of these functional groups are present. The spectral pattern is also influenced by further physical rules, i.e., Fermi resonance (Siesler 2007; Siesler et al. 2002).

With regard to extrusion, this means that the spectrum changes along with the varying melt composition, temperatures (or pressure) and with everything that influences these parameters, i.e., the screw speed and the screw filling ratio. As such, the spectrum provides a lot of information, but it is harder to interpret without knowing other process parameters.

In pharmaceutical extrusion NIR has been used for studying the API content (Schenck et al. 2011; no accuracy stated) and the polymer-API interactions directly in the die (Saerens et al. 2012; accuracy of 1.5 % API, determined by error of prediction) and the API content in film extrusion (Tumuluri et al. 2004); accuracy of 3.5 % API, determined by HPLC reference).

15.4.1.2 Raman Spectroscopy

Raman spectroscopy is based on frequency shifts in reflected light versus incident monochromatic laser light. These shifts are caused by inelastic scattering of photons on a molecule, with induced transitions between vibrational states of the irradiated molecules. The incident laser light is in the VIS to NIR region, but most commonly 785 nm diodes or 1,064 nm Nd:YAG lasers are used.

Radiation excites molecules to a virtual state, which does not correspond to electronic or vibrational levels. The virtual state is not stable, hence the molecules relax quickly. When relaxing, they may return to an excited vibrational state, whereby a part of the incident energy is converted and the scattered light shifts to lower frequencies (so-called Stokes radiation). At room temperature, it is the most commonly observed effect in Raman spectroscopy (Siesler 2007). At higher temperatures a certain fraction of molecules is already in an excited vibrational state, according to the Boltzmann distribution (Hertel and Schulz 2010a). Therefore, they can relax back to the ground state whereby the rotational energy is converted and the reflected light shifts to higher frequencies (Anti-Stokes radiation). The transition from Stokes to

Anti-Stokes radiation is a temperature effect and thus, the spectrum will be affected by temperature changes. Another possible influence is a pressure change, e.g., by undesired process effects like a pulsating melt stream through the die or plugging of the die. These effects may cause a model to be valid only in a small process parameter range.

A molecule is Raman active, if the polarizability changes with interatomic distance of a bond (Hertel and Schulz 2010b). Polarizability a is the proportionality constant between an external electric field E, such as the irradiating laser light, and the induced dipole moment P.

$$P_{Induced} = aE$$

High polarizability implies high mobility of the electrons. Thus, Raman is well suited to measure nonpolar bonds, i.e., C–C, C = C, S–S, N = N (Reichenbächer and Popp 2007), like carbon chains and aromatic rings. The spectra are not affected by the presence of water, which is highly polar. In contrast to Raman the NIR signal is strong for polar bonds and the two spectroscopic techniques can be used complimentary. Especially for symmetric molecules a vibration can be seen either in Raman or NIR. This is caused by different selection rules for the visibility of a vibration in the spectrum (Reichenbächer and Popp 2007).

15.4.1.3 Chemical Imaging

For chemical imaging primarily Raman and NIR sensors are used to obtain spatially resolved (2D) information of the chemical structure, i.e., the API distribution on the tablet surface. The 2D image cannot be taken in one shot, but rather via several techniques: (1) Filter techniques: 2D images are taken by a detector chip using optical band-pass filters for different wavelengths. Although this method is inexpensive and fast, it involves only a limited number of wavelengths. (2) Push broom principle: A 1D line is projected onto a detector chip. The second dimension of the chip is used for wavelength separation, which is achieved by reflecting the light at a grating. If the sample is moving in relation to the detector, the surface can be mapped as a function of time. The "surface" can also refer to the time evolution of the process. For this technique NIR camera systems are used. (3) Rasterizing the surface: The surface is scanned point by point (0D) in x and y direction, resulting in a 2D image. As the sample should not move during rasterizing, this method is impractical for inline measurements. Moreover, measurements can be time consuming: using Raman to map a tablet with a 100 × 100 pixel image 10,000 single point measurements are necessary, which can easily last overnight. One major advantage of such systems is the high resolution, which is superior to the push broom principle. Both Raman and NIR systems exist.

Depending on the complexity of the system either filter or push broom techniques are the method of choice in a manufacturing environment. For rather simple tasks, if few wavelengths are sufficient, i.e., moisture detection, filter techniques are used.

For complex formulations and also during development push broom is necessary. It can be used for film extrusion, for studying the drug content uniformity over time and the die cross-section followed by a shaping calander or for testing the final product after injection molding.

15.4.1.4 Particle Size Analysis

There are many techniques for particle size analysis. Typically, these methods provide information regarding PSD and the particle's parameters, such as sphericity, i.e., Feret-Min and Feret-Max. In a continuous extrusion process, PSD and sphericity can be important for quality control of downstream processes. Typical examples are die-face pelletizing or strand cutting. The following techniques have been used online or inline to study particulate matter:

- *Image analysis*: Images of dispersed individual particles or of the product stream are analyzed by advanced computational algorithms. Image analysis has the distinct advantage of delivering information about the particle shape and can measure in very dense systems (i.e., powder streams). Novel systems use special lighting (i.e., Eyecon by Innopharma Labs uses red, green, and blue LEDs from different angles) for better particle-edge detection.
- *Focused beam reflectance measurement*: A focused laser beam, rotating at high frequency, illuminates the product stream. If the laser beam hits a particle, the light is scattered back to the detector. Thereby, the chord length can be calculated, which is related to the mean particle diameter.
- *Laser diffraction*: A broad laser beam is diffracted from the particles, and the angle of diffraction is detected. Large particles exhibit a small diffraction angle, as opposed to small particles. From the diffraction pattern the volume equivalent sphere diameter is calculated. An assumption of the Mie theory, which is used to analyze the diffraction pattern, is the existence of spherical particles. Therefore, nonspherical particles, like needles or flakes, will result in incorrect measurements.
- *Spatial filter velocimetry*: A laser beam focuses on a detector, consisting of a fiber-optical array. If a particle passes the beam, it casts a shadow on the detector. Results are the chord length and notably the particle velocity at the probe position, which is not accessible with other techniques.

15.4.2 Integration in the Extruder

Selecting an appropriate measurement position to integrate the analyzer probes into the extruder is critical for obtaining accurate results and for developing a reliable control strategy. Therefore, possible measurement positions must be evaluated, according to the theory of sampling (TOS). Temperature and pressure probes for melt characterization have been used for many decades. Still, achieving repeatable and accurate results remains a challenge. Melt pressure probes usually detect pressure differences, thus require a pressure calibration at the beginning of each operation to set the pressure baseline correctly. Appropriate conditions during the calibration procedure are essential, as otherwise an incorrect absolute pressure, i.e., a baseline offset between different extrusion runs can occur. The calibration procedure has to be defined in the manual and usually clean probes and steady temperature conditions are required. The probes can be very sensitive to inexact calibration conditions, for example during contact with small amounts of material and temperature gradients due to incomplete heating of the machine.

During temperature measurement, the measured melt temperature is strongly influenced by the barrel temperature, more so for low throughput extrusions due to direct contact with the barrel. Thus, the melt temperature measured is always some value between a local barrel and the actual melt temperature.

In addition, there are challenges associated with spectroscopic measurements of the API concentration in the melt. Inline measurements of melt composition with spectroscopic sensors during extrusion are discussed in detail.

15.4.2.1 General Considerations Regarding the TOS

Sampling is critical for correct measurements. This applies to classical thief sampling with offline analysis as well as to inline PAT sensors. There are practical rules regarding correct sampling defined by the TOS (Esbensen and Paasch-Mortensen 2010) that need to be applied to evaluate different measurement positions for the extrusion process. For this purpose the material stream in an 8-0 plate (i.e., the plate between the 8-shaped extruder barrel and the circular, 0-shaped die, see Fig. 15.9) shall be approximated as a pressure-driven flow through a pipe.

Modeling the die section as a cylinder with a laminar flow of a Newtonian liquid, the steady-state velocity profile u(r) is described by the Hagen–Poiseuille equation

$$u(r) = -\frac{1}{4\eta}\frac{\Delta P}{\Delta x}(R^2 - r^2),$$

Where u depends on the viscosity η, the pressure drop ΔP, the length of the cylinder Δx, the cylinder radius R, and the radius r where the flow velocity shall be calculated. Figure 15.8 compares the velocity profiles of a perfect plug flow (red) and a Newtonian fluid (black) in a tube in the steady state, without slip. A perfect plug has the same velocity μ_{av} at every point in the tube, whereas the Newtonian profile exhibits a parabolic shape with the maximum velocity μ_{max} in the center. For non-Newtonian flows, as typically seen in extrusion, the inhomogeneity may be even more pronounced. However, the velocity profile of a shear thinning non-Newtonian liquid is somewhere between the ideal plug flow and the Hagen Poiseuille profile. The shape of the real profile in the die section of the extruder depends on the rheology (Rauwendaal 2001; Kohlgrüber 2007), the temperature field and the pulsation of the inlet stream.

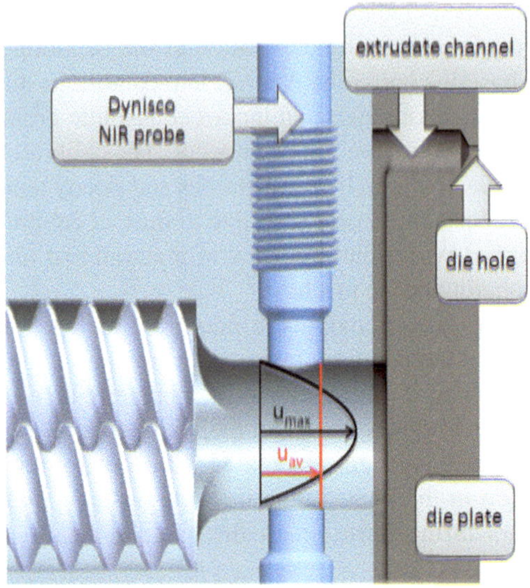

Fig. 15.8 A section of the extruder where the NIR probe is situated. The figure shows the velocity profile of (*red*) a perfect plug with constant velocity μ_{av} and (*black*) a Newtonian melt with a parabolic shape and the maximum velocity μ_{max} in the *center*

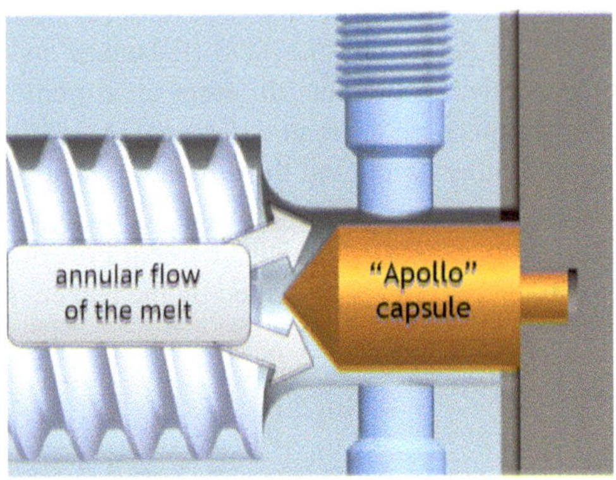

Fig. 15.9 Modification of a die section by inserting a tapered metal cylinder, thus creating an annular channel

As it is observed from Fig. 15.8, the material exchange close to the barrel wall (i.e., the region sampled by a spectrometer with a small penetration depth) is small compared with the core region. Therefore, a sample taken close to the barrel wall does not necessarily represent the overall concentration, but rather the local concentration in the vicinity of the barrel wall. If for some reason fluctuations in the API concentrations occur, sampling of the wall region does not provide an accurate representation of the products quality.

15.4.2.2 Application to Extrusion

The TOS rules for thief sampling can be applied to inline PAT measurements. For inline spectroscopic sensors the sample amount depends on several factors: (1) the irradiated volume, as defined by the spot size of the sensor and the penetration depth in the material, (2) the integration time per spectrum, and (3) the velocity of the material in the irradiated volume (Scheibelhofer et al. 2012).

The sample volume is often significantly smaller, compared with thief sampling methods. If the volume is too small compared to the unit dosage form, there may be subsampling issues. Subsampling can be avoided by choosing a longer integration time or adapting the process, to achieve higher velocities of the material. In addition, with a small spot size only a fraction of the product stream can be monitored. For extrusion purposes, one way to overcome these problems is using a smaller inner piping diameter (Esbensen and Paasch-Mortensen 2010), ideally in the order of the irradiated volume. However, there is one more issue: the barrel diameter is defined by the extruder specifications and cannot be freely changed. As such, another approach must be taken.

Special care must be taken to avoid accumulation of the material in front of the sensor, in which case the sensor signal can be biased, preventing a correct measurement of temporal evolution of the material's properties. This could happen, for example, if the probe's sapphire window is not in plane with the inner barrel wall for measurements in an 8-0 plate. Two measurement positions will be presented below, which fulfill the above criteria for measuring the actual melt composition.

Modified 8-0 Plate The 8-0 plate may be modified to perform measurements directly in the plate. The necessary modifications depend on the actual design of the extruder. The example below involves a Coperion ZSK 18 connected to a novel die-face pelletizer developed by Automatik Plastics Machinery GmbH.

The original pipe flow was changed to an annular flow by placing a tapered metal cylinder (we termed it "Apollo" capsule) in the 8-0 plate. The capsule forced the melt in an annular flow close to the barrel walls, where the probe was located. Figure 15.9 shows the sampling region with the modified geometry. The benefit of the annular flow is the increased average velocity due to a smaller cross-sectional area and higher shear rates in the vicinity of the probe. Thus, the material exchange in the volume sampled by the spectrometer is much increased and more representative of the total mass flow. The disadvantage of the reduced cross-section is a higher pressure drop and increased shear. However, in the case under consideration these effects are negligible compared with the condition in the extrusion die (annular gap dimension: inner diameter is 15.8 mm, outer diameter is 18 mm and length 19 mm; die plate configuration: 2 holes with diameter 1 mm and length of 1 mm).

Custom-made Die or Bypass In several studies a custom-made slit-die (Alig et al. 2005, 2010), a flow cell directly attached to the die (Fischer et al. 1997; both inline) or a bypass for at-line measurement (Coates et al. 2003; Fischer et al. 2011) were used. The advantages included good interface possibilities and a good exchange of the material in front of the sensor. These setups allow a fast and easy comparison of different measurement techniques (i.e., spectrometers) to select the most suitable one (Alig et al. 2010). However, it cannot be used in conjunction with downstream processes like die-face pelletizing or direct shaping methods. Still, if the extrudate is milled to powder for further downstream processing, a custom-made slit die may be the method of choice.

15.5 Process Integration into Computerized Systems

Process understanding requires the identification and explanation of all critical sources of variability, thus offering an accurate and reliable prediction of the product quality (Bakeev 2010). Ideally, it should be based on a mechanistic understanding of formulations and process factors, which in turn comprises (1) the identification of key parameters and effects, (2) real time and continuous measurement of selected key parameters, and (3) a control strategy based on selected univariate and multivariate real-time measurements (Koller et al. 2011).

Clearly, these tasks require for an IT infrastructure that can aggregate real-time process data from multiple unit operations, raw material data, PAT data, and equipment status (Schenck et al. 2011), which may also be used for process control. Moreover, the pharmaceutical industry intends to move away from paper and towards electronic systems, again requiring computerized systems.

15.5.1 Introduction to Computerized Systems

In response to the growing industry requirements, the regulatory authorities developed guidelines and regulations for computerized systems, including:

- EU GMP Annex 11 Computerized Systems (European Commission 2011a)
- EU GMP Chapter 4 Documentation (European Commission 2011b)
- 21 CFR (Code of Federal Regulations) Part 11 (US Department of Health 2011).

Moreover, the Good Automated Manufacturing Practices (GAMP5) Guidelines (ISPE Headquarters 2008) of the International Society for Pharmaceutical Engineering (ISPE) present a suitable approach for validation of computerized systems. GAMP has become a state-of-the art guideline for computer validation throughout the pharmaceutical industry (Linz and Seeger 2006; Fig. 15.10).

Requirements for all forms of computerized systems (needed for an automated continuous manufacturing setup) are as follows:

- Computerized systems should not result in a decreased product quality or process control.
- The overall risk of the process should not be increased.

In the context of continuous manufacturing a subcategory of computerized systems, namely data acquisition and process control systems, are important. Moreover, data acquisition and process control systems are a prerequisite for successful continuous process applications. Specifically, for the implementation of real-time quality control

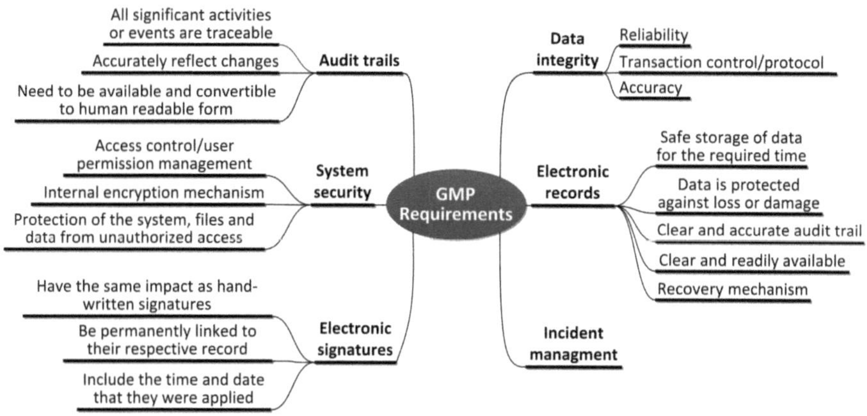

Fig. 15.10 GMP requirements for a computerized system that focus on data acquisition and process control systems. (For more details on GMP requirements see publications of the regulatory authorities)

as well as process control, a certain IT infrastructure is essential. With respect to the previously stated requirements the developer of the IT infrastructure has to consider the GMP requirements:

- Data integrity: A data acquisition and process control system must have built-in checks for the correct and secure data entry and processing. Data integrity and system security (e.g., access authorization for the operator) must be considered during the design stage of the system (European Commission 2011a).
- Electronic records: Electronic records must be readable, accurate, and accessible throughout the retention period (Roemer 2011).
- Electronic signatures: Many situations in our paper-based world legally require a signature or initials. One of the main reasons to promote and establish computerized systems in the pharmaceutical industry is avoiding a vast amount of archived paper. Thus, electronic signatures must be equivalent to handwritten signatures, initials, etc., required under the regulations. Depending on the country-specific legislation, electronic signatures may need to be linked to their respective records and include the time and date when they were entered (European Commission 2011a; Linz and Seeger 2006).
- Audit trails: A record of all GMP-relevant data and changes to the configuration data (e.g., creation, change, and cancellation of access authorizations) must be documented. Thus, audit trails need to accurately reflect the changes and must be convertible to a human readable form (Roemer 2011; O'Neill et al. 2011; Hiob 2011).
- System security
- Incident management

This basic implementation guideline is summarized in Fig. 15.10. For further information regarding the implementation and validation of computerized systems from a regulatory perspective refer to publications and guidelines of the regulatory agencies, such as GAMP 5.

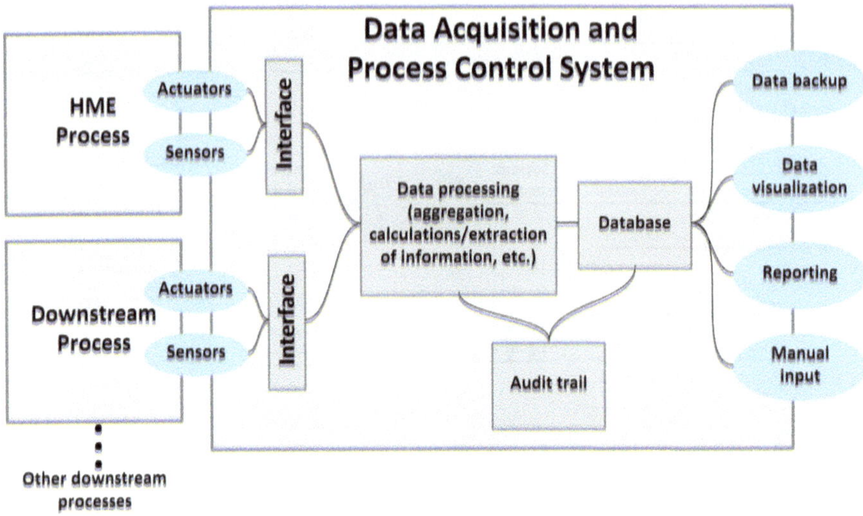

Fig. 15.11 Schematic of a basic data acquisition and process control system for a continuous manufacturing line consisting of several processes (i.e., feeding, extrusion, downstream, etc.)

15.5.1.1 Architecture of Supervisory Control Systems

The basic implementation of the data acquisition and process control system for a continuous HME system contains the following hardware and software units in order to assure the functionality and to meet the GMP requirements:

- Processes units
- Interfaces
- Data processing unit
- Databases
- Audit trail unit

A basic architecture of a supervisory control system is shown in Fig. 15.11. In the following the basic units (cf. gray blocks in Fig. 15.11) are discussed in more detail:

Processes Several processes (e.g., HME, pelletizer, etc.), each equipped with univariate (e.g., temperature) and multivariate (e.g., spectrometers) sensors and actuators from different manufacturers, must be monitored and controlled.

Interfaces The interface from the sensors and actuators to the computerized system has to be selected carefully with the consideration of the GMP requirements. Especially data integrity has to be guaranteed. An easy integration of the sensor systems and actuators into the software can be accomplished via standardized interfaces. This increases the flexibility and ease of use of the sensor systems, actuators, and software. A common interface in the automation industry is OPC (Object Linking and Embedding for Process Control) technology.

OPC technology is used to distribute data from different univariate sensors to OPC clients and from OPC clients to actuators. An OPC server determines the interface between the sensors/actuators and several different software packages that implement an OPC client. Thus, each OPC client can access the data provided by the OPC server simultaneously. In continuous manufacturing several OPC servers (e.g., one OPC server per process or sensor system) have to be merged. In the context of OPC, different interface types are available. For example, OPC Data Access (OPC DA) specifications define communication protocols for real-time communications between sensors/actuators and computerized systems. As such, OPC DA deals only with real time and not with historical data. However, the next generation of software automation solutions belongs to the concept of OPC Unified Architecture (UA). It overcomes a drawback of OPC DA, i.e., the communication is based on COM (Component Object Model) and DCOM (Distributed Component Object Model) and thus tightly depends on Windows operating systems. COM and DCOM are standards introduced by Microsoft that allow the communication between different applications on the same machine (COM) as well as the communication across network boundaries (DCOM). Specifically in continuous manufacturing the data exchange between different machines and one central unit has to be reliable, secure and fast. Thus, OPC UA offers an enhanced interoperability with other platforms and guarantees data reliability and security. OPC UA has the additional advantage of allowing the exchange of higher-level structured data, e.g. PAT data.

To integrate PAT tools (e.g., NIR spectrometer) into a computerized system, a large amount of data must be collected, analyzed, visualized, transferred, processed, and stored within a short timeframe. Many manufacturers of PAT tools provide an OPC DA server. This entails the limitations as discussed earlier. OPC UA would provide a reliable, efficient way to transfer higher-level structured data, but this will take some time until it is established in the industry.

Data Processing Unit One of the most important tasks of the data processing unit is the aggregation of sensor data from several different interfaces. Each sensor system has its own sampling rate, which is a trade-off between noise reduction, relevance to the dynamics and restrictions from a computational performance perspective. However, the data processing unit has to timely align the data from several sensor systems in an appropriate way in order to allow real-time process analysis. Additionally, the integration of multivariate (MVDA) models and basic calculations in the system should be realized by the data processing unit.

Database Data from several processes and sensors, independent of their interfaces to the computerized system, must be time-aligned stored in a database. The operator should have access to the data from the database for offline analysis via a graphical user interface.

Audit Trail As discussed in the previous section, all significant activities and events must be traceable, with a certain service responsible for an accurate audit trail.

Moreover, the system should provide some basic functionality, such as data visualization and the capability to export reports and to backup data frequently. Additionally, the system needs an interface for the operator to enable electronic signatures and set inputs manually.

15.5.2 Monitoring the HME Process via SIMATIC SIPAT

In response to the growing interest in PAT, Siemens among other companies developed a PAT software solution, SIMATIC SIPAT (Siemens AG, Brussels, Belgium), that was designed to meet the basic GMP requirements discussed in the previous section (i.e., data integrity, electronic records, incident management, electronic signatures, system security, audit trail).

SIPAT collects data in real time from various monitoring sources and performs an aggregation function to guarantee time alignment (De Frenne 2011). Observations, including measurements, manipulated variables (MVs), and PAT data (spectra, images, etc.) are available in real time. SIPAT supports external calculation engines to transform process data into information that can further be used to detect abnormalities in the process. For example, in our work external calculation engines used are MATLAB (Mathworks, Inc., USA) and SIMCA-Q (Umetrics, Sweden). The real-time prediction engine SIMCA-Q facilitated the integration of models developed in SIMCA-P+ (Umetrics, Sweden), from which multivariate data analysis (MVDA) models (e.g., PCA, PLS) were developed. MATLAB is a programming environment for algorithm development, data analysis and numerical computation, and integrating MATLAB functions extended SIPAT's capability to perform complex calculations (e.g., multivariate calculus).

In order to monitor and control a continuous process, appropriate univariate and multivariate sensors and actuators must be introduced in SIPAT. On the one hand, SIPAT offers common communication interfaces (i.e., OPC technology), making it possible to use sensors from different manufacturers. On the other hand, it allows the developer to import PAT tools (e.g., spectrometer) by providing customized communication interfaces.

In the context of SIPAT, a base station, collector stations, clients, and central database exist. A collector station is responsible for acquiring data of a certain sensor system, which means that each spectrometer and univariate sensor (e.g., temperatures, feed rates, etc.) provided by the extruder represents a collector station. A base station unifies several collector stations belonging to one or several unit operations, e.g., extrusion, blending, tableting, etc. (De Frenne 2011; De Tandt 2011). A central database is mandatory for the SIPAT system, where all of the system's configuration information and historical and real-time data are stored.

A certain network architecture (Fig. 15.12) based on the software design of SIPAT and the given hardware was required. The main components were (1) a spectrometer PC as a collector station, (2) an industrial PC as a collector station for all extrusion data, (3) a Coperion PC with an operator monitor (independent from the SIPAT system), (4) an HME process, (5) a server PC with a central database and a base station, and (6) a number of clients. Communications between the SIPAT components were based on Ethernet technology.

The HME system had two collector stations: (1) the spectrometer was integrated into SIPAT via its own collector interface. Sentronic provides a driver that integrates the SentroPAT FO spectrometer into SIPAT. The configuration of the spectrometer

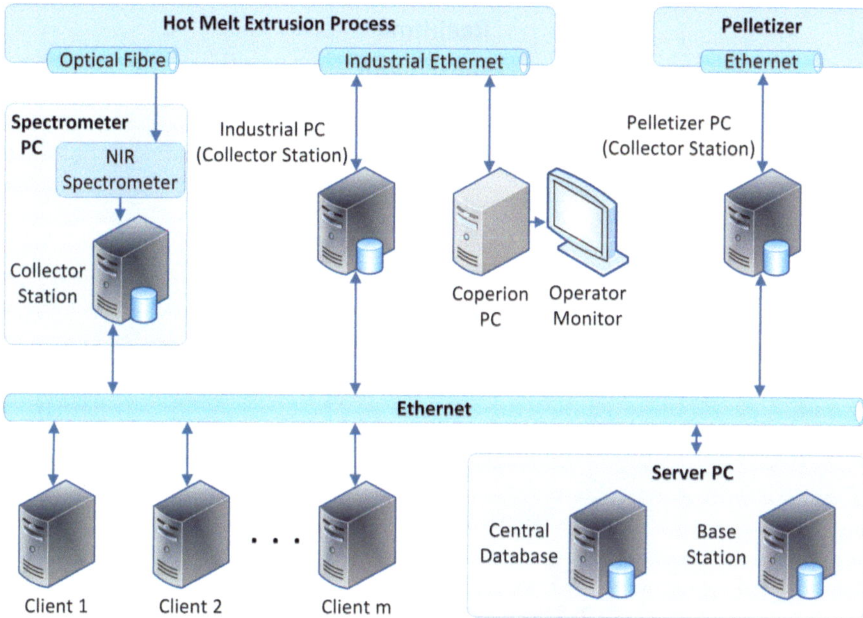

Fig. 15.12 Network architecture of the hot melt extruder and the pelletizer

(e.g., definition of the integration time, the number of spectra that were averaged, etc.) was predefined by an inhouse software from Sentronic and loaded from the SIPAT collector interface. (2) Based on OPC technology, an OPC server and the SIPAT collector station (acting as an OPC client) were installed on an industrial PC. Furthermore, the Coperion PC and the operator monitor were used to display SIMATIC WinCC. Windows Control Center (WinCC) served as supervisory control, a data acquisition system, and a human machine interface (HMI). This allowed the operator to monitor the process via either WinCC or SIPAT, the latter processing the spectra in real time.

Figure 15.13 shows interactions between SIPAT, SIMCA-Q, MATLAB, and the process itself. A predefined input sequence (e.g., reference feed rate) was generated offline in MATLAB. After acquiring a new set point of the reference feed rate, SIPAT distributed the scalar to the extruder via OPC technology. The extruder forwarded the value to the feeder that modified the feed rate.

In order to investigate the influence of input parameters on the product quality, the observed spectrum was interpreted in real time using SIMCA-Q by projecting the data onto a reduced dimensional space. After the experiment, the observed data were analyzed directly in SIPAT or exported in a standard file format. A similar procedure was applied to gather the required data for the development of MVDA models, which will be discussed in the next section.

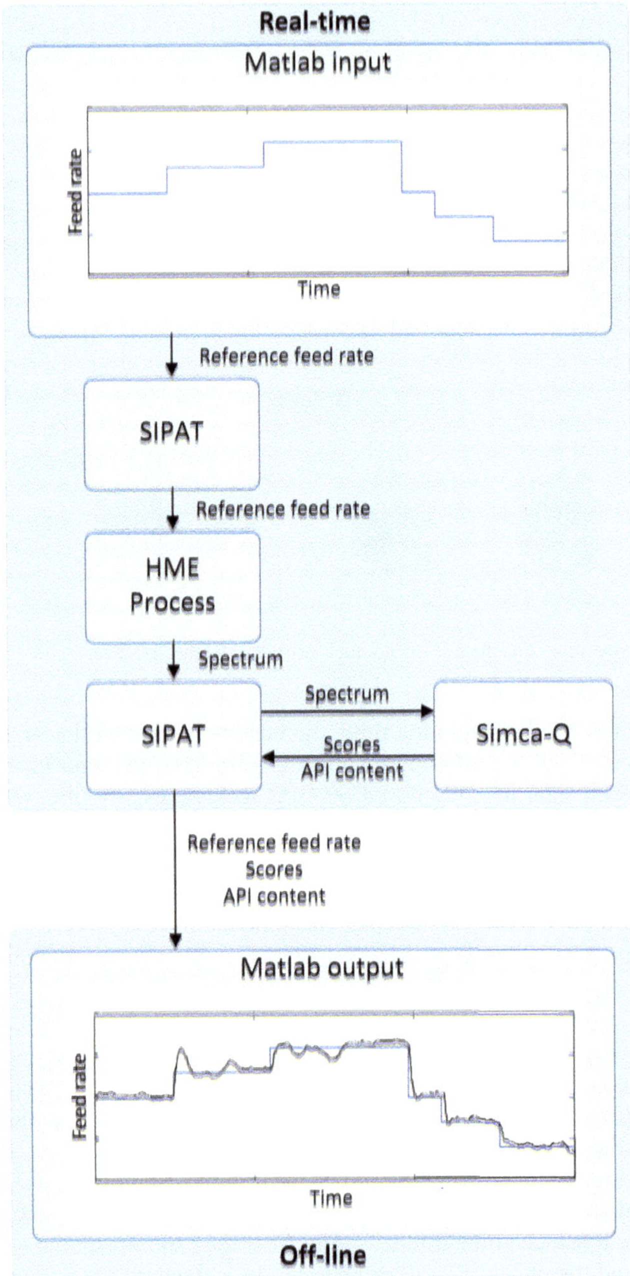

Fig. 15.13 Manipulation of the reference feed rate and observation of the spectrum that can be analyzed in real time using SIMCA-Q. Input and output data can be analyzed offline, i.e., in MATLAB

15.6 Continuous Process Analysis

Process analysis is important throughout the whole life cycle of a drug, from process development to production monitoring and continuous improvement. In all phases rational decisions depend on good data. Especially during continuous manufacturing, the state of the process needs to be known. For example, during extrusion if the melt temperature increases, the rheological behavior changes, impacting the processability by downstream operations, i.e., die-face pelletizing. Temperature can also influence the polymorphic forms of the API, or if the API dissolves in the matrix or remains crystalline. Thus, temperature needs to be tightly controlled. However, by displaying all measured process parameters, e.g., all temperature readings along the barrel, it is hard to grasp important information, as raw data may overwhelm operators. Additionally, computerized fault detection also needs to summarize multiple process parameters. Therefore, methods for condensing information are critically needed and are generally described as *process analysis*. Process analysis helps providing the necessary understanding of the interdependencies of various factors.

In this section, examples of process analysis approaches are presented, including:

- *Monitoring of process parameters*: Here data are collected, to develop and optimize a process and to avoid all kinds of process upsets (i.e., pressure build up). This facilitates the development of a knowledge space, in which the process is yielding consistent results.
- *Qualitative analysis*: An important utility to summarize the information contained in the large amount of process parameters or spectra is qualitative analysis, i.e., principal component analysis (PCA). Additionally, it can be useful for detection of deviations from the desired state during continuous extrusion.
- *Quantitative analysis*: Often critical quality attributes (CQAs) cannot be measured directly by a sensor. Therefore, the impact of the process state on the CQAs has to be predicted. Exemplarily, the API content prediction with a PLS model is presented.

In the following examples, a corotating twin-screw extruder (ZSK 18, Coperion GmbH, Germany) was used, and the raw material was fed with two twin-screw LIW feeders (K-PH-CL-24-KT20 & K-CL-KT20, K-Tron, Switzerland) at the first barrel. The pellets were obtained by a novel pelletizer, developed by Automatik Plastics Machinery GmbH. For spectral acquisition a diode array-based NIR spectrometer, SentroPAT FO (Sentronic GmbH, Germany), with a fiber-optic Dynisco NIR probe was used. The Dynisco probe is a special probe for 1/2″ UNF thread, as used for extruders. The spectrometer covers the wavelength range of 1100–2200 nm with a resolution of 2 nm. NIR spectra were collected in the 8-0 plate as described earlier. The 120 spectra were averaged and an integration time of 0.014 s per spectrum was used. SIPAT (Siemens AG, Belgium) was applied as data acquisition software to collect, timely align, and store the process and spectral data.

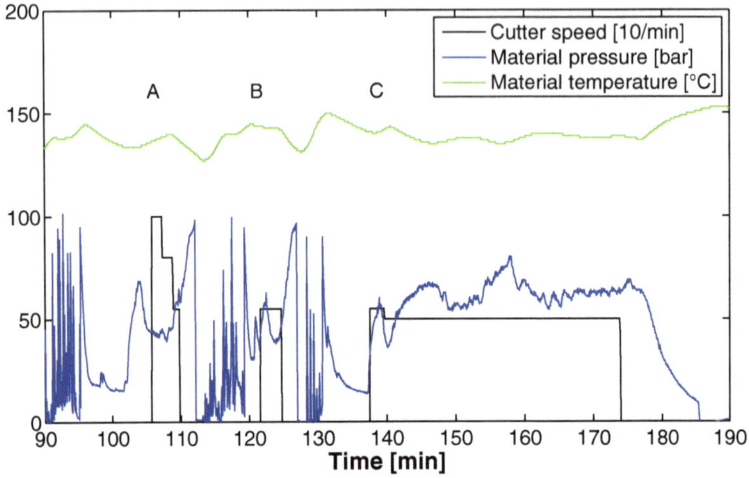

Fig. 15.14 Qualitative Analysis: instable process (early stage of process development)

15.6.1 Monitoring of Process Parameters

The main driving force behind process analysis, as part of the QbD development, is to develop a routine process, which produces a consistently high product quality. This requires process parameter settings which provide a stable process. A stable process means that constant input variables (such as temperature profile, knife speed, etc.) lead to constant output variables (product quality), for example, die pressure, material temperature. A theoretical steady-state process is defined by constant process parameters over the entire process time. In reality, however, measurement accuracy or process fluctuations lead to varying output variables within a process window. As long as all variables stay within a defined range, the process is assumed to be stable. However, first this stable process needs to be developed on the basis of an analysis of process parameters that have a strong impact on the CQAs. Evaluation and validation of the root causes of a stable or an unstable process yields important information about the relation between the process parameter and the stability of a process.

Figure 15.14 shows data of an early development stage of the extrusion line coupled with a hot die-face cutter. Here, process monitoring was used to investigate the interaction between extruder and die-face cutter. For illustrative reasons only three variables (over 50 variables are recorded), cutter speed, material pressure, and temperature were chosen for a simple analysis. The material temperature and pressure showed significant dependence on the cutter. Each start of the die-face cutter (a, b, c in Fig. 13) caused an increase of the material pressure. The reason for the increase was a cooling of the die by the air stream of the cutter, which was used to quench the hot-melt strand and the cut pellets. Subsequently, the melt temperature was reduced and the viscosity was increased, resulting in a higher pressure drop. The first two

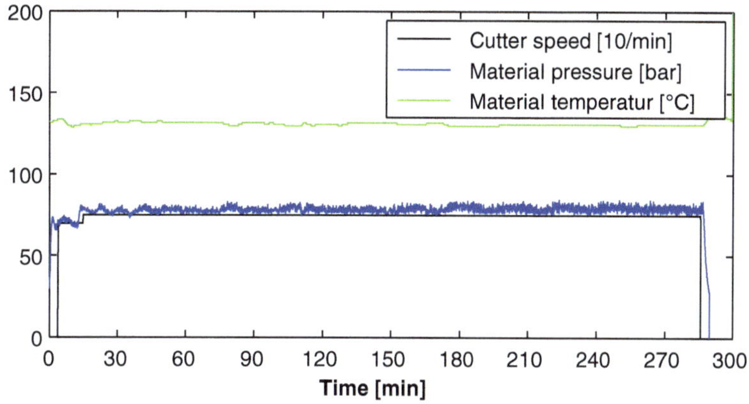

Fig. 15.15 Stable process performance over 5 h (stable process)

starts (A and B) of the hot die-face cutter caused such a high increase (100 bars) that the emergency shutdown of the extruder was triggered. After the second shutdown the energy balance at the extrusion die was changed by lowering the cooling air flow rate and the third start (C) was running for a longer period. Such information can be used to develop a better understanding of the interaction between the individual process parameters. Thus, optimal process settings can be easier defined or the insight triggers a modification of the experimental setup. In this example a better thermal decoupling of the melt temperature in the die zone and the hot die-face cutter was enough to achieve an improved process.

Figure 15.15 shows data of the same extrusion line but for an optimized process. As can be seen the fluctuations of the process variables are within a small range around a steady mean value. The data of such a stable process may provide the basis for the development of an advanced PAT strategy. In the next sections the development of PAT models for qualitative and quantitative analysis based on a stable process is shown.

15.6.2 *Qualitative Process Analysis*

The online monitoring of chemical (e.g., assay, morphology) and physical parameters (e.g., the moisture content, particle size, or temperature of an intermediate material) is even more important in continuous manufacturing, as processes are connected with each other. For example, our downstream process requires an intermediate product (i.e., the hot strand) with consistent physical parameters (temperature and viscosity) in order to process the material properly.

In general, a large number of process variables can be monitored and are available for offline and real-time analysis. Specifically in continuous manufacturing, where several processes are combined to a production line (i.e., feeding, extrusion, and pelletization), the number of variables is even larger. Thus, it is hard to evaluate

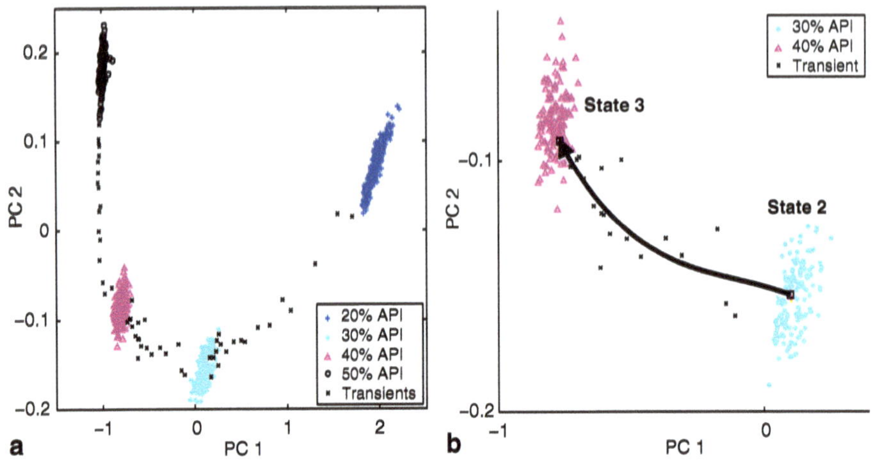

Fig. 15.16 **a** Different API concentrations (20, 30, 40, and 50 %) result in the clustering of samples in the score plot (PC 1 versus PC 2). **b** The trajectory from state 2 (30 % API concentration) to state 3 (40 % API concentration) was monitored in real time via SIPAT. The mean settling time from one process condition to the next stable state was approximately 100 s

trends based on these data. Moreover, many process parameters are highly correlated (e.g., neighboring barrel temperatures of the extruder), and therefore a much lower number of independent variables can be used to monitor the process status.

PCA is an excellent method to extract the essential information of large data sets (process parameters or spectral data). Here, PCA was employed for analysis of NIR spectra, collected while adjusting the feed rates to obtain different API concentrations. The projection of the spectra to a reduced space (i.e., the principal components) was applied to determine trends and variations of the physical and chemical parameters of the intermediate product and to identify process upsets in real time. Figure 15.16a illustrates the score plot of the principal components 1 versus 2. Adjusting the feed rates of the feeders resulted in different API concentrations in the melt, which appear in the score plot as clusters of observations. Applying PCA to the data could thus be seen as classification of the main events affecting the process. The four clusters represented 20 % (blue), 30 % (cyan), 40 % (magenta), and 50 % (black) API content in the melt in the die section (the observations in gray related to transitions between stable process conditions). A closer inspection of the trajectory from state 2 (30 % API concentration) to state 3 (40 % API concentration) allowed real-time monitoring of the process stages, as illustrated in Fig. 15.16b.

15.6.3 Quantitative Spectral Analysis

As mentioned above, a quantitative method is useful to predict a CQA, for example, the API content. This can be done (among other methods) by analyzing the NIR spectrum. Unlike qualitative analysis, the prediction of the CQA with quantitative methods should be independent of changes in physical properties (i.e., melt temperature and material pressure). It should be a soft sensor for this specific attribute. The development of such models is only feasibly for a robust process, with no major process upsets.

For API content analysis the method chosen was partial least squares regression (PLS-R), which is also known as projection to latent structures (Wold et al. 2001). To develop a PLS model, spectra of samples with known API content are taken. The PLS algorithm analyzes the spectrum, and latent variables are statistically computed, which provide the best achievable linear relationship to the known API content. Thereby, the information contained in the spectrum, which commonly consists of > 500 variables (wavelength), is projected to only a few latent variables. Quality measures of the model are R^2 and Q^2, which account for the explained variance within the model data (R^2) and for validation (Q^2). $R^2 = 1$ means, that the model can accurately predict the API content of spectra that are included in the model. But this could also imply that the model fitted noise. This is called overfitting and results in an unstable model. Therefore, a cross validation technique is commonly used. If there is a large difference, greater than 0.2 between R^2 and Q^2 the model is unstable. A model with high Q^2, close to 1, suggests a robust model. Using the PLS method, the API content was predicted from the measured NIR spectrum.

15.6.3.1 Development and Validation of a PLS Model to Determine the API Content

For model building a premix of each API concentration was prepared, ranging from 0 to 60 % API with 10 % increments. These premixes were extruded and the according spectra recorded. Based upon these data, a PLS model was established. The PLS model consisted of the first four latent variables, with $R^2 = 0.999$, $Q^2 = 0.974$, and RMSEP = 0.53 % (root mean square error of prediction). This suggested an accurate model. The validation was determined by excluding a premix from the model, i.e., 10 % API, and predicting the excluded premix.

The resulting model is presented in Fig. 15.17, in which the model data are shown in black and the validation data in blue. The validation data for 0 % API, with a predicted API content of around −9 % was notably off the correct value. When including 0 % API in the model, the predictions were accurate. The main reason was found in the optical properties of the melt. At the extrusion process temperatures, the pure matrix was more translucent compared with a matrix with embedded drug crystals. By adding 10 % crystalline API the melt became significantly more opaque. As such, in the case of a solid dispersion with a translucent matrix in the low API regime the NIR signal was very sensitive to concentration changes. This feature could be exploited to detect low dosage APIs, which is useful, for example, for high-potential APIs.

The model was validated in subsequent continuous extrusion runs. Samples were taken directly at the die, analyzed with HPLC, and compared with the NIR determined

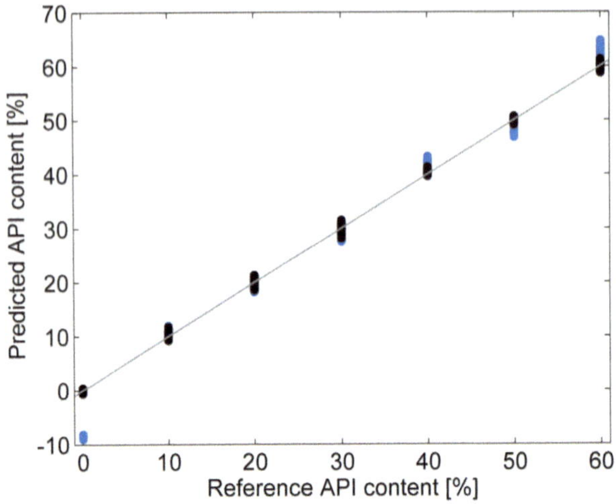

Fig. 15.17 The chemometric model with 0–60 % API content in the melt with $R^2 = 0.999$ (*black*), $Q^2 = 0.974$ (*blue*), RMSEP = 0.53 % and four latent variables. Due to the translucent matrix, cross validation (*blue*) of the 0 % group failed with a predicted API content of about –9 %. Yet, the predictive capability of the model, including 0 %, is high

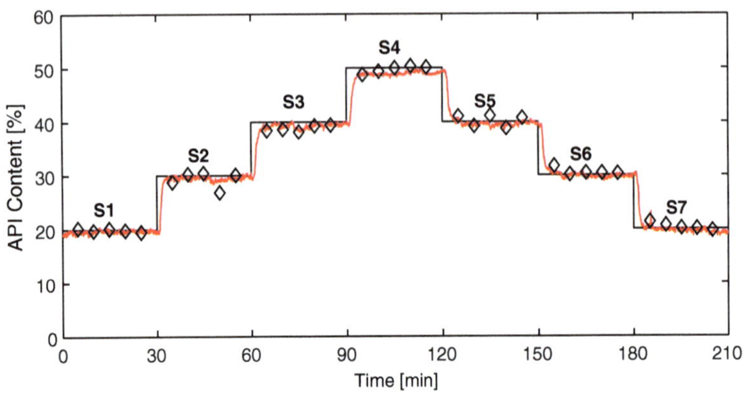

Fig. 15.18 Data of one continuous extrusion run with seven steps in API content, 20–50 %. *Black* preset concentration profile, *red* NIR determined API content, *diamonds* offline HPLC measurements

drug content value. The results are presented in Fig. 15.18, with the NIR determined API content shown in red. The results of the offline samples are represented by black diamonds. The points showed a good agreement with the preset profile and the NIR predictions. One point out of 35 strongly deviated (10.66 %) from the preset concentration and was presumed an outlier. The reason for this point being an outlier is probably due to the sample preparation for HPLC analysis. The average deviation of all samples was –0.47 % and without the outlier –0.16 %. Therefore, the model can be used for drug content monitoring during production and for studying the influences upon it.

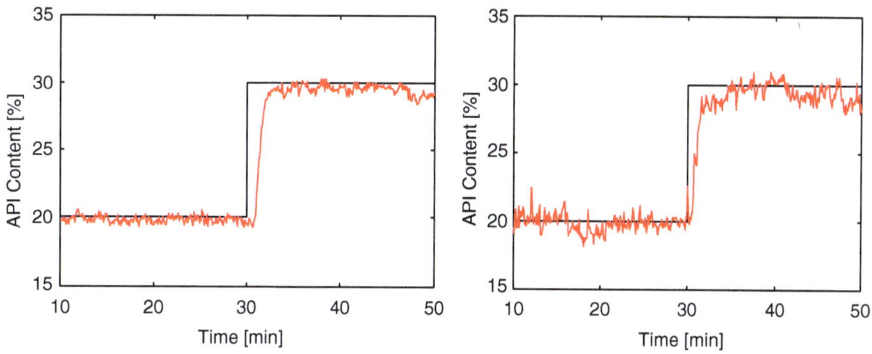

Fig. 15.19 Comparison of different screw designs. *Left*: screw with three kneading zones. *Right*: screw with conveying elements only. Note the differences in the drug content fluctuations. The kneading screw produces a more homogenous melt

15.6.3.2 Comparing Different Screw Designs

One of the major factors that influence uniformity of the drug content is the screw design, which can change the mixing capabilities of an extruder (mixing is the ability to compensate for feeder dosing fluctuations, i.e., pulsating streams). Figure 15.19 compares two-screw designs: one with three kneading zones and the other only with conveying elements. The differences in the drug content uniformity between the screws can be easily assessed by visual inspection. As can be seen the kneading screw produces a much more homogenous API concentration. Obviously, strong high-frequency fluctuations of the feeder exist, which are transported through the extruder, if no kneading elements are present. Thus, kneading elements are critical for obtaining a robust continuous process.

Deviations of NIR determined API content and feeder settings in the transition region between subsequent steady states were due to the residence time in and the mixing behavior of the extruder. Thus, these curves can be used to obtain a good understanding of the mixing and the residence time distribution (RTD) in the extruder. Mixing can be assessed by analyzing the standard deviation (SD) during the steady state between step changes. A low SD indicates a more homogenous melt. The RTD can be calculated by the first derivative of the response by the NIR determined API content to a change of feeder settings. The mean residence time in the extruder is approximated by the delay between the step change of feeder settings and the corresponding concentration increase by half of the step height. The difference in delay times confirm a longer residence time for the kneading screw (Fig. 15.19, left), compared with the conveying screw (right). The average residence time in the extruder is in the order of 1–3 min.

For the purpose of controlling the feeder, only low-frequency fluctuations can be handled. The measurement setup in the 8-0 plate does not allow controlling (i.e., mitigating) high-frequency fluctuations, since the material has already passed the extruder. Therefore, for high-frequency fluctuations this setup can be used as quality control, while long-term feeder deviations can not be controlled.

15.7 Conclusion

Continuous manufacturing is not yet widely used in the pharmaceutical industry. However, there is significant interest. The unit operations of a continuous manufacturing line are similar to other industries. However, the requirements concerning purity, choice of materials, and GMP documentation for pharmaceutical production are typically more stringent. In this chapter the requirements and challenges for continuous manufacturing and process monitoring have been discussed. The key results of our work can be summarized as follows:

Continuous Extrusion

- Development of a stable process is a prerequisite for continuous manufacturing.
- Feeding, extrusion (screw design, etc.), and downstream processes need to be tailored to fit a specific formulation.
- Strong relationship between feeding performance and content uniformity exists.

Downstream

- Numerous products and shapes can be manufactured.
- There is a strong coupling between upstream and downstream processes (i.e., material temperature at the extrusion die due to an altered energy balance).
- Control strategies for process disturbances are needed.

PAT Analyzers

- Suitable sensor positions need to be determined.
- Inline monitoring systems need to be adapted to the process and carefully tuned.
- Robust inline API content determination is required to ensure quality.

Computerized Systems:

- GMP requirements for computerized systems need to be met.
- Integration of unit operations and PAT analyzers into a central system is beneficial (e.g., via SIPAT or other software).
- Data acquisition and real-time monitoring of process parameters is mandatory.

References

Alig I, Fischer D, Lellinger D, Steinhoff B (2005) Combination of NIR, Raman, ultrasonic and dielectric spectroscopy for in-line monitoring of the extrusion process. Macromol Sym 230(1):51–58

Alig I, Steinhoff B, Lellinger D (2010) Monitoring of polymer melt processing. Meas Sci Technol 21(6):062001

Bakeev K (2010) Process analytical technology: spectroscopic tools and implementation strategies for the chemical and pharmaceutical industries. Wiley, Hoboken

Bialleck S, Rein H (2011) Preparation of starch-based pellets by hot-melt extrusion. Eur J Pharm Biopharm 79(2):440–448

Coates PD, Barnes SE, Sibley MG, Brown EC, Edwards HGM, Scowen IJ (2003) In-process vibrational spectroscopy and ultrasound measurements in polymer melt extrusion. Polymer 44(19):5937–5949

Crowley MM, Zhang F, Repka MA, Thumma S, Upadhye SB, Battu SK, McGinity JW, Martin C (2007) Pharmaceutical applications of hot-melt extrusion: part I. Drug Dev Ind Pharm 33(9):909–926
De Beer T, Burggraeve A, Fonteyne M, Saerens L, Remon JP, Vervaet C (2011) Near infrared and Raman spectroscopy for the in-process monitoring of pharmaceutical production processes. Int J Pharm 417(1–2):32–47
De Frenne F (2011) User manual; simatic sipat version 3.1.1. History. Siemens AG, p 230
De Tandt P (2011) Exception handling; simatic sipat version 3.1.1. Siemens AG, p 21
Engisch WE, Muzzio FJ (2012) Method for characterization of loss-in-weight feeder equipment. Powder Technol 228:395–403
Esbensen KH, Paasch-Mortensen P (2010) Process sampling: theory of sampling—the missing link in process analytical technology (PAT). In: Bakeev KA (ed) Process analytical technology, 2nd edn. Wiley, Hoboken, pp 37–80
European Commission (2011a) Annex 11: Computerized Systems. Good manufacturing practices for medicinal products for human and veterinary use, vol 4. European Commission, Health and Directorate General (Public Health and Risk Assessment Pharmaceuticals), Brussels
European Commission (2011b) Documentation Good manufacturing practices for medicinal products for human and veterinary use, vol 4. European Commission, Health and Directorate General (Public Health and Risk Assessment Pharmaceuticals), Brussels
Fischer D, Bayer T, Eichhorn KJ, Otto M (1997) In-line process monitoring on polymer melts by NIR-spectroscopy. Fresenius J Anal Chem 359(1):74–77
Fischer D, Müller J, Kummer S, Kretzschmar B (2011) Real time monitoring of morphologic and mechanical properties of polymer nanocomposites during extrusion by near infrared and ultrasonic spectroscopy. Macromol Sym 305(1):10–17
Hertel IV, Schulz C-P (2010a) Zweiatomige Moleküle. Atome, Moleküle und optische Physik 2. Springer, Heidelberg, pp 1–88
Hertel IV, Schulz C-P (2010b) Molekülspektroskopie. Atome, Moleküle und optische Physik 2. Springer, Heidelberg, pp 247–327
Hiob M (2011) Dokumentation. GMP-Berater. Maas und Peither AG – GMP-Verlag, Schopfheim
ICH Harmonised Tripartite Guidline (2009) Pharmaceitical development Q8(R2) international conference on harmonisation of technical requirements for registration of pharmaceuticals for human use
ISPE Headquarters (2008) GAMP® 5: a risk-based approach to compliant gxp computerized systems. p 352
Khinast JG, Baumgartner R, Koscher G, Roblegg E (2012) R1142GB: System for producing a solid preparation from a suspension.
Khinast JG, Baumgartner R, Roblegg E (2013) Nano-extrusion: a one-step process for manufacturing of solid nanoparticle formulations directly from the liquid phase. AAPS PharmSciTech. doi:10.1208/s12249-013-9946-0
Kipping T, Rein H (2013) A new method for the continuous production of single dosed controlled release matrix systems based on hot-melt extruded starch: Analysis of relevant process parameters and implementation of an in-process control. Eur J Pharm Biopharm 84(1):156–171
Kohlgrüber K (2007) Co-rotating twin-screw extruder. Carl Hanser Verlag, München, p 369
Koller DM, Posch A, Hörl G, Voura C, Radl S, Urbanetz N, Fraser SD, Tritthart W, Reiter F, Schlingmann M (2011) Continuous quantitative monitoring of powder mixing dynamics by near-infrared spectroscopy. Powder Technol 205(1–3):87–96
Linz T, Seeger S (2006) 21 CFR Part 11 Revesited. In: Swarbick J (ed) Encyclopedia of pharmaceutical technology, third. Informer Healthcare USA Inc New York, p 4370
Mezger T (2011) The rheology handbook, 3rd revise. Vincentz Network, p 432
Mürb R-K (2012) Kunststoff granulieren und/oder pelletieren? Chem Ing Tech 84(11):1885–1893
O'Neill S, Grout B, Diehl B, Garsthein E, Hammond S, Moshgbar M, Maris S, Timmermans J, Redl K, O'Sullivan J (2011) Prozessanalytische technologien. GMP-Berater. Maas & Peither AG GMP-Verlag, Schopfheim

Peinado A, Hammond J, Scott A (2011) Development, validation and transfer of a near infrared method to determine in-line the end point of a fluidised drying process for commercial production batches of an approved oral solid dose pharmaceutical product. J Pharm Biomed Anal 54(1): 13–20

Rauwendaal C (2001) Polymer Extrusion, 4th ed. Hanser, Munich, p 781

Reichenbächer M, Popp J (2007) Schwingungsspektroskopie. Strukturanalytik organischer und anorganischer Verbindungen, B.G. Teubner Verlag / GWV Fachverlage GmbH, Wiesbaden Teubner, pp. 61–118

Repka MA, Majumdar S, Battu SK, Srirangam R, Upadhye SB (2008) Applications of hot-melt extrusion for drug delivery. Expert Opin Drug Deliv 5(12):1357–1376

Roblegg E, Jäger E, Hodzic A, Koscher G, Mohr S, Zimmer A, Khinast J (2011) Development of sustained-release lipophilic calcium stearate pellets via hot melt extrusion. Eur J Pharm Biopharm 79(3):635–645

Roemer M (2011) Computervalidierung. GMP-Berater, Maas & Peither AG GMP-Verlag, Schopfheim

Roggo Y, Chalus P, Maurer L, Lema-Martinez C, Edmond A, Jent N (2007) A review of near infrared spectroscopy and chemometrics in pharmaceutical technologies. J Pharm Biomed Anal 44(3):683–700

Saerens L, Dierickx L, Quinten T, Adriaensens P, Carleer R, Vervaet C, Remon JP, De Beer T (2012) In-line NIR spectroscopy for the understanding of polymer-drug interaction during pharmaceutical hot-melt extrusion. Eur J Pharm Biopharm 81(1):237–230

Scheibelhofer O, Koller DM, Kerschhaggl P, Khinast JG (2012) Continuous Powder flow monitoring via near-infrared hyperspectral imaging. IEEE international instrumentation and measurement technology conference proceedings (I2MTC)

Schenck L, Troup GM, Lowinger M, Li L, McKelvey C (2011) Achieving a Hot melt extrusion design space for the production of solid solutions.In. In: am Ende DJ (ed) Chemical engineering in the pharmaceutical industry: r & d to manufacturing. Wiley, Hoboken, pp 819–836

Siesler HW (2007) Basic principles of near-infrared spectroscopy. In: Burns DA, Ciurczak EW (eds) Handbook of near-infrared analysis, 3rd edn. CRC Press Taylor & Francis Group, NW, pp 7–20

Siesler HW, Ozaki Y, Kawata S, Heise H (eds) (2002) Near-infrared spectroscopy: principles, instruments, applications. Wiley-VCH, Weinheim

Tumuluri SVS, Prodduturi S, Crowley MM, Stodghill SP, McGinity JW, Repka MA, Avery BA (2004) The use of near-infrared spectroscopy for the quantitation of a drug in hot-melt extruded films. Drug Dev Ind Pharm 30(5):505–511

US Department of Health (2004) Guidance for Industry. PAT—A framework for innovative pharmaceutical development, manufacturing, and quality assurance, no. September. FDA, Rockville

US Department of Health (2011) Guidance for industry: 21 cfr part 11; electronic records; electronic signatures, glossary of terms, vol. 1. FDA, Rockville

Vankeirsbilck T, Vercauteren A, Baeyens W, Van der Weken G, Verpoort F, Vergote G, Remon JP (2002) Applications of Raman spectroscopy in pharmaceutical analysis. Trends Anal Chem 21(12):869–877

Wold S, Sjöström M, Eriksson L (2001) PLS-regression: a basic tool of chemometrics. Chem-Eur J 58:109–130

Zema L, Loreti G, Melocchi A, Maroni A, Gazzaniga A (2012) Injection molding and its application to drug delivery. J Control Release 159(3):324–331

Chapter 16
Hot-Melt Extrusion Process Design Using Process Analytical Technology

Andreas Gryczke

Abstract Emerging analytical technologies have facilitated a range of testing approaches to assess critical material attributes of pharmaceutical products. As the industry moves toward a continuous manufacturing paradigm, unit operations capable of supporting this approach to drug production will continue to gain importance. Hot-melt extrusion (HME) is one such operation, conducted in a continuous nature, which can be streamlined with the implementation of process analytical technology (PAT) to function as next generation technology. By incorporation of in-line probes and spectral detectors it is possible to determine compositional and process aspects of the production. This chapter describes the underlying principles of extrusion and associated monitoring technologies for implementation of a PAT-based development approach.

Abbreviations

HME	Hot-melt extrusion
PAT	Process analytical technology
QbD	Quality by design
GI tract	Gastrointestinal tract
FDA	(US) Food and drug administration
NIR	Near-infrared
MFI	Melt flow index
FT-NIR	Fourier transformation near-infrared spectrometry
QA	Quality assurance
DoE	Design of experiments
CQA	Critical quality attribute
SPC	Statistical process control
SE	Specific energy
SME	Specific mechanical energy
SMEC	Specific mechanical energy consumption
API	Active pharmaceutical ingredient

A. Gryczke (✉)
BASF SE, Lampertheim, Germany
e-mail: andreas.gryczke@basf.com

M. A. Repka et al. (eds.), *Melt Extrusion*,
AAPS Advances in the Pharmaceutical Sciences Series 9,
DOI 10.1007/978-1-4614-8432-5_16,

n	Extruder screw speed
τ	Extruder torque
$\dot{m}$	Extruder throughput
P	Extruder motor power
O	Extruder engine loading in percent
n_{max}	Extruder maximum adjustable screw speed
ΔP	Pressure at extruder die
SFL	Specific (screw) feed load
VSFL	Volume-specific (screw) feed load
ρ	Melt density
RT	Residence time
RTD	Residence time distribution
$\bar{t}$	Mean residence time
SLR camera	Single-lens reflex camera
Px	Pixel
ANOVA	Analysis of variance
S/N	Signal to noise ratio
$RT_{mean} = \bar{t}$	Mean residence time
Rev	Revolution
V_{free}	Free volume of extruder (inner barrel volume minus screw volume)
T_g	Glass transition temperature
Mw	Molecular weight
C_p	Specific heat capacity
D_O/D_I	Outer screw diameter to inner screw diameter ratio
DSC	Differential scanning calorimetry
QC	Quality control
OOS	Out of specification
R & D	Research and development
DF	Degree of freedom

16.1 Introduction

Hot-melt extrusion (HME) is the most frequently used process in the plastics industry and has gained significant attention over the last 30 years in the pharmaceutical space, with many companies including BASF contributing to the development (Goertz and Klimesch 1987). Today, the technology is used to support the production of amorphous dispersions, controlled release products, and shaped systems for several different routes of administration. Solid dispersions are manufactured using melt extrusion to reduce drug particle size, thereby increasing the particle surface for better dissolution. They also disperse the drug molecularly in a polymeric carrier so that it is already in a dissolved state when it is released into the GI tract. This can have major advantages for poorly soluble compounds.

These systems can be manufactured with soluble, insoluble, or insoluble but permeable polymers to control the drug release of the drug substance. Material selection can also be used to produce brittle or soft extrudates. As the viscosity can be adjusted later by the addition of, e.g., plasticizer, its adhesiveness can be easily controlled at the application temperature. This allows the manufacture of easy-to-grind extrudates in the same manner as it allows the manufacture of adhesive films. Beyond material selection, melt extruders can also be used to manufacture dispersions with different physical characteristics. Due to the high pressures generated and compounding of materials in the molten state, dense extrudates are frequently prepared. Through the use of supercritical fluids it becomes possible to modulate product densities of the product to alter performance.

The characteristics of a particular extrudate are defined by:

a. The ingredients of the extrudate.
b. The process parameters and thereby the process setup required to manufacture the extrudate.

This means that an extrudate is defined by:

a. Its formulation.
b. Its constitution such as shape, density, etc.

Pharmaceutical formulations manufactured by a melt extrusion process often contain (besides the drug molecule itself and carriers such as polymers, lipids, or sugars/sugar alcohols) further ingredients such as plasticizers, glidants, fillers, antioxidants, antistatics, bioadhesives, disintegrants, pH-modifiers, pigments and colorants, solubilizers, and pore formers. Practically any substance or excipient which is used in pharmaceutical formulations can be used for the manufacture of melt-extruded pharmaceutical formulations. But, even more exotic materials can be extruded like clays, ceramics, food, metals, and of course, plastics (Rauwendaal 2010).

An extruder is a machine that develops pressure to pump a material or composite through a die and can be considered a type of pump. These systems are used to produce a material flow where the product can be shaped in further steps, such as in the following examples:

- Strands, which are cooled and cut into small cylinders (for the manufacture of implantable stents and filaments)
- Hot-face pelletization for making pellets to fill into capsules
- Sheets for breaking into flakes for subsequently milling to powder for further processing into tablets
- Films which can be cast on a carrier or can be laminated between several layers, e.g., for manufacturing transdermal therapeutic systems
- Blown foils, e.g., for coating tablets or other dosage forms
- Calenders, where the melt can be shaped into the desired form
- Chill-roll, where the melt gets cooled under controlled conditions to allow a specified relaxation of the extrudate
- Injection molding for shaping into tablets or capsules

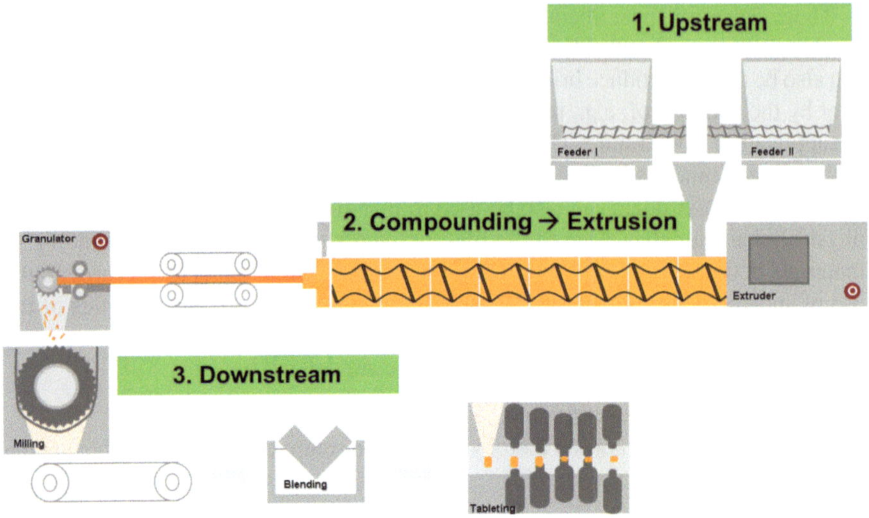

Fig. 16.1 Overview of the melt extrusion process

- Tubes and profiles
- Foaming for higher porosity in the extrudate or for improved milling properties
- Catheters
- Compression molding to shape into final dosage form.

As discussed earlier, a melt extruder is used to manufacture solid dispersions. The extruder is principally used for the uptake of substances, melting of substances, distributing and dispersing substances into others, homogenizing, degassing, and finally discharging (extrusion). The substance can be fed into an extruder as a solid, a liquid, a paste, a melt, or a gas. Successful development of extrusion systems requires a complete understanding, which can be achieved through the application of process analytical technology (PAT) as part of a quality by design (QbD) development program. This chapter discusses the application of PAT and QbD for HME.

16.2 Characteristics of Hot-Melt Extrusion Processes

This section describes process attributes of melt extrusion, rather than specific advantages so as to allow the reader to judge for themselves which points can be considered favorable and which points are to be avoided. The melt extrusion process as shown in Fig. 16.1 is a continuous process since it feed raw materials without interruption which are then processed to an extruded intermediate and potentially directly shaped. Several process steps are connected on-line, including the feeding of materials (the upstream), the compounding/extrusion part and the necessary downstreaming (cooling/shaping). The downstream process can be continued on-line to a final product such as a tablet packed in a blister and packaged.

In the extrusion line, all the material particles to be processed pass through the machine more or less one after the other. Compared to other processes such as fluid-bed systems or blenders which are batch systems, extrusion is a continuous small volume process which processes material in a longitudinal fashion along the length of the process section. While the overall motion of material can be viewed to proceed in one direction, the complex flow within the system allows for a range of internal patterns for material motion, including sections where flow is in the opposite direction.

The compounding step comprises a distributive mixing step, where the materials are thoroughly mixed and a dispersive mixing step, where agglomerates are broken down and particle size is reduced, in some cases to the molecular level. The mixing or blending process, called compounding, can be carried out with or without the aid of solvents.

The required energy, necessary for the desired transformation of material (blending, dispersing, melting, etc.), can be dosed very precisely since in the best case the energy is supplied via the extrusion screws, with material melted primarily through viscous heat generation. This allows fine-tuning of the energy consumption by varying the screw speed, throughput, and the configurations of the screws and the barrel.

As the extruder is meant to be a closed system like a pipe with forced conveying by the extruder screws, the residence time distribution (RTD) can be adjusted and controlled very well. The mean residence time (RT_{mean}) can be very short (less than 30 s) but it can also be very long (> 5 min). The extruder offers a broad adjustable range which is also same for energy consumption.

The extruder has a small footprint, especially in a production scale, and compares very favorably with other equipment which has to process individual batches and therefore requires a large space. The process also offers the possibility for in-line and on-line analytics. The majority of pharmaceutical processes can be equipped with analytical systems; however, attaching these systems to the melt extrusion process is interesting since it is a continuous process. During manufacture not only the material will pass a certain position only once, but also all material will pass through a given region. This allows for the potential of full batch sampling and analysis. The extrusion process can produce very large batches by running without interruption up to 24 h a day, 7 days a week. Given the high output, the risk of not monitoring the critical process steps might lead to high cost if a large amount of material turns out to be out of specification (OOS).

Extruders equipped with PAT should be seen as one component in a series of different process devices connected on-line to continuously manufacture a product. The next section will show the possibilities and the specifics of PAT as applied to the melt extrusion process.

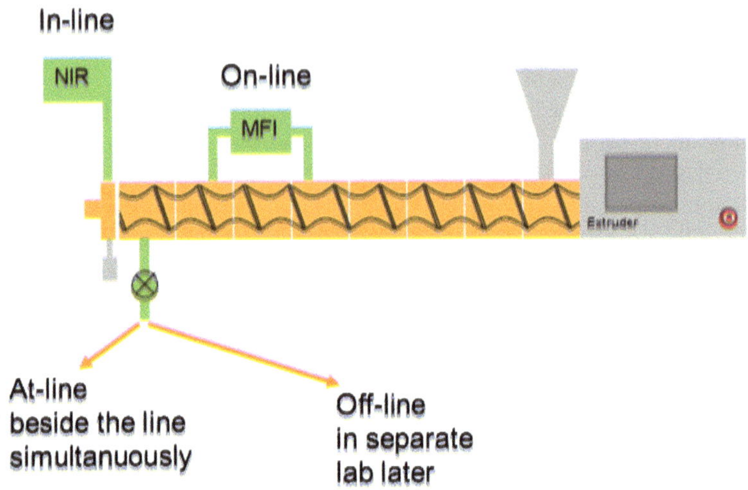

Fig. 16.2 In-line, off-line, at-line, and on-line monitoring. *NIR* near-infrared, *MFI* melt flow index

16.3 Definition of Process Analytical Technology

The scope of this section provides a definition of PAT and its associated tools for characterization. This begins with a definition of the key terms, as defined in the Food and Drug Administration (FDA) Guidance, related to testing of products during the production cycle (US FDA 2004):

- In-line: Measurement where the sample is not removed from the process stream and can be invasive or noninvasive.
- On-line: Measurement where the sample is diverted from the manufacturing process and may be returned to the process stream.
- At-line: Measurement where the sample is removed, isolated, and analyzed in close proximity to the process stream.
- Off-line: Measurement disconnected in a timely and local manner from the manufacturing process.

These concepts are further illustrated in Fig. 16.2.

Depending on the nature of the equipment and the technique, it is possible to use analytical systems in different modalities. This explains why a spectrometer can be used both in-line and on-line. An example of an on-line measurement is an on-line rheometer which is fed by a melt stream diverted from the extruder's main stream. In-line sensors allow measurement with no delay but they might be influenced by the process conditions (temperature, pressure, melt flow). On-line methods require a diverted sample melt stream (which can be withdrawn from the main process whether continuous or discontinuous), which reaches the measuring device with a certain delay. On-line sensors are isolated from the main process, which allows their maintenance and exchange by other sensors during the production run.

Regulatory authorities have specific definitions of and views on how these systems should be applied in the development and manufacturing settings. Specifically, the guidances state

> The Agency considers PAT to be a system for designing, analyzing, and controlling manufacturing through timely measurements (i.e., during processing) of critical quality and performance attributes of raw and in-process materials and processes, with the goal of ensuring final product quality. It is important to note that the term analytical in PAT is viewed broadly to include chemical, physical, microbiological, mathematical, and risk analysis conducted in an integrated manner. The goal of PAT is to enhance understanding and control the manufacturing process, which is consistent with our current drug quality system: quality cannot be tested into products; it should be built-in or should be by design. Consequently, the tools and principles described in this guidance should be used for gaining process understanding and can also be used to meet the regulatory requirements for validating and controlling the manufacturing process. (US FDA 2004)

Further, the FDA sees PAT to be conducive to the design and development of well-defined and well-understood processes that ensure a predefined quality for the manufactured product. Beyond the regulatory benefits, application of a structured development using PAT can allow organizations to realize tangible benefits in the product lifecycle.

1. Speed

Live feedback for process control is possible. While the production line is running, the process can be monitored and adjusted to ensure product specifications are maintained and no material is lost.

2. Automated sample handling

This is important from the safety point of view. There is no operator contact with the processed material in in-line and on-line analytics. Also, in the sense of sampling, the possibility of errors can be reduced.

3. Laboratory test replacement

Data which can be generated in real time do not need to be determined off-line later. This can also save lab space and allow for the potential of real-time release.

4. In situ analysis

Measuring, e.g., with a Fourier transformation near-infrared spectrometry (FT-NIR) in-line in the running process, one can measure the material being processed at the processing temperature, at the melt flow speed, and under the actual pressure in the system. This would be difficult to obtain off-line in an analytical lab. Especially, the possibility of monitoring morphology changes in the material in the process is an important opportunity. By having several different analytical tools running in parallel in-line or on-line, more complex analytical problems can be addressed.

5. Economics

By continuously monitoring the process, the product yield can be optimized. In the sense of real-time release (as envisaged by the FDA), produced material would not have to be put in storage after manufacture to wait for release based on off-line analytics. This can reduce both warehouse and lab costs.

In summary, this shows just how broad the impact of PAT is. The following section will adapt this to the melt extrusion process.

Rauwendaal (2010) lists the requirements for efficient extrusion which he intended for plastics extrusion but it is even more relevant in pharmaceutical melt extrusion:

- Efficient machinery
- Preventive maintenance
- Quality materials
- Trained work force
- Efficient troubleshooting
- Design of experiments (DoE)
- Statistical process control (SPC)
- Instrumentation and control
- Data acquisition system
- Total line control.

The last five points especially have to be considered in the context of process analysis and interplay with PAT. Every modern stand-alone extruder is already equipped with a certain amount of instrumentation, including:

- Temperature measurement and controls for the extruder barrels
- Speed measurement and control for the screw(s)
- Pressure measurement at least at the extruder die with feedback control as emergency switch-off of the extruder in the case of overpressure
- Torque measurement (at least relative) by measuring the power consumption of the extruder motor and feedback control for stopping the extruder screws or slowing them down in the case of too high torque being applied to the screws.

Rauwendaal (2010) further recommended measuring and monitoring at least the following parameters in an extrusion process with standard instrumentation:

- Melt pressure
- Melt temperature
- Motor load
- Barrel and die temperatures
- Screw speed
- Power consumption in each temperature zone (heating and cooling)
- Ambient temperature
- Relative humidity
- Temperature of feedstock entering the extruder
- Moisture level of feedstock entering the extruder
- Flow rate of cooling water
- Temperature at cooling water inlet
- Temperature at cooling water outlet
- Vacuum level at venting port (when applicable).

By monitoring these parameters many potential problems in the extrusion process can be avoided. The process can thus be maintained, optimized, and characterized by certain important aspects. Extrusion system parameters, such as specific energy (SE), can be determined by these parameters. These system-specific parameters are discussed in more detail in Chap. 2.

The aforementioned extruder instrumentation allows the operator to control the equipment in an optimal way. Some critical quality attributes (CQAs) can be linked via the extrusion system parameters to the values monitored with the extruder instrumentation.

Beyond the standard measurements and process controls used during extrusion, more advanced analytical techniques can be coupled with the technology. Similarly, the term PAT is often used in conjunction with FT-NIR or other spectroscopic tools. This suggests a potential to track material levels of components as well as impurities using advanced measurement technologies. The chemical constitution of the product is an important part of the product specification. However, this applies more to the pharmaceutical industry than to the plastics industry. In a pharmaceutical product, small amounts of impurities can lead to an OOS result where the same impurity level for a plastic product might well be acceptable. Hence, in pharmaceutical quality assurance (QA) the chemical constitution is analyzed and confirmed. Spectroscopic methods allow measuring the chemical information at a molecular level. Quality attributes, such as drug concentration in extrudates, are often determined using spectroscopic tools where the spectra are used as data for comprehensive regression analysis. This method of measurement is called "soft sensor analysis." A soft sensor uses available measured data (e.g., FT-NIR spectra) to predict parameters such as drug concentration. The required process model for the soft sensor can be based on a process simulation or on a regression model from a performed set of experiments (e.g., from the DoE). It is not a requirement that a soft sensor is fed with complex raw data such as a spectrum, but could be values from extruder instrumentation such as speed, throughput, pressure, temperature, etc. Hence, it is conceivable to design a soft sensor to "measure" attributes like impurity level, if the impurity level can be linked to extrusion system parameters, such as SE consumption and residence time (RT) and if these extrusion system parameters can be determined based on values from the extruder instrumentation as previously described.

To avoid unnecessary cost and effort, it is important to develop a comprehensive understanding of the data the basic extruder instrumentation supplies. Within the context of process and formulation development, it can be useful to use PATs (including spectroscopic tools) right from the start to identify potential risks in the process (for the formulation as well as for process stability). If the process is well designed, PATs other than the standard extruder instrumentation might be used at process stages where the risks could not be reduced or where monitoring of product quality is not possible with standard instrumentation (Fig. 16.3).

Guenard and Thurau (Bakeev 2010) give a comprehensive and excellent introduction to the topic of implementation of PAT, including definitions.

Process analytics is not an invention by the pharmaceutical industry but has experienced increasing interest in recent years for good reasons. In 2004, the US FDA

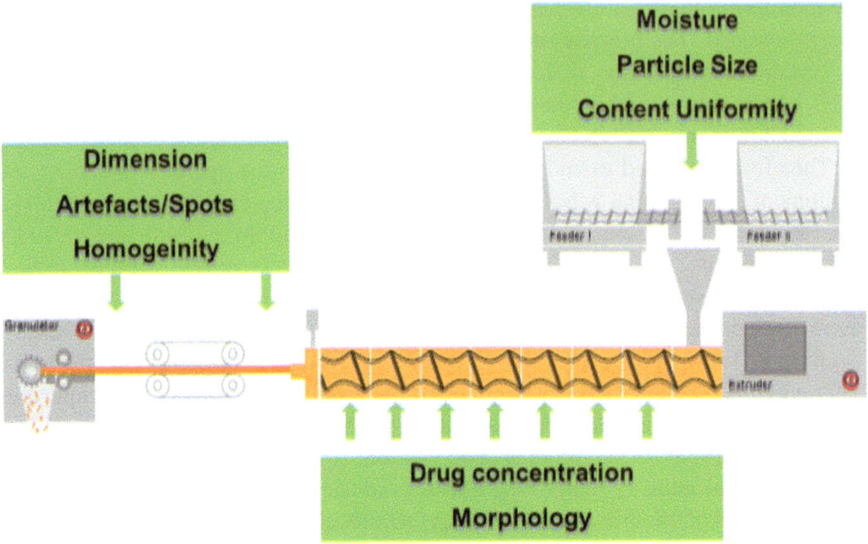

Fig. 16.3 Sampling locations for process monitoring

showed a regulatory initiative in its guidance (US FDA 2004), which has motivated the pharmaceutical industry to apply such innovative systems.

16.4 Process Analysis Elements

Regardless of whether one thinks about univariate sensors, such as a speedometer for measuring and controlling the screw speed, or if one thinks about a complex multivariate system, such as FT-NIR spectrometry, several key elements should be considered which together can be seen as a closed (sub)system. The key elements are:

1. *A process instrument* is necessary. It can be a simple sensor or a more complex device such as a spectrometer, a gas chromatograph, or an HPLC.
2. *An instrument control and data acquisition system* is also required. This is usually hardware which can be a controller for a pressure transducer including an analog–digital converter or it can be an instrument with all the electronics necessary to operate a spectrometer.
3. *Method development* is important to ensure that, e.g., the electrical signal coming from a temperature sensor is producing the right signal and is not influenced by other factors such as melt flow.
4. *Data processing and chemometrics.* Raw data usually need to be processed to result in information. In the case of a temperature probe, the electrical signal will be correlated to temperature. In the case of spectrometers like FT-NIR, complex

chemometric models are required to make use of the raw spectra collected and to correlate changes or specific characteristics in a spectrum to physical or chemical product attributes or even to process these.

5. *An implementation approach* is very important. In the case of a temperature probe to monitor and control barrel temperature, it is common to use feedback control based on this temperature probe. In this way, the temperature is automatically adjusted by the extruder control system to the set value so that the operator need not switch on or off a heating cartridge manually for a certain period of time. But for other quality attributes like the concentration of a drug in a formulation, it is not that easy and one needs to develop a strategy as to what extent feedback control is desired. Assume a control strategy for OOS events when monitoring the drug concentration in the extrudate at the extruder die. It is important to define not only an allowed deviation in concentration but also the unit of time for which this deviation is acceptable. And, it still has to be decided whether the extrusion line discards material automatically in the case of an OOS event or whether the operator makes the final decision.

16.5 Basics of Process Characterization

Before running the extruder, the system must be correctly assembled. Based on experience, simulation, or estimations, one can make a decision on the screw and the barrel design. It is also possible to determine the need for a degassing section and whether certain mixing and melting sections are required. Information about feeder requirements can also be established based on the overall process design. For example, depending on the nature and number of feed streams it is possible to design split feeding or single feeding of a homogeneous blend. Having decided on a specific temperature profile, one can choose a screw speed and finally start the extruder having set a feed rate.

It is assumed here that the process targets the best possible quality with the chosen setup in the sense of the important CQAs. However, determination of which parameter and to what level they must be adjusted to achieve the desired product attributes may remain unclear. Questions may still remain regarding how to change these parameters to optimize quality and to what degree are the CQAs linked to the extrusion input parameters.

Figure 16.4 shows the link of CQAs to extrusion input variables via the route of extrusion system parameters. Why is this the recommended way? Take the example of low impurities as a CQA. If the impurities are higher than specified and the cause is supposed to be in the process, one would consider reducing, e.g., the barrel temperature. What would this mean? The thermal energy might be reduced by this step, but the viscosity of the material increases due to the lower temperature and this will cause a higher torque, which will result in higher mechanical energy consumption. The result of lowering the barrel temperature might not produce a satisfactory outcome in this case.

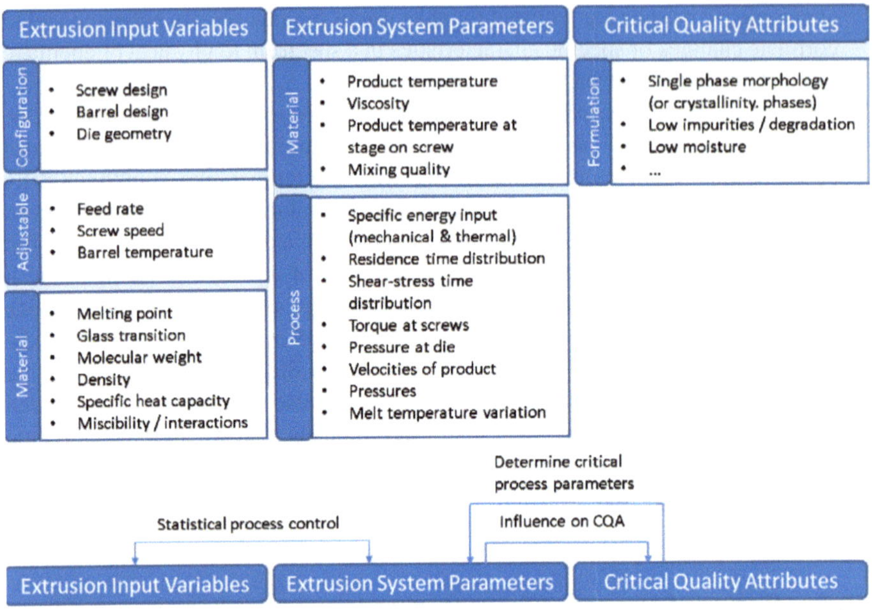

Fig. 16.4 Link of critical product attributes to input variable via process system parameters

The recommended way would be to consider very high energy consumption as a possible cause of degradation of the active pharmaceutical ingredient (API) or excipients. Energy is supplied mechanically and thermally to the material over a certain period of time. It is worth examining the SE consumption, especially at the specific mechanical energy consumption (SMEC) and RTD. In regard to SE consumption, in most cases one will aim to have the extrusion running close to autogenous conditions. The point of autogenous condition is reached if the entire energy required for conveying the mass inside the extruder, for melting and mixing at a desired temperature and for discharging the mass, is provided by the extruder screws only. This means that the heating and cooling systems would not have to contribute to maintain the process at the desired temperature level. In practice, it is often not possible to match the point of autogenous extrusion perfectly; however, most of the required process energy, generally more than 80 %, should be provided by the screws. There are certain reasons for the importance of a good process design in small scale processes as follows:

1. If the screws provide too much energy, surplus energy would have to be withdrawn via the barrel-cooling system. The surplus energy would lead to at least partial degradation of the extruded material and hence increases in impurity levels.
2. If the process on a small scale is not developed to run autogenously, problems can occur at scale-up. This is because of the increasing volume to surface ratio inside the extruder when scaling up. If on a small scale process the barrel is cooled

100 % of the time, the area available to withdraw surplus energy in a large scale process might be insufficient.

3. The material in the extruder is melted or softened mainly due to viscous heat generation, which is the product of viscosity and shear rate squared (Rauwendaal 2010). Viscosity depends highly on the molecular weight (Mw) of the polymer; the higher the Mw, the higher the viscosity and the higher the viscous heat generation. In some cases, excessive viscous heat generation occurs and the extruder has to be cooled to maintain the melt temperature at the desired level (Rauwendaal 2010). The shear rate has a much stronger impact on viscous heat generation. The shear rate depends on screw geometry and screw speed as well as throughput.

To summarize the aforementioned points:

If one considers melt temperature as a cause of impurities, melt temperature is a function of viscous heat generation, which, in turn, is a function of viscosity and shear rate. This means that linked input variables to check or to optimize could be the polymer, the screw geometry, the screw speed, and the throughput. This explains why it might be worthwhile not just to lower the barrel temperature level, since the melt temperature (via the viscous heat generation) would be still the same, but surplus energy would be withdrawn by the extruder cooling system. Although this would probably not be the most efficient way to reduce impurities, changing screw configuration (e.g., less mixing or kneading sections), reducing screw speed, or even considering another polymer with a lower Mw might well be the right way.

It is also necessary for SE consumption and RTD during development as both are important characteristics of the extrusion system. The SE consumption has to be considered as the most characteristic value. With most commercially available extruders only the SMEC is measurable. The SMEC can be calculated using the following equations:

$$\mathrm{SMEC} = \frac{2 \cdot \pi \cdot n \cdot \tau}{\dot{m}} \left[\frac{\mathrm{kWh}}{\mathrm{kg}}\right], \tag{16.1}$$

where n is the screw speed, τ is the torque on the screw shafts and $\dot{m}$ is the throughput (Kohlgrüber and Wiedmann 2008).

Taking n in revolutions per minute (rpm) and τ in Nm, the power P in kW can be calculated using the following equation:

$$P = \frac{2 \cdot \pi \cdot n \cdot \tau}{60} \ [\mathrm{kW}]. \tag{16.2}$$

The SMEC can also be calculated using the following equation (Levine 1997):

$$\mathrm{SMEC} = \frac{n \cdot P \cdot O}{n_{\max} \cdot \dot{m}} \left[\frac{\mathrm{kWh}}{\mathrm{kg}}\right], \tag{16.3}$$

where O is the engine loading in percent and $n_{\max}$ is the maximal screw speed in rpm of the extruder.

Finally, we can calculate the SMEC by using the following equation:

$$\mathrm{SMEC} = \frac{\tau \cdot n}{\dot{m}} \left[\frac{\mathrm{kJ} \cdot \mathrm{min}^{-1}}{\mathrm{kg/min}} = \frac{\mathrm{kJ}}{\mathrm{kg}} \right], \tag{16.4}$$

where τ is the torque in kJ (Villmow et al. 2010).

The SMEC depends much on the torque. The torque applied to the screws by the motor is used for:

1. Turning the screws (τ_{empty})
2. Pumping the material through the die (τ_{pumping})
3. Shearing the material (τ_{shearing}; Liang et al. 2002)

Hence, we can write the total torque τ_{total} applied to the screws as:

$$\tau_{\mathrm{total}} = \tau_{\mathrm{empty}} + \tau_{\mathrm{pumping}} + \tau_{\mathrm{shearing}}. \tag{16.5}$$

Further, we can write:

$$\mathrm{SMEC}_{\mathrm{total}} = \mathrm{SMEC}_{\mathrm{empty}} + \mathrm{SMEC}_{\mathrm{pumping}} + \mathrm{SMEC}_{\mathrm{shearing}}. \tag{16.6}$$

To calculate τ_{pumping} we can use:

$$\tau_{\mathrm{pumping}} = \frac{\mathrm{SFL} \cdot \Delta P}{2 \cdot \pi \cdot \rho_{\mathrm{true}}}, \tag{16.7}$$

where SFL means the specific feed load and is given by:

$$\mathrm{SFL} = \frac{\dot{m}}{n} \tag{16.8}$$

and ΔP is the measured pressure [Pa] at the extruder die.

This means that the determined total SMEC does not describe the energy consumed by the material by shear only. This is important to consider while linking CQAs such as impurities to SMEC. The energy portion which is applied for simply turning the screws is unlikely to be a cause for impurities. The $\mathrm{SMEC}_{\mathrm{empty}}$ can be measured directly and also includes the loss of energy on transmission in the extruder gear box.

Another important characteristic which has a significant influence on product quality is the RTD. This is important, for instance, for heat-sensitive products and for dispersion and melting (Kohlgrüber and Wiedmann 2008). The RTD allows conclusions to be drawn such as the mixing process inside the extruder.

A short RT is often termed one of the attributes of an extrusion process. However, this is unspecific as long as the term "residence time" is not defined. Often, RT is measured with an optical tracer, which has to be of a kind that does not influence the process.

The RTD is represented by the exit age distribution (Levenspiel 1972):

$$\int_0^\infty E(t)dt = 1, \tag{16.9}$$

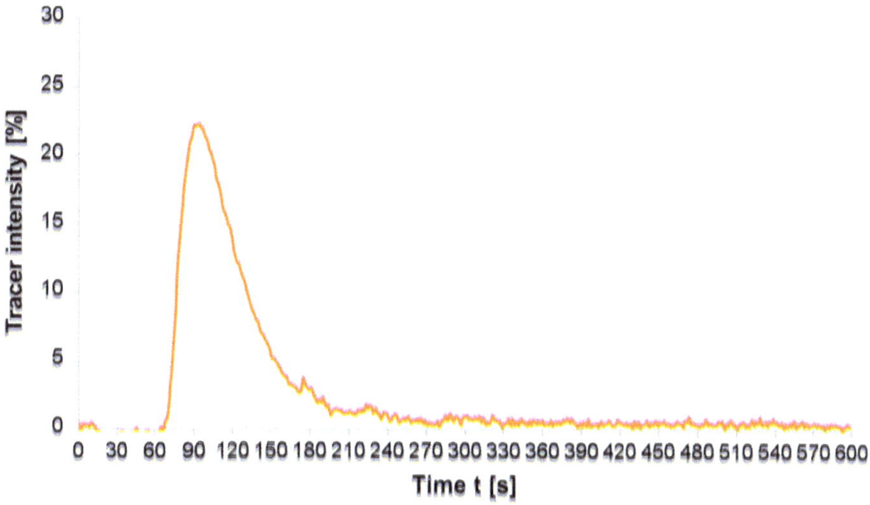

Fig. 16.5 Residence time distribution (RTD) following tracer pulse at time $t = 0$ s

$$E(t) = \frac{c}{\int_0^\infty c dt} = \frac{c}{\sum_0^\infty c \Delta t}, \tag{16.10}$$

where c is the tracer concentration (intensity) at time t. The exit age function has units of inverse time.

The cumulative RTD function can be obtained from (Levenspiel 1972):

$$F(t) = \int_0^t E(t)dt = \sum_0^t E(t)\Delta t = \frac{\sum_0^t c\Delta t}{\sum_0^\infty c\Delta t}. \tag{16.11}$$

The mean residence time is given by the first moment of the exit age function:

$$\bar{t} = \int_0^\infty t \cdot E(t)dt \tag{16.12}$$

The mean residence time can be calculated by using:

$$\bar{t} = \frac{\int_0^\infty tcdt}{\int_0^\infty cdt} = \frac{\sum_0^\infty tc\Delta t}{\sum_0^\infty c\Delta t}. \tag{16.13}$$

Figure 16.5 shows an RTD (plot of raw data, not transformed into exit age distribution). It shows the tracer intensity measured at the die versus time. The time gives the

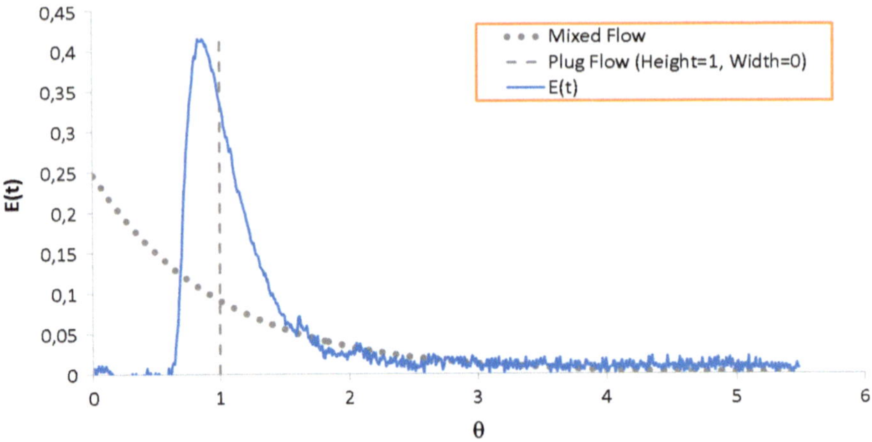

Fig. 16.6 Exit age distribution, also called normalized residence time distribution (RTD)

information on how long it took from the injection of the tracer to appearance at the die. The tracer is injected as a pulse once at the beginning ($t = 0$ s) to a running formulation in the extruder. It visualizes the different velocities inside the extruder for the given extrusion input variables such as screw configuration, throughput, screw speed, etc.

The RTD provides information on:

- How the material is axially distributed inside the extruder under the given circumstances. The spread of the distribution can be seen as a measure of the axial mixing capability of the extruder. Axial mixing has an impact on the dispersive mixing in the extruder.
- How close the mixing comes to either an idealistic plug flow or to an idealistic mixing flow. Taking the reciprocal of the absolute delta in the area under the F(t) curve gives a mixing number, where a larger mixing number means closer to mixing flow. This corresponds mainly to the distributive mixing in the extruder. Figure 16.6 shows, for a 16-mm corotating twin-screw extruder from Thermo Fisher Scientific, that a very low throughput will result in poor mixing even if the SMEC is increased by increasing the screw speed. A certain degree of fill in the extruder is required to improve the distributive mixing.

The RT_{mean} indicates the average time a fluid element remains in the extruder (Kohlgrüber and Wiedmann 2008). Figure 16.7 shows the cumulative RTD, which is often used for better understanding and better evaluation of data. Figure 16.8 shows the relevance of RTD for the homogeneity of the product. Kohlgrüber and Wiedmann (2008) state that the RT_{mean} indicates the average time a fluid element remains in the extruder. Further, they describe the distance between the inflection points in the tracer concentration curve in Fig. 16.8 as a measure of axial mixing where a larger value of “a” means a more pronounced back-mixing and hence the mixing ability of the extruder. To ensure a homogeneous mixture in terms of time while fluctuations

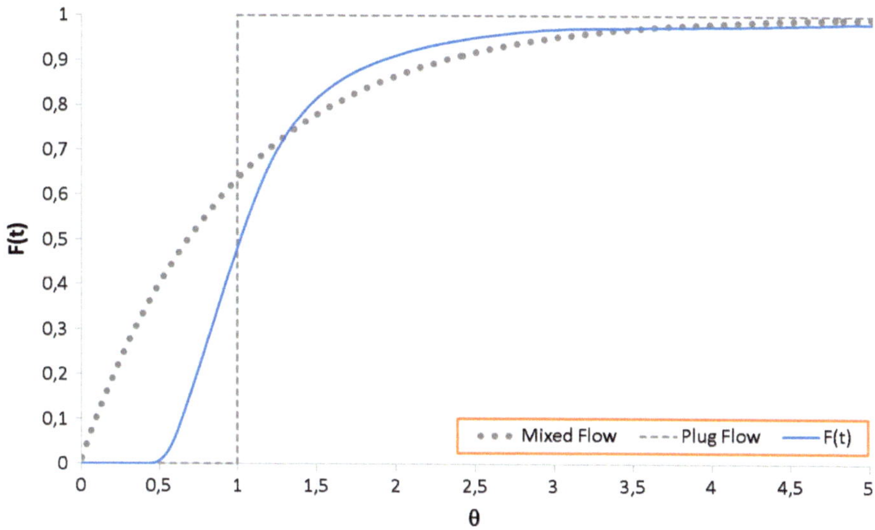

Fig. 16.7 Cumulative function of residence time distribution (RTD) over normalized time

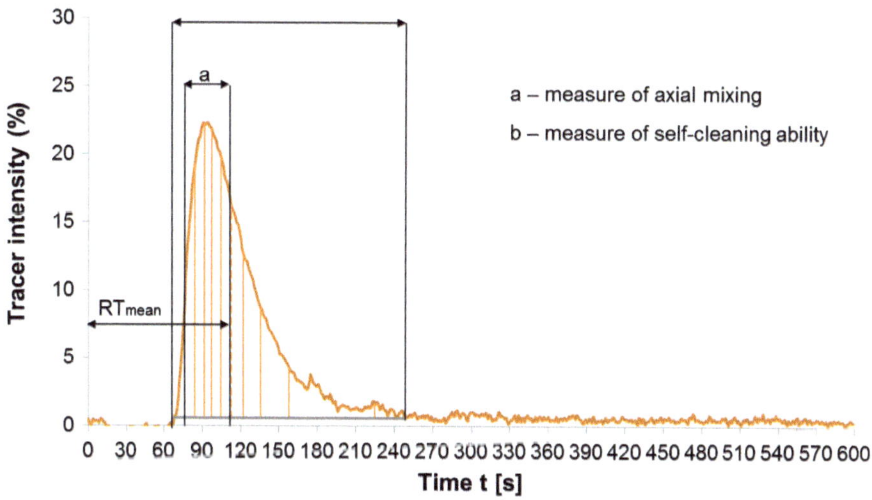

Fig. 16.8 Relevance of residence time distribution (RTD) for homogeneity

in throughput occur, the period of the throughput rate fluctuation must always be less than the RT_{mean} and, if at all possible, less than the time interval "a" illustrated in Fig. 16.8. The time interval "b" is a measure of the self-cleaning ability of the extruder and represents the minimum time one has to wait when changing the product (drug molecule, concentration, etc.) in research trials.

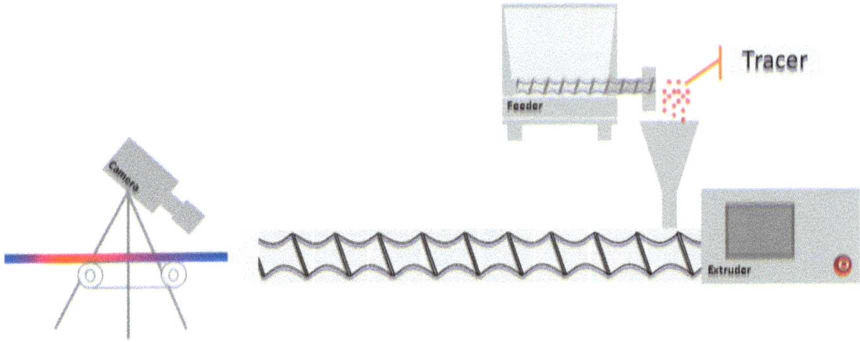

Fig. 16.9 Setup for determining residence time distribution (RTD) and mixing quality

16.6 Experimental Methods to Determine Residence Time Distribution

Optical methods can be used to determine the RTD. This section describes a method which is efficient, easy to implement, easy to adapt, and suitable for low budgets.

Figure 16.9 shows the experimental setup. A camera (e.g., single-lens reflex (SLR) camera) is positioned at the end of the extruder. The camera should be equipped with a strong light source to avoid deviation caused by environmental light fluctuations. The camera takes photos at predetermined intervals, for instance every second. By means of a simple image analysis, the tracer intensity can then be read off the resulting pictures, as shown in Fig. 16.10. A colorant is normally used as tracer to give a significant signal at very low concentration. A low tracer concentration is important if the extrusion system parameters are not to be changed by the tracer itself. The RTD is plotted and the descriptive numbers previously explained for the RTD, such as RT_{mean}, can be determined.

This optical method is not only a good one for determining the RTD but can also be used to determine mixing quality as expressed in homogeneity of concentration over time. This can be done by using, e.g., a film die or mounting a dedicated cuvette through which the melt flows and measuring the tracer intensity not only over time but also over the width of the die or cuvette, respectively. By combining individual line pictures, a picture can be generated showing the entire coloration and discoloration process as illustrated in Fig. 16.11.

In Fig. 16.11, some 600 pictures were taken at a speed of 1 picture per second. Each experiment was set to 10 min. At a fixed predetermined position, a line was extracted by image analysis software (see Fig. 16.11) and, according to the timeline, a new picture was generated by combining the line pictures. This picture (the middle picture in Fig. 16.11) showed the whole extrusion process over a period of time. By reading off the tracer intensity over the width of the die, the homogeneity and mixing quality can be determined in addition to the RTD (see Fig. 16.11, the right-hand picture). Such pictures also allowed the process to be reviewed for parameter shifts which

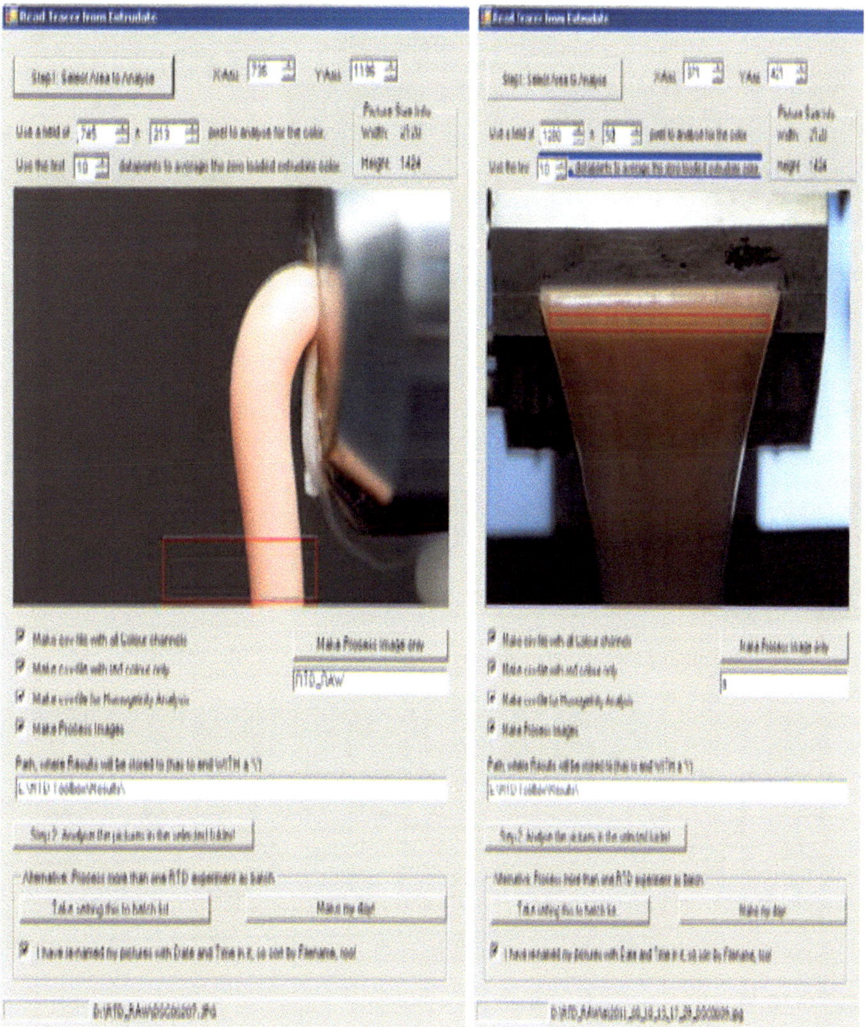

Fig. 16.10 Determination of tracer concentration by image analysis on (*left*) extrusion strand and (*right*) an extruded film

might have been overseen in real time. An example is given in Fig. 16.12 where, shortly after the tracer washout phase, the right-hand part of the film was frozen. The plugging material then shifted over time from the right to the left-hand side of the die. This took place within a time frame of several minutes. However, it shows that such image analysis can be also used to determine the deviation in the dimensions of the extruded film or strand and to analyze for black spots, etc. Figure 16.13 shows an example where a strand extrusion process was monitored for 10 min.

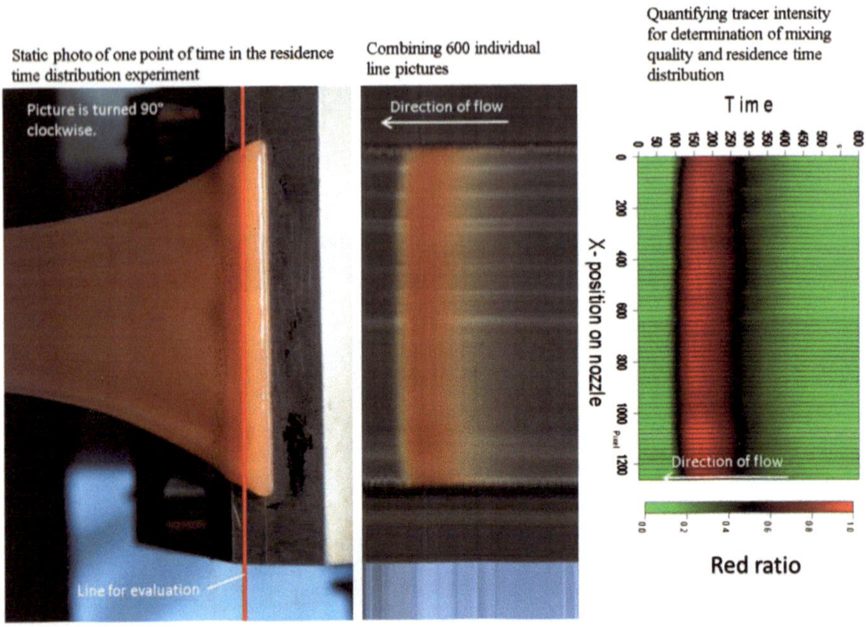

Fig. 16.11 Determining the mixing quality by image analysis

Fig. 16.12 Visualization of slow changes in the process

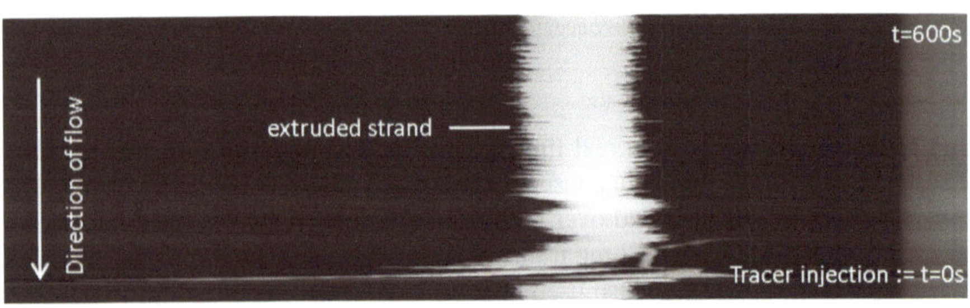

Fig. 16.13 Visualization of 10-min strand extrusion to monitor dimension of strand

The author used a consumer SLR camera in the examples. Such a camera is sensitive and reliable enough. Assuming the camera sensor has a resolution of 24.6 million pixels, the picture taken by this camera shows 6,048 × 4,032 pixels. A total of 6,048 pixels correspond to a sampling length of 36 mm. Assuming that for the picture analysis, a field of 20 × 20 pixels would be averaged to 1 pixel to compensate for noise in the image sensor signal, the sensor would deliver 36,000 μm/(6,048 Px/20 Px) = 119 μm resolution for an image scale factor of 1:1, which standard macro lenses deliver. The use of bellows can produce an image scale factor of 5:1, where the resolution would be improved to 24 μm. This example demonstrates that, with low-cost equipment, a mid-resolution optical analyzer can be realized. There is development taking place to use unmodified consumer cameras, e.g., for spectral imaging; these would measure a spectrum at up to 0.8 nm spectral resolution, thus rivaling commercial spectrometers (Habel et al. 2012).

The previous sections show the importance of process characterization in optimizing and scaling up an extrusion process by linking CQAs not directly to the input variables but to the extrusion system parameters. In this way, optimization of the process can be achieved based on knowledge of the extrusion system parameters and being able to readjust the process and the formulation by adjusting the input variables.

SPC is a useful manufacturing tool to determine whether a variation is inherent to the process or due to special causes of variation (US FDA 2004). To consider a process a controlled one, these causes have to be removed. SPC is used to ensure that a process works well continuously after being established. Helpful tools are DoE but also multivariate analysis which can look at multiple variables simultaneously. In SPC, the process capability can be determined by measuring the variation of the process and comparing it with the specification limits.

By using the results obtained from analysis of the CQAs, the critical process parameters can be identified; in the case of a melt extrusion process, these are the extrusion system parameters.

16.7 Case Study for Process Characterization—Determination and Evaluation of Residence Time Distribution

This section explains, using a real case as an example, the characterization of an extrusion process using a mechanistic approach on how to determine the RTD and the SMEC as these two extrusion system parameters are critical to the design of extrusion processes.

The following study was performed on a Polylab Extruder from Thermo Fisher Scientific. It was set up as a corotating 16-mm twin-screw extruder with a functional length of 40 L/D. The screws were set up with conveying characteristics and had two mixing sections, each of 4D length. The extruded material was fed into the first barrel. No other feeding port was used. The extruder had ten temperature zones. Zone number 1, closest to the gear box, was cooled with ambient temperature water and served as the feed intake zone. Zone number 8 was equipped with a venting port

Table 16.1 Results of ANOVA and main absolute effects

							Absolute effect of varied parameters		
Spalte 1	R^2	R^2_{adj}	Q^2	W^2	RMS	RMS/ Ym	Feed rate	Screw speed	Barrel temperature
RT_{mean}	0.972	0.961	0.931	0.986	12.13	0.079	− 140 s	− 54 s	+ 14 s
SMEC	0.987	0.977	0.941	1.000	0.21	0.13	− 2.32 kWh/kg	+ 1.76 kWh/kg	− 1.76 kWh/kg
Mixing number	0.941	0.906	0.796	0.961	0.003	0.14	0.0161	0.009	− 0.0063

R^2 coefficient of determination, R^2_{adj} adjusted coefficient of determination, Q^2 prediction coefficient, W^2 reproducibility at constant parameter levels, *RMS* root mean square, *RMS/Ym* relative standard deviation in terms of the mean response, RT_{mean} mean residence time, *SMEC* specific mechanical energy consumption, "+" indicates that an increasing parameter will result in ascending change of the target variable, "−" indicates that an increasing parameter will result in descending change of the target variable

open to atmospheric conditions. A film sheet die of dimensions $5 \times 0.3\,\text{cm}^2$ was mounted at the end of the extruder.

Soluplus® from BASF SE was used for the extrusion. Soluplus has a glass transition temperature (T_g) of 70 °C and is used as a polymeric solubilizer for poorly soluble drugs. For this study, no drug was used. A red pigment was used as a tracer and was added in 100-mg doses. All machine configuration parameters, such as screw design, were kept constant for this study. In a DoE, the feed rate, the screw speed, and the barrel temperature were modified. The DoE was chosen as the central composite design with an orthogonal factor α of 1. The parameter range for the DoE was chosen for:

- Feed rate between 500 and 1,500 g/h
- Screw speed between 100 and 500 rpm
- Barrel temperature between 120 and 180 °C.

Table 16.1 shows the DoE, resulting RT_{mean} determined, and SME input.

A camera was installed at the end of the extruder taking a picture every second. Every experiment was run for 10 min (600 s). Prior to introducing the tracer into the feeding zone of the extruder, the camera had already collected ten pictures in 10 s. Using these pictures, the signal to noise (S/N) ratio and the baseline for tracer determination were determined. Thus, a total of 610 pictures were taken per extrusion run. The tracer intensities were read off the pictures with a small self-written software tool. In a spreadsheet, the raw data were then analyzed for RT_{mean}, for the resulting mixing number and the SMEC, from which the SMEC for the empty running extruder was subtracted. An ANOVA was then performed using a Visual X-Sel 11.0 (Ronniger 2012). Also, the main effects of the varied parameters on the RTD and on the SMEC were investigated using Visual X-Sel.

As can be seen, the RT_{mean} varied in a range between 72 and 267 s. As we will see later, this might be the result of insufficient utilization of the machine in terms of throughput. This is a lab-scale extruder for most users and, therefore, the throughput might be reduced to save material during formulation development. For production scale, a higher degree of fill would be utilized to maximize throughput. The SE consumption varied between 0.16 and 6 kWh/kg. This shows the tremendous range for the energy consumption. To understand importance of the energy consumption by the extruded material visible, the following example is given: It requires about 0.055 kWh/kg of energy to heat Soluplus® from ambient temperature to 120 °C and about 0.08 kWh/kg from ambient temperature to 180 °C. This shows that in all performed experiments the energy provided via the screws was sufficient to heat the polymer to the desired temperature by viscous heat generation. The surplus energy was used for shearing and pumping the material.

The minimum RT_{mean} distribution was obtained at the maximum investigated throughput and maximum screw speed but minimum barrel temperature; whereas the maximum RT_{mean} was obtained in the experiment where the throughput and the screw speed are at their minimum but temperature is at its maximum. The minimum SMEC was obtained at the maximum throughput and maximum barrel temperature but at minimum screw speed. The maximum SMEC was obtained at minimum throughput and minimum barrel temperature but at maximum screw speed. This means that both the RT_{mean} and the SMEC can be controlled via the throughput in the same manner but that screw speed and barrel temperature will act conversely to both. This is an important first result and we will discuss what this means later.

The results of the ANOVA are noted in Table 16.1. Here, one can also find the main effects of feed rate, screw speed, and barrel temperature on the investigated variables such as RT_{mean} and SMEC. Table 16.1 indicated (Ronniger 2012):

- R^2, the coefficient of determination, indicated how well the regression model can explain the data.
- R^2_{adj}, the adjusted coefficient of determination, which considered the degree of freedom (DF) via the number of observations in the dataset and via the number of independent variables. The DF of residuals is equal to the DF of the model.
- Q^2, the prediction coefficient, indicates the accuracy of explanation for further data. The calculation was made by leaving out every data row step-by-step (cross validation). Q^2 is always less than R^2.
- W^2, the reproducibility at constant parameter levels, indicates the response stability disturbed through random errors. Best case is $W^2 = 1$. Measuring errors are included as other deviations as in, for instance, modifications.
- RMS, the root mean square, is the standard deviation as a measure of the scatter of the model. Using RMS, the confidence range can be determined.
- RMS/Ym is the relative standard deviation in terms of the mean response.

The absolute effects are equipped with a "+" or "−" indicating whether the effect will result in ascending or descending change of the target variable such as SMEC. A "+" indicates that an increasing parameter will result in an ascending change of, e.g., SMEC. It can be seen in Table 16.1 that all of the investigated response variables, such as RT_{mean}, can be described well by the obtained regression model.

For the RT_{mean} distribution, it can be seen in Table 16.1 that throughput and screw speed have significant influence. Barrel temperature did not significantly influence RT. As the throughput in a corotating twin-screw extruder is independently adjustable from screw speed, the throughput should have a major impact on the RT_{mean}. This

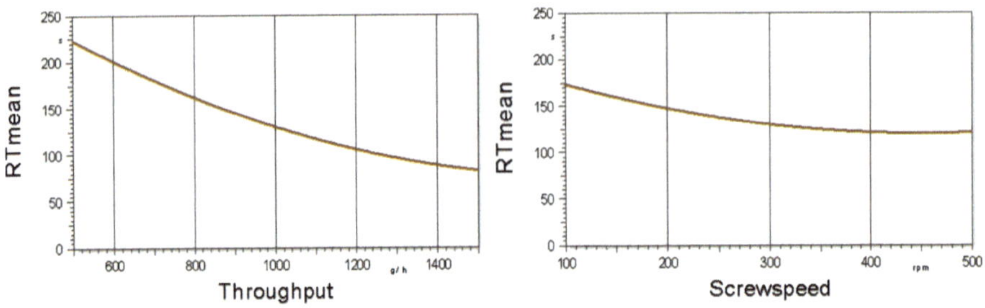

Fig. 16.14 Trace plots for the mean residence time (RT_{mean})

can indeed be shown in Table 16.1 with an absolute effect of 140 s for the throughput compared to 52 s for the screw speed. An increase in throughput will decrease the RT_{mean}. Also, an increasing screw speed will result in a decreased RT. Both parameters work in the same direction with respect to the RT. The dominance of throughput over screw speed in terms of their effect on the RTD can be seen in Fig. 16.14, where process line plots are shown for the different throughput and screw speed settings at a barrel temperature of 120 °C. Visually, it is obvious that the throughput has the major influence on the RTD. In Fig. 16.14, the influence can be seen in a scatter plot demonstrating nicely how an increase in throughput changes the RTD and hence the RT_{mean}. Low screw speed will result in a prolonged RT. The barrel temperature does not show an impact at all. To understand the relationship between throughput, screw speed, and RT_{mean}, Fig. 16.14 shows the traces of the two parameters and their influence on the RT_{mean}. The effect of throughput is stronger than the effect of screw speed and, hence, the slope of the trace plot is larger for the throughput than for the screw speed. Thus, a change in throughput will have a stronger impact on the RT_{mean} compared to a change in screw speed.

For the SMEC, the results are different. All three investigated parameters—throughput, screw speed, and barrel temperature—show a significant impact on the SMEC as expected. If the throughput is increased, the SMEC will decrease because the process energy is provided to a larger quantity of material. If the screw speed is increased, the SMEC will also increase because of increased shear stress of the material by the faster screws. If the barrel temperature is increased, melt temperature will increase and this will decrease the viscosity, resulting in a lower power consumption (lower torque) and hence lower SMEC. Figure 16.15 shows the trace plots for the three independently investigated parameters. Of interest is the fact that throughput and screw speed act against each other in regard to the SMEC. This allows the introduction of the SFL, which is the quotient of throughput and screw speed and gives the amount of material handled by the screws per revolution. The fourth trace plot in Fig. 16.15 shows the influence of the SFL on the SMEC. It makes sense to work with this variable for process analysis. If the amount of impurities in the extruded product is too high and the barrel temperature cannot be changed because it is already at the lower limit for melting the API, the trace plot for the SFL shows how much the SFL needs to be increased to reduce the SMEC significantly. After this

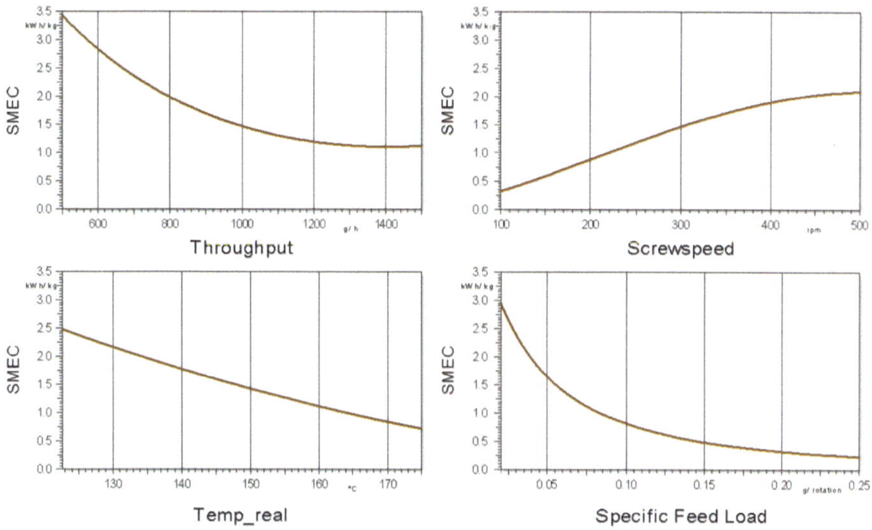

Fig. 16.15 Trace plots for the specific mechanical energy consumption (SMEC)

decision has been made, two parameters can be adjusted: throughput (increase) and screw speed (decrease). Reducing screw speed would reduce the SMEC but would not dramatically influence the RT. However, bearing in mind that a long RT means that the material is exposed longer to shearing of the mixing or kneading elements (they produce the long RT and spread the RT significantly), then an additional reduction of the RT_{mean} could be useful. In this case, increasing the throughput might be the better trigger. Reducing the barrel temperature could reduce thermal energy stress but would increase the mechanical energy stress and would result in higher viscous heat generation. Hopefully, this example will show how relevant it can be to understand the relationship between the CQAs, the extrusion system parameters, which describe what happens in the process, and the extrusion input variables. It depends much on the specific case and the causes of a problem as to how a particular solution will look like.

16.8 Process Parameter Chart

For process optimization and visualization, a process parameter chart can be useful. There is a rationale for looking into such charts as they:

- Enable high-performance extrusion
- Enable higher yields to be obtained
- Determine the suitability of changes in both the equipment and the formulation
- Define design space for the process.

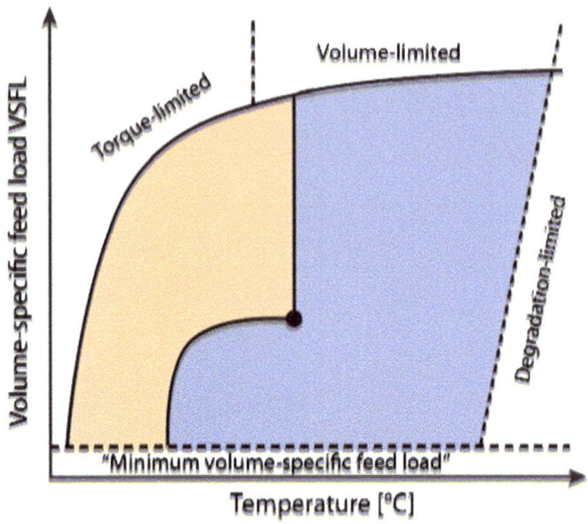

Fig. 16.16 Process parameter chart

A process parameter chart should display not only the technical aspects but also the economic effects of changes made in the process design. Further, a process parameter chart should be scale independent if it is to be used in scale-up.

A new process parameter chart which captures all the aforementioned specifications will be introduced here. The proposed process parameter chart is shown in Fig. 16.16. At first glance, it does not look complex. It is an XY-diagram, displaying the volume-specific feed load (VSFL) versus the extrusion temperature (mean barrel temperature). In the graph, there is a curve representing the determined maximum achievable VSFL values and there is a vertical line. The vertical line indicates the process boundary. Only the position on the temperature axis is relevant for reading the line; the line height does not matter. Every VSFL value to the left of the line is limited by the maximum torque of the extruder (configuration), while every VSFL value to the right of the line is limited by maximum intake (volume). The right vertical dotted line indicates the process boundary given by maximum allowed temperature, which can be determined by degradation points of either the active substance or any nonactive ingredient in the processed formulation.

The VSFL is given by:

$$\mathrm{VSFL} = \frac{\dot{m}}{V_{\mathrm{free}} \cdot n} = \left[\frac{g}{\mathrm{cm}^3 \cdot \mathrm{rev}}\right], \tag{16.14}$$

where V_{free} is the free volume of the extruder.

If the process curve is determined at maximum achievable throughput for the given machine configuration, the process design space is given by the area under the curve. Its boundaries are the process curve to the left and to the top, the temperature to the right, and a certain minimum VSFL, below which a process is no longer realistic.

Using VSFL includes all process parameters which can be changed. Figure 16.17 lists the most relevant influencing parameters. It shows that any change in formulation or machine configuration will result in a change in the process parameter chart.

Material	Machine	Process
• Particle size	• Power of motor	• Temperature
• Particle shape	• D_a/D_i	• Throughput
• Density	• Free volume	• Screw speed
• Tg	• Screw design	
• Mw	• Starved or forced feeding	
• C_p	• Feeder	
• Flow properties	• Length/Diameter	
	• Die	

Fig. 16.17 Parameters influencing volume-specific feed load (VSFL)

With regard to the extrusion system parameters it can be said that:

- A higher distributive mixing is linked to a higher feed load, but also to the machine configuration (screw design, barrel length, etc.). Distributive mixing is important to gain the required homogeneity of the extrudate.
- A higher dispersive mixing is linked to lower regions of the feed load. Even if the screw is redesigned to have more dispersing capability, the achievable feed load usually decreases and hence the process curve reaches lower regions. But, also having a lower throughput per screw rotation will yield a higher degree of dispersion. Dispersive mixing destroys any agglomerates and reduces particle size of, e.g., drug substance, up to a molecular level in case of glassy solid solutions.
- RT will increase for a given machine configuration when lowering the feed load (mainly triggered via the throughput).
- Shear stress will decrease if the feed load is increased. Shear stress is the product of shear rate and viscosity and hence will be also influenced by machine configuration, e.g., screw design.

The economic impact of changes in the extruder setup or in the formulation is mainly shown by the maximum achievable throughput on the extruder. Figure 16.18 shows that this information is represented by the height of the curve. The higher the curve, the higher the achievable throughput on a machine which results in the most relevant process cost. A shift of the curve on the x-axis indicates the temperature range where the chosen setup can be operated and hence the technical feasibility for a setup.

Generally, there are two important numbers which can be directly derived from the chart. One is the minimum extrusion temperature (lower boundary extrusion temperature in Fig. 16.19) below which no extrusion is possible. The other is the temperature which can be expected to produce a stable extrusion process (see Fig. 16.19).

A higher VSFL yields a higher mixing number. The mixing number mainly describes the distributive mixing. Hence, a higher filling degree usually leads to higher distributive mixing. A higher filling degree also leads to a shorter RT_{mean}. RT is one influencing factor for dispersive mixing. Longer RTs often correlate with higher dispersive mixing. Changing the screw design can impact both distributive and dispersive mixing. RT is taken into account in the curve via the height since RT is mainly

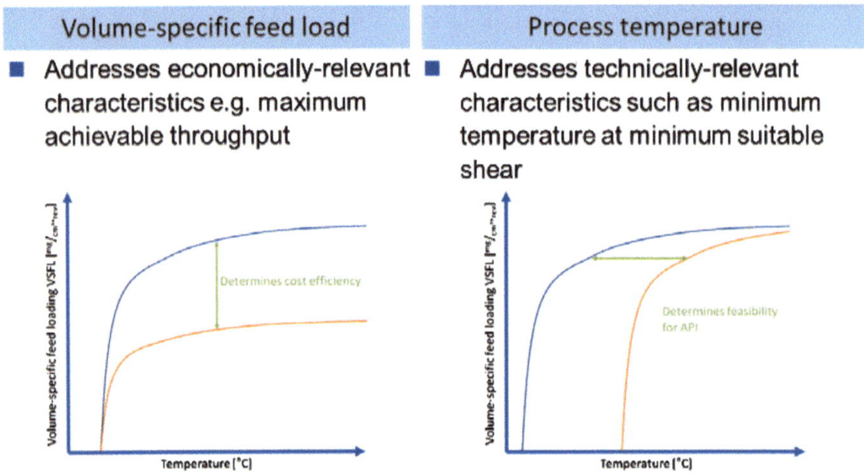

Fig. 16.18 Process parameter chart showing economic and technical characteristics of the setup

Fig. 16.19 Lower boundary temperature indicated by process parameter chart

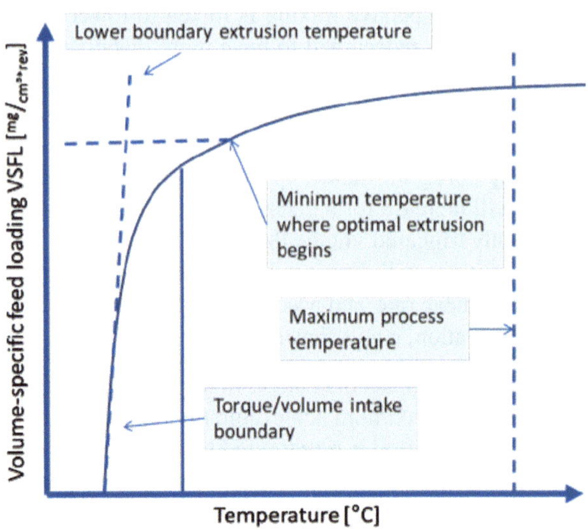

a function of throughput and, to a minor extent, the screw speed. SMEC is also taken into account in the curve via the height. The higher the curve, the lower the SMEC will be at a particular point of temperature. The process parameter chart becomes more powerful when two or more curves are compared, as illustrated in Fig. 16.20.

Figure 16.20 can be used to discuss different scenarios. In general, any value below the curve can be used. There are, however, very low values of VSFL which might not be practicable.

1. Assume a formulation is running on a fixed extruder configuration. The only parameter being changed is the screw speed. The upper curve corresponds to the higher screw speed, the lower curve to the lower screw speed. The range between the two curves thus indicates the adjustable range of SMEC, indicating

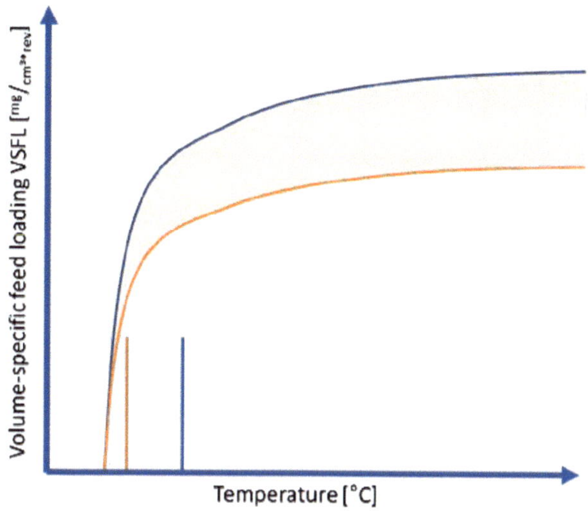

Fig. 16.20 Process parameter chart with two curves for comparison

the available range of energy for dispersive mixing. Both curves might be obtained at maximum achievable throughputs but both curves could also be obtained at other throughputs.

2. This is assuming a scale-up case. Both curves correspond to the same formulation. The upper curve corresponds to the lab scale machine while the lower curve corresponds to the large scale machine. Both curves represent the maximum achievable throughput for the machines. The two curves do not match, indicating that the available process window is different. In this case, the range between the two curves should be avoided during the development of process and formulation since it cannot be adjusted later during large scale processing. This is important information for process design when aiming at maximum yield from an extruder. If the left pin in the chart belongs to the large scale process and the right pin belongs to the lab scale process, it indicates that there is room for improvement in material intake in the large scale process. The target would be to get the pin in the large scale process shifted to the right because in the current plotted situation the process is limited not by the extruders power but by intake issues caused, e.g., by density, particle size, or screw design (free volume).
3. This case is similar to case number 2, a scale-up, but this time the upper curve corresponds to the large scale extruder and the lower curve corresponds to the lab scale extruder. This time, the range between the two curves indicates a range which cannot be reached on the small scale machine. When developing the formulation and designing the process on a small scale, it is thus important to consider that the large scale offers some more room for improving throughput and hence yields.

In conclusion, the process parameter chart has advantage of simplistic visualization while containing all the extrusion data of a formulation regarding process parameters. This allows easy investigation of any changes in formulation, machine setup, or parameters. Another advantage is its capability to directly show the economic impact of changes made to the process.

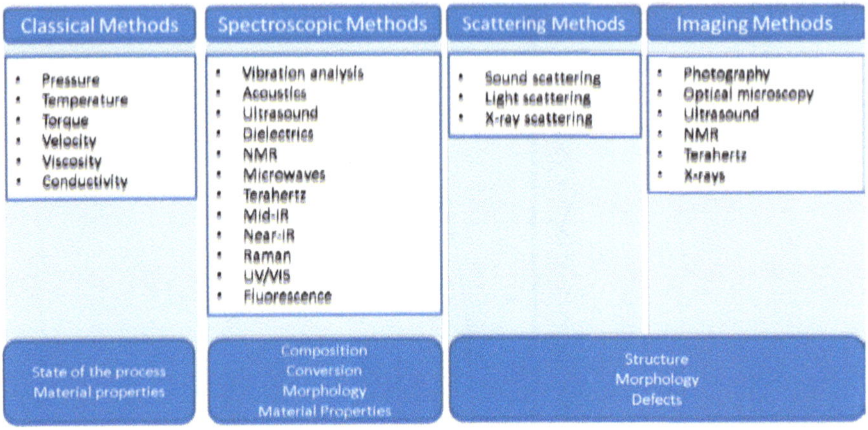

Fig. 16.21 In-line and on-line monitoring methods for the melt extrusion process. (Adapted from Alig et al. 2010)

16.9 Process Analytical Technology Applied to Melt Extrusion

The intention of this section is to provide an overview of common analytical tools used in-line and on-line in extruders. The examples given here are not necessarily from the pharmaceutical industry but provide an overview of what is possible today. An excellent review of the current status of PAT as applied to melt extrusion is given by Alig et al. (2010).

Figure 16.21 is adapted from (Alig et al. 2010) and gives an overview on already applied methods for in-line monitoring. All the methods shown are currently being used in polymer processing but not all of them are necessarily being used or have been used in pharmaceutical melt extrusion processes. Hence, it should provide inspiration, perhaps, to utilize an appropriate method for specific analytical problems. The number of methods being used shows that many analytical questions can be addressed in this way.

If one is missing an analytical method, which can be used for the characterization of melt extrudates off-line such as, for instance, differential scanning calorimetry, then one could consider which other off-line method could at least be a partial substitute. In the case of differential scanning calorimetry (DSC), different analytical questions are addressed. If the question is to determine T_g by DSC then it may be worthwhile considering an oscillating rheometer can determine α-, β- and γ-relaxations with high resolution. If the analytical question addresses relaxation processes, then dielectric spectroscopy might be a suitable alternative. Both oscillating rheometry and dielectric spectroscopy are now available as on-line tools.

16.9.1 Rheology

For process characterization, process optimization, and process control, access to precise rheological data is important. Rheological behavior depends very much upon temperature, process history, and pressure. Off-line measurements are very precise but they do not take the process conditions into account. It is thus useful to have an on-line possibility of determining rheological data.

If the throughput in an extrusion process becomes too high, the process can become unstable. One possible result can be a turbid, mat, or rippled extrudate surface with a sharkskin-like pattern or can even result in a gross melt fracture which occurs at the die exit to the molten extrudate. One possible reason for this is the inhomogeneous temperature distribution between the screw channel center and the barrel surface, where better removal of heat can take place if the barrel is cooled than in the screw channel center where the screw is insulated by the melt with its poor heat convection properties. Nonuniform melt temperatures mean different viscosities which lead to nonuniform melt flow. Kolnaar and Keller (1997) performed experiments with small-angle x-ray scattering to investigate polymer structures inside the die.

Rheology provides information on the macromolecular structure of the melt and its morphology. In this regard, it is interesting to study the rheological properties not only at the end of an extruder but also along the extruder. Covas et al. (2004) and Maia (2001) present concepts for on-line monitoring of the extrusion process along the extruder. They describe concepts to collect samples from the extrusion process at different positions over the screw length. They also describe methods to perform on-line rheology and determination of RTD. An on-line capillary rheometer is discussed as well as an on-line oscillating rheometer. Both get samples from an on-line sample-collecting device. Maia (2001) has conducted a study on the peroxide-induced thermal degradation of polypropylene performed with on-line rheology showing the influence of the process conditions and the peroxide content on the viscosity along the extruder at different process conditions.

Palza et al. (2010) performed on-line rheological studies to correlate these data with melt flow instabilities such as "sharkskin" and melt fracture. They determined on-line pressure oscillations arising from melt flow instabilities by using piezoelectric pressure transducers. By using these transducers they obtained precise pressure signals indicating the frequency and amplitude of pressure oscillation, which correlate with melt flow instabilities. They opened up the vision of developing an intelligent extruder which would be able to detect and control melt flow instabilities automatically, allowing for automatic optimization of the process at the maximum extrusion capacity.

16.9.2 Ultrasound

Ultrasound and other acoustic on-line and in-line methods are mainly used in relation to density. If liquids such as water or gases such as supercritical CO_2 are used to obtain a porous extrudate, ultrasound is a suitable method for obtaining quality information. However, data treatment is not that easy. Usually, an artificial neural network is used and this requires training before information can be derived from the data. The advantage of ultrasound is that it can be sent through opaque melts and is also not dependent on whether a melt is electrically nonconducting.

Abu-Zahra and Karimi (2002) found a good correlation of the density of foamed PVC when measured with an on-line ultrasound sensor. They noted that melt temperature and pressure have to be included in the data model as these affect the ultrasound in the melt. In another work (Abu-Zahra et al. 2002), they were able to characterize the morphology of a polypropylene polymer blend and additionally determine the relative concentrations of the two fillers.

16.9.3 Terahertz

The terahertz (THz) frequency range is considered from 0.1 to 10 THz. Krumbholz et al. (2009) used THz time-domain spectroscopy to in-line monitor a polymeric compounding process. They found this method suitable for determining fillers since most polymers hardly absorb THz waves, which allows discrimination from fillers. At lower THz frequencies, polymers behaved transparent to the THz waves. The authors postulated a possible use of THz spectroscopy to detect bubbles and contaminations as well as the degree of damage to the filler. For their study, they used a newly developed fiber-coupled antenna, which allowed in-line coupling of the THz spectrometer setup to the extruder. They found a free-space setup sealed in a $90 \times 90 \times 30\text{-cm}^3$ aluminum box mounted on a shock-absorbing optical bench. Wietzke et al. (2007) used THz time-domain spectroscopy to measure the content of different fillers such as magnesium hydroxide, glass fiber, calcium carbonate, and silicon dioxide in polymers such as polypropylene.

16.9.4 Dielectric Spectroscopy

Dielectric spectroscopy determines the chemical and physical characteristics of organic materials by measuring their dielectric properties such as dipole moment. Dielectric spectroscopy uses frequencies between 10 and, approximately, 100 MHz. Abu-Zahra (2004) combined ultrasound and dielectric measurements to correlate the data with the melt density of the PVC melt. An oppositely mounted dielectric sensor was used to measure the dielectric properties, which are correlated to mechanical viscosity.

Lee et al. (2005) used dielectric relaxation spectroscopy to monitor the relaxation of nylon/clay composites in a polymer melt. They were able to discriminate between two different nylon grades (nylon 6 and nylon 12) as the α-relaxation showed a different characteristic frequency, different relaxation time distribution, and different dielectric intensity.

16.9.5 Near-Infrared Spectroscopy

Tumuluri et al. (2004) used in-line FT-NIR spectroscopy to determine clotrimazole in a hot-melt extruded polyethylene oxide film. They referenced the FT-NIR data to off-line HPLC data. Gryczke et al. (2010) discussed the development of an FT-NIR method which enables reliable drug content in melt to be determined. NIR spectroscopy is a versatile tool for in-line monitoring since NIR is sensitive to chemical and physical changes in the sample. On the one hand, this is an advantage as it can measure numerous properties; on the other hand, it makes the development of a reliable method more complex. With respect to the extruder, NIR is sensitive to the changes in melt temperature, melt flow, and pressure. Thus, melt temperature, throughput, and screw speed are process variables which need to be considered when building a chemometric model. Unless these factors are included in the chemometric model, a model designed to determine drug content would deliver changing values even if there were no changes in drug content but changes in temperature, throughput, or screw speed.

16.9.6 Raman Spectroscopy

Tumuluri et al. (2008) used Raman spectroscopy on-line and off-line to determine the content of clotrimazole and ketoprofen in hot-melt extruded polyethylene oxide films. They varied the drug content within a range of 1–20 % and were able to discriminate accurately between the different tested concentrations. The authors detected changes in the crystallinity of the drugs. It has also been mentioned that method transferability can be an issue that can be solved. Also, changing the instrument can lead to different results. Saerens et al. (2011) used in-line Raman spectroscopy to determine the metoprolol content in EUDRAGIT® RS melt. The purpose was to investigate the suitability of Raman spectroscopy for in-line polymer drug solid state characterization. They found broadened Raman peaks in the case of amorphous extrudates compared to crystalline drug containing extrudate and they found peak shifts indicating interaction between drug and polymer as hydrogen bonds.

16.10 Conclusion

The intention of this chapter was to provide an overview and introduction to the topic of PAT and the related tools as applied to pharmaceutical melt extrusion processes. It has been shown that PAT is not just quality control (QC) tools but important methods leveraging available state-of-the-art analytical tools for characterizing, developing, optimizing, and scaling the process. Further, it was discussed that the implementation of PAT is an important step which requires some expertise and should be performed by a team consisting of analytical experts in the corresponding methods, the operators of the process line, the galenic R & D experts, and the engineers planning the process. In terms of implementation, the discussion also has to involve which cases and up to what degree a full loop-back control with fully automated readjustment of the extrusion process is possible and preferable in the future. Such automation would surely make systems much more complex than they already are and one has to bear in mind that such algorithms then also have to be validated for use in production. More importantly, at least in the short term, would be to utilize PAT to mitigate risks in the extrusion process, bearing in mind that PAT is not required to a large extent during manufacturing. Further, the importance of understanding extrusion system parameters has been shown as these are the response variables to changes in the process. The extrusion system parameters can be linked to CQAs; the extrusion input variables, however, should not be directly linked to these. By introducing and discussing SMEC and RTD, two powerful variables were discussed that allow for a much better process understanding. Of course, there are other important response variables but a start should be made with these two fast and positive experiences. Finally, different analytical methods were discussed. Not all of them are already being used for pharmaceutical purposes yet, but this is a field of research where the pharmaceutical industry should have examining experiences in other industries where extensive in-line monitoring has been routinely implemented. Finally, the author hopes that readers will now be motivated to use PAT and explore the possibilities for their own applications.

References

Abu-Zahra N (2004) Real-time viscosity and density measurements of polymer melts using dielectric and ultrasound sensors fusion. Mechatronics 14(7):789–803. doi:10.1016/j.mechatronics.2003.11.001

Abu-Zahra NH, Karimi S (2002) On-line monitoring of PVC foam density using ultrasound waves and artificial neural networks. Int J Adv Manuf Technol 19(8):618–622. doi:10.1007/s001700200067

Abu-Zahra N, Fedek W, Neyfeh T, Salem A (2002) Ultrasound measurement of two-filler concentrations in polypropylene compounds. Part 2: On-line calibration. Int J Adv Manuf Technol 20(11):812–816

Alig I, Steinhoff B, Lellinger D (2010) Monitoring of polymer melt processing. Meas Sci Technol 21(6):062001. doi:10.1088/0957–0233/21/6/062001

Bakeev KA (2010) Process analytical technology spectroscopic tools and implementation strategies for the chemical and pharmaceutical industries. Wiley, UK

Covas JA, Carneiro OS, Costa P, Machado AV, Maia JM (2004) Online monitoring techniques for studying evolution of physical, rheological and chemical effects along the extruder. Plast Rubber Compos 33(1):55–61. doi:10.1179/146580104225018300

Goertz H-H, Klimesch RG (1987) Verfahren zur Herstellung von festen Pharmazeutischen Formen, Patent EP0240904

Gryczke A, Heil C, Leister D (2010) Inline monitoring of a hot melt extrusion process by near infrared spectroscopy. AAPS-APV HME Expert Symposium 2010 Tarrytown USA, Conference proceedings

Habel R, Kudenov M, Wimmer M (2012) Practical spectral photography. Imaging 31(2). http://www.cg.tuwien.ac.at/research/publications/2012/Habel_2012_PSP/

Kohlgrüber K, Wiedmann W (2008) Co-rotating twin-screw extruders fundamentals, technology and applications. Hanser Gardner, Munich

Kolnaar JWH, Keller A (1997) A singularity in the melt flow of polyethylene with wider implications for polymer melt flow rheology. J Non-Newtonian Fluid Mech 69:71–98

Krumbholz N, Hochrein T, Vieweg N, Hasek T, Kretschmer K, Bastian M, Mikulics M et al (2009) Monitoring polymeric compounding processes inline with THz time-domain spectroscopy. Polymer Testing 28(1):30–35. doi:10.1016/j.polymertesting.2008.09.009

Lee Y-H, Bur AJ, Roth SC, Start PR, Harris RH (2005) Monitoring the relaxation behavior of nylon/clay nanocomposites in the melt with an online dielectric sensor. Polym Adv Technol 16(2–3):249–256. doi:10.1002/pat.576

Levenspiel O (1972) Chemical Reaction Engineering, 2nd edn. Wiley, New York

Levine L (1997) More on extruder balance. Cereal Foods World 42:22

Liang M, Huff HE, Hsieh F-H (2002) Evaluating energy consumption and efficiency of a twin-screw extruder. J Food Sci 67(5):1803–1807. doi:10.1111/j.1365–2621.2002.tb08726.x

Maia JM (2001) On-line theometry for twin-screw extrusion (along the extruder) and its applications. Appl Rheology 12(1):18–24

Palza H, Naue I, Wilhelm M, Filipe S, Becker A, Sunder J, Gottfert A (2010) On-line detection of polymer melt flow instabilities in a capillary rheometer. KGK-Kautschuk Gummi Kunststoffe 63(10):456–461

Rauwendaal C (2010) Understanding extrusion. Hanser Gardner, Cincinnati

Ronniger C (2012) Software Visual-X-Sel® 11.0 Multivar, Release 11.1112. http://www.crgraph.de

Saerens L, Dierickx L, Lenain B, Vervaet C, Remon JP, De Beer T (2011) Raman spectroscopy for the in-line polymer-drug quantification and solid state characterization during a pharmaceutical hot-melt extrusion process. Eur J Pharm Biopharm: offiArbeitsgemeinschaft für Pharmazeutische Verfahrenstechnik e.V 77(1):158-163. doi:10.1016/j.ejpb.2010.09.015

Tumuluri SVS, Prodduturi S, Crowley MM, Stodghill SP, McGinity JW, Repka MA, Avery BA (2004) The use of near-infrared spectroscopy for the quantitation of a drug in hot-melt extruded films. Drug Dev Ind Pharm 30(5):505–11. doi:10.1081/DDC-120037481

Tumuluri VS, Kemper MS, Lewis IR, Prodduturi S, Majumdar S, Avery Ba, Repka Ma (2008) Off-line and on-line measurements of drug-loaded hot-melt extruded films using Raman spectroscopy. Int J Pharm 357(1–2):77–84. doi:10.1016/j.ijpharm.2008.01.036

US FDA (2004) Guidance for industry PAT—a framework for innovative pharmaceutical development, manufacturing, and quality assurance. http://www.fda.gov/downloads/Drugs/GuidanceComplianceRegulatoryInformation/Guidances/UCM070305.pdf. Accessed 22 Dec 2011

Villmow T, Kretzschmar B, Pötschke P (2010) Influence of screw configuration, residence time, and specific mechanical energy in twin-screw extrusion of polycaprolactone/multi-walled carbon nanotube composites. Compos Sci Technol 70(14):2045–2055. doi:10.1016/j.compscitech.2010.07.021

Wietzke S, Jansen C, Rutz F, Mittleman D, Koch M (2007) Determination of additive content in polymeric compounds with terahertz time-domain spectroscopy. Polymer Testing 26(5):614–618. doi:10.1016/j.polymertesting.2007.03.002

Part V
Regulatory and Commercialization Considerations

Chapter 17
Consistency of Pharmaceutical Products: An FDA Perspective on Hot-Melt Extrusion Process

Abhay Gupta and Mansoor A. Khan

Abstract Hot-melt extrusion represents an efficient technology to formulate drugs with poor aqueous solubility into safe and effective drug delivery systems. The interest in this technology continues to grow as it is suitable for both high dose and potent low dose compounds. It is amenable to real-time monitoring and control of the consistency of the product with respect to content uniformity and crystalline conversion. Quality-by-design (QbD) has become an essential part of modern pharmaceutical quality systems since it incorporates an enhanced product and process understanding. It encourages real-time monitoring and control of critical material attributes and process parameters. The QbD paradigm can be easily incorporated into hot-melt processes. It has the potential to replace traditional batch processes due to its continuous nature and ease of scale-up from laboratory scale to commercial scale.

17.1 Introduction

In the USA, almost every aspect related to the over-the-counter and prescription drugs are regulated by the Food and Drug Administration (FDA), an agency within the US Department of Health and Human Services. FDA derives its authority from the US Federal Food, Drug, and Cosmetic Act (FD&C), which was enacted by the US Congress in 1938 (Federal Food, Drug, and Cosmetic Act 2012). In addition to enforcing the FD&C Act and the associated regulations, FDA routinely issues guidance documents that represent FDA's current thinking on various aspects related to the regulated products. Guidance documents are not regulations or laws but are issued to provide consistency in regulatory approach for the evaluation and approval of applications and establish inspection and enforcement procedures.

Globalization of the pharmaceutical industry during the second half of twentieth century led to a need for a greater harmonization of regulatory assessment processes for drug applications across the USA, Europe, and Japan. The International

M. A. Khan (✉) · A. Gupta
Division of Product Quality Research, Office of Pharmaceutical Science,
US Food and Drug Administration,
Silver Spring, MD, USA
e-mail: mansoor.khan@fda.hhs.gov

M. A. Repka et al. (eds.), *Melt Extrusion*,
AAPS Advances in the Pharmaceutical Sciences Series 9,
DOI 10.1007/978-1-4614-8432-5_17,

Conference on Harmonisation of Technical Requirements for Registration of Pharmaceuticals for Human Use (ICH) was launched in 1990 to rationalize, streamline, and harmonize the technical requirements for reporting and evaluating the data on safety, quality, and efficacy of pharmaceutical product registration in the ICH region, spanning the USA, Europe, and Japan. The harmonization is being achieved through development and issuance of the ICH Tripartite Guidelines, which are based on a scientific consensus among the regulators and research-based industry representatives from the ICH regions. The ICH guidelines are primarily divided into Safety, Quality, and Efficacy topics, reflecting the three criteria that form the basis for drug assessment and approval process. Guidelines that do not fit into one of these topics are considered multidisciplinary guidelines and include the ICH medical terminology (MedDRA), the Common Technical Document (CTD) and the development of Electronic Standards for the Transfer of Regulatory Information (ESTRI). Among the Quality topics, ICH guidelines of particular relevance to product and process understanding are ICH Q8 (Pharmaceutical Development), ICH Q9 (Quality Risk Management), and ICH Q10 (Pharmaceutical Quality System) (ICH Harmonised Tripartite Guideline 2005, 2008, 2009). The overall goal of these guidelines is that the end user of a pharmaceutical product gets a *quality* product and the state of pharmaceutical manufacturing leads to the *desired state*. Pharmaceutical quality has been defined in ICH Q6 as the suitability of either a drug substance or a drug product for its intended use and includes such attributes as the identity, strength, and purity (ICH Harmonised Tripartite Guideline 1999). The *desired state* of the pharmaceutical manufacturing may be achieved by the following (US Food and Drug Administration 2004):

- Designing effective and efficient manufacturing process to ensure product quality and performance
- Using the mechanistic understanding of the effect of formulation and process factors on the product performance to set the product and process specifications
- Continuous *real time* quality assurance

When hot-melt extrusion (HME) is used as a unit operation, it is expected that a quality product is produced consistently and the process is moving toward the desired state.

17.2 Hot-Melt Extrusion

HME is the process of applying thermal or mechanical energy to heat a material under controlled conditions, followed by mixing in the molten state and then pushing the molten mass under controlled conditions through an orifice of fixed cross-sectional profile to create a new material. Industrial application of HME dates back to the middle of last century, especially in preparation of thermoplastic materials. Lately, the process is becoming popular among the pharmaceutical industry due to its solvent free, continuous nature with only a few processing steps. These processes have

Table 17.1 The Biopharmaceutics Classification System (BCS)

BCS class	Solubility	Permeability
1	High	High
2	Low	High
3	High	Low
4	Low	Low

been used to prepare solid solutions, in which the drug is molecularly dispersed in the crystalline carrier forming a one-phase system at the molecular level, or solid dispersions, in which the drug is dispersed as amorphous clusters, crystalline particles, or at the molecular level in the amorphous carrier forming a two-phase system. Since material in their amorphous state exhibit higher solubility as compared to their crystalline states, the process has been used to improve dissolution and oral bioavailability of a number of poorly water-soluble drugs, e.g., ritonavir, ibuprofen, nifedipine, etc. (Klein et al. 2007; Gryczke et al. 2011; Maniruzzaman et al. 2012; Zajc and Srčič 2004). The aqueous solubility, along with the intestinal permeability and dissolution, are the most important factors that govern the rate and extent of drug absorption from immediate release solid oral dosage forms.

The aqueous solubility and intestinal permeability are the basis for classifying drug substances into the four classes of the Biopharmaceutics Classification System (BCS) (Table 17.1) (US Food and Drug Administration 2000). Substances for which the highest dose strength is soluble in 250 ml or less of aqueous media over the pH range of 1–7.5 are classified as highly soluble. The volume estimate of 250 ml for the solubility determination is derived from typical bioequivalence study protocols that prescribe administration of a drug product to fasting human volunteers with a glass of water. Similarly, substances which show 90 % or more absorption of an administered dose based on a mass balance determination or in comparison to an intravenous reference dose, are classified as highly permeable. The permeability class boundary is based indirectly on the extent of absorption of a drug substance in humans and directly on measurements of the rate of mass transfer across the human intestinal membrane. The BCS classification plays an important role in the selection of formulation components and the product dosage form.

The drug present in solid dosage forms must dissolve upon administration to become bioavailable. A growing number of new drug substances, going into clinical trials, exhibit low aqueous solubility. HME is increasingly being utilized to enhance bioavailability, reduce dose and side effects, and increase the duration of action of these drugs (Klein et al. 2007). The process is also being used to mask the bitter taste of some drugs and to formulate modified or targeted release dosage forms (Gryczke et al. 2011; Maniruzzaman et al. 2012). HME has been used to prepare a variety of pharmaceutical dosage forms, such as, immediate and modified release tablets, transmucosal and transdermal systems, ophthalmic implants, etc. (Repka et al. 2007). All materials used for HME process must be thermally stabile under the HME processing conditions. The drug must have sufficient solubility in the carrier to ensure a thermodynamically stable system and carriers with high solubilization capacity are more suited for the HME process. The carrier could be hydrophilic polymers or low melting point waxy materials. Although hydrophilic polymeric materials are preferred as the carrier matrix, e.g., povidone, copovidone, etc., the final selection depends on the miscibility of the drug in the carrier matrix. In some formulations, the carrier may also act as a binder, stabilizer, or release-controlling agent. In addition to the drug and the carrier, other excipients may also be added to impart specific properties to the HME formulation. Plasticizers are typically added to lower the glass transition (T_g) of the polymer lowering the processing temperature. The addition of plasticizers also reduces the drug and polymer degradation and lowers the shear forces needed to extrude the material, thereby improving the processability of certain high molecular weight polymers. Antioxidants may be added to prevent oxidative degradation, while release-modifying agents may be added to alter the drug release profile. Other components, such as acid receptors, light absorbers, thermal lubricants, etc., may also be added depending on the properties of the drug and the final dosage form.

Only excipients that are generally recognized as safe (GRAS), have an official compendial monograph, are listed in the FDA's inactive ingredients database (US Food and Drug Administration 2012), or have documented human use in the proposed level may be used to prepare drug products. Hence, most excipients used for pharmaceutical HME are those that have been used to prepare conventional dosage forms. In addition, generic drug products, if not intended for parenteral, ophthalmic, or otic use, are not required to contain same excipients as those present in the reference listed drug product. However, the sponsor of these drug products must demonstrate that the inactive ingredients do not affect the safety or efficacy of the proposed drug product (Code of Federal Regulations 2012). An excipient is not considered new if it has appeared in an approved drug product for a particular route of administration. Otherwise, all new excipients require extensive evaluation and review for pharmacological activity using a battery of standard testing, including tests for chronic toxicity and carcinogenicity if needed, before being approved for use as an ingredient in a drug product (US Food and Drug Administration 2005). This ensures that all potential new excipients are safe for use in human pharmaceuticals at the concentrations being used in the HME formulation when administered via the proposed route of administration. This is especially important because the amount of excipient, particularly polymer, present in HME formulations is typically far more than that present in conventional dosage forms.

The uniform distribution of the drug in the carrier matrix is of paramount importance in HME formulations. Hence, the pharmaceutical HME process is typically carried out using a screw extruder as they provide more homogeneous products as compared to the ram extruders. Extrudes with twin-screws are preferred over single-screw machines as they have shorter transit time, lower tendency to over-heat, and provide better mixing due to higher dispersing capacity. Corotating twin screws with intermeshing design are the most popular as they can be operated at high speeds, provide good mixing and high throughput, and are self-wiping, which prevents localized over-heating. These extruders are typically characterized by the length-to-diameter ratio (L/D), which expresses the ratio between the length and the diameter of the screw, outer diameter (OD), which is the outer diameter of each screw for a twin-screw extruder, and the ratio of the OD to inner diameter (OD/ID). The residence time of the material in the extruder may range from a few seconds to over a couple of minutes depending upon the type of extruder, feed rate, screw speed, screw length, and the L/D ratio. The barrel temperature, screw speed, and feed rate are among the most important process parameters. The barrel temperature is dependent on the thermal stability and the glass transition temperature of the drug/polymer. It should be high enough to ensure complete solubilization or dispersion of the drug in the polymer melt without degrading them. The screw speed and the feed rate together determine the shear stress and the dwell time.

The continuous and constant delivery of drug, polymer, and other excipients, and the steady flow of these materials through the barrel are essential to ensure uniformity in the output material. These extruder parameters also play an important role in scale-up of the process. During scale-up, the geometries of extruders, the screw profiles, and the peak shear at different scales should be nearly identical. The equipment limits at the larger scale may not permit the use of conditions that were theoretically predicted based on the smaller scale equipment affecting the uniformity and/or stability of the drug in the extruded material. Hence, the downstream processing capabilities, e.g., feeding system, heating/cooling systems, cutting/collecting equipment, etc., should be considered as part of the scale-up process. All extruders used for pharmaceutical HME processes must also confirm to the cleaning and validation requirements of the current good manufacturing practice (cGMPs) and all surfaces that come in direct contact with the materials or finished product must be nonreactive, nonabsorptive, and nonadditive.

A number of analytical methods have been used to evaluate the HME products, including differentiating between solid solutions, solid dispersions, and physical mixtures of drug and polymer. X-ray diffraction (XRD) and differential scanning calorimetry (DSC) are widely used to characterize the crystalline properties of the HME products (Figs. 17.1 and 17.2). Absence of crystalline peaks in the fingerprint region of the drug in the XRD profile and lack of melting transition in the DSC profile indicates absence of crystalline drug. XRD and DSC profiles also allow for differentiation between solid dispersions, in which the drug is present in both crystalline and amorphous states and solid solutions, in which the drug is present in an amorphous state, i.e., absence of crystalline drug. DSC is also used to determine thermal properties, such as melting point, degradation temperature, etc., and to study drug–excipient incompatibility. Other analytical methods include thermogravimetric analysis (TGA) to determine thermal stability of the materials by studying their desolvation and decomposition behavior, light/electron microscope to detect the presence of drug crystals, and near-infrared and Raman spectroscopy for qualitative and quantitative analysis of the drug in the HME product.

17.3 Quality-by-Design

In March 2004, FDA launched the Critical Path Initiative to modernize the scientific processes through which drug products are developed, evaluated, and manufactured (FDA's Critical Path Initiative, http://www.fda.gov/ScienceResearch). The initiative

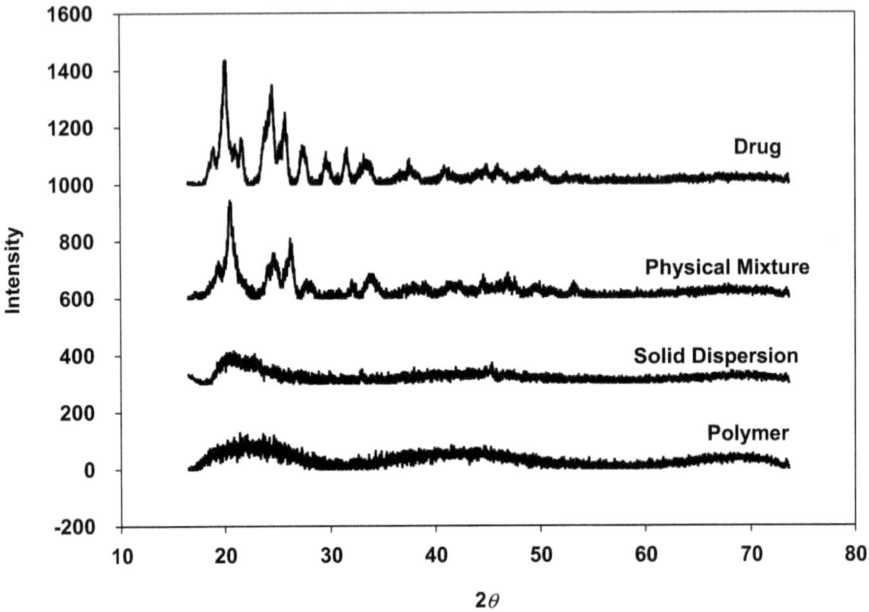

Fig. 17.1 X-ray diffraction profiles of pure drug, pure polymer, their physical mixture, and their solid dispersion

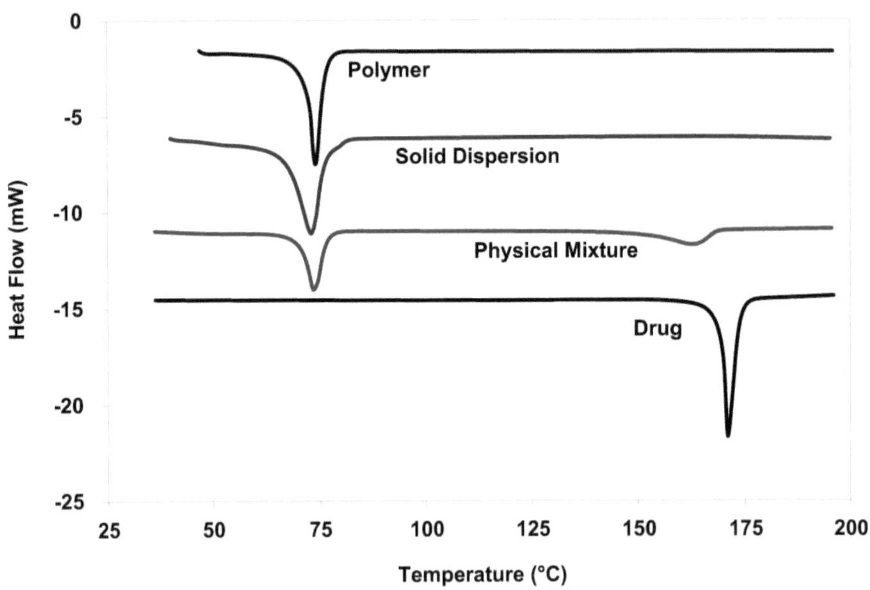

Fig. 17.2 Differential scanning calorimetry profiles of pure drug, pure polymer, their physical mixture, and their solid dispersion

encourages use of scientific and technical tools to evaluate and predict the safety, effectiveness, and manufacturability of drug products by identifying critical attributes during product development. QbD is an important element of this initiative. Under the QbD paradigm, quality is built into the final product through a comprehensive scientific understanding and management of formulation and manufacturing variables. QbD is defined by the FDA and the ICH as a systematic approach to product development that begins with predefined objectives and emphasizes product and process understanding and process control, based on sound science and quality risk management (QRM). It utilizes a multivariate model based on prior knowledge and experimental assessment to link product and process measurements to the desired quality attributes, and provides a framework to transfer product knowledge and process understanding during the lifecycle of the product. An important element of the QbD paradigm is the concept of design space, which is the multidimensional relationship linking the critical quality attributes (CQAs) of the drug product to the critical material attributes and critical process parameters (CPPs). The design space is proposed by the applicant and is subject to regulatory assessment and approval. Working within the design space is not considered a change, while a movement out of design space typically initiates a regulatory postapproval change process, which is dependent on the nature of the change.

The knowledge and the understanding gained about the product and manufacturing process is presented as part of the Pharmaceutical Development section (Sect. 3.2.P.2) of the drug product application, if submitted in the ICH M4 CTD format. Demonstration of greater understanding of pharmaceutical and manufacturing sciences creates a basis for flexible regulatory approaches based on the level of relevant scientific knowledge provided. This flexibility can be realized by demonstrating an enhanced knowledge of product performance over a range of material attributes, manufacturing process options, and process parameters thorough an application of prior knowledge, formal experimental designs, and/or process analytical technology (PAT). PAT is a system for designing, analyzing, and controlling manufacturing through timely measurements (i.e., during processing) of critical quality and performance attributes of raw and in-process materials and processes with the goal of assuring final product quality (FDA's Critical Path Initiative, http://www.fda.gov/ScienceResearch). A process is considered to be well understood when (1) all critical sources of variability are identified and explained; (2) variability is managed by the process; and (3) product quality attributes can be accurately and reliably predicted over the design space established for materials used, process parameters, manufacturing, environmental, and other conditions. HME being a continuous process is ideally suited for real-time monitoring and control of the process, thereby ensuring consistent product quality. Use of off-line and on-line near-infrared and Raman spectroscopy for the quantitative and qualitative analysis of active ingredients in drug loaded hot-melt extruded film formulations has been reported (Tumuluri et al. 2004, 2008; Saerens et al. 2011, 2012). Other applications of PAT included identification of polymorphic forms, crystalline/amorphous state, presence of impurities/degradation products, moisture content, uniformity of drug, etc., via on-line, in-line, or at-line measurements of the CQAs.

Over the last few decades, a number of patents have been issued on the pharmaceutical applications of the HME process, some of which have led to commercialization of HME formulation of such drugs. In some instances, e.g., protease inhibitor combination drug Kaletra® when formulated as a HME product showed significantly higher bioavailability, leading to reduced dosing frequency and a drug product that did not require refrigeration due to improved product stability. Due to the critical nature of the disease being treated by this product, the drug approval for this product was fast tracked by the FDA, with the approval being granted in less than half the time it would have taken if reviewed as per the standard timeline. Some of the other commercial products prepared using HME process includes oral dosage forms of ritonavir (Fort 2000) and etravirine (Baert and Verreck 2006).

Recrystallization of the amorphous drug, over the product shelf-life, may lead to a change in the product dissolution profile. Therefore, it is important to determine and control the crystallinity of the drug as it may be present in an amorphous state in these products. In addition, the water content of these products should also be controlled within tight specifications as water, due to its plasticization properties, may facilitate crystallization of the amorphous substances over the product shelf-life. If needed, QRM principles may be used to manage the risk to the product quality over the lifecycle of a product. QRM is a systematic process for the assessment, control, communication, and review of the risks to the quality of the drug product across the product lifecycle. (ICH Harmonised Tripartite Guideline 2005) It allows for effective and consistent risk-based decisions, by both regulators and industry, across the product lifecycle, by providing documented, transparent, and reproducible methods for managing the risk. Risk is commonly defined as the combination of the probability of occurrence of harm and the severity of that harm. A successful QRM begins with identifying the potential for risk, followed by risk analysis, risk evaluation, risk control, and documentation and communication of risk among all stake holders. Some of the risk management tools that are commonly used include basic risk management facilitation methods such as flowcharts, check sheets, etc., Failure Mode Effects Analysis (FMEA), Fault Tree Analysis (FTA), Preliminary Hazard Analysis (PHA), Risk ranking and filtering, etc., which can be used during different stages of product and process development/implementation processes. Application of these tools helps in identifying potential variables that may be deemed critical to the HME process (Fig. 17.3). An effective QRM approach provides a proactive means to identify and control potential quality issues during product development and manufacturing process, thus ensuring a high quality drug product for the patients.

A QbD approach to drug product development begins with defining the quality target product profile (QTPP) that would reproducibly ensure desired product quality, and deliver the labeled therapeutic benefit of the drug product in a safe and efficacious manner. Examples of QTPPs include identity, purity, dissolution, bioavailability, crystallinity, and stability of drug in the final dosage form. The desired product QTPP is ensured by identifying, studying, and controlling the CQAs of the input material, e.g., drug substance, excipient, etc., intermediates and the drug product within appropriate limits. Special consideration should be given to the physicochemical properties of the drug substance, e.g., particle shape and size, polymorphism,

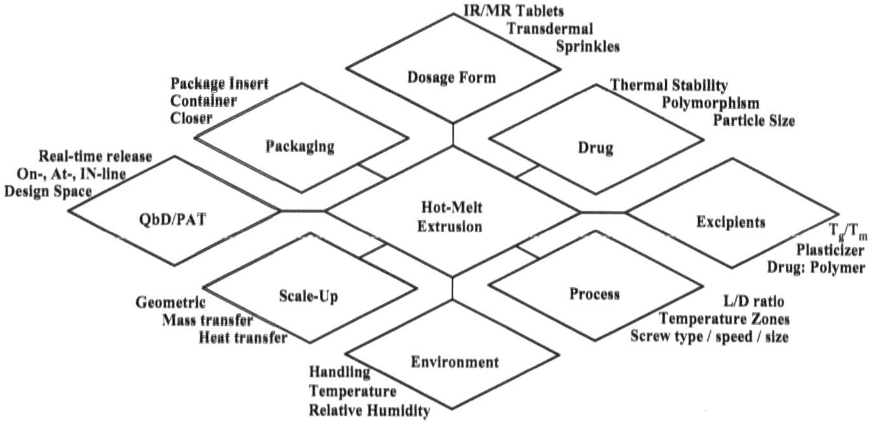

Fig. 17.3 An example of variables influencing the hot-melt extrusion process

aqueous solubility as function of pH, pKa, photolytic and oxidative degradation, partition coefficient, membrane permeability, and/or oral bioavailability. Gaps in knowledge on drug substance, potential excipients, and process operations should be identified. Risk assessment tools may be used to identify hazards requiring further investigation and to find CQAs of the final product that must be controlled to meet the product QTPPs. An iterative process of experimentation, mathematical models, and QRM may be used to identify and rank critical material attributes, CPPs, and sources of variability that can adversely affect the product CQAs. The information about the enhanced product and process understanding can be used to create a design space describing the relationship between the process inputs and the drug product CQAs. If the design space is established at a smaller scale, it must be verified at the commercial scale. A control strategy for the entire process, encompassing input material controls, process monitoring and controls, design spaces around individual or multiple unit operations, and/or final product specification should also be established. PAT tools may be incorporated into the control strategy for real-time monitoring and control of the process.

17.4 Concluding Remarks

QbD has become an integral part of modern pharmaceutical quality systems as it incorporates an enhanced product and process understanding. It encourages real-time monitoring and control of critical material attributes and process parameters. The paradigm can be easily incorporated into hot-melt processes. HME represents an efficient technology to formulate drugs with poor aqueous solubility into safe and effective drug delivery systems. The interest in this technology is growing as it is suitable for both high dose and very potent low dose compounds. It is amenable to real-time monitoring and control of the consistency of the product with respect to content uniformity and crystalline conversion. It has the potential to replace traditional batch processes due to its continuous nature and ease of scale-up from the laboratory scale to commercial scale.

Disclaimer The findings and conclusions in this article have not been formally disseminated by the Food and Drug Administration and should not be construed to represent any Agency determination or policy.

References

Applications for FDA Approval to Market a New Drug, Title 21 Code of Federal Regulations. Pt 314. 2012 ed. Available: http://www.gpo.gov/; Accessed: 02/02/2013.

Baert L, Verreck G (2006) Antiviral compositions. US Patent 7,887,845, 3 Feb 2006

Federal Food, Drug, and Cosmetic Act, Title 21 U.S. Code. 2006 ed. Suppl V, 2011. Available: http://www.gpo.gov/; Accessed: 02/02/2013.

Fort JJ, Krill SL, Law D, QiuPorter Y, Porter WR, Schmitt EA (2000) Solid dispersion pharmaceutical formulations. US Patent 7,364,752, 10 Nov 2000

Gryczke A, Schminke S, Maniruzzaman M, Beck J, Douroumis D (2011) Development and evaluation of orally disintegrating tablets (ODTs) containing Ibuprofen granules prepared by hot melt extrusion. Colloids Surf B Biointerfaces 86:275–284

ICH Harmonised Tripartite Guideline (1999) ICH Q6A: specifications: test procedures and acceptance criteria for new drug substances and new drug products: chemical substances

ICH Harmonised Tripartite Guideline (2005) ICH Q9: quality risk management

ICH Harmonised Tripartite Guideline (2008) ICH Q10: pharmaceutical quality system

ICH Harmonised Tripartite Guideline (2009) ICH Q8 (R2): pharmaceutical development

International Conference on Harmonisation (1999). Specifications: Test Procedures and Acceptance Criteria for New Drug Substances and New Drug Products: Chemical Substances. Q6A. 1999. http://www.ich.org/; Accessed: 02/01/2013.

International Conference on Harmonisation (2005). Quality Risk Management. Q9. http://www.ich.org/; Accessed: 02/01/2013.

International Conference on Harmonisation (2008). Pharmaceutical Quality System. Q10. http://www.ich.org/; Accessed: 02/01/2013.

International Conference on Harmonisation (2009). Pharmaceutical Development. Q8(R2). http://www.ich.org/; Accessed: 02/01/2013.

Klein CE, Chiu YL, Awni W, Zhu T, Heuser RS, Doan T, Breitenbach J, Morris JB, Brun SC, Hanna GJ (2007) The tablet formulation of lopinavir/ritonavir provides similar bioavailability to the soft-gelatin capsule formulation with less pharmacokinetic variability and diminished food effect. J Acquir Immune Defic Syndr 44:401–410

Maniruzzaman M, Boateng JS, Bonnefille M, Aranyos A, Mitchell JC, Douroumis D (2012) Taste masking of paracetamol by hot-melt extrusion: an in vitro and in vivo evaluation. Eur J Pharm Biopharm 80:433–442

Repka MA, Battu SK, Upadhye SB, Thumma S, Crowley MM, Zhang F, Martin C, McGinity JW (2007) Pharmaceutical applications of hot-melt extrusion: part II. Drug Dev Ind Pharm 33:1043–1057

Saerens L, Dierickx L, Lenain B, Vervaet C, Remon JP, De Beer T (2011) Raman spectroscopy for the in-line polymer-drug quantification and solid state characterization during a pharmaceutical hot-melt extrusion process. Eur J Pharm Biopharm 77:158–163

Saerens L, Dierickx L, Quinten T, Adriaensens P, Carleer R, Vervaet C, Remon JP, De Beer T (2012) In-line NIR spectroscopy for the understanding of polymer-drug interaction during pharmaceutical hot-melt extrusion. Eur J Pharm Biopharm 81:230–237

Title 21, Code of Federal Regulations, Part 314. Sec 314.94(a)(9)(ii). 2012 ed
Title 21 UStateC, Chapter 9—Federal Food, Drug, and Cosmetic Act. 2012 ed
Tumuluri SV, Prodduturi S, Crowley MM, Stodghill SP, McGinity JW, Repka MA, Avery BA (2004) The use of near-infrared spectroscopy for the quantitation of a drug in hot-melt extruded films. Drug Dev Ind Pharm 5:505–511
Tumuluri VS, Kemper MS, Lewis IR, Prodduturi S, Majumdar S, Avery BA, Repka MA (2008) Off-line and on-line measurements of drug-loaded hot-melt extruded films using Raman spectroscopy. Int J Pharm 357:77–84
US Food and Drug Administration (2000) Guidance for Industry. Waiver of In Vivo Bioavailability and Bioequivalence Studies for Immediate-Release Solid Oral Dosage Forms Based on a Biopharmaceutics Classification System
US Food and Drug Administration (2004) Guidance for Industry. PAT—A Framework for Innovative Pharmaceutical Development, Manufacturing, and Quality Assurance
US Food and Drug Administration (2005) Guidance for Industry. Nonclinical Studies for the Safety Evaluation of Pharmaceutical Excipients
US Food and Drug Administration (2012) Inactive Ingredients Database. http:/www.accessdata.fda.gov/scripts/cder/iig/index.cfm; Accessed: 02/01/2013
Zajc N, Srčič S (2004) Binary melting phase diagrams of nifedipine-PEG 4000 and nifedipine-mannitol systems. J Therm Anal Cal 77:571–580

Chapter 18
Melt Extrusion: A Commercial Perception to Practicality

Lessons Learned from Meltrex® Technology

Sejal P. Shah and Jörg Breitenbach

Abstract Over the past two decades, Meltrex® has evolved from being a novel processing technique into a comprehensive manufacturing technology. This chapter discusses the successful application of Meltrex® in the commercial development of delivery systems for varied drugs such as lopinavir (Kaletra®), ritonavir (Norvir®), ibuprofen, and verapamil hydrochloride (Isoptin®). Besides the commercial applications, Meltrex® can be successfully applied in design of solid dispersions (bioavailability enhancement of poorly soluble drugs), alter or control dissolution profiles (sustained release), reduced dosing frequency, deter substance abuse, improve safety, and efficacy of drugs. Moreover, continuous processing, rigorous and comprehensive regulatory documentation, and use of generally regarded as safe (GRAS) material are some of the striking features which make Meltrex® a popular drug delivery technology.

18.1 Introduction

Improving solubility, rate of dissolution, and hence the absorption and bioavailability of newly discovered drugs remains a challenging task ahead of the current research- and innovation-driven pharmaceutical industry. A considerable portion of the new molecular entities (NMEs) discovered today have low aqueous solubility and this portion may be increasing with the advent of combinatorial chemistry, progress in synthetic chemistry, and high throughput methods in drug discovery.

Currently, most pharmaceutical organizations observe a rapid decline in the sales value of their respective key products following patent expiry and competition from generic manufacturers. Introduction of new formulations of out-of-patent molecules

J. Breitenbach (✉)
Abbott GmbH & Co. KG,
Ludwigshafen, Germany
e-mail: joerg.breitenbach@abbott.com

S. P. Shah
Department of Pharmaceutics, School of Pharmacy,
The University of Mississippi,
University, MS, USA

M. A. Repka et al. (eds.), *Melt Extrusion*,
AAPS Advances in the Pharmaceutical Sciences Series 9,
DOI 10.1007/978-1-4614-8432-5_18,

with significant improvement in their pharmacokinetic profiles remains an important strategy to benefit patients largely.

Although, solubilization approaches are available, the challenge exists in incorporating them in an appropriate dosage form due to issues with reproducibility, scale-up, physical and chemical stability of the drug, and excipients used for formulation, in particular when solid dispersion or amorphous embeddings are used.

Melt extrusion, a thermally intensive process of converting a thermoplastic raw material into a product of defined shape, offers one possible solution (Crowley et al. 2007; Repka et al. 2008). Melt extrusion has been successfully applied as a solubility enhancement technique for poorly soluble drugs (Lakshman et al. 2008; Tho et al. 2010; Feng et al. 2011; Liu et al. 2012; Sakurai et al. 2012). Solubilization in the melt extrusion process occurs through dispersion of a poorly soluble drug(s) in a polymeric carrier matrix.

Over the past 15 years, several products have been successfully formulated through melt extrusion technology, Isoptin® SRE-240, fast-acting ibuprofen, Norvir®, and Kaletra® are examples of successfully developed tablet formulations using the proprietary Meltrex® technology. Abbott GmbH & Co. KG has adapted and developed Meltrex® (proprietary melt extrusion-based technology) into a process for the production of pharmaceuticals, which can meet the industry needs for both enhanced bioavailability of NMEs and improved formulations of existing products.

Kaletra®, containing the anti-HIV protease inhibitors lopinavir and ritonavir, is the first coformulated pharmaceutical composition to be successfully tableted using Meltrex® technology. More importantly from a therapeutic perspective, in comparison with the soft gelatin capsule (SGC) formulation of Kaletra®, Meltrex® tablets require fewer dose forms to be taken each day, can be stored at room temperature (SGC needs to be refrigerated), and do not need to be taken with food. The melt extrusion tablet formulation of Kaletra® represents a unique example of patient-centric HIV therapy.

18.2 Melt Extrusion: A Solid Dispersion Technique

Melt extrusion is a methodology wherein drug(s) are dispersed in a polymeric (or lipid) carrier matrix, thereby forming a solid dispersion (Shah et al. 2012). As defined by Chiou and Riegelman, a solid dispersion is a dispersion of one or more active pharmaceutical ingredient (API) in an inert carrier in the solid state, prepared either by melting, solvent, or the combined melting–solvent method. They present a classification of solid dispersions as eutectic mixtures, solid solutions, and glass solutions based on the physical state of the drug and the carrier, respectively (Chiou and Riegelman 1971).

The solid dispersions prepared by melt extrusion-based processing can be classified as crystalline solid dispersions, amorphous solid dispersions, and solid solutions (a subgroup of amorphous solid dispersions; Serajuddin 1999; Leuner and Dressman 2000; Janssens and Van den Mooter 2009; Shah et al. 2012).

Although pharmaceutical solid dispersions are prepared with an aim to enhance the solubility and hence bioavailability of drugs, equal importance should be given to studies focused at understanding the physical form of drug in the carrier matrix (recrystallization behavior, stability, etc.).

18.3 The Meltrex® Technology

Meltrex® technology has been developed with the specific aim to combine the preparation of solid dispersions with defined controlled-release characteristics in a single manufacturing process (Breitenbach 2002, 2006; Breitenbach and Lewis 2008). The drug can ultimately exist in either an amorphous or a crystalline form. Meltrex® technology can be used to formulate crystalline drug substances into solid dispersions independent of their polymorphic form or the particle-size distribution.

Meltrex® represents a process free of volatile solvents for the production of pharmaceuticals, which can meet the industry needs for both enhanced bioavailability of NMEs and improved formulations of existing products. In essence, rather than using polymers simply as binders or excipients, as in the conventional tableting process, Meltrex® uses their intrinsic matrix-forming properties as a fundamental part of a drug delivery system (Figs. 18.1 and 18.2).

In the Meltrex® process, the drug substance itself (or as a premix, depending for example on the flow characteristics of the drug) together with a pharmaceutically acceptable polymer and any additional excipients (if required) are fed continuously through a loss-in-weight hopper system into the twin-screw extruder. The ingredients are thoroughly mixed by the corotating screws where most of the energy input is provided by the shear forces of the corotating screws. In addition, the chambers can be heated briefly during the mixing process.

The heating step softens the polymer to enhance mixing and subsequent flow through the slit or nozzle; often the drug itself may act as a plasticizer, which further reduces the need for high temperatures during processing. The actual temperature can depend on a variety of factors such as the thermal stability of the drug, the thermoplasticity of polymer being used, and the pharmacokinetic profile that is required for the final product.

Typical operating temperatures are at around 90–120 °C but, for example, peptides have been successfully extruded at much lower temperatures. Extrusion through a slit creates a ribbon that passes through two calendar rollers, which can result in tablets either in a variety of shapes and sizes or in a thin film (Xellex™). The tablets require a minimum of further processing and can be coated for cosmetic or taste-masking purposes and the films can be used as a coating or, for example, for buccal administration. Extruding through a nozzle produces spaghetti-like strands, which can be cut into pellets for filling into capsules. Granules can also be produced from the extruder and these can then be milled and compressed, often directly, into conventional tablets. As the drug substance and the polymer are thoroughly mixed during the extrusion process, the release profile is an intrinsic property of the resulting formulation and is not dependent on, for example, any additional coating.

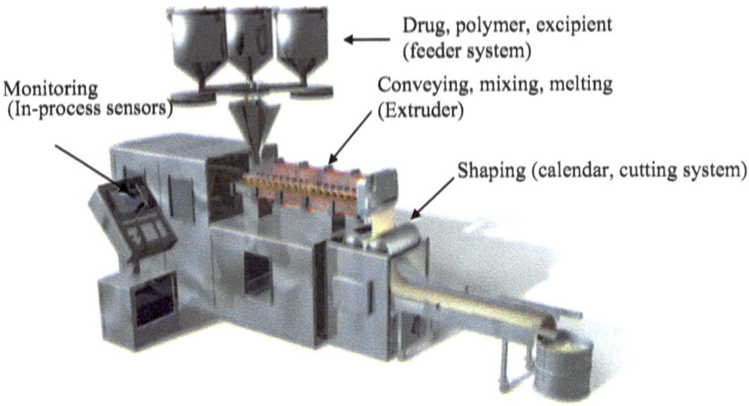

Fig. 18.1 Schematic representation of a typical pharmaceutical twin-screw extruder that represents the core element of the Meltrex® manufacturing process. The basic components of this equipment are the hopper, barrels, screws, kneaders, dies, and a kneading device. (Adapted with permission from Breitenbach (2002))

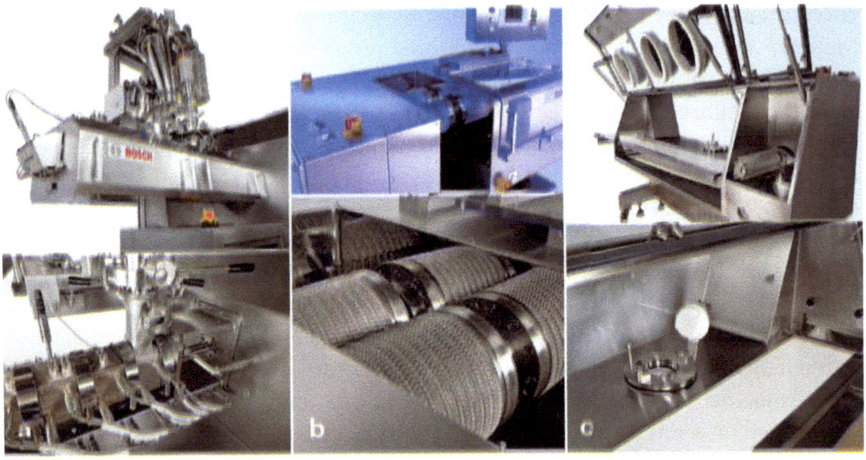

Fig. 18.2 **a–c** Depiction of the individual components of a BOSCH Hot Melt Extrusion Line WCF, BPK, BCK (extruder, calendar, cooling tunnel). **a** BOSCH pharma extruder WCF 0040. **b** BOSCH pharma calendar BPK0050. **c** BOSCH pharma cooling tunnel BCK0050. (Used with permission of BOSCH (2012))

Often the resulting formulation gives a clear, transparent tablet. Solid dispersions produced by processes like melt extrusion may have the additional advantage of being able to combine both enhanced bioavailability and a specific controlled-release profile in the same formulation. The drug load for solid dispersion formulations is typically in the range of 30–60 %, depending on the solubility of the drug in the polymer. Clearly, it is advisable to avoid a supersaturated product, which may then recrystallize and real-time stability data results demonstrate with a variety of different compounds that no crystal formation in the solid dispersion formulations occurred for more than ten years in some cases.

Following are the engineering or built-in design advantages of the twin-screw extruder:

1. *Short-residence time application to thermolabile substances*: The average residence time for drug polymer mix in the twin-screw extruder is around 2 min. The extrusion process moves the molten mass continuously through the extrusion channel, thereby avoiding heat stress on the drug-polymeric matrix blend. This time duration is sufficient enough to ensure the plasticity of extruded material and simultaneously also avoid any major decomposition or unwarranted changes in the physicochemical properties of melt-extruded material. Several drugs and drug carriers are thermally unstable and undergo major physical and even chemical changes on prolonged exposure to high temperatures required to melt or fuse the material.
2. *Shaping*: Another core element of the Meltrex® process is the ability to shape the molten strand that is expelled from the extruder. The mass is forced through two rollers, which can then produce tablets of various shapes or a thin film (Xellex®). The process as such can also produce granules for subsequent milling and compression into conventional tablets. The final drug release profile is an intrinsic property of the extrudate and this profile remains unhampered during milling, compression, or any subsequent downstream processing step and the extrudate can be simply used for the manufacture of a specific delivery system.
3. *Self-wiping screws*: The root of the self-wiping screw, wipes on the shaft of adjacent screw, thereby ensuring near-complete emptying of the extruder at the end of process. Also, this built-in design makes the extruder amenable to Good Manufacturing Practice operations wherein cleaning and sanitization can be easily performed at the end of a batch process.
4. *Versatility*: The extruder is designed with maximum versatility. Die plates can be changed to alter the configuration and final form of the extruded material. In addition, the rate of extrusion and mixing action could be easily controlled. These features allow versatile formulations to be processed on a single machine, thereby optimizing equipment utilization.

A further advantage resulting from the formation of solid dispersions with technologies like the Meltrex® technology is that it may produce stable amorphous final dosage forms, thus eliminating many of the formulation and bioavailability problems associated with drug compounds, which exhibit several polymorphic forms.

18.4 Overview of Kaletra® Tablets

Highly active antiretroviral therapy (HAART) is an intensive treatment regimen for HIV that employs a combination of antiretroviral drugs (protease inhibitor and nucleoside analogue; von Hentig 2007). Although HAART is an effective therapy, it is limited by the patient-unfriendly treatment regimen. The treatment regimen requires frequent administration of several pills. Further, the pills need to be administered around mealtimes. The fact that some of the drugs require refrigeration further reduces the convenience of a regimen. The problem is further compounded in the HIV-affected developing parts of world where maintaining a cold chain for shipment and refrigerated storage during subsequent use of the medication is a major challenge.

A recent development, which illustrates the potential benefits achievable by solid dispersions, has been the reformulation of the leading protease inhibitor drug combination product, Kaletra® as part of Abbott Laboratories' on-going product improvement strategy (Breitenbach 2006). Kaletra® is a preparation of two protease inhibitors, lopinavir and ritonavir; it is the only coformulated protease inhibitor combination dosage form available in the market. (Cvetkovic and Goa 2003; Oldfield and Plosker 2006; von Hentig 2007; Croxtall and Perry 2010). The improvement of pharmacokinetic properties of lopinavir is attributed to the inhibitory effect of ritonavir on cytochrome P450 enzyme (primarily responsible for the metabolism of lopinavir).

Active ingredients, lopinavir and ritonavir, are poorly soluble in water and have negligible absorption from the gut when administered as unformulated solids. The first-marketed Kaletra® product was a soft-gel capsule (SGC) formulation, each capsule containing 133 mg lopinavir/33 mg ritonavir, with an adult regimen of 6 capsules per day.

The new Kaletra® tablets, made using Abbott's Meltrex® technology, contain 200 mg lopinavir/50 mg ritonavir per tablet in a "solid solution" with a resulting adult regimen of only 4 tablets per day. Abbott received US Food and Drug Administration fast-track approval of Kaletra® tablets for marketing in the USA in October 2005, followed by EU approval in July 2006.

The new Kaletra® tablets not only reduce the daily pill burden (6 SGC vs. 4 tablets), but also provide several other benefits. The newly developed Kaletra® tablets did not require refrigeration and could be stored at ambient temperature. Further, the tablets did not require administration around mealtime, thereby providing a more flexible medication schedule. Overall, the melt-extruded formulation significantly improves HIV therapy management as demonstrated by providing reduced dosage frequency (4 tablets vs. 6 SGCs), flexible regimen (independent of mealtime) and ambient temperature storage of the tablets (Fig. 18.3).

18.5 The Kaletra® Tablet: Patient Benefit

Introduction of the Kaletra® Meltrex® tablet formulation reduced the daily pill count from 6 SGCs per day to 4 tablets per day. Meltrex® tablets could be stored at room temperature (SGC needs to be refrigerated). Unlike the SGC, the tablet contains nonanimal-sourced excipients and does not use unsaturated fatty acids as formulation excipients. Further, the Meltrex® tablet need not to be taken with food and provides

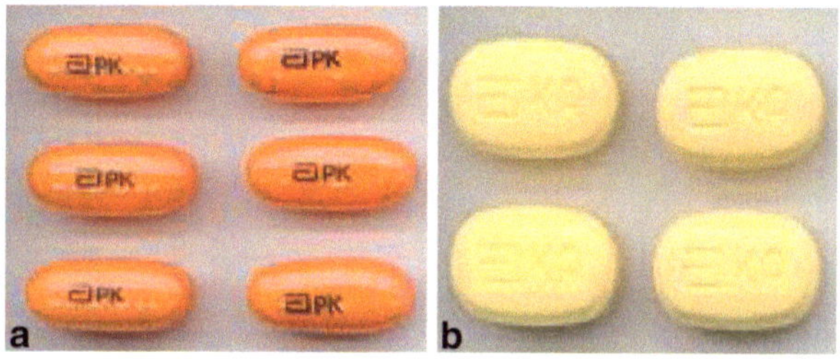

Fig. 18.3 An illustration of Kaletra® Meltrex® tablets and the original softgel capsules. **a** 6 Softgel capsules. **b** 4 Kaletra® Meltrex® tablets. (Used with permission from AbbVie Deutschland GmbH & Co. (2013))

consistent plasma levels of both drugs on a day-to-day basis independent of the varying meal conditions. Lower variability of lopinavir levels compared with those with the SGC.

In addition, it was demonstrated that the Meltrex® tablets were bioequivalent to the preexisting SGC formulation (Klein et al. 2007). Bioavailability of the melt-extruded tablet formulation of Kaletra® was assessed in three studies wherein tablet formulations of lopinavir/ritonavir at 800/200 mg or 400/100 mg under different meal conditions were compared with equal doses of the SGC after a moderate–fat meal. The tablet was found to be bioequivalent to the SGC after a moderate–fat meal based on the areas under the concentration-time curve (Fig. 18.4). Overall, the tablet formulation resulted in consistent plasma concentrations of lopinavir independent of the food intake, providing better therapeutic control.

18.6 Application of Meltrex® Technique in Development of Other Pharmaceutical Formulations

In addition to Kaletra®, Abbott has also developed a sustained release formulation of verapamil hydrochloride in cellulose ether matrix (Isoptin® SR-E 240 mg) that was the first directly shaped melt extrusion product on the market, as well as a fast-onset ibuprofen system (Andrews et al. 2010). Also, in 2010 Abbott announced that the US FDA approved a new thermally stable tablet formulation of the protease inhibitor Norvir® (ritonavir). This new formulation does not require refrigeration and can be stored at room temperature making it more convenient for patients.

Ibuprofen is known to be poorly soluble in aqueous acidic media. Though ibuprofen is completely absorbed upon oral administration (nearly 100 % bioavailability), the onset of absorption depends mainly on its dissolution and hence the type of formulation being administered. Meltrex® has been successfully applied to develop fast-onset ibuprofen, wherein ibuprofen is formulated as a solid dispersion in hydrophilic polymer matrix of polyvinyl pyrrolidone (PVP; Klueglich et al. 2005). The melt-extruded material is finally formulated as a film-coated tablet wherein the drug release (ibuprofen release) is dependent on the disintegration time of the tablet and the dissolution of the polymer and is independent of the solubility of ibuprofen. Such a formulation modification imparts fast-release properties to ibuprofen.

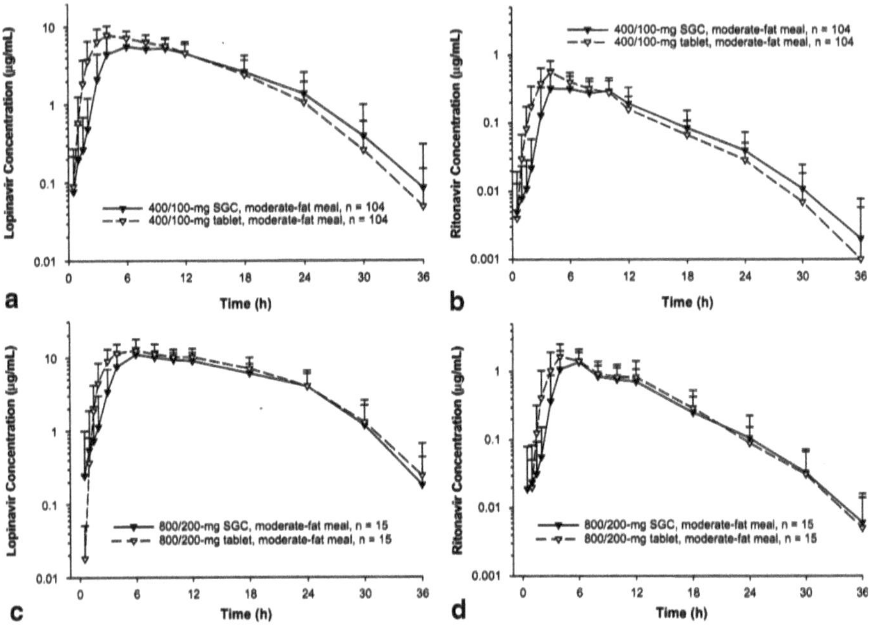

Fig. 18.4 The average ± SD concentrations of lopinavir and ritonavir after SGC (▼) or tablet (▽) administration of a single lopinavir/ritonavir dose of 400/100 mg with a moderate–fat breakfast [lopinavir (**a**) and ritonavir (**b**)] or 800/200 mg with a moderate–fat breakfast [lopinavir (**c**) and ritonavir (**d**)]. (Used with permission from Klein et al. (2007))

As shown in Fig. 18.5, under fasting conditions, ibuprofen lysinate and the novel ibuprofen extrudate tablet are bioequivalent on the basis of area under curve (AUC), C_{max}, and t_{max}. The pharmacokinetic profile of the novel extrudate tablet is different from that of regular ibuprofen under fasting conditions, with the C_{max} for extrudate tablet being approximately 20 % higher and t_{max} almost 1 h earlier than that for regular ibuprofen. The average total exposure as determined by AUC is almost identical though. Thus, although the extent of ibuprofen absorption is equivalent for the novel extrudate and regular ibuprofen, the rate of absorption is much faster for the novel extrudate (C_{max} and t_{max}). In conclusion, the pharmacokinetic profile of novel ibuprofen extrudate formulation is similar to ibuprofen lysinate and it is more rapidly available in plasma than regular ibuprofen (fast onset of action).

Meltrex® has been applied to verapamil hydrochloride, a marketed antihypertensive, and antianginal drug which can interact with alcohol. During extrusion, verapamil hydrochloride gets dispersed in the polymer melt (similar to those of ibuprofen and lopinavir described in the previous paragraphs) and the melt is subsequently shaped in the form of a tablet.

Roth et al. determined the influence of ethanol on the in vitro rate of release of marketed verapamil hydrochloride (240 mg) Meltrex® (Isoptin SR-E) and three marketed verapamil hydrochloride (240 mg) sustained-release (SR) formulations (Roth et al. 2009). Dissolution was tested under standardized conditions, with mediums

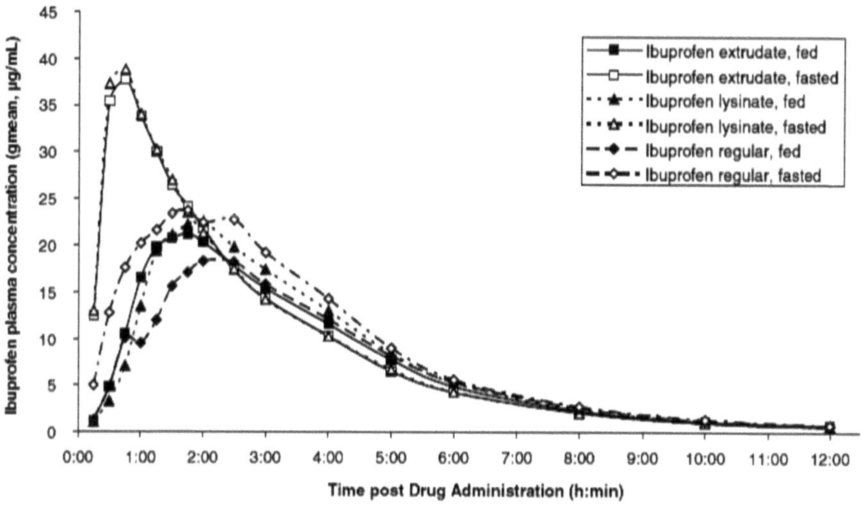

Fig. 18.5 Geometric mean ibuprofen plasma concentrations following single oral administration of a 400-mg ibuprofen extrudate (*square*), lysinate (*triangle*), and regular (*diamond*) tablet under fed (*filled*) and fasted (*open*) conditions, respectively. (Used with permission from Klueglich et al. (2005))

containing ethanol concentrations of 0, 5, 20, and 40 %. The dissolution profiles for verapamil hydrochloride Meltrex® showed no differences between 5 and 40 % ethanol vs. 0 % ethanol ($P > 0.05$). The mean dissolution percentage (%) was identical at 1 h (19 %) in 0 % vs. 40 % ethanol. In contrast, the three verapamil SR formulations showed significant increases in dissolution in 20 and 40 % ethanol versus 0 % ethanol ($P < 0.001$). An initial rapid release (within 2 h) was observed in 20 and 40 % ethanol, with a mean dissolution of 99 % (range 73–107 %). Therefore, unlike the three SR verapamil formulations tested, verapamil Meltrex® was found to be resistant to in vitro dose dumping when combined with readily accessible ethanol concentrations. The aim of their investigation was to determine the influence of ethanol on the in vitro release rate of verapamil hydrochloride from Meltrex® melt-extruded tablets (Form A) in contrast to three other direct compressed verapamil hydrochloride sustained release formulations (Forms B–D; Fig. 18.6).

18.7 Summary and Future Perspective

Over the years, Kaletra® melt-extruded tablets have become an important treatment option for patients with HIV. The application of Meltrex® technology has led to the reformulation of the product into a solid preparation that has significantly improved the convenience of Kaltera® treatment for patients with HIV. Application of the Meltrex® technology led to the overall product improvement (room temperature storage, reduction of pill count, dose schedule independent of meals, etc.) of

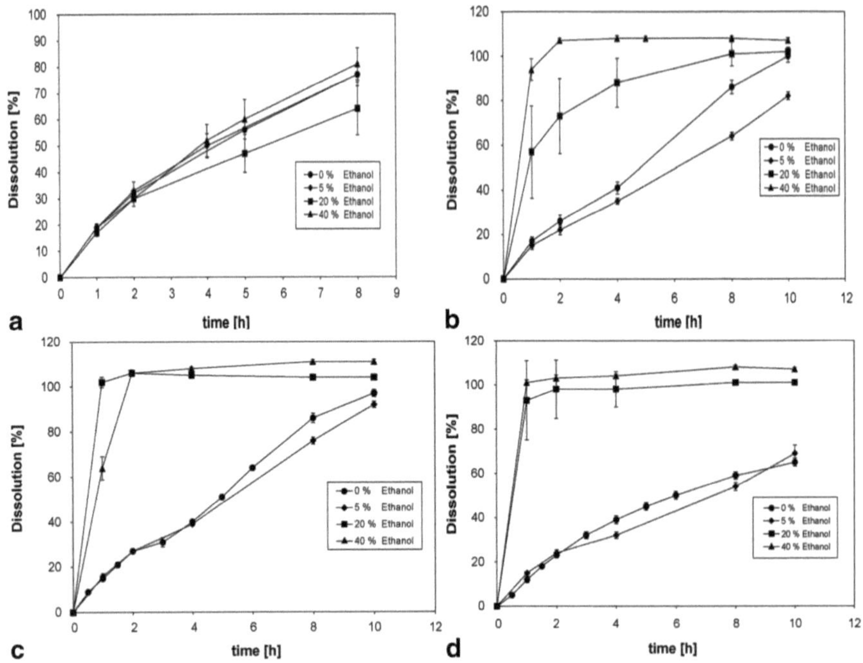

Fig. 18.6 Dissolution profiles (mean dissolution % [± SD]) of verapamil hydrochloride release from Form A (Meltrex®) (**a**), Form B (SR) (**b**), Form C (SR) (**c**), and Form D (SR) (**d**) over time (h), with increasing ethanol concentrations. (Used with permission from Roth et al. (2009))

Kaletra®. In addition, the Meltrex® technology has also been successfully applied to other drug formulations (ibuprofen and verapamil hydrochloride).

Taken together, Meltrex® is a highly efficient manufacturing technology that can be successfully applied for processing poorly soluble drugs as solid dispersion and can also be used to alter and control dissolution profiles of drugs. Meltrex® can be applied to a plethora of pharmaceutical systems with varied benefits such as reduced frequency of daily dosing, aid to deter substance abuse (e.g., opoid analgesics), improve safety, and provide sustained release profile. Meltrex®-based products have been approved in both the EU and USA. Moreover, Meltrex® has steadily evolved from an engineering technology into a comprehensive manufacturing platform amenable to current regulatory guidelines. Finally, since all the materials (polymers, processing aids, etc.) used in melt extrusion in general are pharmaceutically acceptable (GRAS approved), several existing formulations (already marketed) will undergo modification and develop into more efficient dosage forms to be approved in the near future.

References

AbbVie Deutschland GmbH & Co (2013) SOLIQS Global Drug Delivery Business, Germany. www.soliqs.com. Accessed Nov 2012

Andrews GP, Abu-Diak O et al (2010) Physicochemical characterization and drug-release properties of celecoxib hot-melt extruded glass solutions. J Pharm Pharmacol 62(11):1580–1590

Bosch (2012) Technical literature on BOSCH hot melt extrusion line WCF, BPK, BCK. Robert Bosch GmbH, Packaging Technology Division, Germany. http://www.boschconfectionery.com

Breitenbach J (2002) Melt extrusion: from process to drug delivery technology. Eur J Pharm Biopharm 54(2):107–117

Breitenbach J (2006) Melt extrusion can bring new benefits to HIV therapy: the example of Kaletra® Tablets. Am J Drug Deliv 4(2):61–64

Breitenbach J, Lewis J (2008) Two concepts, one technology: controlled-release solid dispersions using melt extrusion (Meltrex®). In: Rathbone MJ, Roberts MS (eds) Modified-release drug delivery technology. Informa Healthcare, New York, pp 179–186

Chiou WL, Riegelman S (1971) Pharmaceutical applications of solid dispersion systems. J Pharm Sci 60:1281–1302. doi: 10.1002/jps.2600600902

Crowley MM, Zhang F et al (2007) Pharmaceutical applications of hot-melt extrusion: part I. Drug Dev Ind Pharm 33(9):909–926

Croxtall JD, Perry CM (2010) Lopinavir/ritonavir: a review of its use in the management of HIV-1 infection. Drugs 70(14):1885–915

Cvetkovic RS, Goa KL (2003) Lopinavir/ritonavir: a review of its use in the management of HIV infection. Drugs 63(8):769–802

Feng J, Xu L et al (2011) Evaluation of polymer carriers with regard to the bioavailability enhancement of bifendate solid dispersions prepared by hot-melt extrusion. Drug Dev Ind Pharm 38(6):735–743

Janssens S, Van den Mooter G (2009) Review: physical chemistry of solid dispersions. J Pharm Pharmacol 61(12):1571–1586

Klein CE, Chiu YL et al (2007) The tablet formulation of lopinavir/ritonavir provides similar bioavailability to the soft-gelatin capsule formulation with less pharmacokinetic variability and diminished food effect. J Acquir Immune Defic Syndr 44(4):401–410

Klueglich M, Ring A et al (2005) Ibuprofen extrudate, a novel, rapidly dissolving ibuprofen formulation: relative bioavailability compared to ibuprofen lysinate and regular ibuprofen, and food effect on all formulations. J Clin Pharmacol 45(9):1055–1061

Lakshman JP, Cao Y et al (2008) Application of melt extrusion in the development of a physically and chemically stable high-energy amorphous solid dispersion of a poorly water-soluble drug. Mol Pharm 5(6):994–1002

Leuner C, Dressman J (2000) Improving drug solubility for oral delivery using solid dispersions. Eur J Pharm Biopharm 50(1):47–60

Liu X, Lu M et al (2012) Improving the chemical stability of amorphous solid dispersion with cocrystal technique by hot melt extrusion. Pharm Res 29(3):806–817

Oldfield V, Plosker GL (2006) Lopinavir/ritonavir: a review of its use in the management of HIV infection. Drugs 66(9):1275–1299

Repka MA, Majumdar S et al (2008) Applications of hot-melt extrusion for drug delivery. Expert Opin Drug Deliv 5(12):1357–1376

Roth W, Setnik B et al (2009) Ethanol effects on drug release from Verapamil Meltrex, an innovative melt extruded formulation. Int J Pharm 368(1–2):72–75

Sakurai A, Sakai T et al (2012) Polymer combination increased both physical stability and oral absorption of solid dispersions containing a low glass transition temperature drug: physicochemical characterization and in vivo study. Chem Pharm Bull (Tokyo) 60(4):459–464

Serajuddin AT (1999) Solid dispersion of poorly water-soluble drugs: early promises, subsequent problems, and recent breakthroughs. J Pharm Sci 88(10):1058–1066

Shah S, Maddineni S et al (2012) Melt extrusion with poorly soluble drugs. Int J Pharm S0378–S5173(12):00990–00998

Tho I, Liepold B et al (2010) Formation of nano/micro-dispersions with improved dissolution properties upon dispersion of ritonavir melt extrudate in aqueous media. Eur J Pharm Sci 40(1):25–32

von Hentig N (2007) Lopinavir/ritonavir: appraisal of its use in HIV therapy. Drugs Today (Barc) 43(4):221–247

Chapter 19
Future Trends

Michael A. Repka

Abstract Hot melt extrusion has been utilized in the plastic, food, and metal industries for over eight decades. Pharmaceutical applications using this technology began in the 1970s and gained significant traction within the last 20 years. Melt extrusion has been considered by many to provide a platform for drug delivery, which has sustained and increased its interest by pharma. *Future Trends* provides a synopsis of melt extrusion's achievements, its continued relevance by the pharmaceutical industry evidenced by patents approved, as well as its potential to change the paradigm for continuous manufacturing.

Hot melt extrusion (HME) has been employed for a number of nonpharmaceutical applications for approximately 80 years. These applications include metals, plastics, and foods, as well as devices (sutures), before extending to pharmaceutical dosage forms. These pharmaceutical applications expanded beginning in the 1970s from formulation/product development to manufacturing. Although there was a "lag time" for HME from then, it has become a foundation technology for research and manufacturing within the pharmaceutical industry over the past two decades—propelling HME as an alternative "platform" for a variety of solid dosage forms.

Innovation in formulation, polymer manufacturing science, and equipment processing and technology has revolutionized the advancement of HME techniques. Industrial acceptability and adaptability continues to influence the development of improved and scalable equipment, serving batch sizes as small as 5 g to 1,000 kg/h. This book describes recent developments in immediate release formulations, including solubilization by transforming drugs into their amorphous state using a solid solution/solid dispersion approach assisted by novel polymer matrices such as Soluplus® and low molecular weight HPC, taste masking of bitter drugs using pH-dependent polymers, as well as effervescent dosage forms. Moreover, exploitation of polymer properties encourages their use in varied modified and targeted release delivery systems. HME for the production of films enables future opportunities to

M. A. Repka (✉)
Department of Pharmaceutics, Pii Center for Pharmaceutical Technology,
School of Pharmacy, The University of Mississippi,
University, MS, USA
e-mail: marepka@olemiss.edu

M. A. Repka et al. (eds.), *Melt Extrusion*,
AAPS Advances in the Pharmaceutical Sciences Series 9,
DOI 10.1007/978-1-4614-8432-5_19,

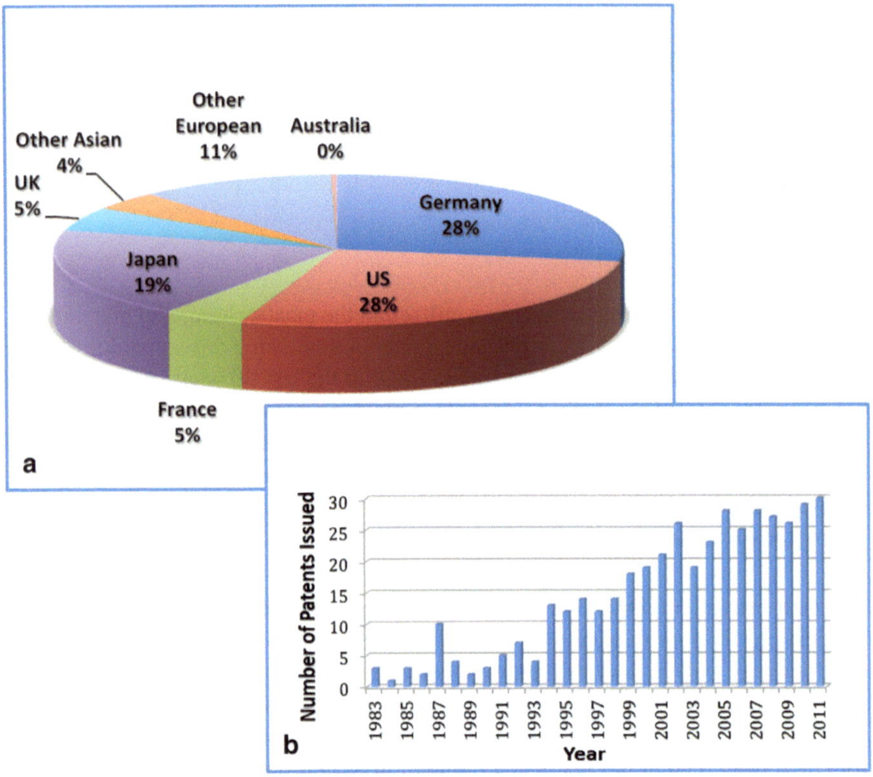

Fig. 19.1 **a** The percentage of hot melt extrusion patents issued by country since 1983 for pharmaceutical applications. **b** The number of hot melt extrusion patents issued for pharmaceutical applications from 1983–2011. (Adapted from Crowley and Repka 2007)

develop immediate and prolonged release patches and multilayer systems to modulate drug release for oral and transdermal applications. The breadth of research and publications in the field of melt extrusion demonstrates this technology's versatility and applicability for transformation of pharmaceutical industry development and manufacturing.

Over the last 15–20 years, hundreds of research papers and extensive reviews have been published on this subject (Breitenbach 2002; Crowley et al. 2007; Repka et al. 2007; Repka et al. 2008). Moreover, issued HME patents have continuously increased worldwide (Fig. 19.1).

This figure as well as other data is evidence of the increasing pharmaceutical interest in melt extrusion. It has been demonstrated that HME technology is an innovative and feasible approach in the preparation of various pharmaceutical drug delivery systems such as pellets (Bialleck and Rein 2011); granules (Liu and Zhang 2001 Liu, X.); immediate and modified release tablets (Crowley et al. 2002; Gryczke et al. 2011); oral fast dissolving systems (Gryczke et al. 2011); transdermal

(Repka et al. 1999; Repka and McGinity 2001); transmucosal, and transungual delivery systems (Mididoddi et al. 2006; Mididoddi and Repka 2007); and implants (Ghalanbor et al. 2010). The physical and chemical stability of hot melt extrudates has been shown related to the nature of the polymer and excipients, the physical state of the drug in the final extrudates, as well as storage and packaging conditions. Although many melt extruded dosage forms tend to demonstrate good long-term stability, at times recrystallization of the actives has been observed during their storage. This kinetic or thermal physical instability is one of the common problems observed with crystalline drugs processed via HME wherein the drug is first converted to a high energy amorphous state during extrusion and reverts back into its crystalline state upon storage (Prodduturi et al. 2004). To overcome such issues, several crystallization inhibitors such as hydroxypropylmethylcellulose acetate succinate (HPMCAS) polycarbophil, polyvinyl pyrrolidone K25 and hydroxypropyl methylcellulose have been incorporated into formulations as problem-solving excipients. These additives have been shown to either significantly prevent or reduce the recrystallization of drugs in HME formulations (Bruce et al. 2007).

In addition to a continuous process, process analytical technology (PAT) has already been introduced for HME techniques, as well as quality-by-design (QbD) principles, to optimize design, analysis, and control within manufacturing processes. Many analytical technologies have been incorporated into the systems including Raman spectroscopy and near-infrared spectroscopy.

Despite the discussion till this point, what is the driving force that could change the paradigm of the development and manufacturing process of pharmaceutical products and devices? One major answer is the enhancement of solubility and bioavailability of poorly soluble drugs. As has been discussed in several chapters of this book, new chemical entities (NCEs), reported in the range of 40–70 %, as well as many generic drugs are classified as BCS Class II or IV compounds—which many times can be successfully produced by HME formulation and processing. *Enhancing solubility*, an ever-increasing topic for seminars, workshops, and meetings, coupled with this same trend for melt extrusion forums, validates the present and future of this enabling technology. This author, as well as a multitude of pharmaceutical scientists, knows that literally every multinational pharmaceutical company has at least purchased melt extrusion equipment with an intent to train and develop their scientists in this growing technology. An ever-increasing number are implementing melt extrusion as a solution to the solubility issues of the majority of NCEs. However, one must ask, "is melt extrusion the only answer to these types of issues?" Certainly not. Other technologies, such as spray drying, have shown to be applicable in some cases (please see Chap. 14 for spray drying discussion). No technique can solve all of the problems that our industry faces. Innovations in formulation, exploiting new polymers, and properties of existing polymers and lipids, as well as increased technical expertise by equipment manufacturers has generated equipment modifications for PAT initiatives, as well as scalability have driven HME into the forefront to change the paradigm of pharmaceutical manufacturing.

Additional advantages of melt extrusion, such as the small equipment footprint, environment friendly (solvent-free) and continuous processes, are becoming even more attractive to pharma scientists and executives. Successful products have been marketed using this technology and many others are in the pipeline. FDA is cognizant of this technology's unique adaptability to PAT and QbD. Process engineers are working together with formulation scientists to enhance this already proven technology. Contract manufacturers with melt extrusion capability are increasing in number across the USA and the world. These examples, and many others, are evidence that HME will be a workhorse within the pharmaceutical industry within the next decade.

Currently, there are only a few commercialized products using HME technology in the pharmaceutical market, which is a contrast to the rapidly increasing interest in HME. The economy over the last several years has most likely played a role in this "wait-and-see" attitude by many. And yes, resistance to change has hampered innovation in every industry.

Regardless, many pharmaceutical companies are embracing melt extrusion technology, moving forward and optimizing the technology—innovating—introducing HME manufactured products. In 2010, a heat stable Norvir® (Abbott Laboratories, USA) tablet (ritonavir) was approved by the US FDA. The product does not require refrigeration and is manufactured using proprietary Meltrex® technology (please see Chap. 18). This product was preceded by Kaletra®, also a melt extruded tablet produced by Abbott, which has significant advantages over the older soft gelatin capsules in terms of dosing frequency and stability. Other melt extruded products include NuvaRing®, Implanon® (Organon, a subsidiary of Merck & Co., USA) and Isoptin SR (Abbott Laboratories, USA). There are a number of HME produced pharmaceutical products and medical devices in the pipeline. Indeed, melt extrusion technology is forging ahead as Amgen and other companies target various markets. Cross-pollination utilizing HME between the pharmaceutical industry, medical device, and biotechnology companies are evidence that melt extrusion technology has earned its place in our industry.

In summary, this author has chosen to paraphrase the Conclusion and a quote from Chap. 17 (FDA perspectives). Melt extrusion represents an efficient technology to formulate poorly soluble drugs into safe and effective drug delivery systems. The technology is growing as it is applicable for both high dose and very potent low dose compounds. HME is amenable to real-time monitoring and control of the consistency of the product (PAT) with respect to content uniformity and crystalline conversion. "It has the potential to replace traditional batch processes due to its continuous nature and ease of scale-up from the laboratory scale to commercial scale." Need we say more about this game-changing technology?

References

Bialleck S, Rein H (2011). Preparation of starch-based pellets by hot-melt extrusion. Eur J Pharm Biopharm 79(2):440–448

Breitenbach J (2002) Melt extrusion: from process to drug delivery technology. Eur J Pharm Biopharm 54(2):107–17

Bruce C, Fegely KA et al (2007) Crystal growth formation in melt extrudates. Int J Pharm 341(1–2): 162–72

Crowley MM, Zhang F et al (2002) Stability of polyethylene oxide in matrix tablets prepared by hot-melt extrusion. BioMaterials 23(21):4241–8

Crowley MM, Zhang F et al (2007) Pharmaceutical applications of hot-melt extrusion: part I. Drug Dev Ind Pharm 33(9):909–26

Ghalanbor Z, Korber M et al (2010) Improved lysozyme stability and release properties of poly(lactide-co-glycolide) implants prepared by hot-melt extrusion. Pharm Res 27(2):371–9

Gryczke A, Schminke S et al (2011). Development and evaluation of orally disintegrating tablets (ODTs) containing Ibuprofen granules prepared by hot melt extrusion. Colloids Surf B Biointerfaces 86(2):275–284.

Liu J, Zhang F et al (2001) Properties of lipophilic matrix tablets containing phenylpropanolamine hydrochloride prepared by hot-melt extrusion. Eur J Pharm Biopharm 52(2):181–90

Mididoddi PK, Prodduturi S et al (2006) Influence of tartaric acid on the bioadhesion and mechanical properties of hot-melt extruded hydroxypropyl cellulose films for the human nail. Drug Dev Ind Pharm 32(9):1059–66

Mididoddi PK, Repka MA (2007) Characterization of hot-melt extruded drug delivery systems for onychomycosis. Eur J Pharm Biopharm 66(1):95–105

Prodduturi S, Manek RV et al (2004) Water vapor sorption of hot-melt extruded hydroxypropyl cellulose films: effect on physico-mechanical properties, release characteristics, and stability. J Pharm Sci 93(12):3047–56

Repka MA, McGinity JW (2001) Influence of chlorpheniramine maleate on topical hydroxypropy-lcellulose films produced by hot-melt extrusion. Pharm Dev Technol 6(3):297–304

Repka MA, Gerding TG et al (1999) Influence of plasticizers and drugs on the physical-mechanical properties of hydroxypropylcellulose films prepared by hot melt extrusion. Drug Dev Ind Pharm 25(5):625–33

Repka MA, Battu SK et al (2007) Pharmaceutical applications of hot-melt extrusion: Part II. Drug Dev Ind Pharm 33(10):1043–57

Repka MA, Majumdar S et al (2008) Applications of hot-melt extrusion for drug delivery. Expert Opin Drug Deliv 5(12):1357–76

Index

M. A. Repka et al. (eds.), *Melt Extrusion*,
AAPS Advances in the Pharmaceutical Sciences Series 9,
DOI 10.1007/978-1-4614-8432-5,

MIX
Papier aus verantwortungsvollen Quellen
Paper from responsible sources
FSC® C105338

If you have any concerns about our products,
you can contact us on
ProductSafety@springernature.com

In case Publisher is established outside the EU,
the EU authorized representative is:
Springer Nature Customer Service Center GmbH
Europaplatz 3, 69115 Heidelberg, Germany

Printed by Libri Plureos GmbH
in Hamburg, Germany